Between scientists & citizens:

Proceedings of a conference
at Iowa State University,
June 1-2, 2012

edited by Jean Goodwin

GPSSA

Great Plains Society for the Study of Argumentation
Ames, IA
2012

ISBN-13: 978-1478152347
ISBN-10: 1478152346
LCCN: 2012911916

Table of Contents

Preface   vi
Acknowledgements   vii

KEYNOTE ADDRESSES

Black Box Arguments and Accountability of Experts to the Public   1
Sally Jackson, University of Illinois

Nonsense on Stilts about Science: Field Adventures of a Scientist-Philosopher   19
Massimo Pigliucci, City University of New York

SELECTED CONFERENCE PAPERS

Public Understanding of Climate Science and the Ethics of Expertise   29
Ben Almassi, College of Lake County

The Peer-Review Certification Label: 39
A Shortcut to Assessing Expertise and Consensus by the Necessarily Uninformed
Bill Anelli, Modesto Jr. College

Evaluating Experts: Understanding Citizen Assessments of Technical Discourse   53
Lauren R. Archer, University of Washington

Distrusting Climate Science: A Problem in Practical Epistemology for Citizens   63
Thomas C. Atchison, Metropolitan State University

The Responsibility of Authority: When Should a Physician Seek a Further Opinion?   75
J. Anthony Blair, University of Windsor

The Cycle of Deliberative Inquiry: Re-conceptualizing the Work of Public Deliberation   85
Martin Carcasson, Colorado State University

Co-Evolving Expertise in Environmental Policy Debates: 99
Rethinking Values and Participants through an Ecological Model of Rhetoric
Piper Corp, University of Pittsburgh

Framing Science: The Influence of Expertise and Jargon in Media Coverage   109
Deserai Anderson Crow & J. Richard Stevens, University of Colorado-Boulder

The Problem of Communicating Beyond Human Scale   121
Michael Dahlstrom & Raeann Ritland, Iowa State University

Testimony Traces in Appellate Review: 131
Expertise Extension in Cases of Domestic Abuse and Eyewitness Identification
Per Fjelstad, University of New Hampshire

Reason, Values and Evidence: Rational Dissent from Scientific Authority   141
Bruce Glymour & Scott Tanona, Kansas State University

What is "Responsible Advocacy" in Science? Good Advice.   151
Jean Goodwin, Iowa State University

Using Delphi to Track Shifts in Meanings of Scientific Concepts in a Long-term,   163
Expert-lay Collaboration on Sustainable Agriculture Research in the Midwest
Nancy Grudens-Schuck & GL Drake Larsen, Iowa State University

Examples, Illustrations, Inductions, Anecdotes, Analogies, Precedents,   173
Narratives, and Personal Testimonies: Are They Essentially Different?
Dale Hample, University of Maryland

Analysis of Arguments Favoring Vaccine Resistance   183
Jessica M. Hample, Western Illinois University

"Accommodating Science": A New Way of Thinking about Rhetorical Dynamics   195
Thierry Herman, University of Lausanne, & Camillia Salas, University of Neuchâtel

Authority in an Age of Expertise   209
Catherine E. Hundleby, University of Windsor

Deliberative Systems View of Efforts to Democratize Energy in Arizona   219
Travis Johnson, Arizona State University

A Pragmatic Paradox Inherent in Expert Reports Addressed to Lay Citizens   229
Fred J. Kauffeld, Edgewood College

The Ambiguous Relationship between Expertise and Authority   241
Moira Kloster, University of the Fraser Valley

Professors and Scholars as Experts: Problem Setting and Methodological   253
Considerations for the Examination of Newspaper Articles
Alain Letourneau, Université de Sherbrooke

Should Scientists Communicate Uncertainty to the Public in Health Controversies?   263
The Case of Endocrine Disrupters' Effects on Male Fertility
*Laura Maxim, Martine Cadot, & Pascale Mansier, Centre national de la
recherche scientifique*

The Ethos of Expertise: How Social Conservatives Use Scientific Rhetoric   275
Jamie McAfee, Iowa State University

Examining News Coverage and Framing: 285
The Case Study of Sea Lion Management at the Bonneville Dam
Tess McBride & Cynthia-Lou Coleman, Portland State University

Scientists, Other Citizens, and the Art of Practical Reasoning   297
Gitte Meyer, Aalborg University

Mork and Mindy, Canola Oil and Mustard Gas: The Dilemma of Scientific   307
Illiteracy in Decisions about Food and Health
Mary L. Nucci & William K. Hallman, Rutgers, The State University of New Jersey

Stephen Jay Gould and McLean v. Arkansas: 315
Scientific Expertise and the Nature of Science in America Culture 1980–1985
Myrna Perez, Harvard

Assessing Bias Charges against Collaborative Expertise, with an   325
Application to the IPCC
William Rehg, Saint Louis University

Objectivity vs. Advocacy: Newspaper Rhetoric during the "Bemis Affair"   335
and the "Oleomargarine Controversy"
David Seim, University of Wisconsin-Stout

The Reasonableness of Argumentation from Expert Opinion in Medical   345
Discussions: Institutional Safeguards for the Quality of Shared Decision Making
A.F. Snoeck Henkemans & J.H.M. Wagemans, University of Amsterdam

Do Experts Help or Hinder? An Empirical Examination of Experts and Expertise   355
during Public Deliberation
Leah Sprain, Andy M. Merolla, & Martín Carcasson, Colorado State University

Scrambling on Defense: An Anatomy of Anthropological Responses to   365
the Mead/Freeman Controversy
Robert Strikwerda, St. Louis University

Analyzing GM Food Risk Arguments through an Online, Multi-media Case Study   379
Tosh Tachino, University of Winnipeg, & David R. Russell, Iowa State University

Dismantling Expertise: Disproof, Retraction, and the Persistence of Belief   393
Christopher W. Tindale, University of Windsor

Signatures and Spinoffs: Sequences of Ignorance in the Theory/Practice Split   403
of the Ecological Society of America, 1917–1950
Kenny Walker, University of Arizona

Balancing Substance and Style on a Budget: How North Carolina Sea Grant   413
Communicates Science (Part 1)
Heather Ward

Expertise and Inauthentic Scientific Controversies: What You Need to Know   427
to Judge the Authenticity of Policy-Relevant Scientific Controversies
Martin Weinel, Cardiff University

The Explanatory Value of Cognitive Asymmetries in Policy Controversies   441
Frank Zenker, Lund University

Preface

We are increasingly dependent on advice from experts in making decisions in our personal, professional, and civic lives. But as our dependence has grown, new media have broken down the institutional barriers between the technical, personal and civic realms, and we are inundated with purported scientific results from all sides. Further, we remain attached to an ideal of critical thinking that envisions citizens as making their own, independent judgments on public affairs—an ideal that places appeals to expert authority among the fallacies.

The general inquiry into the place of knowledge in politics goes back to the image of the philosopher king in Plato's Republic, and in the American tradition, to the exchange between John Dewey and Walter Lippmann on how to balance the contributions of ordinary citizens and experts in contemporary mass democracy. Today scholarship on this question is driven in part by a sense that science is contributing less than it could to our policy debates.

Grappling with these issues requires conversations across disciplines. The conference *Between Scientists & Citizens* brought together eighty scholars from diverse fields: rhetorical and communication theorists studying the practices and norms of public discourse and science communication, philosophers interested in the informal logic of everyday reasoning and in the theory of deliberative democracy, and science studies scholars examining the intersections between the social worlds of scientists and citizens. The essays in this volume include case studies, experimental works and conceptual analyses; they focus on historical and current controversies, on diverse communication practices and institutions, and on the difficulties both experts and layfolk face when deliberating together. Together, they represent a significant advance towards understanding and addressing the normative and practical challenges of communication between scientists and citizens.

ACKNOWLEDGEMENTS

These proceedings, and the conference *Between Scientists & Citizens*, were made possible by generous support from the Center for Excellence in the Arts and Humanities at Iowa State University, the Strengthening the Professoriate@ISU initiative, the Departments of English and Philosophy, and the Program in Speech Communication.

The local arrangements for the conference were undertaken by the Science Communication@ISU project, and special thanks are due to my colleagues, project team members Michael Dahlstrom and Kevin deLaplante. Throughout, Chitra Rajan has remained our project's mentor and motivator.

I also gratefully acknowledge the help of Eric Abbott, James Andrews, Amy Bix, Kevin Blankenship, Jean McGuire, Lulu Rodriguez, Greg Wilson, and Clark Wolf of Iowa State University, and the additional assistance of Robert Craig (University of Colorado), Hans Hansen (University of Windsor), Beth Innocenti (University of Kansas), Melanie Roberts (Emerging Leaders in Science & Society), Dane Scott (University of Montana), Assimakis Tseronis (University of Amsterdam) and Frank Zenker (Lund University).

The volume you hold benefited from the careful attention of Mary Speckhard and the design talent of Erin Zimmerman. Thanks, "peons"!

Jean Goodwin
Ames, IA

Keynote Address

Black Box Arguments and Accountability of Experts to the Public

SALLY JACKSON

Department of Communication
University of Illinois at Urbana-Champaign
Urbana, Illinois 61801
USA
sallyj@illinois.edu

ABSTRACT: Contemporary deliberation depends on an infrastructure of expertise the way travel depends on transportation infrastructure. Like other forms of infrastructure, the inner workings of expertise become less visible to its users, even as expertise itself becomes more indispensable. Accountability is an essential design requirement for any such system.

KEYWORDS: argumentation, blackboxing, design, normative pragmatics, expertise, infrastructure, L'Aquila earthquake trial.

1. INTRODUCTION

Argumentation is among the most powerful tools humanity has for building societies and solving problems. Argument stimulates deeper thought, motivates the search for more information, and at least in some cases produces alignment of belief. We learn through argumentation, both as individuals and societies. We also learn *about* argumentation, and as we learn about it, we change it profoundly. Argumentation in our day is technical not just in the content of some propositions or the qualifications of some participants but in the argument techniques that have evolved over millennia of practice.

Among the most visible of the evolutionary directions of argumentation are the increasingly disciplined bases for opinion-formation and the increasingly complex designs we have for incorporating disciplined opinion into decision making. This conference weighs one of those trends a bit more heavily than the other: focusing more on the growing dependence of citizens on experts than on the growing dependence of expertise itself on layers of technical achievements. However, both trends are important and need theoretical attention, and they are intimately interconnected. I am particularly pleased to be part of a conference that draws together such diverse disciplinary perspectives, because understanding the role of expertise in contemporary society is going to require all the disciplines with a piece of the problem to share their insights.

Despite disciplinary differences in which questions interest us most, what we all share as citizens is a need to make the best possible use of expertise, while reserving the possibility of holding experts accountable for their overall beneficence. Surveying the diversity of research topics at this conference suggests to me that it may be time to attempt a clean break from a notion that has had a powerful hold on rhetoric for centuries: the idea that appeal to authority is a *class* of arguments, some members of which are fallacious while others are sound. Contemporary argumentation studies inherited this notion, and it continues to drive a

Jackson, S. (2012). Black box arguments and accountability of experts to the public. In J. Goodwin (Ed.), *Between scientists & citizens: Proceedings of a conference at Iowa State University, June 1-2, 2012* (pp. 1-17). Ames, IA: Great Plains Society for the Study of Argumentation. Copyright © 2012 the author(s).

search for a standard set of tests of acceptability of appeals to authority that can be applied within the natural limits of a critical discussion—for example, lists of critical questions that can be asked about an argument (Walton, 1989). Plenty of clear cases of argument from authority can be found, but so can plenty of cases where the same epistemic problems are expressed quite differently. Many contemporary practices involving dependence on expertise resist assimilation to the concept of an argument from authority or an appeal to expertise and are not illuminated by the critical questions approach. For example, there are system-level issues to consider, such as the overall quality of the information environment (at this conference, Shanahan and others) and its lack of aids to citizen judgment (at this conference, Anelli; Kloster; and others); institution-level issues, such as management of known sources of bias in deliberative inquiry (at this conference, Rehg; Snoeck-Henkemans & Wagemans; and others), and individual-level issues having to do with citizens' capacity to judge information (at this conference, Pigliucci; Tachino & Russell; Weinel; and others) and with experts' own attitudes toward their expertise (at this conference: Blair; Zenker; and others).

So let's start with what any of us can see when we look around for expertise in play within variously scaled disagreement spaces. Unless we have already decided to restrict our attention to a certain class of arguments, the phenomena we see are quite heterogeneous. Consider an initial set of prototypes that will be familiar to everyone as settings where expertise is or may become a significant part of a disagreement space.

Prototype 1. Individual people purposefully searching for answers to questions, not knowing at first whether there are any experts in the subject. What they want is knowledge to guide their own actions and beliefs—not specifically an expert to trust. But questions about the expert basis for knowledge and about which experts may be trusted often come up when multiple sources have been found, and when the information sources do not all agree on the answer to the question. Most of us probably share the experience of searching the worldwide web for information on health either for ourselves or for family members. Some of us probably have family members who cherry-pick from among experts those whose advice happens to justify what they want to do and then appeal to these at need. In contemporary discourse, there is ample material that can be analyzed in terms of appeal to expert authority and ample opportunity to improve information literacy through critical questioning (from this conference, see McAfee; Nucci; and others).

Prototype 2. Individual people projecting their expertise, either because public information is their business or because they are hoping to build prestige, accumulate social capital, or attract jobs. Projection of expertise has become a small industry, supported by professional networking applications like LinkedIn and by more ambitious projects for mapping scientific expertise like Cornell's VIVO project (see http://vivo.cornell.edu/). The growth of interest in professional networking suggests that the problem ordinary citizens face in choosing among experts is mirrored by the problem experts face in positioning themselves to be the ones chosen from among all possible sources of advice (see especially Tindale, this conference).

Prototype 3. Individual experts responding to questions of public interest. This pattern occurs daily in countless news stories, and it is well known that journalists choose sources for reasons that have little to do with actual expertise. Journalists often have favorite quotemeisters, and in the search for balanced coverage of controversies, they may select for diversity of opinion rather than for presenting the public with a good basis for forming its own opinions.

Prototype 4. Panels of experts chartered to respond to requests for advice on specific points. Typically, this occurs as part of a larger deliberative process, with a subset of contested issues delegated to the experts while use of the answers is left to other participants such as policy makers or the public. The Intergovernmental Panel on Climate Change is a theoretically and practically significant example (Rehg, 2011, and related remarks at this conference), but lower profile efforts of the same kind have become common in policy formation and have given rise to a new form of expertise in design of policy deliberation (Carcasson, this conference).

Prototype 5. Expert bureaucracies that actually make decisions and carry out actions. I have in mind here the myriad agencies that develop and enforce various standards, like the U.S. Food and Drug Administration, but also the many entities that make decisions about how to allocate research funds among possible research topics, like the National Science Foundation and the National Institutes of Health (NIH). For situations of this type, knowing who is and is not an expert is nowhere near sufficient to safeguard public interest. People who know what they are talking about may still overvalue their own interests over others' interests, and people who have few checks on their behavior can take unfair advantage of that fact. The public has little to no say, for example, in how NIH allocates public funds among all of the diseases it might choose for funding.

I have arranged these prototypes along a continuum from the most ad hoc to the most institutionalized. All are important to understanding the role of expertise in what we believe and what we do. And to some extent, as we move up this ladder, the more institutional forms incorporate aspects of the ad hoc forms. For example, expert bureaucracies must have a steady supply of experts to perform specific judgmental tasks, including evaluation of the work of other experts. My own interest is increasingly focused on the more institutionalized end of this continuum, where questions about which individual experts to trust become less interesting and questions about institutional design become more interesting.

Notice that different disciplines involved in this conference have different strengths and weaknesses in dealing with the facets of the problem. Contemporary argumentation theory, especially pragma-dialectics, has greatly improved our understanding of how otherwise reasonable argument schemes may generate fallacious moves in discourse (van Eemeren & Grootendorst, 1992; Snoeck-Henkemans & Wagemans, this conference). Normative pragmatics adds theorizing about the obligations inherent in different speech acts (Goodwin, 2011 and this conference; Kauffeld, this conference). Informal logic suggests new ways of thinking about argument soundness (Johnson & Blair, 1994; Walton, 1989, 2008; represented at this conference by Blair; Tindale; and others). New directions in science studies offer powerful new concepts for thinking about varieties of expertise and about the meta-expertises required for how we live now (Collins & Evans, 2007; Collins & Weinel, 2011; Weinel, this conference). Other perspectives from communication, sociology, political science, and philosophy, too numerous and heterogeneous to cite, help us to understand institutions that emerge from practice and direct our attention to the integrity of our institutions and the problem-validity of their designs for making scientific knowledge actionable.

To the mix of perspectives that are already familiar in scholarship on expertise and expert-based arguments I will also add perspectives from design theory and from a newer line of work known as infrastructure studies. No amount of scientific literacy among the citizenry can eliminate the need for designing participation formats, so design theory is needed to guide how society takes advantage of expert knowledge. And particularly for the expressions of expert dependence that move up the ladder toward institutionalization, it may be helpful for us to think of certain products and practices as infrastructural.

2. ARGUMENTATION DESIGN THEORY

To address argumentation as it now occurs rather than as it once occurred requires that we acknowledge the role of design—and acknowledge as well the fact that design has changed the practice of argumentation over time. Argumentation design theory is not one cohesive strand of work but a set of design concepts and design practices that are implicit in many distinct strands of work. Design themes have been central to my own work, much of which has been done in collaboration with Scott Jacobs.

Jacobs' and my work is grounded in the natural design of conversational argument (Jackson & Jacobs, 1980; Jacobs & Jackson, 1989, 1992), and it recognizes a blurry and reflexive distinction between natural and built environments for argumentation (Jackson, 1998, 2008; Jackson & Jacobs, 2006). Most of our studies have been qualitative empirical discourse analyses, but with a strong theme around attempting to theorize design. I have proposed an explicit design methodology that could provide a framework for a version of translational work in our field (Jackson, 1998).

The natural design of argumentation in conversation is expansion around disagreement. In the normal course of conversation, argument starts with the participants becoming aware that they have a disagreement that makes some sort of difference to one or the other of them. Instead of moving forward with the business at hand, the participants reorganize their efforts around exploring and repairing the disagreement. They do this with varying levels of skill and with varying degrees of emotion.

Jackson and Jacobs have advanced three key propositions concerning the natural design of argumentation.

First, argumentation is about drilling down from disagreements, not building up from agreements. People do begin interactions with considerable bodies of common belief, but they do not and cannot enumerate these as the foundations from which to reach further agreements. Instead, people assume alignment until actions give evidence of misalignment.

Second, unless the misalignment is of no significance to anyone, it expands a speech act exchange by routing conversation into an attempt at repair. In ordinary conversation, the resolution process is commonly just more conversation. If the misalignment can be repaired, the conversation can return to its main business.

Third, sometimes, no resolution is possible within the natural limits of the conversation, and each new expansion simply exposes new disagreement space. Any conversational move provides some set of opportunities for disagreement, including not only what is actually asserted but also any belief or commitment the speaker reveals without actually stating it. Appeals to expert authority do not typically appear without context; they appear, as most argument content does, through challenge and response.

Argument expands around points of disagreement that the participants themselves choose from a constantly reshaping disagreement space. Sometimes participants accept standpoints simply because their co-participants have expressed commitment to these standpoints; other times they question the standpoint, but accept the thinnest possible backing; and other times, they attempt more extensive inquiry. Some have interpreted Grice's (1989)

Cooperative Principle as implying that without this willingness to accept a great deal of content simply on one another's say-so, conversation would be impossible.

So far I have been reviewing what we know about the natural design of argumentation and have said nothing about the problems presented by appeals to technical expertise. An appeal to expert authority occurring in ordinary conversation may look very much like the pattern familiar from informal logic texts: challenged for the basis for some exposed belief, a person may point to any form of support, including support from outside sources of all kinds. Sometimes the fact that authority can be cited for a standpoint is sufficient to end a line of argument. If not, asking critical questions about whether the outside source is truly authoritative is just one way to expand further. Another obvious way to expand the disagreement is for the challenger to demand to know how the outside source came to its conclusion, and yet another is to insist on referring the disagreement to someone outside the discussion itself. When we think about argument as a process of drilling down from disagreements, it is the challenger who must make some strategic calculation about how the discussion proceeds: whether to accept the authority's standpoint, or to challenge the credentials of that authority, or to introduce a competing authority, or to explore the expert basis for the authority's standpoint, or even to literally outsource the decision. These are various ways the sequence can unfold, not fundamentally different ways of reaching conclusions (and as Zenker pointed out at this conference, the challenger's choice among these alternatives may be based on calculation of which offers the greatest strategic advantage).

I want to extend these key propositions with a fourth: there is no limit in principle to how deep participants can go in drilling down from disagreement. But there are limits in practice, and participants sometimes find that it is simply not feasible to continue with ever more fundamental objections. And moreover, sometimes we make judgments about how likely it is that drilling down further will produce anything useful. This characteristic of natural argument is inherited by technical argumentation and by argumentation among experts themselves.

My task now is to show how these key findings about the natural design of argumentation might help us in thinking about technical argumentation and its role in practical affairs of citizens. In the US, we have an ongoing controversy over how schools should educate young teenagers about sex. Ideologically based, the main division of opinion is between those who believe that young people should be taught to abstain from sex until marriage and those who believe that young people should be taught to practice safer sex throughout their lives. This is a very complex controversy, one that has played out between parents and school officials in countless school districts as well as between politicians and public health agencies. A large empirical research literature has built around this controversy, and it is available for invocation by participants.

Figure 1 presents a partial concept map of some of the issues in the controversy as it had developed by late 2007. This is not an argument diagram but an abstraction from many individual arguments to a set of themes that connect to one another in various ways, such as support versus opposition. Whether any of these particular themes appear in any text or in any bounded discussion of sex education depends on what opinions participants have going in to the discussion and how they choose to expand from any disagreements that appear. The materials that are available for invocation from participants include artifacts remaining from earlier contributions. In the lower right hand corner my concept map shows just one of many such artifacts: a research report published in *BMJ* and widely covered in the popular press

(Underhill, Montgomery, & Operario, 2007). Most citizens who encountered this artifact saw it mentioned in the news, in stories announcing that abstinence-only sex education does not work, or in quasi-official websites with information on sex education programs.

In my own prior analysis of this controversy (Jackson, 2008), I suggested that expert arguments enter public debate as "black boxes" that are not really open to expansion without assistance of experts. In actor-network theory (from which I have borrowed this concept), a black box is a stabilized practice that has ceased to require explanation and defence within an expert community (Latour, 1987). An ordinary educated citizen can make a few inroads into evaluating the contents of the box, but to go very far the explorer will have to actually build expertise. Sooner or later, without specialty expertise, argument drills down to something that must be taken on expert authority.

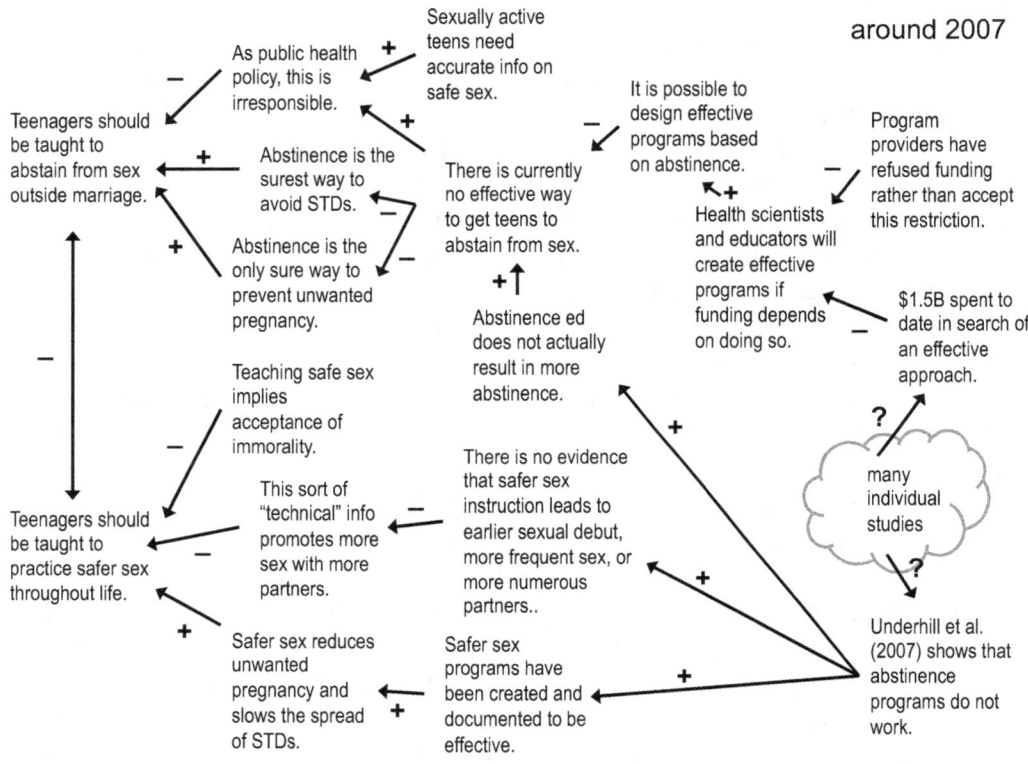

Fig. 1. Partial concept map of abstinence-only sex education controversy.

Consider the plight of the intrepid school board member who wants to understand how scientists could possibly conclude that abstinence education does not work—for it is the citizen who disagrees with this conclusion who is most likely to want to find its flaws. This intrepid soul can find the Underhill et al. (2007) article without much difficulty, since it has been mentioned in the news. With only the ubiquitous expertise of a literate adult, the citizen can identify a first level of backing for the conclusion: something like Figure 2 from the original article, reproduced below, showing a summary of all identifiable studies comparing abstinence-only sex education to "usual care" (whatever curriculum the school district had in place prior to introduction of a curriculum designed to teach abstinence from sex outside of marriage).

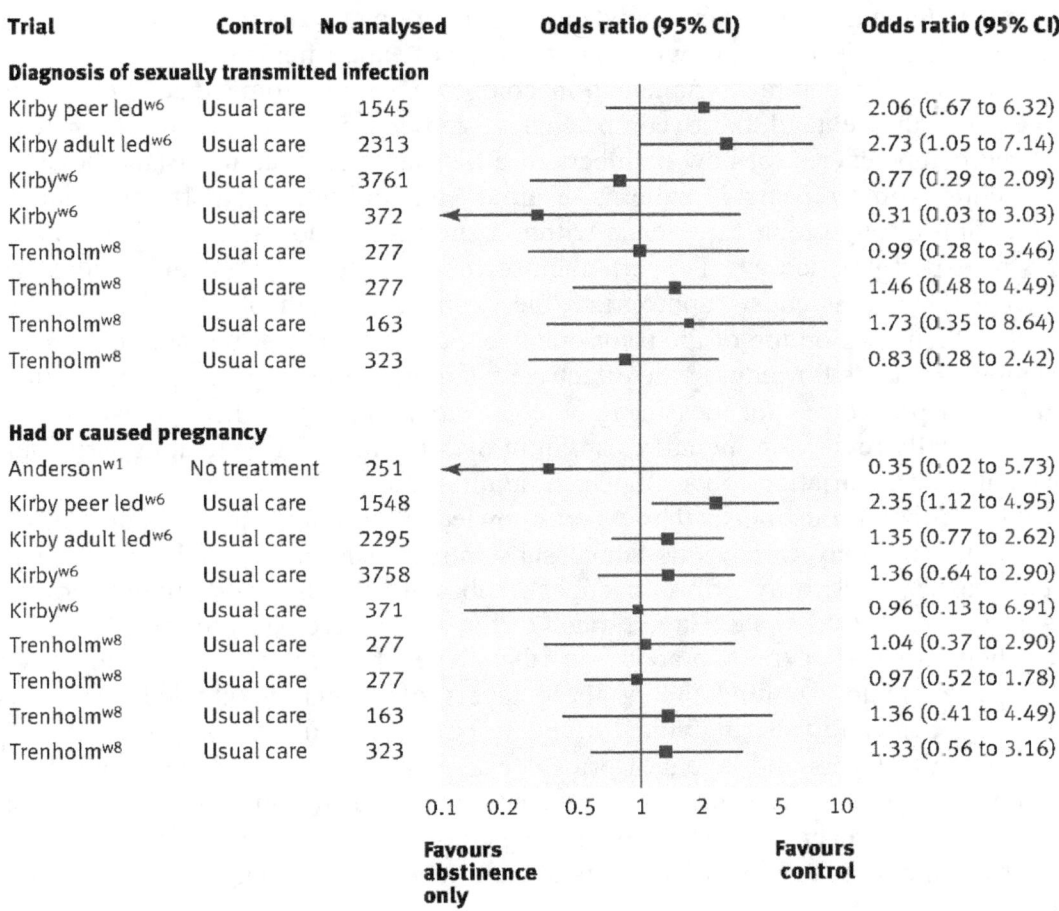

Trial	Control	No analysed	Odds ratio (95% CI)	Odds ratio (95% CI)
Diagnosis of sexually transmitted infection				
Kirby peer led[w6]	Usual care	1545		2.06 (0.67 to 6.32)
Kirby adult led[w6]	Usual care	2313		2.73 (1.05 to 7.14)
Kirby[w6]	Usual care	3761		0.77 (0.29 to 2.09)
Kirby[w6]	Usual care	372		0.31 (0.03 to 3.03)
Trenholm[w8]	Usual care	277		0.99 (0.28 to 3.46)
Trenholm[w8]	Usual care	277		1.46 (0.48 to 4.49)
Trenholm[w8]	Usual care	163		1.73 (0.35 to 8.64)
Trenholm[w8]	Usual care	323		0.83 (0.28 to 2.42)
Had or caused pregnancy				
Anderson[w1]	No treatment	251		0.35 (0.02 to 5.73)
Kirby peer led[w6]	Usual care	1548		2.35 (1.12 to 4.95)
Kirby adult led[w6]	Usual care	2295		1.35 (0.77 to 2.62)
Kirby[w6]	Usual care	3758		1.36 (0.64 to 2.90)
Kirby[w6]	Usual care	371		0.96 (0.13 to 6.91)
Trenholm[w8]	Usual care	277		1.04 (0.37 to 2.90)
Trenholm[w8]	Usual care	277		0.97 (0.52 to 1.78)
Trenholm[w8]	Usual care	163		1.36 (0.41 to 4.49)
Trenholm[w8]	Usual care	323		1.33 (0.56 to 3.16)

Favours abstinence only — Favours control

0.1 0.2 0.5 1 2 5 10

Fig. 2. Summary of evidence from Underhill et al. (2007).

Anyone who is not too easily discouraged can see that the study-by-study evidence divides unevenly between results that show abstinence programs more effective than the comparison and results that show abstinence programs less effective. More expertise is needed to avoid faulty reasoning at this level: sociological discrimination is required to understand that this pattern of findings is, overall, strong evidence against the success of abstinence programs judged against the outcomes measured in the studies, and that it is not evidence of scientific controversy (Weinel, this conference). Additional skeptical questions can be asked at this level. One might suspect, for example, that the authors of the study selected evidence in a biased way to support their own preferred conclusion; but to make any reasonable judgment about this question the individual must understand not only systematic review but also certain standard criteria scientists use to judge experimental design. If the individual is satisfied that the authors selected studies in an unbiased way and that the evident heterogeneity in the body of studies can still yield a consistent picture, a next recourse is to consider whether the individual studies have been fairly and correctly categorized, and this requires still more expertise, most particularly in statistics. Few citizens who are not themselves social scientists could get this far, but for those who reach this point and find nothing to object to, there are still avenues for critique. For example, before giving up entirely on the idea of abstinence education, one might want to examine the actual curriculum materials used in the individual studies to make qualitative judgments based on knowledge of child psychology, instructional design, or persuasive message design, all of which are themselves expert fields.

Among our natural argumentation competencies are some that allow us to invent improvements that extend these competencies, and these extended competencies become naturalized competencies for new members of a discourse community. Black boxes assemble these inventions into repeatable patterns of argumentation within expert communities. Black boxes develop through long experience within a substantive domain, and they become stable features of expert practice when experts themselves come to regard them as the usual way to draw a conclusion. Scientists appropriate these practices from other fields and build new practices of their own on top of the prior practice. So in the case of research on sex education, for example, methods for judging the effectiveness of educational programs are built on a pre-existing set of procedures for measuring outcomes and comparing them statistically. Methods for systematically reviewing literatures are built over methods for generating large numbers of studies that answer variations on a single central question.

For citizens, this means that expert knowledge introduced into practical discussion or policy making is likely to have its actual substantive basis in a very large number of very different expert fields. Any effort to engage substantively involves digging down through *layers* of expertise. And at each layer, much of the substantive basis for conclusions has been left implicit because expert practice at that level has long since been black-boxed.

How should we think about these layers of specialty knowledge? Set aside the question of whether to trust individual scientists. Can we really trust knowledge claims that come wrapped in layers and layers of prior decisions about how to draw conclusions? Every one of those layers involves some substantive content, but also some historical decision by one person or group to accept a way of doing things based on the expert guidance of some other person or group, and these historical decisions introduce path-dependence into a current array of decisions.

Although concepts and methods familiar in argumentation studies have much to contribute within layers, and something to contribute between layers, I suggest that theorizing about layering as a characteristic of contemporary argumentation practice requires some new concepts and methods. It is this notion that leads me to suggest that we look to infrastructure studies for ideas about how to theorize our heavily layered practices and about how to design better interfaces between one layer and another.

3. INFRASTRUCTURE THEORY

Does it make sense to talk about expert organizations and institutions as infrastructure for decision making? I think it does. At first, it may be very difficult to see where I am heading in this discussion of infrastructure theory. It will help to remember that I am arguing that argumentation itself has evolved many technical systems with device-like components. And it may also help if I say up front that the natural competence we all have for arguing is connected to these technical systems in the same way that people's natural ability to walk is connected to technical systems that aid in transportation and the way people's natural ability to talk is connected to technical systems for telecommunication. Technical systems do not usually replace natural competencies, but they do often change our options for how to do things.

The formal study of infrastructure is very new. The concept itself is really a twentieth-century concept, and scholarly interest in infrastructure has been rising since about the 1980s, with studies of large technical systems. Much of the work in infrastructure studies involves analysis of the evolution of particular systems, such as emergence of the power distribution infrastructure (Hughes, 1983). Lately, though, general theory has been emerging of what it is to be infrastructural, and it is also increasingly understood that skills and competencies can be infrastructural (Bietz, Baumer, & Lee, 2010).

An increasingly standard account of what makes a technical system infrastructural is that offered by Star and Ruhleder (1996), based on the observation that infrastructure is not a category of technical systems but a kind of relationship between a technical system and the work it supports. Systems are infrastructure for particular users or uses when they are embedded in ("sunk into") some practice, when they operate transparently and unproblematically, when they have extension in time and space beyond any one occasion of use, and when their use is "naturalized" in the learning of a practice built over them. They depend on standards (so as to interoperate with other systems), are built on the installed base of other and earlier infrastructure, and become visible when they "break."

Technical systems that become infrastructural for a set of users are those that can be taken for granted, incorporated as layers beneath other technical systems. The more infrastructural they become, the less visible they become to the user. Highway systems and electricity distribution systems are infrastructural for contemporary, developed economies, allowing individuals and communities to build other services that take these systems for granted.

As Edwards, Bowker, Jackson, and Williams (2009) point out, infrastructures cannot just be designed and built; they must *grow*, becoming infrastructural not as a matter of prior engineering but as a matter of incorporation into work. For a technical system, becoming *infrastructure* is an achievement that occurs slowly over time if it occurs at all (S. J. Jackson, Edwards, Bowker, & Knobel, 2007). Not all efforts to build infrastructure have been successful. Some have simply failed to attract users (Star & Ruhleder, 1996) and others have failed to achieve "can't-do-withoutness" (Edwards et al., 2009).

Technical systems are infrastructural for people who take them for granted and construct other activities that assume the technical capabilities as facts about the environment. A technical system may be considered infrastructure when its capabilities have been embedded so deeply in work that the work cannot continue if the technical capability fails, and especially when learning to do the work includes learning to rely on the technical system as though it were a natural fact.

A key insight of Star and Ruhleder (1996; also Bowker, Baker, Millerand, & Ribes, 2010, and others working in infrastructure studies) is that infrastructure must be understood relationally. A system is not born infrastructural; that status is earned slowly, as people take advantage of a system capability to create other layers of function over that capability.

One feature absent from prior accounts of infrastructure is recognition of the moral responsibilities that grow along with infrastructure. People who build and maintain technical systems do not necessarily envision the level of responsibility that might follow from other people coming to depend on the system. But as a technical system becomes embedded in people's work and daily lives, and as other systems and associated practices are built over them, a generalized duty of care begins to build as the system becomes more infrastructural. A system builder may not have had any ambition to become so essential to others' welfare, but once people come to depend on a system, responsibility for beneficence begins to attach to everyone who has a hand in the system.

For the study of how the public comes to depend on scientific expertise, much can be learned from infrastructure studies. The first and most important lesson is that, pervasively in contemporary human society, systems and associated practices are built over other systems and their associated practices, without necessarily any explicit attention to the terms under which this occurs. Explicit agreements about who is responsible for what tend to emerge late in the process of infrastructuring. First comes dependency of one system on another, then comes recognition that one system is vulnerable to failure in the other, and only then comes a desire for assurances that the embedded system will not be allowed to fail.

A second lesson is that social responsibility and duty of care are not elements of a job description, but obligations that follow automatically with how much consequence one person's actions may have for another's well-being. Building a bridge connected to a highway obligates someone (a government, usually) to see to the continued safety of the structure.

A third lesson is that infrastructuring occurs when trust is invested in a technical system, including in a system that is not (at first) worthy of trust. We put trust in physical infrastructures and build other technical systems over them even though we know that the infrastructure may fail. Often, the fact that other technical systems have been layered over a prospective infrastructure speeds the hardening of that system into something truly dependable—by raising the stakes for failure of the system.

Now I believe we are prepared to test a conjecture: that many dilemmas of contemporary practice will benefit from thinking about how to make expert advice as infrastructural as possible—designing them for dependability and then just depending on them until they need our attention again. This will shift our focus from evaluating individual experts or individual contributions of expert advice toward actively designing devices that organize scientific and technical knowledge for use.

4. THE L'AQUILA EARTHQUAKE TRIAL

As a context for testing my conjecture I have chosen events around a criminal trial that has been in progress since September 2011—the trial of seven scientists for alleged negligence prior to the 2009 L'Aquila earthquake (see "Scientists in the Dock," 2011, for a good overview). If we wanted to choose one controversy as the focus of a cross-disciplinary multi-perspectival study of expertise in public deliberation, we could hardly do better than the events surrounding the L'Aquila earthquake trial. This controversy demonstrates in the most tragic way what is at stake for both citizens and scientists when expertise is mismanaged. Depending on how we slice the events occurring before and after the earthquake, the case seems to fall right within the focal strength of any of several prominent lines of work. For pragma-dialecticians, there are derailed strategic manoeuvres to be untangled; for social epistemologists, there are conflicting sources of expert advice to the citizenry and conflicting claims about the track records of expert methodologies; for Third Wave science studies, there are disputes over who should be participating in the decisions; for informal logicians, there is a trial involving heavily contested expert testimony. I will of course concentrate on what this case might teach us about design and infrastructuring—but by no means do I think that we can understand this complicated set of events without many other theoretical perspectives.

So to review the history, on April 6, 2009, a devastating earthquake killed 308 people in the Italian city of L'Aquila. For several months prior to the large quake, the area had been experiencing smaller tremors. A researcher at a nearby laboratory had predicted that a major quake would occur, based on unusual emissions of radon gas, and had been denounced by the local authorities. On March 31, 2009, civil authorities convened a panel of scientists in L'Aquila to evaluate the level of risk. Discussion at that meeting focused on the impossibility of accurately predicting when earthquakes will occur. Nevertheless, just one week before the quake, public officials reassured the people in the region that there was no special danger despite the ongoing swarm of smaller quakes. One official stated that, in fact, the swarm of smaller quakes was actually a good sign. He claimed to have gotten this opinion from "the experts." Citizens who had been sleeping out of doors returned to their homes, and many died a week later as a result of that decision.

A little over a year after the quake, Italian prosecutors brought criminal charges of manslaughter against seven scientists (the meeting convenor Bernardo de Bernardinis and six experts in seismology), provoking a worldwide outcry by other scientists (Hall, 2011). The trial has been underway since September 2011, and the seven defendants made their first statements just two days before our conference opened.

Let us return to five prototype problems I mentioned earlier. In many controversies, all of these prototypes appear in a tangle, and this is especially noticeable in the events surrounding the L'Aquila earthquake.

Prototype 1. Individual citizens were alarmed by the swarm of small tremors and were highly motivated to seek out expert advice on whether to sleep outdoors. Two very different opinions were available to them: at one extreme, the highly alarmist opinion of Giampaolo Giuliani, who claimed to be able to predict that a large quake was imminent; and at the other extreme, the highly reassuring opinion of Bernardo de Bernardinis, who claimed that the small tremors were actually a favorable indication that there would be no large quake in the near future. Here citizens were forced to actually choose between sources claiming expertise. Neither, however, was himself a seismologist. Giuliani based his claims on his own observations, using a nonstandard method for prediction. De Bernardinis based his claims (so he said) on consultation with other scientists.

Prototypes 2 and 3. Active projection of expertise was at work as well, with Giuliani attempting to build his own credibility by explaining the nature of his experience and the novel basis of his prediction and other scientists attempting to discredit him. Journalists of course played their usual role, seeking out commentators on Giuliani's predictions, both before and after the deadly quake.

Prototype 4. A now infamous hearing of the Major Risks Commission brought together six highly qualified experts for the specific purpose of determining whether the series of smaller quakes indicated elevated risk for L'Aquila and the surrounding region. The convening of an expert panel is the sort of thing that commonly happens when an expert community has not reached consensus on an issue and when much is at stake. There is now evidence (from a recorded telephone conversation) that the expert panel was convened for the specific purpose of squelching the alarmist message of Giampaolo Giuliani (Cartlidge, 2012; Nosengo, 2012), and that is in fact what happened as a result. Here we see the deep need for looking beyond questions of who may participate as an expert. There is no doubt whatsoever that the six earthquake scientists brought together at L'Aquila "knew what they were talking about," but somehow they seem not to have known what to say to the people of L'Aquila before, during, or after the meeting (Hall, 2011).

Prototype 5. The scientists assembled at L'Aquila were not random choices from the expert community but members of the Major Risks Commission, an institutionalized component of the national public safety bureaucracy. They were also members of global research communities whose research topics and methods carry institutional history of all kinds, ranging from path-dependent scientific modeling techniques to the highly contingent professional values that defend boundaries between scientific knowledge and the sociopolitical uses of that knowledge. Public officials had a legitimate institutional interest in suppressing Giuliani's message if they believed it to be scientifically invalid; allowing his alarmist message to be spread without countering it risked losing the public's confidence in the public agencies responsible for risk information.

Later it may be possible to do much more with this tragic and complicated case, but a few observations are possible even now.

First, in circumstances like this, citizens cannot take individual responsibility for evaluating competing expert claims. They must delegate this work. There was a formally composed body whose remit was to do this work. The meeting convened to do this work had an appropriate and usually reliable design: a consultation with a panel of the most highly qualified scientists available. Other experts reviewing the transcript of the meeting have affirmed the scientific correctness of statements made by the participants, especially on the impossibility of accurately forecasting earthquakes. However, this normally reliable device was invoked not by mutual agreement among participants, but as a coercive move by government officials who already knew what answer they hoped to get from the inquiry and who had also already tried other methods of repressing Giuliani's alarmist message. The meeting lasted less than an hour.

Significantly, what came out of the meeting was a statement to the press by a nonspecialist rather than a formal statement authored by the specialists (Nosengo, 2010). Whenever one technical system must interact with another, the interface becomes important. There were opportunities for compromise and distortion at multiple points in the process of consulting the scientists about the danger of a major earthquake, and this was most evident in the interface between the panel of earthquake scientists and the press.

Second, an integral part of the controversy has been who should and who should not be contributing to decision making in virtue of their expertise, but in addition to the ground-level dispute over participants' legitimacy, there are undisputed experts involving themselves on issues beyond their expertise.

Consider Giampaolo Giuliani first. He is not a seismologist, but he is a working scientist who holds a patent on a device for measuring radon emissions and had been working on the use of his predictive technique for some time. Authorities attempted to suppress his contributions (Sample, 2009), referring to him as a scaremonger and even (indirectly) as an imbecile (Cartlidge, 2012). However, the idea of predicting earthquakes from gas emissions is not in itself implausible, and Giuliani believed that his own investigations were validating his methods. One might say that Giuliani is in the "should not contribute" category, because his ideas had been considered and rejected by the relevant experts, specifically by those who later played a part in the L'Aquila meeting of the Major Risks Commission (Dollar, 2010). But remember that Giuliani was working on a new method based on this idea. A person's participation in decision making cannot be dependent on employment of techniques that are already blackboxed; blackboxing is something that occurs after multiple successful uses of a technique, not something achieved in advance of successful use.

Next consider Bernardo de Bernardinis. He also is a scientist but not an expert in earthquakes. He convened the March 31 hearing, and he gave statements to the press on behalf of the Commission. The gist of his statement was that the swarm of small tremors did not indicate any elevated risk of a large quake, and that in fact the swarm of small tremors was a favorable sign. A transcript of the hearing was published in an Italian magazine, and most capable readers find that the co-defendant scientists were scientifically correct in their discussion, saying nothing that would have justified de Bernardinis in this conclusion (Jordan, 2011; Nosengo, 2010). Based on a recorded telephone message disclosed during the trial, it appears that de Bernardinis was instructed to convene the hearing with the specific purpose of finding a basis for reassuring the citizens (Nosengo & Nature News Blog, 2012), and he may well have felt that his idea was confirmed by the specialists' review of the situation. His six colleagues undoubtedly knew that what de Bernardinis said was false and dangerous; that is, they undoubtedly knew that people should remain vigilant. But they did not contradict de Bernardinis, then or later. The members of the Major Risks Commission definitely belong in the "should contribute" category, but what now becomes apparent is that *what* people should contribute is more important than *which people* should contribute.

I will not discuss the contributions of the other six defendants in the trial. Each of them made their first appearance in court as I prepared these remarks. But let us consider, finally, the participation of the broader scientific community, especially after the announcement that the seven commission members would be indicted and tried for manslaughter. About the immediate protest from within the scientific community, it should be noted that geophysicists and other scientists are no better positioned than any other citizen to judge charges of negligent conduct, if these are not based on the quality of the science but on the performance of a public duty. Among the range of possible bases for a charge of negligence would be focusing on issues peripheral to the assessment of risk, collaborating in a sham proceeding with a predetermined outcome, allowing a spokesperson to misrepresent the scientific conclusions, or simply failing to anticipate public response to an overly reassuring message. Any of these, if they occurred, would fail ethical standards such as those suggested by Hardwig (1994), but whether these failures occurred, and whether their occurrence constitutes criminal conduct, is clearly not a scientific question that can be decided by other scientists.

The third and most important element of this case is what it reveals about the importance of accountability for anything the public is expected to treat as infrastructure. From the perspective of infrastructure theory, earthquake measurement and monitoring make up an infrastructure-in-waiting. These activities are already highly standardized, and they are invisible to the public except as reports of how powerful a given tremor was. But these capabilities have not yet achieved infrastructure status because people have not quite figured out what to do with them. A technical system becomes infrastructural when practices are built on top of it, and what can be built on top of earthquake science is still a source of tension between citizens and scientists. Despite scientists' widespread agreement that earthquakes cannot be accurately predicted, citizens want and expect predictions of when and where earthquakes will occur. This may never be successful, but it is certainly the case that people will continue searching for ways to predict earthquakes. Scientists themselves envision a different infrastructural use of earthquake science: as a justification for construction standards that make earthquakes survivable. Unfortunately, public officials and citizens alike are far more interested in earthquake warnings than in long-term investment in safer buildings, a point that some scientists have made about L'Aquila (Hall, 2011; "Science in the Dock," 2011).

Before this or any other variety of expertise can become infrastructural, someone must accept the duty of care that comes from building something that others build over. On this point, I want to examine two formal statements issued by scientific organizations, one issued by the American Geophysical Union and another written by Alan Leshner on behalf of the American Association for the Advancement of Science. These statements trouble me, because they not only fail to acknowledge the expert community's duty of care but actively reject it for scientists engaged in important public decisions. Specifically the two statements suggest that scientists may refuse to share their expertise unless they themselves are held harmless:

> The criminal charges against these scientists and officials are unfounded. Despite decades of scientific research in Italy and in the rest of the world, it is not yet possible to accurately and consistently predict the timing, location, and magnitude of earthquakes before they occur. It is thus incorrect to assume that the L'Aquila earthquake should have been predicted. The charges may also harm international efforts to understand natural disasters and mitigate associated risk, because *risk of litigation will discourage scientists and officials from advising their government or even working in the field of seismology and seismic risk assessment.* (Anonymous, 2010, italics mine)
>
> Years of research, much of it conducted by distinguished seismologists in your own country, have demonstrated that there is no accepted scientific method for earthquake prediction that can be reliably used to warn citizens of an impending disaster. To expect more of science at this time is unreasonable. It is manifestly unfair for scientists to be criminally charged for failing to act on information that the international scientific community would consider inadequate for issuing a warning. Moreover, *we worry that subjecting scientists to criminal charges for adhering to accepted scientific practices may have a chilling effect on researchers, thereby impeding the free exchange of ideas necessary for progress in science and discouraging them from participation in matters of great public importance.* (Leshner, 2010, italics mine)

At issue is whether de Bernardinis and the six earthquake scientists exercised the due diligence demanded by the situation. Until the trial is concluded, no one is really in a very good position to say whether anyone failed in their duty of care for the citizenry. However, for citizens to depend on expert communities in decision making, representatives of those expert communities must be accountable for performance failures, just as any infrastructural system includes accountability for individuals who build and operate the infrastructure. The Major Risks Commission clearly had a duty of care with respect to the citizens of L'Aquila, and the criminal trial is about whether this duty was shirked when scientific opinion about the predictability of earthquakes was invoked as support for public misinformation—the reassuring official statement that the small tremors were a favorable sign rather than a cause for continued vigilance. As Hardwig (1994) observed, the public depends on experts to blow the whistle in cases like this one.

My point here agrees with Douglas (2003), who argues that scientific practice is subject to the same moral responsibilities as apply to any other social action: the responsibility to consider and weigh risk associated with one's own choices, and especially with the unintended consequences of one's own choices. Scientists who take actions without thinking through potential consequences may create harm through negligence, regardless of whatever scientific purpose may be served by the action. While Douglas's argument is focused on choice of scientific programs rather than on advice-giving, her position that scientists must accept that there can be negligence in their performance as scientists is potentially relevant to this case.

5. CONCLUSION

How scholars deal with the theoretical problems of understanding expertise will naturally affect (and be affected by) how the public deals with the practical problems of evaluating expert opinion. Especially throughout the later half of the 20th Century, citizen groups and government agencies have been tinkering with designs for tapping into our vast stores of expertise. But the design process has been guided by trial-and-error more than by theory. What might a well-developed argumentation design theory offer?

First let me clarify what I mean by design theory. Whatever its specific content, the theory should be capable of supporting design, including proposals for novel designs. Reasoning from analogy with other design enterprises, we should expect that each unique controversy may have unique design requirements—as do unique building sites. Design disciplines have general principles and models of how these principles apply in particular settings, but they assume that any new situation will require its own design based on application of the principles and inspired by the successful models of the past. The characteristics of the controversy should shape the design of a deliberation process the way a building site shapes the design of the structure. The role of theory is to suggest design hypotheses that get implemented as actual designs for the conduct of argumentation.

Argumentation design theory will likely draw content from many sources, the most obvious of which are normative theories of argumentation (van Eemeren, Grootendorst, Jackson, & Jacobs, 1993). I have suggested in these remarks that certain design theory problems may also benefit from what we know about other layered technical systems—that is, from what we know about infrastructure. There is infrastructural potential in society's scientific and technical capacity, but infrastructure isn't infrastructure until its operation can be taken for granted and other activities can be confidently built over it. Good designs for exploitation of technical expertise must let the details of expert judgment sink into invisibility as far as the ordinary citizen is concerned, but both the basis for expert judgment and its social consequences must become a matter of great care for those who work inside the technical system. In other words, instead of focusing circumstantially on whether and when it is reasonable to place confidence in technical experts, argumentation design theory focuses on how to design trustworthiness into relationships between technical experts and citizens.

How, exactly, does trustworthiness get designed into a system? Turning again to comparison with infrastructural systems, a key feature is an acknowledged duty of care. People in the infrastructure business have an inescapable duty of care for those who depend on them. Scientific fields must embrace this duty of care, and while it is possible that what is happening in L'Aquila will discourage scientists from giving advice, it is also possible that individual scientists who engage with public controversy will grow accustomed to accepting the same kinds of personal risk that public officials do.

Our ability to rationally manage our dependence on technical expertise is vulnerable in many ways, but among these, one of the most serious vulnerabilities is the notion that experts themselves should have no accountability for the foreseeable consequences of how they deploy their expertise. But accountability, like other aspects of a sociotechnical system, must be designed with care to achieve intended effects (trust) without unwanted effects (suppression of experts' participation in practical affairs). Accountability may take much less drastic forms than criminal prosecution, and it can incorporate protection for those acting in good faith. The worldwide scientific community has been shocked by the criminal prosecution of these Italian scientists. But this extreme method of calling the experts to account was only attractive because of the absence of any well-proportioned method of doing so.

From an infrastructure theory perspective, the task we face is not deciding whether to trust experts or which experts to trust, but to recognize that expertise is a societal investment that can be shaped for greater trustworthiness. Professional training can be infused with ethical principles (Hardwig, 1994) and with analytic frameworks for considering broader social impacts (Cohen, 2012). Accountability structures can be devised for more institutionalized forms of expert participation. Procedural checks and balances can be incorporated to control for known sources of bias. And other strategies are of course possible. Since we must depend on experts, we must make it as safe as possible to do so.

ACKNOWLEDGEMENTS: Thanks to Scott Jacobs for his many detailed comments and suggestions on this manuscript, and for years of collaboration on argumentation design theory.

REFERENCES

Anonymous (2010). AGU statement: Investigation of scientists and officials in L'Aquila, Italy, is unfounded. *Eos, Transactions, American Geophysical Union, 91*(28), 248. Retrieved from http://www.agu.org/pubs/crossref/2010/2010EO280005.shtml
Bietz, M. J., Baumer, E. P. S., & Lee, C. P. (2010). Synergizing in cyberinfrastructure development. *Computer Supported Cooperative Work: The Journal of Collaborative Computing, 19*(3–4), 245–281.
Bowker, G. C., Baker, K., Millerand, F., & Ribes, D. (2010). Toward information infrastructure studies: Ways of knowing in a networked environment. In J. Hunsinger, L. Klastrup, & M. Allen (Eds.), *International handbook of Internet research* (pp. 97–117). Dordrect, Netherlands: Springer.
Cartlidge, E. (2012, January 26). Italian official added to list of defendants in earthquake trial. *ScienceInsider.* Retrieved from http://news.sciencemag.org/scienceinsider/2012/01/italian-official-added-to-list.html
Cohen, J. (2012, February 7). A seismic crime. *ABC Science.* Retrieved from http://www.abc.net.au/science/articles/2012/02/07/3425182.htm
Collins, H. M., & Evans, R. J. (2007). *Rethinking expertise.* Chicago, IL: University of Chicago Press.
Collins, H., & Weinel, M. (2011). Transmuted expertise: How technical non-experts can evaluate experts and expertise. *Argumentation, 25,* 401–413.
Dollar, J. (2010, April 5). The man who predicted an earthquake. *The Guardian.* Retrieved from http://www.guardian.co.uk/world/2010/apr/05/laquila-earthquake-prediction-giampaolo-giuliani
Douglas, H. E. (2003). The moral responsibilities of scientists: Tensions between autonomy and responsibility. *American Philosophical Quarterly, 40,* 59–68.
Edwards, P. N., Bowker, G. C., Jackson, S. J., & Williams, R. (2009). Introduction: An agenda for infrastructure studies. *Journal of the Association for Information Systems, 10*(5), 364–374.
Goodwin, J. (2011). Accounting for the appeal to the authority of experts. *Argumentation, 25,* 285–296.
Grice, H. P. (1989). *Studies in the way of words.* Cambridge, MA: Harvard University Press.
Hall, S. S. (2011). Scientists on trial: At fault? *Nature, 477,* 264–269.
Hardwig, J. (1994) Toward an ethics of expertise. In D. E. Wueste (Ed.), *Professional ethics and social responsibility* (pp. 83–101). Lanham, MD: Rowan and Littlefield.
Hughes, T. P. (1983). *Networks of power: Electrification in Western society 1880–1930.* Baltimore, MD: Johns Hopkins University Press.
Jackson, S. (1998). Disputation by design. *Argumentation, 12,* 183–198.
Jackson, S. (2008). Black box arguments. *Argumentation, 22,* 437–446.
Jackson, S., & Jacobs, S. (1980). Structure of conversational argument: Pragmatic bases for the enthymeme. *Quarterly Journal of Speech, 66,* 251–265.

Jackson, S., & Jacobs, S. (2006). Designing countermoves to questionable argumentative tactics. In F. H. van Eemeren, M. D. Hazen, P. Houtlosser, & D. C. Williams (Eds.), *Contemporary perspectives on argumentation: Views from the Venice Argumentation Conference* (pp. 83–100). Amsterdam, the Netherlands: International Centre for the Study of Argumentation (SICSAT).

Jackson, S. J., Edwards, P. N., Bowker, G. C., & Knobel, C. P. (2007). Understanding infrastructure: History, heuristics, and cyberinfrastructure policy. *First Monday, 12*(6), 24–24.

Jacobs, S., & Jackson, S. (1989). Building a model of conversational argument. In B. Dervin, L. Grossberg, B. J. O'Keefe, & E. Wartella (Eds.), *Rethinking communication: Paradigm exemplars* (pp. 153–171). Beverly Hills/Newbury Park, CA: Sage.

Jacobs, S., & Jackson, S. (1992). Relevance and digressions in argumentative discussion: A pragmatic approach. *Argumentation, 6*, 161–176.

Johnson, R. N., & Blair, J. A. (1994). *Logical self-defense*. New York, NY: McGraw-Hill.

Jordan, T. (2011, September 21). Don't blame Italian seismologists for quake deaths. *New Scientist, 2831*. Retrieved from http://www.newscientist.com/article/mg21128310.200-dont-blame-italian-seismologists-for-quake-deaths.html?DCMP=OTC-rss&nsref=online-news

Latour, B. (1987). *Science in action: How to follow scientists and engineers through society*. Cambridge, MA: Harvard University Press.

Leshner, A. (2010). Open letter to President Giorgio Napolitano. Retrieved from www.aaas.org/gr/docs/10_06_29earthquakelettertopresidentnapolitano.pdf

Nosengo, N. (2010). Italy puts seismology in the dock. *Nature, 465*, 992. Retrieved from http://www.nature.com/news/2010/100622/full/465992a.html

Nosengo, N. (2012, January 25). Wiretap evidence could aid Italian seismologists' defense. *Nature* News Blog. Retrieved from http://blogs.nature.com/news/2012/01/wiretap-revelation-could-aid-italian-seismologists-defence.html

Nosengo, N., & Nature News Blog (2012, February 16). California seismologist testifies against scientists in Italy quake manslaughter trial. *Scientific American™*. Retrieved from http://www.scientificamerican.com/article.cfm?id=california-scientist-test

Rehg, W. (2011). Evaluating complex collaborative expertise: The case of climate change. *Argumentation, 25*, 385–400.

Sample, I. (2009, April 6). Scientist was told to remove internet prediction of Italy earthquake. *The Guardian*. Retrieved from http://www.guardian.co.uk/world/2009/apr/06/italy-earthquake-predicted

Scientists in the dock: An extraordinary manslaughter trial starts in Italy (2011, September 17). *The Economist*. Retrieved from http://www.economist.com/node/21529006

Star, S., & Ruhleder, K. (1996). Steps toward an ecology of infrastructure: Design and access for large information spaces. *Information Systems Research, 7*(1), 111–134.

Underhill, K., Montgomery, P., & Operario, D. (2007). Sexual abstinence only programmes to prevent HIV infection in high income countries: A systematic review. *BMJ, 335*(248). doi: 10.1136/bmj.39245.446586.BE

van Eemeren, F. H., & Grootendorst, R. (1992). *Argumentation, communication, and fallacies: A pragma-dialectical perspective*. Hillsdale, NJ: Erlbaum.

van Eemeren, F. H., Grootendorst, R., Jackson, S., & Jacobs, S. (1993). *Reconstructing argumentative discourse*. Tuscaloosa, AL: University of Alabama Press.

Walton, D. N. (1989). *Informal logic: A handbook for critical argumentation*. Cambridge, UK: Cambridge University Press.

Walton, D. N. (2008). *Informal logic: A pragmatic approach* (2nd ed.). Cambridge, UK: Cambridge University Press.

Keynote Address

Nonsense on Stilts about Science: Field Adventures of a Scientist-Philosopher

MASSIMO PIGLIUCCI

Philosophy Program,
The Graduate Center at the City University of New York
365 Fifth Avenue
New York, NY 10036
USA
massimo@platofootnote.org

ABSTRACT: Public discussions of science are often marred by two pernicious phenomena: a widespread rejection of scientific findings (e.g., the reality of anthropogenic climate change, the conclusion that vaccines do not cause autism, or the validity of evolutionary theory), coupled with an equally common acceptance of pseudoscientific notions (e.g., homeopathy, psychic readings, telepathy, tall tales about alien abductions, and so forth). The typical reaction by scientists and science educators is to decry the sorry state of science literacy among the general public, and to call for more science education as the answer to both problems. But the empirical evidence concerning the relationship between science literacy, rejection of science and acceptance of pseudoscience is mixed at best. In this chapter I argue that—while certainly important—efforts at increasing public knowledge of science (science education) need to be complemented by attention to common logical fallacies (philosophy), cognitive biases and dissonance (psychology), and the role of ideological commitments (sociology). Even this complex, multi-disciplinary approach to science education will likely only yield measurable results in the very long term. Meanwhile science remains, as Carl Sagan famously put it, a candle in the dark, delicate and in need of much nurturing.

KEYWORDS: science literacy, science education, philosophy, psychology, sociology, pseudoscience.

1. INTRODUCTION: THE ACCIDENTAL PUBLIC INTELLECTUAL

Most academics do not engage in public outreach, for a variety of reasons. First off, it is at least implicitly discouraged by the very structure of the academy, where one's career is advanced (in decreasing order of importance, regardless of what anyone would tell you) by one's success in research or scholarship (measured by peer-reviewed publications and grants), one's teaching, and finally one's service. It would be nice if at least the somewhat vague category of "service" included outreach lecturing and writing, but it is in fact pretty much confined to *internal* (and not really very useful) service, such as on departmental or college committees.

Second, many of us are simply never trained for public outreach (or teaching, for that matter), and do not know how to relate in an understandable and engaging way to laypeople. Contra a popular academic myth, there is no particular reason for there to be a strong correlation between one's scholarly excellence in a highly technical field of expertise and one's ability to communicate in a non-technical matter about the broader context in which that scholarly work flourishes or makes sense (Sperber, 2001).

Consequently, until 1997 I was yet another example of a non-engaged academic, pursuing my first tenure track job at the University of Tennessee, within the specialty field of

Pigliucci, M. (2012). Nonsense on stilts about science: Field adventures of a scientist-philosopher. In J. Goodwin (Ed.), *Between scientists & citizens: Proceedings of a conference at Iowa State University, June 1-2, 2012* (pp. 19-28). Ames, IA: Great Plains Society for the Study of Argumentation. Copyright © 2012 the author(s).

evolutionary biology (and more specifically the study of gene-environment interactions in plant model systems). But then something happened that made me pay attention and turned me into the accidental public intellectual that I have become since. A few months earlier, in March 1996, the Tennessee state legislature had attempted to pass a bill that would have ensured equal time for creationism in science classes throughout that state's public high schools. A *New York Times* article published at the time summarized the idea in this manner:

> "If evolution is true, then it has nothing to fear from some other theory being taught; the truth will prevail," State Senator David Fowler, a Republican from Chattanooga, argued on the Tennessee Senate floor this week. "But if intelligent design is the truth, then God forbid we should not teach it to our children." (Applebome, 1996)

The bill actually died in committee (though, unfortunately, the TN legislators were at it again in 2012, this time successfully (Thompson, 2012)). But during those months I had awoken to a reality that perhaps should not have surprised me to begin with: I was now living (and operating professionally, as an evolutionary biologist!) in the middle of the Bible Belt, in a country characterized by one of the most religious-puritanical attitudes among modern advanced societies (Uhlmann Poehlmanb, Tannenbaumc, & Bargh, 2012). As a reaction, I started one of the very earliest "Darwin Day" events on campus, an outreach program aiming at explaining to the public the nature of science and at exploring the relationship between science and religion. It has since become a major international event, with hundreds of locations worldwide every year (darwinday.org).

Once I got started, I realized that public outreach ought to be an important part of what academics should be doing, in part because they are among the best-positioned individuals in our society to function as public intellectuals, to aid the layperson—in the words of Noam Chomsky—in the pursuit of a course in "intellectual self-defense" (Baillargeon, 2008). This quickly led me to maintaining a blog devoted to science and philosophy for the public (rationallyspeaking.org), as well as to publish a number of advocacy books concerned with pseudoscience (Pigliucci, 2002, 2010). Most recently, and to my own amusement, my scholarship and outreach converged, with the co-editing of a new book on the philosophy, history, and sociology of the so-called demarcation problem (Popper, 1957), i.e. the conceptual divide between science and pseudoscience (Pigliucci & Boudry, 2013). What follows is a summary of the lessons I have learned while putting together the book and during the now more than fifteen years devoted to interacting with the public, initially as a scientist, and more recently as a philosopher.

2. NOT JUST SCIENCE EDUCATION

It is common for scientists and science educators to bemoan the fact that the American public is not scientifically literate, a state of affairs that is often blamed for much misunderstanding of science and embracing of pseudoscience. Indeed, the American Association for the Advancement of Science thinks that science literacy is so crucial that they have launched an ambitious "2061" project to improve it substantially (www.project2061.org/publications/bsl). There are good reasons to think that scientific literacy matters (Hazen, 2002), including so that people can better appreciate the world around us and, most crucially, make intelligent decisions about their lives—including who they vote for when it comes to public issues informed by science.

However, scientific literacy cannot be the whole story. A report by the National Science Foundation pointed out that *both* Americans and Europeans do not "have a firm grasp of basic scientific facts and concepts, nor do they have an understanding of the scientific process," (2004, p. 7-15) which makes it puzzling why, for instance, denial of evolution is a popular stance in the US but not in Europe (Miller, Scott, & Okamoto, 2006).

Indeed, a number of years ago one of my students and I carried out a preliminary study of the relationship between scientific literacy and belief in pseudoscience, and the results were not at all encouraging (Johnson & Pigliucci, 2004). For instance, while we found a (small, but statistically significant) effect of major (science vs. non-science) in the amount of *factual* knowledge of science that undergraduate students were able to master, we uncovered no significant differences between science and non-science majors in either *conceptual* understanding of science or belief in pseudoscience. (Interestingly, and contra popular lore, there was also no effect of gender on either science understanding or acceptance of pseudoscience.)

A more recent and much larger survey confirmed our results (Impey, Buxner, Antonellis, Johnson, & King, 2011): 10,000 students taking astronomy classes as part of their general education requirements were tested, and it turned out that belief in pseudoscience was high in the sample, that students' degree of science literacy was only marginally better than in the general population, and—most crucially—that there was no correlation between their science literacy and their acceptance of pseudoscience. The authors also found that over a period of two decades there was no measurable improvement in students' science literacy.

This is not exactly surprising to people who have spent years "in the field," so to speak, actually talking to (and sometimes debating) proponents of pseudoscientific notions. Contrary to popular belief among academics (and scientists in particular) these people are both intelligent and knowledgeable about science. Take the issue of creationism, for instance. I have debated the likes of Duane Gish (Institute for Creation Science) and Michael Behe and William Dembski (both associated with the Discovery Institute), among others. Gish and Behe have a PhD in biochemistry, and Dembski has academic credentials in both mathematics and philosophy. They know far more about natural selection, the chemistry of life, and so forth than most, and yet they subscribe to untenable notions of young earth creationism (Gish) and Intelligent Design creationism (Behe and Dembski). This observation is not limited to the leaders of these movements either, as a good number of the rank and file creationists (or paranormalists, or climate change deniers, you name it) that I've met during the years are also well-versed in the basics of science, and can discourse on points of scientific method and even philosophy of science.

None of this argues that science literacy is not a worthy goal, but it certainly points out that—in itself—more general science education is not likely to significantly ameliorate the problem. The implication is that there must be other, so far less explored, factors playing into so much misunderstanding of science and acceptance of pseudoscience by the general public. My suggestion is that we need to consider three other, interconnected, spheres of influence: philosophy (particularly as it concerns critical thinking and informal logical fallacies), psychology (pertinent to people's proneness to engage in cognitive biases and cognitive dissonance), and sociology (concerning the strength and dynamics of people's ideological commitments).

3. FROM EDUCATION TO PHILOSOPHY AND PSYCHOLOGY

Professional skeptics—i.e., people who spend their time debunking all sorts of pseudoscientific claims—will quickly point out that believers in pseudoscience commit an orgy of logical fallacies, particularly of the informal kind (e.g., www.theskepticsguide.org/resources/logicalfallacies.aspx). Indeed, a recent project by Richardson, Smith, and Meaden (2012) has cataloged the most common fallacies and allows internet browsers to paste a fallacy-specific link to any blog, article, or commentary they find around the internet (hence the title of the project: "Your Logical Fallacy Is...").

Examples are in abundance. Consider just the following small sample (more in Baillargeon (2006)):

> Hasty generalization: "Acupuncture works; my brother stopped smoking by seeing an acupuncturist."
> *Ad Populum*: "Most people believe in astrology, so there must be something to it."
> False dilemma: "Either medicine can explain how someone was cured, or it is a miracle."
> Appeal to ignorance: "No one has ever proved that UFOs do not exist, so they might exist."
> Red herring: During a discussion on global warming someone says: "What you really have to worry about is a government too prone to regulating the economy, which will keep people from being decently employed."
> *Post hoc, ergo propter hoc*: "I was wearing a red sweater when I won at the casino. If I wear the red sweater again, I will win again." (Richardson et al., 2012)

And the list could go on and on. If we begin with the (empirically testable) assumption that awareness of logical fallacies helps people think more clearly about issues, then it follows that courses in critical thinking and informal reasoning may be significantly more effective than straightforward science education (though, of course, the two are not mutually exclusive; indeed, there is some evidence that inferential skills are affected both by general education and by specific knowledge (Franks, 1998; Zachos, Pruzek, & Hick, 2003)).

However, this cannot be the end of the matter. It has become better known during the last several years of research in psychology that people have a natural tendency toward a number of cognitive biases, several of which actually explain why some of the above mentioned logical fallacies are so common and persistent. Take the *post hoc ergo propter hoc* fallacy, for instance. This is the mistake of inferring a causal connection between two events simply on the ground that one followed the other (typically within a short period of time). Statisticians constantly warn us that correlation is not the same as causation, and yet this type of elementary mistake in logic is arguably at the foundation of much superstitious behavior. A plausible (if difficult to test empirically (Kaplan, 2002)) argument can be made, however, that a cognitive bias favoring quick causal inference evolved because it was fitness-enhancing in ancient humans.

Moreover, Skinner (1948) famously showed that human beings are not the only animals to engage in superstitious behavior triggered by a false causal inference (although in the case of Skinner's pigeons, presumably such inference was unconscious). He reported:

> [O]ne bird was conditioned to turn counter-clockwise about the cage, making two or three turns between reinforcements. Another repeatedly thrust its head into one of the upper corners of the cage. A third developed a 'tossing' response, as if placing its head beneath an invisible bar and lifting it repeatedly. Two birds developed a pendulum motion of the head and body, in which the head was extended forward and swung from right to left with a sharp movement followed by a somewhat slower return. The body generally followed the movement and a few steps might be

taken when it was extensive. Another bird was conditioned to make incomplete pecking or brushing movements directed toward but not touching the floor. (Skinner, 1948, p. 168)

The problem, however, is that human beings persist in such behaviors much longer than experimental animals, who typically abandon their superstitious habits when they repeatedly fail to deliver the goods. For instance, work carried out by G.L. Wolford at Dartmouth (summarized by Gazzaniga, 2003) employed a test in which the subject is guessing whether a light will appear on the bottom or on the top of a computer screen. The setup is such that the light appears on top the majority of the times, but the sequence is random (i.e., there is a bias in the random appearances of the light). Rats quickly learn to maximize their performance, simply by limiting themselves to hit the top button. Human beings, however, think they can infer the rule behind the sequence, performing significantly worse than the rats, on average. So much for *Homo sapiens*. In all fairness, though, if the subjects of Wolford's experiment were *told* what was going on they would quickly adjust their behavior as a result of understanding, an option not available to the rats.

There are a number of other cognitive biases that are well-known to interfere with our critical thinking abilities, and that accordingly are increasingly taught alongside an understanding of logical fallacies. For instance, people habitually over-rely on their memories, so much so that 25% of participants in a survey were convinced that they got lost for long periods of time in a shopping mall when they were children (Baillargeon, 2008). If true, this would be a phenomenon of epidemic proportions, which would scarcely go unnoticed among the security personnel of shopping centers.

Our unjustified trust in our perception abilities has been well-documented as well, for instance in the cases of the unreliability of eye-witness testimony (Vidmar, Coleman Jr., & Newman, 2010) and in that of a number of standard and yet bewildering optical illusions (Ditzinger, 2001). Interestingly, a common cognitive bias that leads to (or at least reinforces) pseudoscientific belief takes place when unreliable perception combines with an innate tendency to see patterns in the world around us (another example of the human hyper-tendency to infer causality on the basis of scant data). In 2004, for instance, the face of the Virgin Mary "found" on a piece of toast fetched a bewildering $28,000 when sold, a type of (expensive!) cognitive mistake that Hadjikhani, Kveraga, Naik, and Ahlfors (2009) traced to an early activation of face-specific areas of the cortex by "face-like" objects (though exactly what made that particular object face-like remains largely unexplored).

A classic example in the pseudoscientific literature is, of course, the famous "face on Mars," allegedly discovered by NASA, but that clear images from the Mars Global Surveyor revealed to be simply an unusual configuration of a natural mesa, initially photographed at an angle that gave the (vague) impression of a giant sculpture of a human-like head on the surface of the red planet (Fraknoi, 2003). Naturally, the solution of the mystery does not seem to have done much to abate belief in a NASA-inspired conspiracy to hide the truth from the unsuspecting public.

One cannot leave even a cursory discussion of the psychological underpinnings of pseudoscience without mentioning a well-known concept in psychology that helps us make sense of a lot of the phenomenology concerning pseudoscience, and which indeed was invoked initially precisely in the context of a pseudoscientific cult: cognitive dissonance. The phrase was famously coined by Leon Festinger and colleagues in 1956, in their *When Prophecy Fails* (Festinger, Riecken, & Schachter, 1956).

MASSIMO PIGLIUCCI

Festinger et al. managed to infiltrate a cult growing around the pronouncements of one Dorothy Martin (initially identified in the book as "Marian Keech"). Martin claimed to be receiving messages from extraterrestrials from the (imaginary, as far as we know) planet Clarion. One day the message said that the world would end on December 21st, 1954. However, a group of flying saucers would pick up Miss Keech's followers. As a direct consequence of their belief in Martin's prophecies, many of those followers cut their ties with their families, quit their jobs, sold their possessions, and joined the wait. Problem was, the appointed date arrived and (predictably) nothing happened. Hours of tension followed, until at 4:45am the following morning, Martin announced the reception of another transmission from the Clarionians: the faith of the believers had saved the world, and no flying saucers rescue mission was needed after all!

Now, one would expect that Martin's followers would be angry and upset, and perhaps would go to the police, or sue Martin for all her worth. Instead, the majority of the cult members began to spread word of the good news, attempting (and succeeding, for a while) to make new disciples. Festinger et al.'s (1956) interpretation of what was going on was that too many of Martin's followers had invested too much at that point, both emotionally and materially. Now they were faced with the choice of either facing up to the fact that they had been gullible fools, or of telling the world (and themselves) that they were valiant heroes whose courageous behavior had actually saved the world. Apparently, for some people this is an easy choice: they avoid the rabbit hole and gladly take the blue pill.

Unfortunately, the Clarionian debacle was far from the last example of the genre. As recently as 2011 a new cultish movement was started by Evangelical Biblical literalist Harold Camping, with similar dynamics (and outcomes) to the classic Martin case (Bartlett, 2012). Just as in the 1950s, followers of Camping firmly believed their leader, who had predicted the end of the world by earthquakes, based on his (unorthodox and unfounded) reading of the books of Daniel and Revelation. Once again, they had to deal with the undeniable fact that the world did not, in fact, end on the announced date of May 21, 2011. And their reactions were remarkably similar to those of the followers of Martin. Here are some excerpts from Bartlett's (2012) article on the aftermath of the cult:

> With less than three months to the day of Christ's return, I desire to spend more time studying the Bible and sounding the trumpet warning of this imminent judgment...
> Based on everything we know, and when you look at the timelines, you look at the evidence—these aren't the kind of things that just happen. They correlate too strongly for it not to be important....
> Even if it's 99.9 percent, that extra .1 percent makes it not certain. It's like the weather. If it's 60 percent, it may or may not rain. But in this case we're saying 100 percent it will come. God with a consuming fire is coming to bring judgment and destroy the world....
> "Of course I'll be disappointed if it doesn't happen, but I feel like God's not going to let us down."

And, most revealingly: "I turned my back on the world. I can't afford to doubt."

Just as was argued by Festinger et al. (1956), however, cognitive dissonance has its limits, and some of Camping's followers apparently exceeded them. One of the people he interviewed told Bartlett (2012):

> After October 22, I said 'You know what? I think I was part of a cult. My wife and I joke that when my kids get older they're going to say that we're the crazy parents who believed the world was going to end.

24

Another wrote: "definitely lost an incredible amount of faith. It makes me wonder just how malleable our minds can be. It all seemed so real, like it made so much sense, but it wasn't right. It leaves a lot to think about." Indeed.

4. THE ROLE OF IDEOLOGY IN PSEUDOSCIENCE

The last aspect of the problem of misunderstanding of science and acceptance of pseudoscience I wish to briefly discuss is perhaps the most difficult to handle: the role of strong ideological commitments in how we filter just about everything else, including what should otherwise be relatively straightforward scientific information.

I will illustrate the issue with a specific example, concerning self-styled professional skeptics and the question of anthropogenic climate change. I am referring to the popular (now no longer airing) television show "Bullshit!," hosted by magicians Penn Gillette and Raymond Teller. The show was funny, if often crass because of Penn's tendency to curse, and very intelligently put together. Each episode examined some pseudoscientific claim and proceeded to debunk it with a combination of investigative journalism and empirical demonstrations. Penn and Teller do not pretend to be doing rigorous science—after all, it is a television program meant to entertain—but their antics also manage to educate, and I actually use them regularly in a class I teach on the nature of science, to provide my students with what turns out to be a very effective combination of laughs and food for serious thought.

Even Penn and Teller, however, sometimes get it spectacularly wrong, and it is instructive to examine one example because it vividly illustrates the role of ideology (in this case, political, though it may just as well be religious) in public science debate. Episode 13 of the first season of Bullshit! aired in 2003 and tackled the problem of climate change (McLaughlin, Moldave, & Small). The choice of topic by Penn and Teller may appear strange at first glance, since—despite the scientific discussions and the sociopolitical controversy— atmospheric physics certainly is no pseudoscience. It takes only a few minutes of background research to begin to guess why they chose to be skeptical of global warming: Penn Gillette is a well-known libertarian and a fellow of the Cato Institute, a think tank that has repeatedly taken positions against the emerging scientific consensus on global warming. The Cato Institute, it should be added, is funded in part by the Exxon-Mobil Corporation, not exactly a neutral player in discussions about energy production and use. Of course, the suspicion of bias is not enough to condemn Penn and Teller's treatment of climate change, but one's baloney detector's alert level should go up a couple of additional notches once a few more things become apparent from the broadcast. To begin with, Penn and Teller set up the episode by pitting oil-industry lobbyists against hippie college protesters to make their case that the climate change movement is a sort of New Age irrational belief. The only credentialed academic to speak on the program is economist Bjorn Lomborg (2001), a notorious skeptic of global warming (and not an atmospheric physicist).

Things became worse as the show transitioned from a clearly imbalanced presentation to outright misrepresentation of the debate. One of the guests was Jerry Taylor (of the above mentioned Cato Institute), who said

> In the mid '70s we were told pollution is going to cause a new ice age . . . The very same scientists who argued an ice age was coming because of industrial pollution then shifted gears and argued industrial pollution will bring on a greenhouse warming world with virtually no breath in between. (McLaughlin, Moldave, & Small, 2003).

This is simply false, as the idea of a temporary cooling of the earth's temperature was advanced in the popular press (not in academic, peer-reviewed journals), prompted by speculations about the massive injection of aerosols in the atmosphere. To compare a few magazine articles with the overwhelming scientific literature on global warming is a joke, and not a particularly funny one, given what is at stake.

While the Bullshit! episode is obviously anecdotal (though, I think, actually representative of ideological distortions of science policy debates), there is also systematic research bearing on the issue of science and ideology. A comprehensive presentation of the cognitive science aspects of this research has been summarized recently for a broader public by Chris Mooney (2012), while Corey Robin (2011) has put together a more nuanced historical and philosophically informed analysis of conservative ideology.

Robin (2011) develops his central thesis by way of a number of case studies (from Thomas Hobbes to Ayn Rand), defending the idea that conservatism is, at core, a combination of reaction against challenges to power hierarchies as well as a strong sense of entitlement about private property. Interestingly, as some commentators have noted, Robin's analysis helps making sense of the otherwise strange association between social conservatives and libertarians in the United States: while the latter endorse all sorts of positions that are abhorred by the former (e.g., legalization of drugs, prostitution, etc.), they share a sense of the inviolability, almost sacredness, of private property. It is this very thing that associates the otherwise progressive thinking and science endorsing Penn Gillette and Raymond Teller with the like of US Senator Jim Inhofe of Oklahoma, who famously refers to climate change as a "hoax."

Mooney's (2012) book is a bit more problematic. The research discussed in it is certainly interesting, if not quite as definitive as the author boldly states, but one wonders what exactly we are learning from the general conclusion that conservatives' and liberals' brains are wired differently. Quite apart from the fact that the terms "conservative" and "liberal" are highly sensitive to their particular cultural and historical setting, and from the fact that they do not capture but a fraction of people's attitudes toward political positions and ideologies, Mooney's book is another example of a recent trend that belongs to the "This is Your Brain on X" genre (or cottage industry, depending on how one looks at it). *Of course* the brains of people who think differently about X will be different. How *else* could distinct ways of thinking and behaving be carried out by the human animal? In fact, I am willing to go so far as to agree with Mooney that there may even be some genetic factors that may bias people's developmental psychology toward the conservative or progressive end of the spectrum (though then we enter the complex and empirically treacherous territory of gene-environment interactions (Pigliucci, 2001)).

The danger with Mooney's and others' approach to the biologization of ideology is that people will take the existence of genetic and/or neurobiological differences among groups as indicative of strong determinism (a position, to be fair, not endorsed by Mooney himself), quickly leading to the conclusion that nothing could therefore possibly change people's minds. This would be a highly unfortunate outcome that would essentially negate the value of public discourse about pretty much anything at all we care about, and this in turn would be a serious blow to the very idea of democracy. Ideological biases are important to explain some people's rejection of science and embracing of pseudoscience, but we need to thread carefully about the implications of research on ideology for the prospects of science education and even political progress.

5. CONCLUSION: THEN WHAT?

Carl Sagan, in his influential *The Demon-Haunted World* (1996) famously referred to science as "a candle in the dark," a precious thing, always in danger of being extinguished by a variety of threats, ignorance and superstition among others, but also ideological demagoguery in the service of political, religious, or corporate interests.

The analysis outlined above shows why Sagan got it right, and why the task confronting us is much more difficult than some may have thought. Rejection of science and belief in pseudoscience are not just the product of science illiteracy, more or less easily fixed by augmenting science teaching (and at any rate, certainly not by doing so primarily at the college level, or chiefly by way of more instruction about science facts as distinct from an understanding of science as a set of methods). It is much more complicated than that. Progress will be slow and will require much effort, and it will not be accomplished without taking seriously the contributions of philosophy (critical thinking, informal fallacies), psychology (sources and types of cognitive biases, cognitive dissonance), and sociology (roles and dynamics of ideological commitments).

It may very well be that the best that science educators, philosophers, and social scientists can do in the short run is to keep the candle lit, and that it takes a long view to remain confident of the possibility of progress (after all, a few centuries ago most people still believed in witches and demons...). Even so, what is at stake is well worth the fight. We just need to keep sharpening our tools along the way.

REFERENCES

Applebome, P. (1996, March 10). 70 years after Scopes trial, creation debate lives. *The New York Times*. Retrieved from http://www.nytimes.com/1996/03/10/us/70-years-after-scopes-trial-creation-debate-lives.html?pagewanted=all&src=pm

Baillargeon, N. (2008). *A short course in intellectual self-defense*. New York, NY: Seven Stories Press.

Bartlett, T. (2012, May 18). A year after the non-Apocalypse: Where are they now? *Religion Dispatches*. Retrieved from http://www.religiondispatches.org/archive/atheologies/5983/a_year_after_the_non-apocalypse%3A_where_are_they_now/

Ditzinger, T. (2011). Optical illusions: Examples of nonlinear dynamics in perception. *Nonlinear Dynamics in Human Behavior, 328*, 179–191.

Festinger, L., Riecken, H., & Schachter, S. (1956). *When prophecy fails: A social and psychological study of a modern group that predicted the destruction of the world*. New York, NY: Harper-Torchbooks.

Fraknoi, A. (2003). Dealing with astrology, UFOs, and faces on other worlds: A guide to addressing astronomical pseudoscience in the classroom. *Astronomy Education Review, 2*, 150–160.

Franks, B. A. (1998). Logical inference skills in adult reading comprehension: Effects of age and formal education. *Educational Gerontology, 24*, 47–68.

Gazzaniga, M. S. (2003). The split brain revisited. *Scientific American, 287*, 26–31.

Hadjikhani, N., Kveraga, K., Naik, P., & Ahlfors, S.P. (2009). Early (N170) activation of face-specific cortex by face-like objects. *Neuroreport, 20*, 403–407.

Hazen, R. M. (2002). Why should you be scientifically literate? *Actionbioscience.org*. Retrieved from www.actionbioscience.org/newfrontiers/hazen.html

Impey, C., Buxner, S., Antonellis, J., Johnson, E., & King, C. (2011). A twenty-year survey of science literacy among college undergraduates. *Journal of College Science Teaching, 40*, 31–37.

Johnson, M., & Pigliucci, M. (2004). Is knowledge of science associated with higher skepticism of pseudoscientific claims? *American Biology Teacher, 66*, 536–548.

Kaplan, J. M. (2002). Historical evidence and human adaptations. *Philosophy of Science, 69*, S294–S304.

Lomborg, B. (2001). *The skeptical environmentalist: Measuring the real state of the world*. Cambridge: Cambridge University Press.

McLaughlin, J., Moldave, R., & Small, E. (Writers), & Price, S. (Director). (2003, April 18). Environmental hysteria [Television series episode]. In S. Price (Producer), *Penn & Teller: Bullshit!*. Showtime Entertainment.

Miller, J. D., Scott, E. C., & Okamoto, S. (2006). Public acceptance of evolution. *Science, 313*, 765.

Mooney, C. (2012). *The Republican brain: The science of why they deny science—And reality*. Hoboken, NJ: Wiley.

National Science Foundation. (2004). Science and technology: Public attitudes and understanding. Retrieved from www.nsf.gov/statistics/seind04/pdf/c07.pdf

Pigliucci, M. (2001). *Phenotypic plasticity: Beyond nature and nurture*. Baltimore, MD: Johns Hopkins University Press.

Pigliucci, M. (2002). *Denying evolution: Creationism, scientism, and the nature of science*. Sunderland, MA: Sinauer.

Pigliucci, M. (2010). *Nonsense on stilts: How to tell science from bunk*. Chicago, IL: University of Chicago Press.

Pigliucci, M., & Boudry, M. (Eds.). (2013). *Philosophy of pseudoscience: Revisiting the demarcation problem*. Chicago, IL: University of Chicago Press.

Popper, K. (1957). Philosophy of science: A personal report. In C. A. Mace (Ed.), *British philosophy in mid-century* (pp. 155–191). London: Allen and Unwin.

Richardson, J., Smith, A., & Meaden, S. (2012). Thou shalt not commit logical fallacies. Retrieved from yourlogicalfallacyis.com

Robin, C. (2011). *The reactionary mind: Conservatism from Edmund Burke to Sarah Palin*. New York, NY: Oxford University Press.

Sagan, C. (1996). *The demon-haunted world: Science as a candle in the dark*. New York, NY: Ballantine Books.

Skinner, B. F. (1948). "Superstition" in the pigeon. Journal *of Experimental Psychology, 38*, 168–172.

Sperber, M. (2001). *Beer and circus: How big-time college sports is crippling undergraduate education*. New York, NY: Holt.

Thompson, H. (2012, April 11). Tennessee 'monkey bill' becomes law. *Nature News*. doi:10.1038/nature.2012.10423

Uhlmann, E.L., Poehlmanb, T.A., Tannenbaumc, D., & Bargh, J.A. (2012). Implicit Puritanism in American moral cognition. *Journal of Experimental Social Psychology, 47*, 312–320.

Vidmar, N., Coleman Jr., J.E., & Newman, T.A. (2010). Rethinking reliance on eyewitness confidence. *Judicature, 94*, 16–19.

Zachos, P., Pruzek, R., & Hick, T. (2003). Approaching error in scientific knowledge and science education. Proceedings from *7th International History, Philosophy of Science and Science Teaching Conference, 947–957*. Winnipeg.

Public Understanding of Climate Science and the Ethics of Expertise

BEN ALMASSI

Department of Philosophy
College of Lake County
19351 W. Washington St.
Grayslake IL 60030
United States
balmassi@clcillinois.edu

ABSTRACT: Public understanding of climate change turns significantly on epistemic trust and distrust of those claiming rational-social authority. Attending to the ethics of expert/non-expert trust relations and to argumentation and rhetoric in popular climate discourse, I argue, illustrates the importance of epistemic trustworthiness for the social propagation of climate scientific knowledge.

KEYWORDS: climate change, credibility, ethics, expertise, social epistemology, trust.

1. INTRODUCTION

Public understanding of climate change is a curious thing for social epistemological scrutiny. No subject of observation and conversation is more generic than the weather. Unlike chemical bonds, alleles, or leptons, many climatological phenomena make themselves apparent to the unaided inexpert eye. Strikingly, 19% of respondents to a 2008 survey identified "personal observations of warmer temperatures in their local communities" as their primary factor for belief in global warming, tied with glacial melting (19%) and changing weather patterns (18%) as most frequent answers (Borick, 2010, p. 33). For skeptics, personal observations (42%) was the most frequent factor identified for disbelief, well ahead of natural explanations (19%) and insufficient scientific evidence (11%) (Borick, 2010, p. 35). Both believing and disbelieving members of the American public are apparently looking to their own climate assessments as especially evidentially significant.

These things may tempt us into misunderstanding climate change as something we each can know on our own; yet the best evidence for climate change available to us doesn't admit of strictly independent individual assessment. As with much scientific knowledge, our grasp on anthropogenic climate change and appropriate responses to it turns on *epistemic dependence*: not only wary reliance on others' empirical or evaluative claims, but also relations of epistemic *trust*. Such dependencies include trust among climate researchers (no one of whom could do all the work alone) and public trust and distrust toward testimonies from those who earn—or claim—authority on global climate.

To be sure, public understanding of climate change is not monolithic but includes many *publics* with varying experiences, capabilities, and commitments. Yet would any of us be wise to pursue strict self-reliance on climatological beliefs? Consider me: even if I have good grasp of the greenhouse effect, this and my observations of local temperatures and weather patterns would be weak justification for any particular beliefs on global climate change. Whether my observations *even constitute* evidence for anthropogenic climate change and not another cause is unclear to my unaided evaluation; I cannot reliably judge its evidential relevance myself. As Hardwig (1994) emphasized in his work on epistemic dependency, it would be positively irrational of me to attempt strictly autonomous assessment, as it would be for one researcher to comprehensively measure global temperatures entirely by herself, neglectful of the rich social-evidential resource of trustworthy testimony.

Epistemic dependency may be recognized as useful, rational and responsible. When the scope of this dependency is obscured, however, expert and non-expert parties to trust relations are made vulnerable to exploitation of this trust. Vulnerability to exploitation of trust is urgent when the knowledge claims are political and controversial, as popular discourse continues to frame anthropogenic climate change. My hope is that by attending to the sometimes-neglected ethical dimensions of citizen-scientist epistemic dependency, understood in terms of trust and distrust, we may better appreciate how public understanding of climate change requires *trustworthiness*. Drawing on Baier and other trust theorists, I model public understanding of climate science as nested/overlapping trust relations, each with potential for promise and exploitation depending on their *moral health*. I find that the rhetoric and argumentation deployed in popular discourse on climate provide many illustrations for how untrustworthiness erodes the moral health of citizen-scientist epistemic dependency. I look to a range of climate-science testimonies including the recent dispute over climate science consensus between the Marshall Institute and the authors of *Merchants of Doubt*. In this way I offer a preliminary explication of role-specific duties of trustworthiness for morally and epistemically healthy public understanding of climate change.

2. TRUST AND CLIMATE SCIENCE

Solomon and Flores observed that trust relationships enable remarkable freedom: "not only the freedom from suspicion and distrust but the freedom to realize all sorts of possibilities … the freedom to engage in projects which one could not or would not undertake on one's own" (2001, pp. 7–8). This applies not only to business and personal relationships, but our collective understanding of global climatological phenomena as well. An enterprise necessarily broadly extended over space, time, and specialization, climate science is just the sort of project these authors recognize as made possible by mutual trust.

Solomon and Flores acknowledged their debt to Baier's work on trust, especially "Trust and Antitrust." Baier (1986) emphasized that trust is messy: it can be thrust upon us without consent, unrecognized by one or more parties to it, rational and irrational, morally healthy and rotten. Perhaps most important is her characterization of trust as a dependency distinctive from what she calls mere reliance. Specifically, Baier saw trust as not simply prediction and reliance on another's steady habits, but reliance by the trustor on the *good will* of the trusted toward her and the object of her trust. Characteristic of genuine trust relationships for Baier is its notable *discretion*: that is, the trusting person allows the other some discretion in determining how to meet her trust, such that the stronger the trust, the more discretion the trusted person is afforded (Baier, 1986, pp. 234–237). Discretion is the source of the powerful freedom noted by Solomon and Flores and also what renders the trustor vulnerable to betrayal as the merely reliant person is not. "One leaves others the opportunity to harm one when one trusts, and also shows one's confidence that they will not take it" (Baier, 1986, p. 235). *Rational* trust, then, requires "good grounds for such confidence in another's good will, or at least the absence of good grounds for expecting their ill will or indifference" (Baier, 1986, p. 235).

Karen Jones's (1996) account of trust as an *affective attitude* is also significantly indebted to Baier, though Jones did criticize Baier for neglecting the full significance of the trustor's *expectation* of the trusted's good will and responsiveness. Jones characterized trust as "an attitude of optimism that the good will and competence of another will extend to cover the domain of our interaction with her, together with the expectation that the trusted will be directly and favorably moved by the thought that we are counting on her" (1996, p. 4). She was less interested in any entrusted object than the situation of the trust relationship. Paul Faulkner (2007) offered a similar portrait in epistemology contrasting *affective* and *predictive* trust. In the latter, Faulkner noted, the listener trusts the speaker to do something in the sense that he knowingly expects *s*he'll be doing it, but doesn't expect anything *of* her. By contrast, trusting affectively means the listener actually expects that the speaker "recognizes his need to know whether *p*, and presumes that the speaker's telling him that *p* is a response to this" (Faulkner, 2007, p. 888).

Just as it is possible to merely rely without genuinely trusting, it is possible to learn from testimony without actually trusting it. We sometimes might find ourselves forced to rely on claims made by those we don't find trustworthy, perhaps because of the paucity of alternatives. In such situations we warily rely, as one might rely on *but not trust in* a damaged car to get to the hospital because it's the only available means. In these cases, we have not exactly opened ourselves to the possibility of *betrayal* since we have no real expectations of trustworthiness. To be sure, even without the vulnerability Baier identified as distinctive of trust, mere reliance on the unreliable can be costly. I may not be surprised when a shrimp cocktail at a bar in Iowa gives me food poisoning, but lack of betrayal may be modest solace.

The ampliative potential of epistemic trust as I here understand it means that trusting non-expert recipients of trustworthy expert testimony may acquire further grounds for belief beyond that available through mere-reliance alone. Consider the social-evidential significance of reflective belief in anthropogenic climate change by a supermajority of climate researchers. Here public trust and distrust by varied publics toward scientific opinion makes a big epistemic difference. Those who trust have reason to find great evidential significance in the percentage of scientists in agreement because we trust these scientists have come to their common beliefs not through deceit, conspiracy or groupthink but through their reflective expert assessments of empirical evidence, models, and fellow scientists' trustworthiness. Those who actively *distrust* climate scientists will place less evidential significance in lopsided expert opinion in favor of anthropogenic climate change. For example, Gardiner explained that novelist Michael Crichton, author of the conspiratorial climate-skeptical *State of Fear*, is dismissive of climate research because he "distrusts the data and methods of the scientists whose work is summarized by the IPCC" (2011, p. 460). Lastly consider a third public that neither trusts nor distrusts: they do not consider climate-scientific expert testimony trustworthy, neither do they dismiss it as untrustworthy. For this third group, lopsided opinion among scientists on climate change is a striking fact requiring explanation. Absent trust and distrust, they neither confidently attribute it to independent epistemic achievement nor to conspiracy.

3. MORAL ROT IN TRUST RELATIONSHIPS

"Most of us notice a given form of trust most easily after its sudden demise or serious injury. We inhabit a climate of trust as we inhabit an atmosphere and notice it as we notice air, only when it becomes scarce or polluted," Baier (1986, p. 234) observed. What is *polluted trust?*

Here Baier proposed what I consider an excellent moral test for trust relationships which we should extend to moral evaluations of citizen-scientist epistemic trust in public understanding of climate change. She articulated a test of the "moral decency" of trust relationships thusly:

> More generally, to the extent that what the truster relies on for the continuance of the trust relation is something which, once realized by the truster, is likely to lead to (increased) abuse of trust, and eventually to destabilization and destruction of the relation, the trust is morally corrupt.
> A trust relationship is morally bad to the extent that either party relies on qualities in the other which would be weakened by the knowledge that the other relies on them. (Baier, 1986, pp. 255-256)

Trustworthy citizen-scientist interdependency on climate change, I submit, should be able to pass Baier's expressibility test.

To illustrate let us consider the "skeptical environmentalist's guide to global warming," Lomborg's *Cool It*. The book opens with a striking claim: "Global warming has been portrayed recently as the greatest crisis in the history of civilization" (Lomborg, 2007, ix). For this claim the author has given no reference or citation. The reader has been immediately asked to trust: specifically to trust that this vivid description accurately captures the arguments of Lomborg's opponents, or perhaps another way of putting the point, to trust that 'greatest crisis historically' even *has* any actually advocating referent and is not just an arresting strawman. Whose view of global warming *is* this: a climate researcher in a peer-reviewed article, IPCC report, newsmagazine, Science *Times* article, anonymous blog post? The reader has been asked to trust the author that this claim has an advocate worth engaging; if that's not the case, the reader's engagement of the author's critique is predicated on ignorance of its strawman status.

A similar requirement of trust in the accurate characterization of unnamed opponents' views undergirded Patrick Michaels's criticism of what he calls "the Popular Vision" of global climate, in his 1992 book *Sound and Fury* published by the Cato Institute. The Popular Vision as Michaels described can be quite radical: for example, "One of the general tenets of those who subscribe to the Popular Vision is that there is a consensus among scientists that the end is at hand" (1992, p. 181). Readers have not been clearly informed what "the end is at hand" means, nor exactly to whom Michaels has ascribed this apocalyptic view; we have been implicitly invited to trust the author that the Popular Vision is not a strawman.

The rhetorical practice of putting a maligned view in quotation marks without reference to an actually identified advocate also turns for efficacy on the author-reader trust relationship. Notice: should these apparent quotes fail to have actually identifiable referents, continuance of this trust relationship is predicated on the reader's ignorance of that fact. In his introduction to *Climate Coup*, published by Cato, the editor said, "We are repeatedly told that 'the science is settled' on global warming (whatever that means) because of what is in our scientific journals" (Michaels, 2011, p. 1). Perhaps Michaels had in mind Oreskes (2004; 2007) on climate science consensus based on peer-reviewed climate research; but he didn't cite Oreskes or anyone else. Notice the rhetorical pull of Michaels's phrasing: 'the science is settled' has been introduced without an actually identified advocate *and belittled* as overly vague. Of course the vagueness criticism only works if readers can trust the author's implication that unnamed people of authority say this.

For another instance of apparent yet unattributed quotation, let us return to *Cool It*. "In public debates, the argument I hear most often is a variant of 'If global warming is going to kill us all and lay waste to the world, this has to be our top priority —everything else you talk about, including HIV/AIDS, malnutrition, free trade, malaria, and clean drinking water may be noble but it is utterly unimportant compared to global warming'" (Lomborg, 2007, p. 124). Note that the argument relayed here (against which the author organized his book) has been given in quotation marks, inviting the reader to imagine actual advocates voicing the very words; indeed, we have been invited to imagine many different people frequently saying them to Lomborg's consternation. But since no reference is provided for these 'quoted' words, the reader must trust the author's implication that they accurately capture his adversaries' views. This is no small matter, since *Cool It* is built around the following thesis:

> That humanity has caused a substantial rise in atmospheric carbon-dioxide levels over the past centuries, thereby contributing to global warming, is beyond debate. What is debatable, however, is whether hysteria and headlong spending on extravagant CO_2-cutting programs at an unprecedented price is the only possible response. Such a course of action is especially debatable in a world where billions of people live in poverty, where millions die of curable diseases, and where these lives could be saved, societies strengthened, and environments improved at a fraction of the cost. (Lomborg, 2007, ix)

Note two dilemmas that have been presented in this passage. First, Lomborg has attributed (what he critiqued as a false dilemma) to unnamed experts the claim that we must commit to hysterical, extravagant, unprecedentedly expensive programs or do nothing. No advocate of carbon-cutting programs as essential would describe them as hysterical or extravagant, so Lomborg has rejected (as-yet uncited proponents') plans even in his very description of them. If the reader is to be moved, she must trust Lomborg's prior assessment that *actually advocated* programs are hysterical and extravagant. The second dilemma, Lomborg endorsed: *either* engage in CO_2-cutting programs *or* address urgent global problems of poverty and disease. Readers have not been given evidence why we must choose between these horns of the dilemma, nor evidence why global climate change should be thought unrelated to problems of global poverty, disease, conflict, and displacement. Again, readers have been implicitly asked to trust Lomborg's assessment of the dilemma as valid. The closest he comes to *arguing* for the dilemma is this generic assertion: "The world lacks the resources and will to solve all its major challenges. Focusing on some issues puts others on the back burner" (Lomborg, 2007, p. 47).

In these rhetorical choices readers are made vulnerable to potential exploitation of trust. Recall Baier's proposed test for moral rot. If in fact the presented dilemma is overly simplistic, if actually advocated emissions-reduction plans are not obviously hysterical or extravagant, if "the greatest crisis in the history of civilization" fails to accurately describe opponents' views or has no actual referent, then continuance of the trust relationship between author and readers is predicated on the author's reliance on readers' ignorance of these things.[*] The next sections consider the further vulnerabilities of trust in climate claims to moral rot: specifically regarding climate consensus, ascriptions of credibility, non-expert expectations, and abdication of good scientific communication skills across epistemic differences.

---

[*] Jean Goodwin argues that Lomborg is engaged in *advocacy*, and as such, does not invite a trust relationship with readers. This is intriguing and worth further attention, but as yet it is not clear to me that advocacy precludes trust.

4. CHARACTERIZATIONS OF CLIMATE-SCIENCE CONSENSUS

The epistemic asymmetry involved in citizen-scientist trust relationship makes them a rich social-epistemic resource but also enables their exploitation. When expert testifiers *depend* on their recipients' lack of expertise in order to propagate their favored interpretation, they fail Baier's moral test. When a trusted speaker or writer relies on trusting recipients' unfamiliarity with a range of alternatives on a disputed issue within a scientific community, for example, or relies on the fact that trusting recipients are unaware of one's own or others' credibility among experts, the attendant epistemic trust is morally corrupt. These problems are nicely illustrated, I think, by contemporary debate over the existence of a climate-science consensus.

Testimony reinforcing or challenging the idea of a climate science consensus need not raise special moral concerns; but it certainly *can,* when presented ambiguously, vaguely, in a way that preys on public ignorance of how consensus is being operationally defined.

Naomi Oreskes (2004; 2007) has defined climate science consensus in a way that is plausible, specific, and transparent. She has grounded her assessment of scientific consensus in comprehensive analysis of peer-reviewed journal articles published 1993–2003: approximately 900 articles on a search of "global climate change." Oreskes found that none were offered as refuting to the notion that "global climate change is occurring, and human activities are at least part of the reason why;" approximately one-fifth explicitly endorsed the view that anthropogenic climate change was the main force behind current climate change, approximately half affirmed this view implicitly (2004; 2007). This finding of climate-scientific consensus as operationally defined by peer-reviewed journal publication informed Oreskes's popular work with Conway (2008; 2010) and has been often cited in climate ethics as evidence against a real debate among climate scientists (cf. Garvey, 2008; Gardiner, 2011; Shrader-Frechette, 2011). We might contrast Oreskes's model of climate consensus with Bray (2010), who directly challenged her analysis by appealing to the results of three surveys, each with approximately 350-550 respondents of alleged climate scientists. Twenty years ago, Fred Singer similarly rejected a climate consensus by appealing to collected survey results (cf. Michaels, 1992, 181). Those who reject the idea of an expert climate science consensus frequently cite petitions, surveys, and private admissions of doubts (cf. Barth, 1998, pp. 8-9; Singer, 2000, p. 39).

For their part, Oreskes and Conway acknowledged and even emphasized in their historical analysis that there have been and still are some scientists like Fred Singer and Fred Seitz that argue contrary to the scientific consensus to build doubt about the extent of human contribution to climate change (2008; 2010). These skeptical claims have been given less through original peer-reviewed climate research, Oreskes and Conway observed, and more through other socially and politically influential channels of communication: newspaper or magazine editorials, letters to the editor, think-tank book publication and white papers, and private conversations with policymakers (2008; 2010). A principal target of criticism in *Merchants of Doubt* is the George C. Marshall Institute, an American think-tank with funding from tobacco, energy and other industry groups. Shortly after publication of *Merchants of Doubt,* the Marshall Institute gave a critical response to Oreskes and Conway's book through its newsletter. Titled "Clouding the Truth," it opened with a quote from Galileo—"In questions of science, the authority of a thousand is not worth the humble reasoning of a single individual" — meant to set the tone.

O'Keefe and Kueter (2010) defended Seitz and other Marshall Institute colleagues as conscientious researchers pushing against scientific hegemony. The issue of consensus is key:

First, Oreskes-Conway assert the importance of consensus – these scientists were on the wrong side of the scientific consensus, they state. Science is not a popularity contest and scientific history is replete with examples of consensus views that were flat-out wrong. Second, Oreskes-Conway say these scientists 'fought the scientific evidence.' That should surprise no one. In fact, if the opposite were true, we all should be very concerned. Challenging the theory, hypothesis, and evidence is after all, the basis of modern science. (O'Keefe & Kueter, 2010, p. 1)

Nevertheless, Oreskes-Conway criticized Seitz, Jastrow, and Nierenberg for rejecting the scientific consensus that anthropogenic factors will cause dramatic climate change. To bolster their support for an alleged consensus, Oreskes-Conway offer a strong defense for the Intergovernmental Panel on Climate Change (IPCC). The recent Climategate revelations should be sufficient to give anyone pause when examining the openness and credibility of the IPCC process...In reality, the only consensus is among those on a [IPCC report] writing team. (O'Keefe & Kueter, 2010, p. 6)

Let us attend carefully to how O'Keefe and Kueter have been framing consensus. In insisting that science is "not a popularity contest," their implication was that Oreskes-Conway understand scientific consensus as what the majority of scientists endorse. While going by the numbers on experts' agreement can be a useful indicator of trustworthiness in facing conflicting testimony (cf. Goldman, 2001), this is not how Oreskes measured consensus, as we have seen. Though they imply that their colleagues "fought the scientific evidence" in noble scientific tradition, they do not remind readers that the fighting occurred outside of peer-reviewed publication. The issue of peer-reviewed research is not broached in this critique of *Merchants of Doubt*, despite the fact that the historians of science criticized for their climate consensus claim explicitly built their analysis from original peer-reviewed climate research.

O'Keefe and Kueter also implied that Oreskes-Conway's consensus claim depends on the IPCC report: if the credibility of the IPCC can be successfully impugned by reference to "Climategate," then presumably Oreskes-Conway's consensus claim is likewise impugned. Yet the force of this defense of the Marshall Institute turns on readers' ignorance of the fact that Oreskes's analysis of climate science consensus was neither a popularity contest nor parasitic on the IPCC anyway.

5. ASCRIPTIONS OF CREDIBILITY

An ethic of trustworthy scientific expertise should include the ways in which testifiers ascribe credibility (or lack thereof) to themselves, their allies, and their opponents.

Kristen Shrader-Frechette observed that "virtually no CC dissenters do peer-reviewed-expert climate research. Most of them are scientifically uninformed, and most are paid by special interests, like the oil lobby" (2011, p. 25). She argued that "scientists like Fred Seitz – who have never done climate research—have no authority from which to disagree with climate scientists who spend their lives doing advanced climate research" (Shrader-Frechette, 2011, pg. 25), described the Cato Institute, American Enterprise Institute, and Heartland Institute as "funded by chemical and fossil-fuel interests," and denounced scientists paid by these groups as decidedly untrustworthy (Shrader-Frechette, 2011, pp. 25–26). Retired physicist Fred Singer, she reminded us, has not published advanced climate research; similarly, despite his visibility as a climate critic biologist Patrick Michaels has not done climate research, and is paid by Cato, an "industry front group" largely funded by coal companies (Shrader-Frechette, 2011, p. 29).

Is this an *ad hominem* attack, distracting readers from engagement with the real issues? I don't think so. As Shrader-Frechette elucidated, the absence of original peer-reviewed climate research and the presence of industry funding are, *taken together*, relevant to ascriptions of expert credibility. Without the fallible yet socially-epistemologically significant filter of peer review, non-expert recipients of these critics' skeptical testimonies must rely more directly on testifiers' good will and responsiveness. In this, testifiers' funding sources are quite relevant to assessing their affective trustworthiness for those of us not providing that funding.

By contrast let us consider Singer's climate-skeptical essay "Cool Planet, Hot Politics." Having cast suspicion on climate researchers funded by government grants, he noted:

> Of course there are think tanks on the other side as well (such as the Cato Institute and the Competitive Enterprise Institute), spreading the message that the best information available from climate science contradicts the alleged need for drastic policies certain to cause great economic harm. Needless to say, these groups don't get any government money. (Singer, 2000, p. 39)

The reader has been invited to regard the Cato Institute as especially trustworthy *because* it doesn't "get any government money." Yet Singer made no acknowledgement of Cato's own industry funding. He cannot consistently insist that Cato's funding is irrelevant to the credibility of its message *and* make the point in the quoted passage. Thus the author relied here for the force of his claim on readers' ignorance of Cato's industry funding.

A retired physicist like Singer, skeptical about anthropogenic climate change, need not be untrustworthy in choosing to testify publicly to his skeptical beliefs. Yet if he does so in a way that depends for its rhetorical force on public ignorance about how well his training and experience in physics prepares him to competently assess climatological research, this fails our test. Likewise, citing one's ideological allies to buttress one's position need not be problematic, but it becomes so when the efficacy of citation turns on readers' ignorance of these allies and their credibility.

6. RECIPIENTS, MEDIA, AND RELUCTANT POPULARIZERS

Shrader-Frechette observed that laypeople can be misled in several ways in misunderstanding climate change. One involves a failure of nonexpert recipients of testimony to appreciate the uncertainty inherent to science: laypeople "may be uncomfortable with uncertainty [and] may erroneously believe good science should be certain" (Shrader-Frechette, 2011, pp. 24–25), and dismiss fallible but reliable climate scientific knowledge. Public misunderstanding may be partially a media failure too, by giving a public platform to climate skeptics lacking real expert credibility in pursuit of superficial balance (Shrader-Frechette, 2011, p. 25; Gelbspan, 2000, p. 25). Public confusion may be a failure, thirdly, of scientists with climatological expertise yet poor communication skills. Shrader-Frechette put the point thusly:

> After all, advanced-scientific researchers are trained to do demanding technical work and make new discoveries, not to popularize science. They are trained to produce knowledge, not disseminate it. Indeed, if scientists become popularizers, they typically become suspect among other experts—who may think that they cannot do technically-demanding work. Poor expert communication thus can leave science open to misrepresentation. (2011, p. 24)

I take this to be an important point not to be overshadowed by the failures of trustworthiness by opportunistic climate skeptics already discussed in detail. Let us recall Jones's expectation criterion for trust. Trustworthy expert testimony means more than just transparency, reliability, and absence of ill will; trustworthy expert facilitation of public scientific understanding means a conscientious responsiveness to trusting non-experts' expectation "that the one trusted will be directly and favorably moved by the thought that we are counting on her" (Jones, 1996, p. 4). When *trusting* scientists, we expect them to recognize themselves to be giving testimony, to recognize that they are making claims employed by us as evidence for our belief and actions. At its most trustworthy, scientific testimony is presented *conscientiously*: sincerely but further with attention to successful recipient uptake. In devaluing good science-popularization, then, otherwise fine scientists abdicate a duty of responsible testimony across epistemic difference and so fall short of trustworthiness.

7. CONCLUSION

To conclude, I have recommended that we be mindful of rhetorical considerations relevant to the moral health of epistemic trust relationships undergirding public understanding of climate. These include but are not limited to

- representations of opponents' commitments and claims
- operational definitions of climate science consensus
- ascriptions of credibility to oneself, allies, and opponents
- non-expert recipients' expectations of certainty
- expert researchers' conscientious communication skills

My remarks are only a partial articulation of what ethics of expertise may look like for morally healthy, mutually conscientious public trust on climate change. I have sought to illustrate the importance of reciprocal, role-specific *trustworthiness* by all parties to our nested, overlapping relationships of social-epistemic interdependency.

ACKNOWLEDGEMENTS: I'm grateful to Moira Gutteridge-Kloster, Monica Aufrecht, Paul Thompson, and Clark Wolf for their recommendations and critical commentaries toward the development of this project. My thanks especially go to Jean Goodwin and her co-organizers of the 2012 meeting of the Great Plains Society for the Study of Argumentation (GPSSA) at Iowa State University. Any and all errors are, of course, mine alone.

REFERENCES

Baier, A. (1986). Trust and antitrust. *Ethics, 96*, 231–260.
Barth, G. (1998). The distorted world of climate models. *The Intellectual Activist*, February 3, 1998, 3–10.
Borick, C. (2010). American public opinion and climate change. In B. Rabe (Ed.), *Greenhouse governance: Addressing climate change in America*, 24-57. Washington, DC: The Brookings Institution.
Crichton, M. (2005). *State of fear*. New York, NY: HarperCollins.
Bray, D. (2010). The scientific consensus on climate change revisited. *Environmental Science and Policy, 13*, 340-350.
Faulkner, P. (2007). On telling and trusting. *Mind, 116*, 875–902.
Gardner, S. (2011). *A perfect moral storm: The ethical tragedy of climate change*. Oxford: Oxford University Press.

Garvey, J. (2008). *The ethics of climate change: Right and wrong in a warming world.* New York, NY: Continuum.

Goldman, A. (2001). Experts: Which ones should you trust? *Philosophy and Phenomenological Research, 63,* 85-110.

Gelbspan, R. (2000). Reality check. *E: The Environmental Magazine, 11*(5) 24–26.

Hardwig, J. (1994). Toward an ethics of expertise. In D. Wueste (Ed.), *Professional ethics & social responsibility,* 83-101. New York, NY: Rowman & Littlefield.

Intergovernmental Panel on Climate Change (IPCC). (2007). *Climate change 2007: The Scientific basis.* Cambridge: Cambridge University Press.

Jones, K. (1996). Trust as an affective attitude. *Ethics, 107,* 4–25.

Lomborg, B. (2007). *Cool it: The skeptical environmentalist's guide to global warming.* New York, NY: Alfred A. Knopf.

Michaels, P. (1992). *Sound and fury: The science and politics of global warming.* Washington, DC: The Cato Institute.

Michaels, P. (Ed.) (2011). *Climate coup: Global warming's invasion of our government and our lives.* Washington, DC: The Cato Institute.

O'Keefe, W., & Kueter, J. (2010, June). Clouding the truth: A critique of *Merchants of doubt. Marshall Institute Policy Outlook,* 1–8. Retrieved from http://www.marshall.org

Oreskes, N. (2004). The scientific consensus on climate change. *Science, 306,* 1686.

Oreskes, N. (2007). The scientific consensus on climate change: How do we know we're not wrong? In J. Dimento & P. Doughman (Eds.), *Climate change: What it means for us, our children, and our grandchildren,* 65-100. Cambridge, MA: MIT Press.

Oreskes, N. & Conway, E. (2008). Challenging knowledge. In R. Proctor & L. Schiebinger (Eds.), *Agnotology: The making and unmaking of ignorance,* 55-89. Stanford, CA: Stanford University Press.

Oreskes, N., & Conway, E. (2010). *Merchants of doubt: How a handful of scientists obscured the truth on issues from tobacco smoke to global warming.* New York, NY: Bloomsbury Press.

Shrader-Frechette, K. (2011). *What will work: Fighting climate change with renewable energy, not nuclear power.* Oxford: Oxford University Press.

Singer, S. F. (2000). Cool planet, hot politics. *American Outlook,* Summer 2000, 38–40.

Solomon, R., & Flores, F. (2001). *Building trust in business, politics, relationships, and life.* Oxford: Oxford University Press.

The Peer-Review Certification Label: A Shortcut to Assessing Expertise and Consensus by the Necessarily Uninformed

BILL ANELLI

Philosophy
Modesto Jr. College
422 Hackberry Ave., Modesto, CA 95354
anellib@mjc.edu

ABSTRACT: Given the well-documented public ignorance of science content and process, it's not so clear what can be done given various distorting mechanisms of large-scale media institutions. This paper explores one avenue for more academy input and meta-epistemic dialogue: an opt-in, client-funded, Academy-enacted peer-review certification label scheme to be located at, say, news dissemination sites, and roughly analogous to shortcut epistemic devices such as food labels. This label scheme might also provide well-intentioned, yet uninformed citizens who are asked to adjudicate apparent expert debates a more reliable epistemic shortcut than the standard, rhetorically vulnerable shortcuts of institutional affiliation or degree.

KEYWORDS: certification, media, argumentation, peer review, social epistemology.

1. INTRODUCTION

> You've got two groups of people who want to be part of history, to dig it up and hold it in their hand. The only difference is I'm doing it to make a living. They're doing it to write papers and make it to associate professor and get tenure. - Entrepreneurial archaeologist Ric Savage of Spike TV's *American Digger*, in response to charges of detrimental amateurism from the American Anthropological Association and Archaeological Institute of America (Carter, 2012)

In this paper I intend to explore the notion of an opt-in, client-funded peer-review certification label performed by appropriate members of the academy and roughly analogous to opt-in, client-funded food labeling such as organic, kosher, or LEED certification labels. My tentative proposal is the peer-review certification label, an epistemic shortcut that could be located at scalable levels: from the footnote to a given media expression to a subsection of a media site or to an entire media site.

In defense of this tentative labeling scheme I will argue the following:

(1) That epistemic shortcuts such as affiliation, degree, and specialty are no longer sufficient and that peer-review certification shortcuts might offer a useful addition to the well-intentioned but uninformed public when seeking to adjudicate apparent disputes between science experts or when trying to evaluate a given factual claim by an expert.

(2) That peer-review certification label shortcuts might improve epistemic standards in journalism, social media, and perhaps public discourse.

(3) That use of a peer-review certification label might enrich public-science dialogues as well as intra-science dialogues regarding the scientific process and related intellectual virtues.

Anelli, B. (2012). The Peer-review certification label: A shortcut to assessing expertise and consensus by the necessarily uninformed. In J. Goodwin (Ed.), *Between scientists & citizens: Proceedings of a conference at Iowa State University, June 1-2, 2012* (pp. 39-51). Ames, IA: Great Plains Society for the Study of Argumentation. Copyright © 2012 the author(s).

(4) That implementing a peer-review certification label scheme in a relatively fair, objective, and accurate manner might be feasible.

2. WHY PEER REVIEW?

David Shatz in *Peer Review: A Critical Inquiry* identifies academic peer review as the system of certification or quality control within the academy:

> The peer review system is often described as a system of certification, and indeed it is, in two senses: acceptance to a journal or publishing house via peer review certifies a body of work, and it also certifies the scholar who produced it ... To say of a published article or book that it was peer reviewed is to say that it is perceived by experts as a contribution to human knowledge ... Peer review is a mechanism, then, for quality control; it protects us from contamination by error and poor argument, and affords us truth or contributions to attaining truth. (Shatz, 2004, p. 1)

That a body of work passes peer review does not mean that the body of work necessarily counts as justified truth or knowledge nor does it mean that the peer-review process was free of politics or bias or that a paper's conclusions were not somewhat skewed towards corporate funders, grantees, or editors' theoretical preferences. It also does not mean that ideas not addressed by peer review or for which there lacks near-consensus in peer review are thereby false claims. It simply means that relevant experts have decided that a given claim has merit and is thus worthy of consideration by other experts whose time and energy are necessarily limited.

One way of understanding academic peer review in the sciences is that there exist multiple communities of scientists, most of whom operate in small core-sets with face-to-face interactions (Shapin, 1994, p. 415) and who are engaged in historically revisable, highly specialized, labor-intensive practices aimed at understanding or discovering factual claims within a fairly specialized domain. Given this context, say a small specialized scientific core-set community that is focused on questions such as "how does Skeletal Development in *Pan paniscus* compare to *Pan troglodytes?*[1]", given the limited time and energy of members of this community, many of whom personally know one another, and given a situation in which the amount of new arguments and claims regarding their area of interest might be relatively large and the competencies varied, some sort of gatekeeping, quality control system is needed.

And while there are many variations of peer review, most peer-review practices enlist expert members of community X to vet new arguments and claims, i.e., new manuscripts submitted to relevant journals (Schatz, 2004; Hames, 2007).

Although criticisms of academic peer review abound (i.e., problems of affiliational, ad hominem, ideological, and aesthetic biases (Shatz, 2004; Chubin, 1990; Hames, 2007)), surveys consistently report that both authors and readers wish to preserve peer review, perhaps with reform as they depend on peer review as a necessary filter and vehicle for professional development (Hames, 2007). Regardless, peer-review flaws are not, I think, fatal to my argument—I only need to assume a) that peer-review certification, or even reformed peer-review certification, as a quality control process is better than the alternatives both within and outside the academy and b) that both the academy and the public are better served if the public has a better understanding of peer review. Finally, peer review in the academy, despite its

---

[1] Bolter, D. R., & Zihlman, A. L. (2012). Skeletal development in *Pan paniscus* with comparisons to *Pan troglodytes*. *American Journal of Physical Anthropology*, *147*, 629–636. doi: 10.1002/ajpa.22025

many flaws, does, I argue, rely on and generate virtues (or internal goods) of courage, honesty, and justice for its participant scholars and scientists. Virtue practices contain community standards of excellence, and considerations of the well-being of larger communities of which a practice is a part (MacIntyre, 1984, pp. 187-190). That academic peer-review practices include these virtues is perhaps evidenced by the following practices that are commonplace in academic peer review:

- A commitment to minimization of bias such as ad hominem, ideological, affiliational, and aesthetic (Shatz, 2004): expert reviewers are assigned who are not themselves part of the editorial staff and assigned in accordance with needed expertise (Weller, 2001).
- A commitment to deliberation and transparency: many journals, and journal editors, publicly disclose their review policies and deliberative procedures (Weller, 2001).
- Different types of peer review have been developed to maximize quality control and vetting of academic work in a variety of contexts: closed and open peer review, single and double-blinded peer review, prepublication peer review, post-publication peer review, journal peer review, conference peer review, foundation/grants peer review and books peer review (Schatz, 2004).
- The decision to publish usually involves consideration of whether a given topic is relevant, interesting, provocative, or controversial to the larger core-set or even the scientific community or public at a given point in time (Weller, 2001).
- Editors and peer reviewers often work with submitters to improve the quality of their work and often give rigorous, lengthy, detailed feedback (Hames, 2007).

Despite its flaws, academic peer review likely incorporates and generates virtues within its core-set communities although I have only sketched the barest outline here. I will now turn to my defense and explanation of the certified peer review labeling scheme.

3. EVALUATING EXPERTS: THE INADEQUACY OF CURRENT EPISTEMIC SHORTCUTS

Given that nearly all members of the public are ill-equipped to independently verify most of the factual claims addressed by the various science and social science disciplines, often divided into small core-set sub-disciplines composed of as few as 15-20 specialists (Shapin, 1994) and given that we are routinely asked to weigh in, via polling and voting, on science-related public policy issues, by necessity we must rely on a set of epistemic shortcuts in order to determine which claim, among competing claims, to accept as justified knowledge. These epistemic shortcuts are usually reduced, given spatial, financial, and temporal limitations in most media expressions to three: 1) institutional affiliation, 2) specialty, and 3) educational degree.[2]

Given the disconnect between public perception versus the reality of scientific consensus, perhaps contributed to by the corrosive effect of what I loosely term *unfriendly zombie expert authorities*, these standard shortcut methods for evaluation are, I argue, inadequate. Other factors that undermine traditional epistemic shortcuts include rapid

---

[2] So "Anthropologist Dr. John Doe of U.C. Berkeley," not "physical anthropologist Dr. John Doe of U.C. Berkeley, author or co-author of 15 publications within the past five years in impact peer-reviewed journals in physical anthropology and reviewer of 25 manuscript submissions for four leading journals…"

digitalization of information (Gasser, 2008) and reformulation of sites of news and media expressions (i.e., the blurring of boundaries between print and online journalism, news producers and news curators, social media sites and blogs (Rosenberg, 2009)).

A zombie expert authority is an authority that, to an uninformed outsider, *behaves* nearly identically to a genuine expert authority yet lacks the necessary properties internal to a genuine expert authority such as an active virtue practice within a scholarly community that includes reviewing and publishing. The zombie authority carries identical symbols such as degree, an institute affiliation[3], a noted specialization, even published papers that *appear* to be peer-reviewed[4] yet the zombie is not in fact a practicing member of the relevant core-set academic community. Given the limited labor-time and financial resources available to journalists, bloggers, news curators, they are vulnerable to zombie expert authorities. Adding a peer-review certification label to the mix might help to retain the advantages of epistemic shortcuts without sacrificing reliability and trust. It might also provide well-intentioned, yet uninformed citizens a more reliable shortcut in adjudicating apparent debates between experts in the sciences and social sciences.

4. A PEER-REVIEW CERTIFICATION LABEL: IMPROVED EPISTEMIC STANDARDS?

Peer-review certification labeling might raise the epistemic justification bar from whether an *individual* institution, by conferring a degree and employment, has properly certified a given scholar's claims (and so "X-claim counts as justified knowledge because Y scholar from Z institution says so") to whether a *community* of scholars certify that claim X counts as justified knowledge (and so "X-claim counts as justified knowledge because a *community* of scholars from multiple institutions say so"). Given that, as described below, the academy will perform much of the heavy lifting of peer-review certification, and given the increasingly limited resources available to journalists for epistemic labor, it seems reasonable that a peer-review certification label might be something that journalists and bloggers could easily utilize in lieu of a recycled rolodex of experts. Finally, repeated reference to, and the branded presence of peer-review certification labels might further highlight, and thus pique, curiosity and interest in the virtues associated with academic peer review.

5. A PEER-REVIEW CERTIFICATION LABEL AND PUBLIC-SCIENCE DIALOGUE

Among the general public there continues to exist significant ignorance regarding both factual knowledge and understanding of scientific inquiry (National Science Foundation, 2012). For example, only 20 to 25 percent of Americans qualify as scientifically knowledgeable (Miller, 2005) and these wide gaps in scientific understanding are not limited to those with limited education as over 60 percent of those with graduate degrees lack basic understanding of what a

---

[3] Often these are ideological think tanks, corporate/industry/non-profit lobby groups, or government public relation institutions.

[4] For example Dr. Arthur B. Robinson, Ph.D. in chemistry, educated at California Institute of Technology and U.C. San Diego and now a clinical researcher on peptides at the Oregon Institute of Science and Medicine and head of the petition project, "a petition (of 31,000 scientists) opposed entirely on scientific grounds published in peer reviewed journals - to the hypothesis of 'human-caused global warming.'" (http://www.oism.org/s32p21.htm)

scientific study is. (National Science Foundation, 2012). Perhaps most disturbing, the public often assumes a lack of expert consensus where in fact there is near expert consensus, for example on key public policy issues such as global warming, medicine, evolution, and even certain fundamentals about our economy (National Opinion Research Center, 2006, 2010; Anderegg, Prall, Harold, & Schneider, 2010). The consequences of public ignorance and increased skepticism about expert consensus on public policy-related issues can be dramatic—as seen in President Obama's recent omission of any discussion of climate change with respect to the Keystone XL Pipeline debate (Song, 2012). Another consequence is that, although the public still places more trust in scientists than in most other expert professions (National Science Foundation, 2012), it is likely that if journalists and educators do not raise the bar with respect to epistemic shortcuts, the public will extend their scepticism and cynicism of expert authorities to scientists—as is already occurring with those who identify politically as conservative (Koebler, 2012).

Given the possibility of increased distrust or confusion about science, a thicker evaluative shortcut such as peer-review certification labeling might a) more effectively counter "external good" cynicism, i.e., that scientists are simply motivated by external goods of power, money, and fame as claimed by Ric Savage of *American Digger* for example, and b) lead to more dialogue between the public and scientists and continued reflection and improvement within scientific communities. For example, more explicit reference to academic peer review by large scale disseminators of factual claims (i.e., textbooks, news organizations, magazines, bloggers, and the like) might contribute to a more informed public, essential to any healthy deliberative democracy according to John Dewey (1954).

Additionally non-scientists (or non-specialist scientists even), due to repeated exposure to the peer-review certification "brand," might be more likely to import valuable academy practices, and even related virtues, to various non-academic sites of public discourse—whether they be civic organizations, city council meetings, political pundit commentary and debate, etc. Also, more discussion of academic peer review might create more openings or spaces for non-scientists to engage with the source practices of knowledge creation in the academy. If we know more about how peer review actually operates in the academy the likelihood is also increased that flaws, for example of ideological bias or lack of availability of raw data, might be improved and the process made more transparent—both to the public at large and within the academy, for example PLoS or the International Journal of Science Innovations and Discoveries. The result might be to increase the likelihood of more dialogue, trust, and understanding between scientists and the public.

Fig. 1. Certification Labels

6. PEER REVIEW CERTIFICATION LABEL SCHEME: A PROPOSAL

But is such a scheme feasible? Below I outline one approach and my aim here is to simply show that a peer-review certification label is a feasible scheme, not that my proposal is the best approach per se.

In my peer-review certification scheme, results and ratings are communicated to the public via a label placed on individual articles, sections of a website, on the home page of a website, as part of the video or even audio portion of a media site. The label should be clickable and aesthetically engaging, interesting, fun, and additionally could incorporate competitive statistics that rank various news organizations' peer-review status as well as methods of measurement by the peer-review certification team—all available in a scalable format that does not overburden the knowledge consumer.[5] Peer-review certification becomes a brand that is ubiquitous, as are labels such as Organic, Kosher, etc., thus rewarding knowledge producers who qualify and pay for the certification labor and gradually setting an expectation for media expression sites. Media companies who ignore peer-review certification would do so at their own peril given that a clear distinction can now be drawn between them and their competitors.

The rough analogy I draw is to food labeling. Just as we cannot be expected to individually test our food products for desired properties of nutritional content, freshness, organic or kosher standards, it is equally unreasonable to expect the public to individually vet factual claims they encounter, or even to vet authorities laying claim to this or that claim (i.e., we cannot expect members of the public to study, for 60 or more years, both the basics of biology, chemistry, physics, psychology, economics, political science or to stay abreast of recent peer review literature in each area). Thus, just as we trust food labels despite our awareness that food labeling process is imperfect and negatively impacted by politics, I assert that we should at least consider various schemes for a standardized, opt-in, client-funded, peer-review certification.

While it is true that government regulatory agencies such as the USDA or FDA are charged with codifying—and enforcing—congressional food regulation legislation, food producers, especially with respect to opt-in food labels such as organic or kosher labels, are required to significantly subsidize the objective, non-based, evaluative costs in certifying a given food producer's product.

Labels bring with them considerations of graphic design, of some implicit notion of property hierarchy in the object in question and a concern to communicate information about important properties to an imagined audience. Labels are also important in that they are affixed directly on the object in question—as sign or text they eradicate distance between a name (of a category or property) and the object referenced by category or property such that the consumer can efficiently make a reliable judgment at the point of sale given our context of large-scale modern food production. Implicit in labels is a relation of trust between consumer and the rating or certifying agency. Given the public's relatively high trust in the academy, it's likely that academy peer-review certification labels would be well-received by the public.

How far can the analogy hold between food labeling practices and peer-review certification labeling? The analogy I wish to make is that both food production and factual claim production or dissemination are highly-specialized, complex, modern activities that contain standardized

---

[5]   For instance one could click on the peer-review certification logo and travel to a webpage that explains the peer-review process in detail as well as a breakdown as to the given website's success.

operations and as such can be evaluated for quality control by independent, competent third parties. Furthermore, consumers of both food and media access these products at centralized points of sale or consumption (supermarkets or cable TV screens or websites or radio stations or newspapers), and finally, a significant number of food or media products (in our case science-related factual claims) are distributed by a relative few producers. Finally, large-scale food producers and large-scale knowledge disseminators possess ample capital to be able to fund organic/kosher or peer-review certification, respectively.

FACTUAL CLAIM DISSEMINATORS, CREATORS, & PEER REVIEW CERTIFICATION

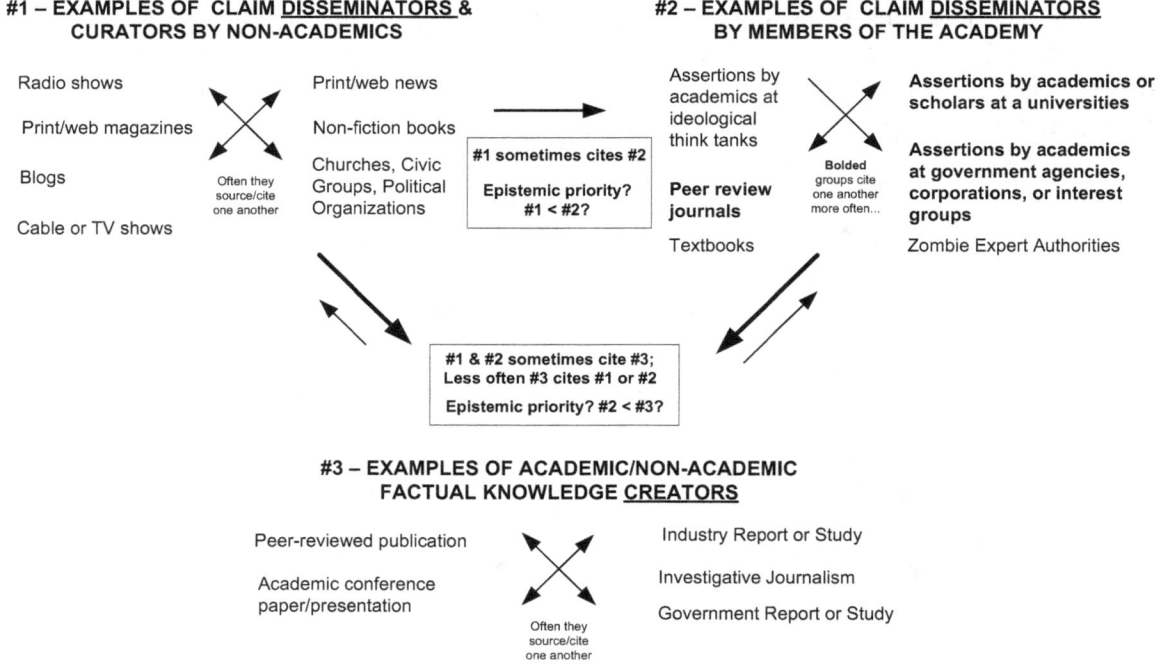

#1 – EXAMPLES OF CLAIM DISSEMINATORS & CURATORS BY NON-ACADEMICS

Radio shows

Print/web magazines

Blogs

Cable or TV shows

Often they source/cite one another

Print/web news

Non-fiction books

Churches, Civic Groups, Political Organizations

#1 sometimes cites #2

Epistemic priority? #1 < #2?

#2 – EXAMPLES OF CLAIM DISSEMINATORS BY MEMBERS OF THE ACADEMY

Assertions by academics at ideological think tanks

Peer review journals

Textbooks

Bolded groups cite one another more often...

Assertions by academics or scholars at a universities

Assertions by academics at government agencies, corporations, or interest groups

Zombie Expert Authorities

#1 & #2 sometimes cite #3; Less often #3 cites #1 or #2

Epistemic priority? #2 < #3?

#3 – EXAMPLES OF ACADEMIC/NON-ACADEMIC FACTUAL KNOWLEDGE CREATORS

Peer-reviewed publication

Academic conference paper/presentation

Often they source/cite one another

Industry Report or Study

Investigative Journalism

Government Report or Study

EARNING PEER REVIEW CERTIFICATION (PRC) FOR A GIVEN MEDIA EXPRESSION OR SITE
(usually from 1 or 2 above)

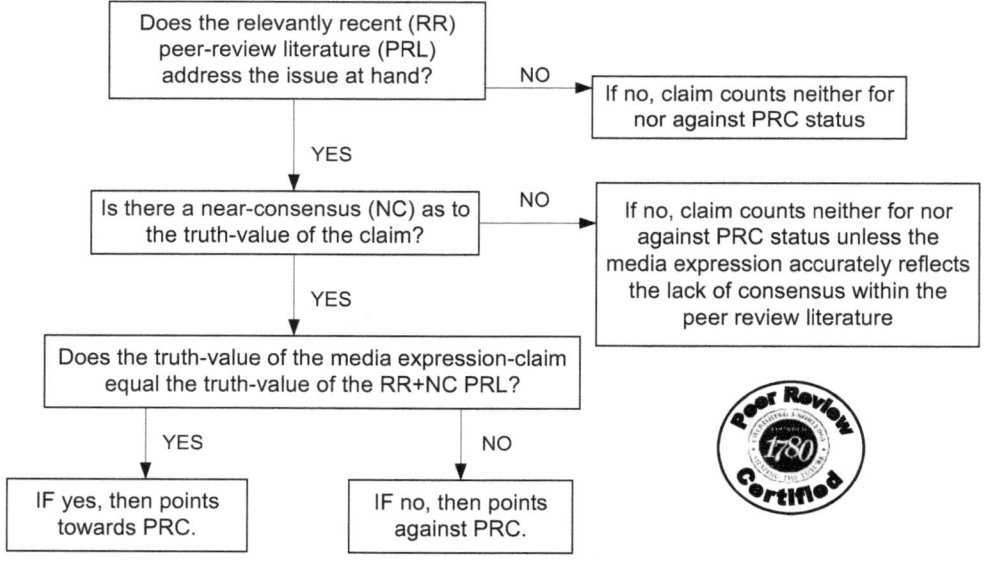

Does the relevantly recent (RR) peer-review literature (PRL) address the issue at hand?

NO → If no, claim counts neither for nor against PRC status

YES

Is there a near-consensus (NC) as to the truth-value of the claim?

NO → If no, claim counts neither for nor against PRC status unless the media expression accurately reflects the lack of consensus within the peer review literature

YES

Does the truth-value of the media expression-claim equal the truth-value of the RR+NC PRL?

YES → IF yes, then points towards PRC.

NO → IF no, then points against PRC.

Fig. 2. Peer Review Certification: Applicants and Decision Tree

The following are the conditions, steps, and rules by which this might take place in one possible scheme are outlined below:

(1) The applicant seeking certification furnishes payment for the labour of an interdisciplinary panel of scholars (or discipline-specific scholars) to certify a given site of knowledge dissemination/production.

(2) Selection of scholars is randomly drawn from a list of scholars who have published in a peer-reviewed academic journal within a relevant recent time period.

(3) Certification status could be quantitatively determined in the following way:
   a. A randomized sample X of prior year media expressions is generated.
   b. From this sample X, an enumerated list of factual claims is created for each media expression of the sample and each claim is paraphrased such that only the factual claims are retained.
   c. Note: By "factual claim" I mean propositional statements that are true or false and whose truth-value can be reliably and consistently determined (i.e., falsified or confirmed) through empirical testing/scientific method and practices.

(4) For the purposes of peer-review certification the following issues are irrelevant:
   a. The form (audio, print, web, blog, essay, report) of the media knowledge expression;
   b. Whether or not the author sources the factual claims;[6]
   c. The identity, background, and experience of the author of the media expression.

(5) Only factual claims that meet the following criteria will be considered candidate factual claims for peer-review certification analysis:
   a. There exists a near-consensus in the relevantly recent peer-review literature with respect to a given factual claim's truth-value or:
   b. There exists a near-consensus in the relevantly recent peer-review literature with respect to the class of factual claims whose truth depends on the given claim being true and vice versa.

(6) Points for or against a given candidate factual claim will be awarded based on whether or not the truth-value of a given claim is aligned with the truth value of the peer review academic community as stated in 5 a or b above.

(7) Factual claims not addressed by the relevantly recent peer-review literature or for which near-consensus is lacking are not counted toward or against a given media expression's peer-review certification status.

(8) Operational definitions for the following would be set by discipline-relevant, or interdisciplinary-relevant academic communities in the sciences and social sciences:
   a. What counts as relevantly recent (RR) with respect to a given topic or set of claims.
   b. What counts as near-consensus (NC) with respect to a given topic or set of claims.
   c. What counts as a legitimate peer-reviewed literature (PRL) with respect to a given discipline.
   d. Thresholds within individual media expressions for counting as peer- review certification (i.e., "a net 70% of candidate factual claims must be aligned with the truth-value of RR-NC-PRL...").
   e. Thresholds that qualify a given media expression for peer-review certification consideration (i.e., minimum conditions that must be met in terms of factual claims vs. nonfactual claims and factual claims vs. candidate factual claims).

---

[6] Of course accurate citations, sourcing, footnotes, etc. makes it much easier for peer-review certification analysis to be completed. However I distinguish between ease of peer-review certification analysis and whether a given media expression is indeed compatible with existing peer-review consensus on a given topic.

    f. Algorithms for determining peer-review certification (PRC or CPR – Certified Peer Review) qualification. For example:

        i. In order for an article to be considered for peer-review status, at least 25% of the article's factual claims must be comparable to the relevantly recent peer-review literature.

        ii. To meet the peer-review threshold, the peer-review percentage in the below formula must be at least 70%:

Peer Review % = # of NET claims whose truth-value equals NC-RR-PRL / # of claims addressed by NC-RR-PRL

(9) Transparency: the operational definitions, selection procedures, evaluation analyses shall be accompanied by rational justifications, be publically available and transparent, and open to ongoing criticism, evaluation, and improvement (i.e., the peer-review certification process will itself be subject to ongoing peer review in a transparent manner).

In sum, the evaluation is not to determine whether the given claim is true or false but whether:

(1) The claim is addressed in recent peer-review literature.

(2) If the claim is addressed then, is the claim accepted as true or false by a rough consensus of the recent peer-review literature? or

(3) If the claim is addressed then, is the claim as yet undetermined as true or false in a review of recent peer-review literature?

    Points toward or against peer-review certification are awarded or subtracted based on whether a given factual claim is presented in such a way that it is consistent with the view on this claim by the recent peer-review literature. So if there is a rough consensus (95% agreement) in the peer-review literature that a given factual claim counts as "true" and the knowledge disseminator presented the factual claim as true, then a point is awarded in favour of certification. If the knowledge disseminator presented the above factual claim as undetermined then a point is subtracted and so forth. Congruence between the knowledge-disseminating organization and the scholarly community (as evidenced by peer-reviewed evidence) results in points toward certification. Non-congruence results in points against. If a given percentage (threshold) of factual claims in a given article are congruent with the peer-review literature then that article has peer-review certification status. If a given percentage of articles or reports in a calendar year have peer-review certification status then that organization is certified as attaining peer-review certification status for that given calendar year

    In this final section of the paper I wish to explore in more detail supporting arguments and then I will conclude by addressing important objections.

7. CONCLUSION

It's most likely the case that with few exceptions, journalists—and their editors—do not, as part of fact-checking protocol, check to see that the truth-value of important factual claims are consistent with the relevantly-recent, near-consensus, peer-reviewed literature, when such near-consensus exists. Indeed there is most likely little awareness among journalists, bloggers, news anchors, and news curators of what academic peer review is and how it works as well as the intellectual practices and virtues that are involved. Thus sourcing obligations are often met with one or two quotes from ready-at-hand university professors, industry research, or ideological think tank fellows.

Often the result is that the public wrongly concludes that on a given issue there is widespread expert disagreement amongst experts or the public fails to gain a properly-nuanced view on a given scientific issue. One possible fallout is that the existing alienation between the academy and the public at large is further exacerbated; this is especially dangerous for those academy institutions that rely on continued taxpayer support. If the academics can't agree on anything, or if they are all "biased" and guided by the sole pursuit of external goods instead of virtuous internal goods, then scientists' knowledge claims are suspect and financial sacrifice on the part of taxpayers for the academy is increasingly unwarranted. I assert that while the general public cannot be expected to understand much of the specifics of a given discipline they can potentially have a rough grasp of academic peer review—after all, everyone relies on peer-review processes for all sorts of consumer decisions—from reading Amazon reviews to Yelp reviews to asking friends for car repair shop advice. Academic peer review is, I assert, simply a more rigorous, formal process of we do informally in our day-to-day lives.

Furthermore the public might learn from peer review were it to become more familiar and widespread. Given the enormity of media images, social media expressions that the public is exposed to each day, something like a peer-review certification label might be an appropriate adaptation in this modern communication environment. It's replicable, is brand-able, is easily recognizable, sound-biteable, is click-able and yet might, in a clever way, redirect knowledge consumers to an overlooked space in which robust intellectual virtue practices are the norm.

I conclude this paper by considering some objections to the notion of a certified peer-review certification label. Objections are stated first followed by my response.

7.1 Some objections and responses

Objection #1: Such a labeling scheme is logistically unworkable.

Response to objection #1: Above I sketched out one possible scheme by which a peer-review certification could be implemented that makes three key assumptions:

(1) That large-scale dissemination sites, such as news organizations (CNN, MSNBC, The Huffington Post, the Wall Street Journal) would be willing to fund and submit to a certification review process in which they might come up short. Why would they pay a substantial fee for less-than-guaranteed certification?

(2) That an objective, fair, and unbiased content analysis, based on sufficient samples, can be conducted with a minimal amount of remunerated scholarly labour-time.

(3) That an algorithm can be developed that would reliably, and objectively, indicate whether or not a given media dissemination site meets a sufficient threshold for peer-review certification.

I applied my peer-review certification formula to three recent stories on yoga injuries, gas prices, and climate change and, I think I have shown that it is logistically possible to carry out such an analysis. As for funding and participation by media companies, the penalty for not participating might outweigh the costs of participating and failing. If just one or two relatively high-quality media sites (for example the New York Times science pages) were to successfully submit to a peer-review certification process by scholars then non-participating competitors would be penalized insofar as the contrast between media products would be quantitatively, objectively evident. The higher-quality site has the label; the "lower-quality" site does not. Just as non-organic food, following the 1995 Organic Consumers Act, is now highlighted as lacking a given social quality ("organic-ness") that it did not lack before, non-peer-review certified media disseminators would be highlighted insofar as they lack peer-review certification. Finally, given that the standards and procedures of peer-review certification would be transparent, presumably media companies would not submit to this process until they were fairly sure, via in-house evaluation, that they would meet peer-review criteria.

Objection #2: That such a scheme would be fatally vulnerable to political manipulation.

Response to objection #2: The weakest two links in my tentative approach that would be vulnerable to explicit political manipulation are:

(1) The selection of scholars or scientists to serve on the peer-review team and

(2) Accusations that the peer-review practices in the academy itself are politically biased toward either political objectives or careerism on the part of scholars as charged by Ric Savage of American Diggers, for example. So comparing media sites to scholarly peer review misses the point. If the gold standard of scholarship is itself fatally corrupt and biased, what sort of gold standard is that?

Use of a lottery system to invite qualifying scholars (those who have recently published or refereed papers) would minimize any political bias or manipulation in choosing the make of an academic peer-review certification team. As for addressing claims that the academy itself is politically biased or concerned solely with external goods (as charged by Ric Savage), labeling schemes such as the peer-review certification label that contribute to more public awareness of, and hopefully more curiosity about, academic practices, that aim at more transparency and openness with the public, might go a long way to pre-empt and proactively defuse charges of political bias within the academy. Deeper familiarization with and exposure to academy peer-review practices, would, I hope, impress upon the public that, despites its flaws, most scientists engage in peer-review as part of a larger set of scholarly practices that include virtues such as courage, justice, and honesty (MacIntyre, 1984).

Objection #3: Given the diversity of peer review practices within and between academic disciplines, no single label is workable.

Response to objection #3: Since much of reporting on science and social science is fairly general and relatively unsophisticated, it should be fairly easy for a review team of scientists to come to easy consensus on whether or not a given set of media expressions accurately reflects consensus in the peer-review literature.

REFERENCES

Anderegg, W. R., Prall, J. W., Harold, J., & Schneider, S. H. (2010). Expert credibility in climate change. *PNAS, 107* (27), 12107–12109.

Annas, J. (2011). *Intelligent virtue.* Oxford: Oxford University Press.

Carter, B. (2012, March 20). TV digs will harm patrimony, scholars say. *New York Times.* Retrieved from http://www.nytimes.com/2012/03/21/arts/television/spikes-american-digger-draws-concern-from-scholars.html

Council, U. G. (2012). U.S. Green Building Council. Retrived from http://www.usgbc.org

Dewey, J. (1926). *Democracy and education.* New York, NY: Macmillan Company.

Dewey, J. (1954). *The public and its problems.* Chicago, IL: Swallow Press Books.

Gasser, J. P. (2008). *Born digital: Understanding the first generation of digital natives.* New York, NY: Basic Books.

Habermas, J. (1981). *The theory of communicative action* (Vol. 1). (T. McCarthy, Trans.). Boston, MA: Beacon Press.

Hames, I. (2007). *Peer review and manuscript management in scientific journals: Guidelines for good practice.* Maiden: Blackwell.

Horster, D. (1992). *Habermas: An introduction.* Philadelphia, PA: Pennbridge Books.

Koebler, J. (2012, March 30). Study: Conservatives' trust of science hits all time low. *Chicago Tribune.* Retrieved from http://articles.chicagotribune.com/2012-03-30/news/sns-201203301004usnewsusnwr201203290329 conservativescimar30_1_science-conservatives-trust-political-debates

MacIntyre, A. (1984). *After virtue* (2nd ed.). Notre Dame, Indiana: University of Notre Dame Press.

National Opinion Research Center. (2006, 2010). General social survey. Chicago, IL: University of Chicago.

National Science Foundation. (2012). Science and engineering indicators 2012. Arlington, VA: National Science Board.

Rogers, M. (2010). Introduction: Revisiting the public and its problems. *Contemporary Pragmatism, 7* (1), 1–7.

Rosenberg, S. (2009). *Say everything: How blogging began, what it's becoming and why it matters.* New York, NY: Three Rivers Press.

Seife, C. (2010). *Proofiness: How you're being fooled by the numbers.* New York, NY: Penguin Books.

Shapin, S. (1994). *A social history of truth: Civility and science in seventeenth-century England.* Chicago, IL: University of Chicago Press.

Shatz, D. (2004). *Peer review: A critical inquiry.* Oxford: Rowman & Littlefield Publishers.

Song, L. (2012, March 26). Climate Change Disappears from Keystone XL Pipeline Debate. *Inside Climate News.* Retrieved from http://www.http://insideclimatenews.org/news/20120326/cushing-obama-keystone-xl-pipeline-climate-change-greenhouse-gas-emissions

Walton, D. (2007). *Media argumentation: Dialectic, persuasion, and rhetoric.* Cambridge: Cambridge University Press.

Weller, A. (2001). *Editorial peer review: Its strengths and weaknesses.* Medford, NJ: Information Today.

Evaluating Experts: Understanding Citizen Assessments of Technical Discourse

LAUREN R. ARCHER

Department of Communication
University of Washington
Box 353740, Seattle WA 98195-3740
USA
archerl@uw.edu

ABSTRACT: Rhetorical analysis of Oregon's 2010 Citizens' Initiative Review (CIR) transcripts provides insight into understanding how citizens evaluate the rhetoric of experts as well as how they process technical discourses. The CIR demonstrates the potential of a Deweyian model of expertise—where *experts inform* and *citizens deliberate*—for improving expert-citizen interactions.

KEYWORDS: citizen engagement, deliberation, John Dewey, *ethos*, expert-citizen interaction, expertise, public understanding of science.

1. INTRODUCTION

In many ways, Philip Wander's (1976) prediction about technical discourse coming to dominate public discussion of policy decisions has come to fruition. Increasingly we rely on "expert opinion" to guide and direct us in what to do and what to think about particular issues (Fischer, 2009). Political actors stall policy decisions by demanding more information from experts and stalemates between differing technical perspectives bog down debate. Alongside this increasing reliance on technical information, growing specialization within academic disciplines has led to an explosion in areas of expertise that might be called on for making these decisions.

Scholars studying issues of expertise often discuss the increasing gap between experts and laypeople and worry that expert discourses are drowning out the voices of citizens (e.g. Goodnight, 1982; Jasanoff, 1989; Wander, 1976). In response to these concerns, some scholars have called for closing that gap and creating space for citizen input (Wynne, 1989). In heeding such calls, various approaches for bringing citizens and experts together in discussion over particular issues or policies have been explored (e.g. Davies and Burgess, 2004; Mitchell and Paroske, 2000). The Oregon Citizens' Initiative Review (CIR) represents one such approach, given its aim to provide citizens with better access to experts with specialized information on state ballot initiatives. Much can be learned from studying these types of events, as examining the interactions between experts and citizens can provide useful information for understanding how publics process and utilize scientific and technical discourses.

The CIR represents a unique opportunity to examine the rhetoric of expertise. This study examines how citizens respond to enactments of expertise as well as how they interpret and evaluate technical information, through rhetorical analysis of CIR transcripts. Analysis reveals a potential disconnect between what experts emphasize and what citizens attend to. Despite an emphasis on *ethos* by experts, citizens focused on assessing support for claims, rather than evaluating experts' character. Encouragingly, within the CIR structure, citizens demonstrated their aptitude for critically examining technical information, rather than simply accepting the information provided unquestioningly. Citizens found their voice alongside experts and demonstrated that although they did not possess technical knowledge, they still had significant contributions to make to the deliberative process. Ultimately, I argue, the Oregon CIR provides an example of a Deweyian model for using experts in public policy decisions where *experts inform* and *citizens deliberate*. Thus, the CIR offers a model for closing the gap between experts and citizens in civic life if its structure and process can be translated into other realms of public policy debate.

I begin with a broad sketch of the rhetorical dimension of expertise. I then provide a history of the CIR, followed by a broad overview of the enactments of expertise that emerged over the course of program. I then analyze the citizens' assessments of expertise, highlighting some general findings from the CIR transcripts. I end with a brief discussion on how the CIR incorporates a Deweyian model of expertise as well as the implications of this study's findings for future efforts to bridge the gap between experts and citizens.

2. EXPERTISE: AN OVERVIEW

As Fischer (2009) notes, "It is increasingly recognized that as societies become more complex so does the importance of expert advice in matters related to governance" (p. 17). With this growing dependence on experts, academic attention has turned increasingly to issues of expertise. According to Lyne and Howe (1990) "an expert is defined by reference to the norms and content of a field" (p. 134). However, expertise is not just a matter of possessing knowledge of a particular field, but also of sharing that knowledge. As Lyne (1990) observes, "Expertise is not only a matter of the relationship of a specialist to a body of knowledge; it is also a matter of the relationship to the audience" (p. 52). Thus, any person with specialized knowledge can be considered an expert, defined as such by the relation between that individual and the larger community.

Experts rely on communication to relay their specialized information to nonspecialists (Carr, 2010; Collins & Evans, 2007; Lyne & Howe, 1990). In the public discourse arena, where healthy competition among experts exists, rhetoric serves as an important tool for persuading listeners to pay attention. Thus, much can be gained by carefully examining the discourses of experts in order to understand which inventional resources they utilize when speaking to nonexperts. Several scholars within rhetoric of science have developed excellent studies on the rhetorical aspects of expertise (Goodnight, 1982; Lessl, 1989; Lyne & Howe, 1990; Miller, 2003; Wander, 1976).

However, much can also be gained from understanding the other side of expert-citizen interactions. Analyzing how laypeople interpret different enactments of expertise can provide insight into perceptions of experts in the public domain, but can also be challenging. While scholars can often easily obtain accounts of expert rhetoric, accounts of audience response are more limited, making it challenging to determine layperson assessments of the communicative practices of experts. Through analysis of the CIR transcripts, which contain citizen discussions about experts, this study aims to provide a more robust understanding of the rhetorical relationship between experts and citizens, and in particular, of citizen assessment of experts and their discourses.

3. OREGON'S CITIZENS' INITIATIVE REVIEW

In 2010, Oregon introduced a pilot program of the CIR. Ultimately the program aimed to provide informed, unbiased evaluations of select ballot measures to guide voters come Election Day (Gastil & Knobloch, 2010, p. 3). In the summer of 2010, the CIR pilot gathered two small groups of Oregon citizens to participate in one of two weeklong panels.[1] During the week, citizens learned about and deliberated on an assigned ballot measure. For the pilot program, two ballot initiatives were reviewed: Measure 73, which called for implementing mandatory minimum sentencing for repeat DUI and sexual offense offenders; and Measure 74, which proposed implementing a dispensary system for the distribution of medical marijuana (Oregon CIR Archive, 2010). Citizens heard presentations from both pro and con initiative advocates and selected background witnesses to testify and fill in information gaps.[2] At the end of each week, citizens engaged in deliberation to sort through all the provided evidence, evaluate the measure, and write a general statement of findings for inclusion in the voter's pamphlet.[3]

The significance of the CIR cannot be overestimated; it represents the first such attempt to change how citizens participate in the initiative process. Although it builds off deliberative democracy theory more generally, and citizen deliberation experiments like citizen juries more specifically, the CIR program remains unique in its commitment to encouraging deliberation amongst a small group of citizens for the benefit of the larger population (Gastil & Knobloch, 2010). Given the success of the 2010 pilot program, the Oregon legislature approved making the CIR a permanent part of the state's initiative process.

The detailed CIR transcripts provide a unique text for studying and understanding the rhetoric of expertise. Small group discussions following expert presentations allowed citizens time to discuss what information they found most useful, but also provided a discursive space for commenting on and evaluating individual speakers as well. Citizen comments reveal how they assess various enactments of expertise while also demonstrating how they receive and interpret complex technical information. Thus, a rhetorical perspective on the exchanges at the CIR provides a better understanding of how audiences of nonspecialists interpret and evaluate the rhetoric of expertise.[4] Additionally, CIR's success in bringing citizens and experts together into conversation carries implications for improving expert-citizen interactions in public policy debates.

4. ENACTMENTS OF EXPERTISE

Carr (2010) argues for attending to enactments of expertise rather than acquisition of expertise (i.e. how expertise is used rather than how expertise is gained), believing that a focus on

---

[1] The first panel met August 9-13 to discuss Measure 73. The second panel met Aug 16–20 to discuss Measure 74. For more background information on the CIR, please see Gastil and Knobloch (2010) or http://www.healthydemocracyoregon.org/citizens-initiative-review.

[2] Citizens selected witnesses from a provided list, which was compiled with the advocates' help.

[3] Please see the Oregon CIR archive website (http://cirarchive.org, 2010) for access to recordings and other background information. All direct quotes in this paper come from transcripts created from audio recordings of the CIR by John Gastil and Katie Knobloch.

[4] I treat both initiative advocates and background witnesses as experts, in part because they all speak about areas of specialized knowledge and in part because citizens did not typically distinguish between them.

communicative interaction will provide new understandings of the role of experts in society.[5] CIR experts relied most heavily on enactments of expertise that emphasized *ethos*. Typically defined as an audience's perception of a speaker's character, a speaker's *ethos* can significantly influence how an audience receives the delivered message. *Ethos* can play a particularly important role in enactments of expertise. As Miller (2003) argues, "The reliance on expertise is an argument from authority, and thus, in rhetorical terms, a signal that *ethos* is an important mode of appeal" (p. 169). The CIR experts seemed to share this understanding and put significant effort into developing their *ethos*.

Aristotle further classified *ethos* into three components: *phronesis* (practical knowledge), *arête* (virtue), and *eunoia* (good will). Within the discourse of CIR, experts employed appeals relating to each component of *ethos*. Presenters established *phronesis* through introductory comments that highlighted credibility by emphasizing professional experience. This included experts not just identifying their professions but also underscoring the length of time spent in that profession. Throughout their presentations, experts relied on anecdotes that spoke to *arête*. Such stories portrayed the speaker as having good intentions and being trustworthy, i.e. by depicting the speaker as a dedicated member of law enforcement or as the victim of a crime trying to keep others from suffering the same experience. Finally, experts' comments on the CIR process overall, and specifically the citizens' efforts, reflected on *eunoia*; positive comments helped convey a sense of good will toward the citizens, while negative comments undermined it.

5. ASSESMENTS OF EXPERTISE

While experts devoted significant time to building *ethos*, citizens paid little attention to such appeals. When citizens did mention the experts, they typically did not refer to them by name, but rather relied on pronouns or occupational references. However, an expert's character did become a topic of discussion for citizens when they perceived a speaker to be violating some aspect of *ethos*, especially *arête* and *eunoia*. Additionally, most anecdotes delivered by experts did not receive explicit attention during small group discussions; instead citizens focused most explicitly on evaluating the relevance, credibility, and objectivity of provided information. In the next sections, I look at these assessments of expertise more closely.

5.1 *Reducing* Ethos *to Occupation*

Regardless of the sometimes elaborate introductions and anecdotes experts used to establish themselves and their credibility, the citizens did not pay much explicit attention to these appeals. However, citizens did focus on *ethos* when they felt an expert to be disingenuous in some regard. Speakers perceived as lacking *arête* received particular attention. For example, one citizen declared, "Sometimes they're feeding us lines of bull." These citizens expressed a clear sense of what they expected of experts and did not take it lightly when speakers failed to meet those expectations, as demonstrated by the following exchange:

---

[5] Space does not permit for a detailed accounting of enactments of expertise. Since the main focus of this paper is on understanding citizen assessments of expertise, this discussion will necessarily be limited and fairly broad.

Panelist 1: You also have to remember that when Dr. Barthwell was speaking yesterday—remember, she represents a pharmaceutical company. That—I took 25 steps back. Come on. Come on.

Panelist 2: One of the first things she does is give us this big pamphlet full of the pharmaceutical drugs that she's been working on, half of which were based on marijuana.

Panelist 1: That's just where I'm coming from. What I heard today [from other background witnesses] was far more concrete. I heard evidence and I heard a solid case for those in need.

Given her professional association, these citizens considered this expert to have a conflict of interest in speaking in opposition to Measure 74, which dealt with establishing medical marijuana dispensaries. This conflict of interest reflected poorly on this expert's *ethos* and negatively impacted the citizens' evaluation of the information she provided.[6] Citizens also commented on experts who, from their perspective, lacked *eunoia*. After a somewhat disorganized opening presentation from the con advocates of Measure 74, a citizen expressed disappointment:

Panelist: My issue is the way the DA presented himself in the beginning. 'I was kind of called in here because so and so's my friend and I'm not really prepared.' Then what are you doing here? That kind of sets the tone so then to me I was questioning in my mind everything that was shared.

Citizens treated their role in the CIR process quite seriously and didn't approve of experts who didn't show enough goodwill toward them or respect for the task at hand. The con advocates on Measure 74 managed to redeem themselves later in the week and re-establish *eunoia*, as evidenced by another citizen's comment: "I found that they actually put in a lot more into it today." Thus, while citizens may not have explicitly discussed an experts' positive *ethos*, they did note those experts whose *ethos* were perceived less positively.

Beyond discussions of negative *ethos*, citizens frequently reduced experts' *ethos* to their profession and forgot the rest. In referencing the various speakers, citizens often used vague pronouns rather than specific names. Citizens struggled to link information with the speaker who provided it and comments such as "individuals that are in law enforcement" or "that lady from yesterday" occurred frequently. Often citizens simply referenced speakers by their occupation. For example, one citizen stated, "I don't think the opponent—the police officer—put out a very good—they don't push out a very good case for it." Another citizen referenced a speaker by saying, "the doctor there." These vague references may be a result of the number of presenters heard over a short span of time. However, this focus on occupation may also result from the emphasis speakers placed on their job experience during introductions. Interestingly, citizens remembered a select few experts explicitly and almost always referenced them by name. For example, citizen participants on Measure 73 identified Craig Prins when he was the source of information being referenced. Citizens may have remembered Prins because they had more interaction with him; he spoke longer than anyone else and even came back a second day for follow up. But advocates also talked up Prins before he spoke to the citizens, and perhaps these recommendations helped frame Prins in a way that made them particularly attentive to him, increasing the likelihood of them remembering his name.

Additionally, citizens often grouped together those presenters with similar occupations. One individual commented, "I thought even the Law Enforcement people were

---

[6]   Another group had a similar exchange over the seriousness of this conflict of interest.

saying that there's really no increase in crime." Another citizen corrected this statement and pointed out that only some, not all, of the law enforcement people had made such a statement, but even then, citizens failed to realize that one of the "law enforcement" individuals appearing at the CIR had spoken as a private citizen, not a police officer.[7] Citizens didn't even use length of service or amount of work experience to distinguish between individuals in the same profession, despite speakers emphasizing such details in introductory comments. Likewise, citizens also did not distinguish between specialities within professions, but rather categorized experts broadly, i.e. as "law enforcement" or "lawyers." This tendency to reduce speakers to a profession may carry implications for public discourses that incorporate expert voices. Lyne and Howe (1990) note how standards of accountability and validity become blurred as technical discourses cross over disciplinary boundaries or from the technical into the public sphere. Within public discourse, especially in media, experts may be asked to speak beyond the scope of their specific area of study. If citizens are likely to think of experts broadly, for example as "scientists," rather than recognize their distinct areas of expertise, they may be ill-equipped to evaluate which experts are best qualified to speak on particular issues and may perceive all expert opinions as equal. This may help explain the ease with which manufactured controversies are able to sustain themselves in the public sphere, often simply by providing an opposing "expert" viewpoint (Ceccarelli, 2011).

5.2 Statistics Not Stories

While experts frequently utilized storytelling, citizens didn't often discuss the stories told. Instead, they expressed a desire for reliable evidence. This emphasis by citizens on information rather than anecdotes might be a result of the overall structure of the CIR and the instructions provided, which directed citizens' attention to evidence. At the beginning of the week, moderators instructed citizens to stay in "learning mode" and to gather as much information as possible with an open mind and without immediately jumping into evaluation of it. Similarly, the moderators framed small group discussions as an opportunity for citizens to select "strong and reliable evidence." Citizens especially valued statistics, as demonstrated when one remarked, "Every time we ask for statistics, they say, 'Oh, we don't have that record.' Or, 'We don't track that.'" This pattern among CIR participants reinforces research by Hornikx (2008), which demonstrates that laypeople perceive statistics as the most persuasive type of evidence while they perceive anecdotes as the least persuasive.

Citizens also exhibited an ability to engage with experts and critically evaluate the quality of claims being made. Several times during presentations citizens asked for source citation or references for the information presented. For example, during a presentation for Measure 74, an advocate relayed a story about a medical marijuana grower being arrested and prosecuted for possessing one plant over the legally allowed limit. At the end of the presentation, a citizen said, "I just really want to know the Washington case by name or date or the cost analysis of $100,000 that was spent for the guy that was over with one plant. I just want to be able to check that out." Rather than immediately accepting the story as accurate because someone with more law enforcement experience shared it, the citizen asked for additional information in order to verify the story. Citizens expressed similar feelings during

---

[7] This confusion may demonstrate the difficulty of separating citizen action from professional identity, an idea discussed by Pielke (2007) specifically in regards to scientists.

small group discussions, which experts were excluded from. One group discussed the reliability of a particular witness based on the information she provided:

> Panelist 1: I'm disagreeing with her on that part. I think she's exaggerating.
> Panelist 2: I don't think we could say evidence's that exaggerating. It's a piece of evidence. I think you would just throw that out and not include that.
> Panelist 3: It sheds doubt on anything else she—
> Panelist 1: Exactly. She was unclear about her information.

This exchange demonstrates not only that citizens did not blindly accept all information given to them by experts, but also, that they were not afraid to express their own perspective alongside that of the experts.

Additionally, group discussions worked quite effectively as a type of quality control. Citizens frequently corrected each other when they misremembered a statistic or confused two pieces of evidence. Also, during discussions, citizens sometimes collectively discovered discrepancies in the data or conflicts in the evidence, as demonstrated in the following interchange:

> Panelist 1: Actually, the thing that we got back from the lady yesterday –
> Panelist 2: Kind of stated the exact opposite.
> Panelist 1: Exactly. And that was from several different Police Chiefs
> Panelist 2: They totally negated that statement.

Furthermore, citizens never forgot that many experts presenting information also had a particular stance on the initiative up for discussion. Throughout the process citizens sometimes expressed scepticism about experts' comments. One citizen commented "They are both trying to pull us onto their side." Another citizen spoke up when he felt that presenters were not properly conducting themselves. He commented, "It seems to me that you guys have been using a lot of fear and exaggeration rather than factual stuff, especially you, sir, the sheriff." The comment calls out these experts on their rhetorical tactics and enforces that the citizens valued credible information over fear tactics.

These observations about citizens' abilities to analyze and evaluate the technical discourse of the CIR experts carry implications for the public understanding of science debate. In America, claims of scientifically illiterate citizens frequently garner headlines (California Academy of Sciences, 2009; Mooney and Kirshenbaum, 2009). However, the CIR transcripts reveal citizens capable of engaging in a dialogue with experts, understanding technical information, and asking for clarification when needed. Although neither initiative relied solely on scientific information for support, with both measures, experts presented some highly technical data and complicated statistics. During this process, citizens depended almost entirely on the experts to provide the information they used in making their recommendation to voters. At the end of the week, citizens spent significant time deliberating in order to sort through all the presented information and select the most pertinent elements for voters. In their evaluation report on the CIR pilot program, Gastil and Knobloch (2010) noted that citizens were successful in recognizing unsupportable claims and selecting accurate information to use in the citizens' statement for the Voter's Pamphlet. This observation speaks to the citizens' ability to not only understand technical presentations but also to evaluate the relevance and credibility of provided information.

6. A DEWEYIAN MODEL OF EXPERTISE

John Dewey envisioned experts as not only possessors of specialized knowledge but also leaders in teaching citizens the skills of deliberation, which he considered essential for improving the strength of democracy in America. Fischer (2009) explains:

> The answer for Dewey was to rethink professional expertise . . . . Dewey called for improvements in the methods and conditions of debate, discussion, and persuasion. The experts would have a special role in such deliberation, but it would take a different form. Instead of only rendering judgments, they would analyze and interpret them for the public. If experts, acting as teachers and interpreters, could decipher the technological world for citizens in ways that enabled them to make intelligent political judgments, the constitutional provisions designed to advance public over selfish interest could function as originally conceived. (p. 28-29)

Dewey's vision of expertise ceded the technical sphere to the professionals but then placed it in service to a political sphere governed by citizens. However, in the post-World War II era, the emergence of a technoscientific sense of expertise steadily gained popularity in America as citizens recognized the role science played in winning the war (LaFollette, 1990). With the continued advancement of both science and technology, the public came to rely more heavily on the advice and guidance of technoscientific experts. As experts became more specialized and citizens became more dependent on technical advice, the separation between experts and citizens grew into the current culture of privileging expert opinion over citizen voices.

However, the overall structure of the CIR reverses the traditional expert-layperson hierarchy by incorporating a Deweyian model of expertise. By mediating interactions between experts and citizens, providing ample time for citizens' questions, and giving citizens time for discussion without experts present, the CIR gave experts the responsibility of informing citizens and placed the onus for deliberation on the citizens themselves. Within such a structuring of expert-citizen interactions, citizens were more likely to speak up rather than simply defer to the experts. Citizens readily pointed out when they felt the experts weren't providing sufficient evidence to support their claims or were demonstrating poor *ethos*. Citizens also demonstrated their ability to assess technical information and hone in on accurate, credible data. Through their role in the CIR, these citizens demonstrated that while they did not possess the same technical knowledge as the experts, they still had plenty to contribute to the discussion and possessed the skills necessary to engage in productive deliberation.

Additionally, citizen comments reveal that they took their responsibility to report back to Oregon voters very seriously. This sense of accountability may have helped motivate citizens to pay close attention to expert presentations, take notes, and carefully weigh all the evidence in ways they might not have done if they weren't responsible to someone else. Citizens' willingness to ask for further explanation or stronger data created a sense of accountability for the experts as well. Replicating this sense of accountability may be an important strategy for raising the quality of interactions between experts and citizens. Citizens' assessments of the experts they encountered provide support for Miller's (2003) claim that face-to-face interactions between experts and citizens can encourage accountability and enable *ethos*—in this case the failure to establish positive *ethos*—to operate as a useful resource in assessing experts. More everyday encounters with expertise, such as reading a news article on the risks of contaminated food or hearing a televised debate on alternative energy options, lack this ability to engage more directly with experts. Thus, events like the Oregon CIR, which can replicate a Deweyian model for utilizing experts, are particularly important for creating opportunities for engaged interaction between experts and citizens.

7. CONCLUSION

Miller (2003) observes, "When our age is characterized as an age of science and technology, what is often meant is simply our deference to expertise in the public realm" (p. 190). This observation reflects the actualization of what Wander (1976) predicted, a society in which technical discourse dominates and "non-expert" citizens are often discounted as too uninformed to have much to contribute. As Myers (2003) notes, "Such divisions between science and non-science, professional and non-professional, divisions that we take for granted, were formed in historical struggles, and are re-formed in everyday practices" (p. 274). Contemporary practices often perpetuate an expert-layperson binary. However, the CIR offers hope for addressing these historical struggles and closing the gap between experts and citizens in civic life. Activities like the CIR, which create a space for citizen engagement and recognize the contributions citizens can make alongside experts, help make that move. Constraining experts to analyzing technical issues and informing while placing the onus of deliberation and debate on citizens holds promise for getting those most affected by policy decisions involved and increasing citizen engagement in politics.

ACKNOWLEDGEMENTS: Thanks to John Gastil and Katie Knobloch for providing access to the Oregon CIR transcripts and to Andrew Waddell for his research assistance. Thanks also to Leah Ceccarelli for her helpful feedback on an earlier version of this paper.

REFERENCES

Aristotle. (2007). *On rhetoric: A theory of civic discourse* (2nd ed.). (G. A. Kennedy, Trans.). New York, NY: Oxford University Press.

California Academy of Sciences. (2009, February 25). American adults flunk basic science: National survey shows only one-in-five adults can answer three science questions correctly. Press Release. Retrieved from http://www.calacademy.org/newsroom/releases/2009/scientific_literacy.php

Carr, E. S. (2010). Enactments of expertise. *Annual Review of Anthropology, 39*(1), 17–32.

Ceccarelli, L. (2011). Manufactured scientific controversy: Science, rhetoric, and public debate. *Rhetoric and Public Affairs, 14*(2), 195–228.

Collins, H., & Evans, R. (2007). *Rethinking expertise.* Chicago: University of Chicago Press.

Davies, G., & Burgess, J. (2004). Challenging the 'view from nowhere': Citizen reflections on specialist expertise in a deliberative process. *Health & Place, 10*(4), 349–361.

Fischer, F. (2009). *Democracy and expertise.* Oxford: Oxford University Press.

Gastil, J. and Knobloch, K. (2010). Evaluation report to the Oregon state legislature on the 2010 Oregon Citizens' Initiative Review. Portland, OR.

Goodnight, G. T. (1982). The personal, technical, and public spheres of argument: A speculative inquiry into the art of public deliberation. *Journal of the American Forensic Association, 18*, 214–227.

Healthy Democracy Oregon. (2010). Oregon CIR Archive. Retrieved from http://cirarchive.org

Hornikx, J. (2008). Comparing the actual and expected persuasiveness of evidence types: How good are lay people at selecting persuasive evidence? *Argumentation, 22*(4), 555–569.

Initiative and Referendum Institute. Oregon. (2011). Retrieved from http://www.iandrinstitute.org/Oregon.htm

Jasanoff, S. (1998). Technocracy and democracy. In *The 5th branch: Science advisors as policymakers* (pp. 229–250). Cambridge, MA: Harvard University Press.

LaFollette, M. (1990). *Making science our own: Public images of science 1910–1955.* Chicago: University of Chicago Press.

Lessl, T. M. (1989). The priestly voice. *Quarterly Journal of Speech, 75,* 183–197.

Lyne, J. (1990). Bio-Rhetorics: Moralizing the life sciences. In H. Simons (Ed.), *The Rhetorical Turn: Invention and Persuasion in the Conduct of Inquiry* (pp. 35-57). Chicago: The University of Chicago Press.

Lyne, J., & Howe, H. F. (1990). The rhetoric of expertise: E. O Wilson and sociobiology. *Quarterly Journal of Speech, 76,* 134–151.

Miller, C. R. (2003). The presumptions of expertise: The role of *ethos* in risk analysis. *Configurations, 11*(2), 163–202.

Mitchell, G. R., & Paroske, M. (2000). Fact, fiction, and political conviction in science policy controversies. *Social Epistemology, 14*(2/3), 89–107.

Mooney, C. and Kirshenbaum, S. (2009). *Unscientific America: How scientific illiteracy threatens our future.* New York, NY: Basic Books.

Myers, F. (2003). Discourse studies of scientific popularization: Questioning the boundaries. *Discourse Studies, 5*(2), 265–278.

Pielke, Roger A. (2007). *The honest broker: Making sense of science in policy and politics.* Cambridge: Cambridge University Press.

Wander, P. C. (1976). The rhetoric of science. *Western Speech Communication, 40,* 226–235.

Wynne, B. (1989). Sheepfarming after Chernobyl: A case study in communicating scientific information. *Environment, 31,* 10–15, 33–39.

Distrusting Climate Science: A Problem in Practical Epistemology for Citizens

THOMAS C. ATCHISON

Department of Philosophy
Metropolitan State University
700 East Seventh St.
St. Paul, MN 55106
USA
Thomas.Atchison@metrostate.edu

ABSTRACT: I briefly present empirical findings suggesting that citizens in contemporary democracies face great difficulties in arriving at an accurate picture of the world and of the relevant policy options and in identifying trustworthy sources of information. Unfortunately, these difficulties do not seem to diminish with more education or with more effort and attention. I argue that a highly polluted information environment can defeat the sorts of strategies generally recommended to individuals under the label 'critical thinking.' Finally, I consider what sort of institutional or systemic conditions would be necessary to provide citizens with a manageable epistemic task.

KEYWORDS: citizens, climate change, cognitive bias, deliberation, expertise, experts, motivated reasoning, practical epistemology, social epistemology, trust.

1. INTRODUCTION

Citizens need to know something, on even the most minimal view of their role in a democratic society. Some would hold, ambitiously, that they ought to know enough to participate meaningfully in self-government. Others would say that it is enough if they can learn that the elites they have been passively allowing to rule (by voting or not voting) are no longer serving them well enough and it is time to 'throw the bums out'. Even this last minimal view, though, requires citizens to know when their interests are no longer being served and to know whether voting for an opposition party is more likely than not to improve the situation. And these things may not be so easy to know.

In this paper I briefly present findings from public opinion research, political psychology and media studies, suggesting that time-constrained citizens in contemporary democracies face great difficulties in arriving at even a minimally accurate picture of the world and of the relevant policy options. The problem is not just ignorance of the kind well documented by decades of political science research (the 'low-information voter'). There is also widespread misinformation and apparent immunity to disconfirming evidence. Unfortunately, these difficulties do not seem to diminish with more education or with more effort and attention. Given the presence in all of us of various cognitive biases, and of the processes lumped together under the label "motivated reasoning," and given the generally 'polluted' character of our information environment, even diligent efforts to inform ourselves are all-too-likely to entrench our prejudices instead of improving the accuracy of our beliefs. Public skepticism about climate science provides a case in point.

I argue that unfavorable epistemic conditions (a highly polluted information environment) can defeat the sorts of strategies generally recommended to individuals under thelabel 'critical thinking,' at least when combined with plausible time and resource constraints and a realistic picture of human cognition. Finally, I consider what sort of changes in the information environment would be necessary to provide citizens with a manageable epistemic task.

2. GROUNDS FOR PESSIMISM (DEMAND SIDE)

I will not provide the standard rehearsal of voter ignorance. Suffice it to say, with Bartels (1996) that "The political ignorance of the American voter is one of the best-documented features of contemporary politics …." Nor is there space here for a rehearsal of the debates over the significance of this ignorance—whether it is seriously disabling or whether most voters manage to make decisions that are 'good enough' despite it. I will just say that I am persuaded by writers like Bartels and Delli Carpini and Keeter (1996) that American voters are uninformed and misinformed to an extent that is consequential in a very clear sense: Election results would be different if voters were better informed.

My main thesis: In the current epistemic environment, no feasible improvement in the information-seeking efforts of individual citizens is going to enable them to acquire the needed knowledge; to improve the level of public knowledge we will need to improve that environment. "Feasible" here refers not to the reluctance of many citizens to engage with politics or to seek out political information. What I mean is that even those who are willing to put a considerable amount of time and energy into investigating candidates and policy issues will find themselves thwarted. Why?

First, "considerable" does not, of course, mean "infinite." None of us has time to learn all that might be useful or relevant, and more time devoted to one issue means less to others. While anyone can learn the rudiments of political processes, parties, and ideologies (and thereby become better-informed than most American voters), many issues of public policy are complex and controverted and could absorb nearly unlimited investigative resources. To become even moderately well-informed on more than a few such issues is impossible. And what might be possible for a person with a good deal of leisure will not be possible for most citizens, who have relatively little time left over after work and family responsibilities are discharged. Inevitably most people will have to rely almost entirely on others (experts and intermediaries) to provide the needed information and analysis in a form that can be assimilated within the available time. And even policy experts will have to rely on others for everything outside their area of expertise. (Some of the consequences of this reliance will be crucial to my argument below.)

Further, more education and paying more attention to the news media don't always seem to help. The data here are not as robust, but we do have examples like:

(1) College-educated Republicans are more likely than less-educated Republicans to be global warming skeptics (Pew polls, 2008, 2010). While less-educated and less-informed Americans do not display much partisan polarization on this issue, more educated and better-informed Americans do. One might hope that higher education would provide better tools and more background information, so that, whatever the truth of the matter, more educated people would be more likely to converge on it. Not so.

(2) For most broadcast media sources, people who watched or listened more frequently were more likely to believe one or more false things about the Iraq war of 2003 (That Saddam Hussein's regime had 'weapons of mass destruction.' That most experts said he did. That Iraq materially supported Al Qaeda. That world public opinion favored the US invasion.) (Kull, et al., PIPA poll, 2003).

(3) Friedman (2005), citing Converse (1964): "Converse's most disturbing and under-remarked finding is that the relatively well informed compensate in dogmatism for their greater knowledgeability" (p. xxii).

Dogmatism may not be the best name for the underlying processes here. One psychological mechanism that may be at work is so-called "belief polarization."[1] This has recently received a careful philosophical analysis by Thomas Kelly (2008). Kelly starts with an empirical finding: Two people, who hold opposite views on a policy issue, can look at the same (mixed) body of evidence, and each come away thinking that their view is better supported by that evidence than the other's is. On the face of it, this is bad from an epistemic point of view, but what seems to be going on here is not simple dogmatism. Each is not simply dismissing the evidence that counts against her view and latching on to the evidence that supports it. Instead she is looking at the disconfirming evidence very carefully—so carefully that she finds good reasons to discount it (methodological flaws in studies, alternative explanations of unwelcome data). And Kelly suggests that it is not unreasonable to look harder at disconfirming evidence. He compares this to scientists who try harder to find explanations of anomalies than of phenomena that fit easily into existing theory. Given scarcity of investigative resources, this seems like a reasonable strategy. Moreover, he says, people pursuing this strategy are not violating the Principle of the Commutivity of Evidence (the anti-path-dependence principle, which says that the order in which you get various bits of evidence shouldn't matter to your epistemic evaluations). They are, instead, arriving at different bodies of evidence (because my total evidence includes, for example, the methodological flaws I have noticed in the studies I have subjected to stricter scrutiny.) My body of evidence is, in fact, biased by this process, but my assessment of the upshot of that evidence is not. So, Kelly concludes, as long as I am unaware of this biasing process, I still count as a reasonable inquirer. When I become aware of this process (as you and I, dear reader, are now aware) then I ought to take account of the fact that my evidence is likely to constitute a biased sample. But the typical case (and the one relevant to my point here) is the case where the subject is unaware of the bias. This is bias, but it is not dogmatism, and it exemplifies a pattern that has other instances: apparently conscientious reasoning that leads nonetheless to epistemically poor results.

This phenomenon shades off into other sorts of cognitive bias and 'motivated reasoning'. The literature on these subjects is vast. Here I will attempt only the briefest of summaries.[2] "Motivated reasoning refers to the unconscious tendency of individuals to process information in a manner that suits some end or goal extrinsic to the formation of accurate beliefs" (Kahan, 2011, p. 18). These goals can be various, but they are all non-epistemic. That is, they are not plausibly related to the goal of discovering the truth whatever it may be. Rather they serve other psychological needs: for self-affirmation, for a sense of group belonging, for

---

[1] The *locus classicus* is a study by Lord, Ross and Lepper (1979). In their terms 'biased assimilation' leads to 'attitude polarization'.

[2] This summary follows Kahan's (2011) summary where a comprehensive set of references may be found.

status, for security, etc.—or simple material self-interest. Beliefs can be an important part of social identity and change of belief can disturb important relationships. (Kahan speaks in this connection of "identity-protective cognition.") Whatever the goal, the distorting processes include biased search (where information is sought selectively to bolster one's beliefs), biased assimilation (where information and arguments are selectively filtered and evaluated) and biased assignments of credibility to sources of testimonial evidence (often on the basis of group membership).

One consequence of these processes is that partisan identification often trumps facts and issue-consistency. People who identify strongly with a particular political party tend to skew their beliefs about quite objective aspects of the economy depending on which party is in power (Achen & Bartels, 2006). They tend to believe, for example, that the unemployment rate or the budget deficit or the rate of inflation are lower (or are falling) when 'their side' is in charge and are higher (or rising) when the other side is in charge. Moreover, partisans will 'agree' with a policy position (more or less regardless of content) if they are told that it is the position of their party (Cohen, 2003). (Part of the picture here is that people will work hard to think of a reason why their party would be supporting an apparently discrepant policy position.)

Crucially, these processes go on below the level of conscious awareness. They are almost entirely unavailable to and denied by the subjects. Consequently, people tend to be far more aware of other people's biases than of their own, and to interpret disagreement in pejorative terms, asking about partisans on the other side "Are they stupid or are they evil?" since it seems incomprehensible that an intelligent, well-motivated person could fail to see such obvious truths. (Kahan uses the term "naïve realism" to refer to this phenomenon.)

Can people be taught to do better than this, to become 'critical thinkers'? It would be nice to think so, but there are reasons to doubt it. Philosophers (and others) who teach critical thinking often recommend a set of strategies of inquiry and evaluation like: consider the arguments on both sides of the question, look for fallacious reasoning and for inductively weak arguments, try to decide which side has the best case. Sometimes this can be effective, but, when the issues are complex and the assessment of evidence requires domain-specific expertise, it is not. As Neil Levy (2006) puts it, reading a book like Bjorn Lomborg's *The Skeptical Environmentalist* is more likely to degrade your epistemic condition than to improve it. Why? Because it contains a large number of superficially plausible arguments which one is in no position to adequately assess (since one lacks the relevant expertise).

Nor should we expect that our determination to think critically and our use of the usual techniques will eliminate motivated reasoning. "Indeed, far from being immune from identity-protective cognition, individuals who display a greater disposition to use reflective and deliberative (so-called "System 2") forms of reasoning rather than intuitive, affective ones ("System 1") can be expected to be even more adept at using technical information and complex analysis to bolster group-congenial beliefs" (Kahan, 2011, pp. 20–21). As Hilary Kornblith argues in "Distrusting Reason" (1999), a rather plausible case can be made for regarding a great deal of apparently sincere and conscientious reasoning as no more than rationalization. Here, intelligence and reasoning skills work against us, by making our rationalizations more plausible to ourselves and to others. Achen and Bartels (2006, p. 44) write, "Most of the time, the voters are merely reaffirming their partisan and group identities at the polls. They do not reason very much or very often. What they do is rationalize. Every election, they sound as though they were thinking, and they feel as if they were thinking, as do we all. The unwary scholarly devotee of democratic romanticism is thereby easily misled." Finally, Kahan (2011, p. 22) reports findings that suggest that exhortations to be objective, to approach issues in a spirit of open-mindedness, and to set aside biases, tend to backfire by triggering the very group identifications one was hoping to defuse. When the issue of bias is put on the table, so to speak, defending the beliefs characteristic of one's group becomes a matter of honor.

3. GROUNDS FOR PESSIMISM (SUPPLY SIDE)

So far I have been describing difficulties that seem to arise from human psychology. But we also have a highly 'polluted' information environment. Politicians and their advocates lie and spread misinformation. Industry sponsored think tanks and institutes, following a playbook initiated by the tobacco industry, try to sow doubt and confusion about, for example, the health effects of formaldehyde or the effects of CFC's on the ozone layer (Jackall, 1988). More recently global warming has received a similar treatment (Oreskes & Conway, 2010; Anderson, 2011). Chain emails spread fabricated or misinterpreted stories. "Push polls" offer tendentious characterizations of issues and candidates in the guise of attempts to measure public opinions. Employers provide distorted information to employees about public affairs and legislation, urging them to contact their elected officials. And, of course, the media environment is saturated with advertising.

It would be possible for the press as an institution to help citizens sort through the smoke and the spin, and news organizations often claim to be aiming to play that role. (Even Bill O'Reilly of Fox News, who strikes many observers as a fairly crude purveyor of conservative propaganda, claims to be providing a "no spin zone," in which politicians are held accountable to the public.) But, like many observers, I see a press that is serving the public rather poorly (Fallows, 1996). For one thing we have a set of media organizations that seem highly partisan (Fox News, the Murdoch-owned press more generally, right wing talk radio). For another, television has adopted an entertainment-focused model in which TV news shows are full of people talking very briefly, emphatically and unreliably about complex issues (evidently on the principle that controversy is more attractive to viewers than content).

But even in the more serious and 'responsible' quarters of the news business, all is not well. There are a number of unhelpful norms of journalistic practice. There is the often lamented (but never abandoned) emphasis on 'horse race reporting,' where the focus is on how statements and decisions will affect the political standing of officials with various groups (or their electoral fortunes), not on the truth of the statements or the effectiveness (much less the wisdom) of the decisions/policies. A closely related tendency is the conflation of perceptions and reality, where the fact that a certain action or policy will be seen in a particular way by some audience is more salient than whether or not that perception is accurate. [3]

But I want to focus mainly on the way that news organizations transmit and mediate the views of experts. There are two standard and contradictory practices, neither of them helpful.

---

[3] An example: "Obama owns the economy" – no distinction is drawn between causal responsibility (his policies have made the economy worse) and political perception (voters will blame the President for the bad economy regardless of why it is bad).

1) In the first sort of case, journalists transmit the claims and judgments of experts and authorities to the public, without comment and without much attention to dissenting views. This seems to have been the case in the run-up to the invasion of Iraq by the US and its allies in 2003. The media did not provide much resistance to the Bush administration's effort to 'sell' the invasion of Iraq to the American public in 2002–2003. The case for war was made to the American people and to the world with a lot of very scary claims about the 'gathering danger' of Iraq—including the possibility that Saddam Hussein would give a nuclear weapon to terrorists who would use it to blow up an American city. As we subsequently learned,[4] the evidence for most of their claims was much shakier than they led us to believe. And for the most part the US media simply relayed these claims to the American people, without much analysis and certainly without giving anywhere near the same kind of prominence to the views of people (many of whom had excellent credentials) who had a different view of the nature of the Iraqi threat and what would be the best way to deal with it.

As reporter Karen DeYoung put it: "We are inevitably the mouthpiece for whatever administration is in power. If the president stands up and says something, we report what the president said" (Kurtz, 2004). And if contrary arguments are put "in the eighth paragraph, where they're not on the front page, a lot of people don't read that far" (Kurtz, 2004).

The result is that even our best newspapers provide reasonably accurate coverage of important issues—at best—only to careful and diligent readers. Even the most casual reader learned from front-page headlines that the Bush administration repeatedly claimed that Iraq had WMD's and ties to al Qaeda. Only a reader who read to the end of the articles and searched the back pages for more could discover that many intelligence experts in and out of the government thought that the evidence for those claims was weak. This suggests that the key to helpful journalism is to make sure that reporting is 'fair and balanced' and gives equal time and equal prominence to opposing views. But my second example cuts the other way.

2) Media coverage of the global warming issue often conforms to a norm of 'balance.' But several researchers have found that this norm is, in this case, doing more to damage public understanding than to improve it. There is growing recognition that the norm of 'balance' as currently understood degenerates too easily into a practice of 'balance as bias' or 'he said-she said' journalism, leaving news consumers confused rather than enlightened. The problem is that readers (or viewers) who are presented with reports of scientific assessments of the nature and significance of global warming and who are then presented with a skeptical or critical response are left with the impression that these views are of roughly equal plausibility, that there is no consensus among the experts. (See Anderson, 2011, for discussion of the research on this point.) Declining public concern about global warming and declining willingness to endorse remedial measures seems to have been produced (in part) by this pattern of coverage.

But now it seems as though journalists are damned if they do (present "both sides of the story") and damned if they don't (by foregrounding the views they find most plausible at the time and ignoring the critics). The consumers of the news are poorly informed in either case. I will try to say something about how to resolve this dilemma in the final section of this paper.

---

[4]   http://msnbc.msn.com/id/5403731/ -- *Senate Intelligence Committee report on pre-war intelligence failures in Iraq*

4. EPISTEMIC DEPENDENCE AND UNAVOIDABLE BUT UNFORTUNATE TRUST

I said above that we will inevitably rely on others for most of what we can claim to know. This is an instance of what Allen Buchanan (2004) calls "epistemic dependence." Unavoidably, we will trust those who nurture us as children. They and others in our immediate social environment will (partly) set the starting conditions for all of our epistemic evaluations. What we can know and what we will fail to know will be affected drastically by those starting conditions and then by contingent facts about the other influences that come our way.

Some examples:

Example 1: Allen Buchanan on growing up racist:

> I grew up in the American South during the 1950s and 1960s in a racist family culture embedded in a society of institutionalized racism. ... I was taught, by explicit dogma and by example, to regard blacks as subhuman. Unlike my mother, I never witnessed a lynching, but I did once see a desiccated, severed black ear of unknown provenance, proudly displayed by a white junior high school classmate. I also recall joking with my friends about the "Tucker telephone," a crank-operated dynamo that was used to deliver electrical shocks to the genitals of black inmates of a nearby penal farm. Largely through luck, I left this toxic social environment at the age of eighteen and came to understand that the racist world view that had been inculcated in me was built on a web of false beliefs about natural differences between blacks and whites. My first reaction was a bitter sense of betrayal: Those I had trusted and looked up to—my parents, aunts and uncles, pastor, teachers, and local government officials—had been sources of dangerous error, not truth. (Buchanan 2004)

Buchanan also claims that he was taught strategies for evading evidence that might have undermined his racist beliefs. Buchanan goes on to argue that the cultural and institutional features of liberal democracy offer our best protection against the possibility that our starting point is so deficient—by drawing us into exchanges with diverse others in a context where there is a presumption of epistemic equality. But what would have happened if he had stayed in his community of origin?

Example 2: Baurmann on "Rational Fundamentalism"

Baurmann's argument (2007) is too complex to summarize here, but the intuitive idea is fairly simple: given the right kind of social environment it is (at least subjectively) rational to become and remain a fundamentalist. A fundamentalist is defined as someone who prioritizes salvation over worldly goods, regards his belief system as certain and not appropriately subject to doubt or criticism, and who divides the world into the good (who can be trusted) and the evil (who should be avoided, if not killed).

Two elements of Baurmann's account are especially relevant to this paper:

> (1) Particularistic trust. The rule is: "distrust everyone who is not a member of your group." When one is the member of a relatively isolated social group that is in hostile and/or conflict-ridden relationships with surrounding groups one will get enough confirmation of the untrustworthiness of members of other groups (they will treat one badly in various ways) that it is (again, subjectively) rational to continue to follow this rule (supposing it has been

taught to one by those in one's "personal trust network"—those one has come to trust on the basis of close personal relationships).

(2) Epistemic seclusion. This is accomplished by the inculcation of norms granting epistemic authority to group leaders and sacred texts and by, as much as possible, keeping group members away from contact with alternative sources of information (e.g., home schooling, ostracism and expulsion of dissenters, forbidding dangerous literature). Particularistic trust helps here, too.

Again, the idea is that in this kind of environment it is subjectively rational for an individual to adopt and maintain a fundamentalist set of beliefs and norms. Baurmann, like Buchanan, emphasizes the fact that those of us who are in a more open-ended and 'enlightened' social environment, where we can extend our trust to the deliverances of science, are (epistemically) lucky, not particularly rational. The superior (objective) rationality "lies in the institutions of science and the culture of an open and liberal society and not in the individual rationality of the single citizen."

Backing off a bit from the totalistic nature of the fundamentalist belief system and its social epistemology, we arrive at:

Example 3: Distrusting Climate Science.

We can think of this as a less drastic version of the previous example. (At least, it's less drastic for the 'climate change deniers' who are not fundamentalists.) The social and epistemic isolation is not as complete and the beliefs are not as extreme. But there is still a version of 'particularistic trust'. One is provided with reasons for disregarding the testimony of scientists, academics, and mainstream journalism. One is persuaded that only Fox can be trusted to give you accurate news reports. Everyone else is said to suffer from liberal bias, anti-Americanism, hostility to markets and to business, moral relativism and other epistemically-disabling conditions. Climate scientists, in particular, cannot be trusted because they have a strong conflict of interest: they want more money for their research, which will only come to them if they alarm the public and thus make it seem that their research is particularly urgent. In this way the denier is inoculated against conflicting evidence. And there is enough hostility, contempt and condescension coming from liberals and scientifically-minded people that the attitude of particularistic trust gets the support it needs.

Examples such as these seem to me to illustrate with particular force the dependence of the individual citizen on his or her epistemic environment. If we know better than the people described in these examples, then we are lucky.

5. MELIORATIVE STRATEGIES

After all this pessimism I wish I had a more persuasive (and less familiar) set of remedies. I do not. And though my main thesis is that we need to adjust our epistemic environment instead of expecting individuals to rise above it, I have no reason to think that the sorts of reforms suggested here (and by other like-minded scholars) have much chance of coming to pass. It might be just as reasonable to hope for a reform of human nature. Nonetheless …
Journalism:

We need more effort to identify and give pride of place to genuine experts rather than flacks, politicians, generic commentators (Brooks & Shields) and horse-race handicappers. We, therefore, need reporters to develop the subject-area knowledge necessary to be reliable guides to who the experts are, what the state of scientific (and other expert) understanding is, and who is likely to be telling us the truth.   We need serious efforts to assess the truth of statements by public officials and others who attempt to shape the public's understanding. (We need more than the sort of quick and perfunctory 'fact-checking' that is so widely practiced. This latter is better than nothing, but only if it employs reasonable standards of evidence. All too often, it does not, as a desire to appear even-handed leads to a refusal to decide which party to a dispute has the better case, and to sometimes-strained efforts to find all parties guilty of something.) This requires (again) more expertise on the part of the journalists, but also a different conception of their role: as fact-finders, not opinion conveyers (Cunningham, 2003.) To play this role, however, more is needed than better-educated and differently-oriented journalists. As noted above, the rationality of science lies in its institutions and practices more than in the virtues of individual scientists. Analogously, journalism cannot rely on the integrity and diligence of individual journalists. It needs some analogue of the institutions of peer review and scholarly debate.

Education:

We need to inculcate scientific and critical habits of mind. Every high school graduate should know enough about the nature of scientific inquiry and the practices and institutions of science to be an intelligent consumer of scientific reports and to have some idea about how to identify real expertise.

Media literacy (and now Web literacy) should be goals of instruction. This would include some awareness of the way journalistic practices (and search engine algorithms) can let us down.

We need to put less emphasis on argument analysis and evaluation and more emphasis on raising consciousness about cognitive and motivational biases. This flies in the face of my own training (as an analytic philosopher), but ironically, it seems to be where the best arguments and evidence lead us. However, there is considerable work left to do before we have a good set of tools for 'de-biasing'. As noted above, it does not seem to be effective to simply make people aware of these biasing processes and then exhort them to do better. Indeed, we have good reason to suspect that people will deploy their knowledge about cognitive and motivational biases in a biased way. What would seem to be required are procedures that will take these biases out of play. But such procedures may not exist for the sorts of knowledge we are worried about in connection with democratic politics.[5]

None of this can work unless the general cultural prerequisites are present: as they are not when religious leaders succeed in de-legitimating all non-religious sources of knowledge, or when conservative 'thought-leaders' teach their followers not to trust academia, the mainstream media, or the institutions of mainstream science. So we need to find way to draw the epistemically segregated and mutually suspicious elements of our society into productive

---

[5] Consider an analogy: It is apparently futile to exhort people who are auditioning prospective orchestra players to be fair with respect to gender. If the evaluators are aware of the gender of the players, then their perception of the quality of the playing will be affected (to the detriment of female applicants). It is however possible and effective to put the auditioning players behind a screen, so that the evaluators cannot see them and are forced to judge purely on the basis of what they can hear. What sorts of procedures could do this kind of job in the realm of politics?

dialogue. Here there are some reasons to be hopeful. Hugo Mercier and Helen Landemore (Landemore & Mercier 2010, Mercier & Landemore 2012) have argued that the problems of motivated reasoning and cognitive biases are most acute for solitary reasoners, and that they can be overcome when people reason together in a properly deliberative way. Building on Mercier's argumentative theory of reasoning, which suggests that the evolutionary basis for reasoning is in competitive social interaction and not in truth-seeking,[6] they argue that people who exchange arguments can compensate for each other's cognitive and motivational biases. Solitary reasoners are likely to construct arguments supporting their pre-existing views, and they are not likely to subject those arguments to rigorous scrutiny. But two or more people reasoning together can check each other and thereby do better. What is necessary for this process to go well, according to Landemore and Mercier, is at least some degree of cognitive diversity among the participants (they cannot be too 'like-minded') and they must actually argue in an adversarial way, putting forward arguments for their views and subjecting the arguments of others to critical scrutiny.

Alas, I must end on a note of caution. A problem with the appeal to the idea of deliberative exchange is that, in the real world, there is no guarantee that any formal rules or procedures will prevent the subtler forms of interpersonal power from corrupting the process. "Discussion is repression" ran a slogan of the German student movement in 1968. The students had a point. Anne Phillips' account of the drawbacks of participatory democracy in the women's movement illustrates the problem (1991, pp. 120–146). In-face-to-face meetings, the emotional relationships of the participants become crucial. Phillips reports many women feeling afraid to voice disagreement, informal and unaccountable patterns of leadership, and the emergence of false consensus. I think these are real problems, and I see no possibility of an easy or permanent solution. It is possible to stipulate 'rules of engagement,' like taking turns and making sure everyone has a chance to contribute to agenda-setting. But nothing of this sort can guarantee that people will not be shamed or seduced in one way or another. Neither the formal features of an 'ideal speech situation' nor any specifiable set of rules can substitute for the (uncodifiable) virtues required for good deliberation.

ACKNOWLEDGEMENTS: I began thinking seriously about these problems a few years ago, when I was struck by the fact that some of the most intelligent and well-informed students in my classes were global warming skeptics, often adamantly so. I presented an early version of this paper at the 3rd Copenhagen Conference in Epistemology in August, 2011. Thanks to Arnon Keren and Klemens Kappel for helpful comments on my talk and to Alvin Goldman for a very useful conversation. Thanks also to George Shulman and Simon Keller for comments on another version presented to the Association for Political Theory Conference at Notre Dame in October, 2011.

REFERENCES

Achen, C. H., & Bartels, L. M. (2006). It feels like we're thinking: The rationalizing voter and electoral democracy. Unpublished manuscript. Retrieved from http://www.princeton.edu/~bartels/thinking.pdf

---

[6] Crudely: the biological function of our reasoning ability is to persuade others and win arguments, not to discover truth.

Anderson, E. (2011). Democracy, public policy, and lay assessment of scientific testimony. *Episteme*, *8*(2), 144–164.

Bartels, L. M. (1996). Uninformed votes: Information effects in presidential elections. *American Journal of Political Science*, *40*(1), 194–230.

Bartels, L. M. (2008). The irrational electorate. *Wilson Quarterly*, *32*(4), 44–50.

Baurmann, M. (2007). Rational fundamentalism? an explanatory model of fundamentalist beliefs. *Episteme: A Journal of Social Epistemology*, *4*(2), 150–166.

Buchanan, A. (2004). Political liberalism and social epistemology. *Philosophy & Public Affairs*, *32*(2), 95–130.

Cohen, G. L. (2003). Party over policy: The dominating impact of group influence on political beliefs. *Journal of Personality and Social Psychology*, *85*(5), 808–822.

Converse, P. (2006). The nature of belief systems in mass publics (originally, 1964). *Critical Review*, *18*(1-3,), 1–78.

Cunningham, B. (2003, July/August). Rethinking objectivity. *Columbia Journalism Review*. Retrieved from http://www.cjr.org/feature/rethinking_objectivity.php

Friedman, J. (2005). Popper, Weber, and Hayek: The epistemology and politics of ignorance. *Critical Review*, *17*(1-2), 1–58.

Goldman, A. (1999). *Knowledge in a social world*. Oxford: Oxford University Press.

Huemer, M. (2005). Is critical thinking epistemically responsible? *Metaphilosophy*, *36*, 522–531.

Kahan, D. M. (2011, November). The supreme court 2010 term - foreword: Neutral principles, motivated cognition, and some problems for constitutional law (Yale Law School, Public Law Working Paper No. 231). *Harvard Law Review 125*, 1-77.

Kelly, T. (2008). Disagreement, dogmatism, and belief polarization. *Journal of Philosophy*, *105*(10), 611–633.

Kornblith, H. (1999). Distrusting reason. *Midwest Studies in Philosophy*, *23*, 181–196.

Kull, S., Ramsay, C., & Lewis, E. (2003). Misperceptions, the media, and the Iraq war. *Political Science Quarterly*, *118*(4), 569–598.

Kurtz, H. (2004, August 12). The Post on WMD's: An inside story. *The Washington Post*, p. A1.

Landemore, H., & Mercier, H. (2010). 'Talking it out': Deliberation with others versus deliberation within. Retrieved from http://ssrn.com/abstract=1660695

Levy, N. (2006). Open-mindedness and the duty to gather evidence; Or, reflections upon not reading the volokh conspiracy (for instance). *Public Affairs Quarterly*, *20*, 55–66.

Lord, C. G., Ross, L., & Lepper, M. R. (1979). Biased assimilation and attitude polarization: The effects of prior theories on subsequently considered evidence. *Journal of Personality and Social Psychology*, *37*(11), 2098–2109.

Mercier, H., & Landemore, H. (2012). Reasoning is for arguing: Understanding the successes and failures of deliberation. *Political Psychology*, *33*(2), 243---258.

Oreskes, N. (2007). The scientific consensus on climate change: How do we know we're not wrong? In J. F. DiMento, & P. Doughman (Eds.), *Climate change: What it means for us, our children, and our grandchildren* (pp. 65–99). Cambridge, MA: MIT Press.

Phillips, A. (1991). *Engendering democracy*. University Park, PA.: Pennsylvania State University Press.

The Responsibility of Authority: When Should a Physician Seek a Further Opinion?

J. ANTHONY BLAIR

Centre for Research in Reasoning, Argumentation and Rhetoric
University of Windsor
Windsor, ON
Canada N9B 3P4
tblair@uwindsor.ca

ABSTRACT: Physicians are normally in a better epistemic position that patients to decide when to seek a further opinion, so when should they do so? The criteria are well known, but how is physician to know when to operationalize them? Training in metacognition and cognitive error avoidance strategies may help.

KEYWORDS: CanMEDS framework, codes of medical ethics, cognitive errors, criteria, expert authority, guidelines, metacognition, second medical opinion.

1. INTRODUCTION

Physicians possess medical knowledge pertinent to diagnoses and treatments that is normally unavailable to patients. It follows that the physician is normally in a better epistemic position than the patient to judge when it is advisable to have a further medical opinion about either a diagnosis or a treatment. Given the physician's duty of care, it also follows that he or she ought to seek or recommend a further medical opinion when the circumstances warrant doing so.

The question initiating this paper was, "In what circumstances should a physician seek a further opinion?" I first look at some national medical association codes of ethics, but those tend to be quite general. Next, I consider the advice that some acquaintances who are physicians gave me. From them I learned that there are well-established protocols for referring a patient for a "consult." However, the notion of "consultation" is ambiguous, and so despite the now common practice of "consulting" in medicine, the initial question remains. Even so, it seems that guidelines, or anything beyond the well-established criteria, are not what physicians need. Instead, they need the ability and disposition to seek further opinions in those circumstances. The hypothesis is proposed that training in metacognition and cognitive error avoidance strategies will help develop that ability and disposition.

2. CODE OF ETHICS

One might think an answer to the question of when a physician should seek a further opinion lies in the codes of medical ethics of professional medical associations. The Canadian, American and British codes of medical ethics endorse a physician's seeking or recommending a further opinion. Here are the pertinent excerpts from the Canadian Medical Association Code of Ethics (2011) (italics added):

Blair, J.A. (2012). The responsibility of authority: When should a physician seek a further opinion? In J. Goodwin (Ed.), *Between scientists & citizens: Proceedings of a conference at Iowa State University, June 1-2, 2012* (pp. 75-84). Ames, IA: Great Plains Society for the Study of Argumentation. Copyright © 2012 the author(s).

General Responsibilities [to the Patient]
15. *Recognize your limitations and, when indicated*, recommend or seek additional opinions and services.
Responsibilities to the Profession
52. *Collaborate with other physicians and health professionals in the care of patients* and the functioning and improvement of health services.

From the American Medical Association Code of Medical Ethics (2011) (italics added):

Opinion 3.04 - Referral of Patients
A physician *may* refer a patient for diagnostic or therapeutic services to another physician, limited practitioner, or any other provider of health care services permitted by law to furnish such services, *whenever he or she believes that this may benefit the patient.* As in the case of referrals to physician-specialists, referrals to limited practitioners should be based on their individual competence and ability to perform the services needed by the patient. A physician should not so refer a patient unless the physician is confident that the services provided on referral will be performed competently and in accordance with accepted scientific standards and legal requirements. (V, VI)
Opinion 8.04 - Consultation
Physicians *should* obtain consultation *whenever they believe that it would be medically indicated in the care of the patient* or when requested by the patient or the patient's representative. When a patient is referred to a consultant, the referring physician should provide a history of the case and such other information as the consultant may need, calling to the attention of the consultant any specific questions about which guidance is sought, and the consultant should advise the referring physician of the results of the consultant's examination and recommendations. (V)
Opinion 8.041 - Second Opinions (8.00 Opinion on Practice Matters)
Physicians *should* recommend that a patient obtain a second opinion *whenever they believe it would be helpful in the care of the patient.*

From the British Medical Association (2011) (italics added):

What is meant by a 'second opinion'?
The CCSC finds it more helpful to use the term 'further opinion' to cover all circumstances. *Consultants often ask for a further opinion from their colleagues when a case is unusually complex or difficult.* The patient should always be kept informed and, unless the patient objects to this, the consent is implicit.

From these texts we can deduce the following prescriptions for when a physician should seek a further opinion. I order them from the more general to the more specific:
A physician should seek a further opinion when…

(1)   it would be helpful in the care of the patient.
(2)   it may benefit the care of the patient.
(3)   it is medically indicated in the care of the patient.
(4)   a case is unusually complex or difficult.
(5)   the physician recognizes the case to be beyond his or her limitations.

(1), (2) and (3) seem to be more or less equivalent, depending on how "helpful," "benefit" and "medically indicated" are parsed. (4) and (5) are more specific, and distinct. Being unusually complex or difficult (4) are two ways of being medically indicated, but there might be others—for example if there is debate within the profession about the diagnosis for the presenting symptoms or for the treatment of the condition in question. And a case might be unusually complex or difficult, yet still be judged by a physician not to lie beyond his or her limits of knowledge and expertise, so (5) is different from (4). Yet like (4), (5) is a special case, since a case's lying beyond the physician's limits is but one way a consultation can be medically indicated or beneficial or helpful in the patient's care.

Notice that only the fifth makes reference to a crucial ingredient if any such prescription is to be taken as an action-guide. *The physician must judge* that a further opinion would be helpful or beneficial in the care of the patient, be medically indicated, be unusually complex or difficult. So the question remains, *on what basis* does or should a physician make this judgment? *When* is it helpful or beneficial or medically indicated in the care of the patent to seek a further opinion? Only (4) "unusually complex or difficult" and (5) "beyond my limitations" (i.e., "I don't know how to explain these symptoms" or "I can't decide what treatment to suggest") provide any action-guiding specifics.

The preceding analysis of these ethical codes might suggest that their guidance regarding the seeking of a further medical opinion is empty or fatuous. However, if the codes are regarded not as action-prescriptions, but instead as generalizations of current best practice, then they cannot fairly be characterized in such negative terms. In emphasizing doing what is best for the care of the patient, the codes may be seen as capturing the idea that physicians following best practice are able to recognize when a diagnosis or treatment plan is problematic and, with the welfare of the patient as the primary concern, will then seek help from qualified colleagues. The alternatives to "the care of the patent" would include such things as the physician's convenience, ego or pride, the interests of insurers, or the convenience of the medical system (the hospital, the laboratory, colleagues' preferences, and so on).

Another way to distinguish these two frames is to see each of them as a different kind of prescription, one being criteria and the other, guidelines. A criterion of right action is a condition of successful performance; a guideline is advice about how to achieve that performance. I conclude that it is more reasonable to read such "Codes of Ethics" not as guidelines or lists of "how to" tips for professionals to consult when in doubt in the practice of their profession, but instead as criteria for the best ethical practices in the profession.

While that way of understanding these codes vindicates their function, it renders them of limited relevance to the motivating question of this investigation. Since these codes of medical ethics are not reasonably to be taken as action-guides, are there to be found elsewhere any useful guidelines for physicians as to when to seek a further medical opinion? Or, indeed, is there a need for such guidelines?

3. WHAT DOCTORS THINK

This paper was originally conceived as a preliminary study to attempt to discover whether practical guidelines can and should be formulated as to when a physician should seek or recommend a further medical opinion. To that end, I sent a brief questionnaire to four old friends who are physicians in Canada with long experience. These included two cardiac surgeons, an orthopedic surgeon, and a pediatric consultant.[1]

The broad thrust of their responses was that seeking further opinions is routine and unproblematic in the practice of medicine in Canada. Presumably the practice is similar in the U.S. and other wealthy countries. The practice of medicine in such countries is so specialized

---

[1]   The names of my sources must remain confidential, since I did not seek their permission to identify them.

that primary care physicians routinely refer patients to specialists for consultations about any presenting symptoms suggesting conditions for which a medical specialty exists. In many cases these specialists work in specialty clinics, often in teaching hospitals where physicians are organized into teams, so that a referred patient's case will routinely be discussed by more than one specialist and/or a group of medical residents who are more or less advanced in their training in that specialty. In the case of diagnoses or treatment recommendations resulting from discussions among such a team, the reference to "further opinion" has in effect already been made.

One of my respondents thought no guidelines are necessary, though he didn't explain why. Quite possibly behind his opinion was the fact that, at least in Canada, physicians are already trained to have internalized the practice of seeking further opinions when it is appropriate to do so. As another respondent pointed out, these already exist in guidelines for the training of medical specialists in Canada. In 1996 the Royal College of Physicians and Surgeons of Canada adopted, for specialist physician training, something called the "CanMEDS" framework, which identifies seven roles leading to optimal health care outcomes, and for each role identifies and specifies a set of "competencies," which are stated in general terms and then amplified in more detail. For the "central role" of "medical expert" in the CanMEDS framework it lists as one of six competencies that a medical expert will:

> 6. Seek appropriate consultation from other health professionals, recognizing the limits of their expertise.

And in particular, medical experts will:

> 6.1 Demonstrate insight into their own limitations of expertise via self- assessment
> 6.2 Demonstrate effective, appropriate, and timely consultation of another health professional as needed for optimal patient care
> 6.3 Arrange appropriate follow-up care services for a patient and their family.

Canadian medical schools are required to train residents in these competencies in order to retain accreditation, and medical specialists are required to exhibit them to become certified and to retain their certification.[2] The CanMEDS competencies are general descriptions of behaviors that presumably will be manifested in the day-in, day-out practice of each specialty.

A couple of my respondents gave some slightly more specific criteria they employ for seeking a further opinion. Here, for instance, are the criteria for seeking a further opinion that were listed by the pediatric consultant:

(A)   when I am questioning whether I have made the correct diagnoses—most of my patients have complex problems;

(B)   when I am prescribing medications which have serious side effects;

(C)   when I have prescribed the range of medications or other interventions that I am comfortable with and the therapeutic response is unsatisfactory to me or the patient/family/school;

---

[2]   "The CanMEDS Roles have been integrated into the Royal College's accreditation standards, objectives of training, final in-training evaluations, exam blueprints and Maintenance of Certification Program." According to the Royal College, the CanMEDS framework has been widely adopted outside Canada as   well.   See: http://www.royalcollege.ca/public/resources/aboutcanmeds.

(D)   when the range of problems that the patient has requires, in my opinion, a sub specialist evaluation (e.g. neurologist, physiatrist, psychiatrist) because of the complexity of the diagnosis.

One of the surgeons added these remarks:

(E)   I ... seek other expert opinions in complex situations in my own specialty, or opinions about other co-morbitities [concurrent diseases] not specifically related to [my specialty], but which may affect investigation, treatment or overall results of treatment.

(F)   As well, mandated are regular "Morbidity and Mortality" Rounds/conferences at which negative outcomes are discussed in detail, alternate options discussed, mistakes identified and lessons learned articulated and shared verbally or in writing.

These are more particular instantiations of the general guidelines listed in the CanMEDS list. They might be summarized, as:

Seek further opinion(s) when (a) you are unsure of the diagnosis, (b) your treatment has effects that may require other medical interventions, (c) the treatment that seems right to you doesn't seem to be working, (d) the patient has problems requiring evaluation by other specialists or sub specialists, (e) the patient has other medical conditions than the one in your specialty that you are treating him or her for that might affect the diagnosis or outcomes of treatment. Moreover, consultations are mandated when (f) in spite of or in the course of the treatment the patient dies.

In sum, and assuming Canada is typical, there is already an appreciation by the medical profession that physicians should seek further opinions, there is a general framework set out as to when to do so, and physicians are trained explicitly to recognize when it is appropriate to do so and to act on that recognition.

4. A TASTE OF REALITY

However, if the above accounts leave a comforting picture of physicians responsibly seeking second opinions when it is medically appropriate to so, that picture is at odds with some realities.

One of my correspondents discussed my query with a fellow physician, whose reaction was to say, "that train has left the station," and that the era of the traditional consult has gone the way of the Dodo. His point was that the situation is no longer one in which a puzzled physician managing a patient's care seeks the professional input of a colleague. Instead, the situation is that medicine has so many specialties and sub-specialties that expertise is highly specialized and medical care is highly compartmentalized. The primary care physician more or less automatically refers patients to specialists for any symptoms falling within the ambit of a medical specialty, and those specialists further refer the patient to other specialists or sub-specialists. The model in which one physician takes responsibility for the care of the patient has for some time no longer been applicable.

The current system comes with benefits and with costs. The obvious benefit is that typically patients receive very high quality medical treatment. Especially in large urban centers where there are medical schools and teaching hospitals, a patient stands to receive treatment grounded in state of the art medical equipment, knowledge and techniques. That is an enormous plus for the present system, not to be discounted. At risk, however, is the treatment of the patient as a whole person and not just as a hip, a tumor, or an X-ray. Moreover, the practice of medicine may tend to become like an assembly line, with each physician focusing exclusively on the immediate medical condition presented to him or her. This system can foster cognitive errors such as diagnostic momentum (initial labels biasing later diagnoses), triage cueing (patient mismanagement due to initial misdirection) and vertical line failure (each specialist thinking within his/her own particular medical silo).

5. ROUTINE AND EXCEPTIONAL CONSULTATION

It may seem that the question with which this paper began can now be seen to be "academic." Practically speaking, physicians now seek the further opinion of qualified colleagues (or "consult," to use the current terminology) as a matter of course in their daily practice. The question of "when" to do so has a more or less automatic answer, namely, when the symptoms suggest a condition or treatment in another medical specialty or sub-specialty, or requiring the services of another medial specialty or sub-specialty—which is a situation that any qualified physician will be well-trained to recognize immediately.

However, in fact the question does not go away. It turns out there is an ambiguity in the notion of medical consultation. On one hand, a consultation is a reference to another physician with the specialized expertise deemed needed to deal with the patient's symptoms. That is the kind of consultation that today occurs routinely in the practice of medicine in many wealthy countries. Call this "routine consultation." On the other hand, a consultation is a reference to a specialist or sub-specialist by a physician who is not confident about the diagnosis or about the treatment plan. This latter kind of consultation will tend not to be routine, for physicians will tend to be confident of their diagnoses or treatment plans. Call this "exceptional consultation." In the first case, the physician judges the case to require someone else's expertise. In the second case, the physician has the case presented or referred to him or her precisely because it is a case requiring his or her expertise, and yet it is so complex or puzzling that the physician judges—or certainly ought to judge—that he or she needs a further medical opinion. So the question of this paper can now be expressed more precisely: "Is there a need for guidelines to help physicians decide when they would be well-advised to make an exceptional consultation to a qualified colleague?" I think the correct answer is a qualified "No," for two different reasons.

6. WHAT'S NEEDED: METACOGNITIVE TRAINING

One reason is that general criteria exist already in such things as the CanMEDS list of general and particular competencies for medical experts, and the more concrete specifications of them that physicians learn during their training or work out for themselves (e.g., those my physician friends mentioned), so that further "guidelines" are unlikely to be more specific or helpful.

The other reason is that what is wanted is not a checklist so much as the ability on the part of the physician to recognize when he or she does not have any business being confident about a diagnosis or a treatment plan and needs advice, and the willingness to make an exceptional consultation at that point. In other words, what is needed is the training that will develop the ability and disposition appropriately to apply the criteria such as those listed in the codes of medical ethics and the CanMEDS competencies. All the guidelines in the world will not make any difference to the sincerely overconfident professional who does not know when he or she is out of his or her depth (see Kahneman 2011).

It is plausible to hypothesize a connection between this requisite self-knowledge and disposition to seek help and an ability and inclination to avoid cognitive errors, for it seems likely that the latter is causally connected to the former. For being alert to the risks of making cognitive errors is at least part of what it is to be aware of one's own limits (of knowledge, of skill, of authority). As a consequence of such alertness, and assuming one is disinclined to make cognitive errors, one will tend to take steps to reduce such errors, and on some occasions the appropriate way to do so is to seek opinions from authoritative colleagues.

It is encouraging to see that in at least some medical schools there are being developed modules on critical thinking in medicine that include exposure to typical cognitive errors in the practice of medicine and the teaching of strategies to avoid them. Groopman's *How Doctors Think* (2007) describes and illustrates many cognitive errors to which physicians are liable, and cites much of the extensive literature on cognitive errors that has grown up over the last 30 years. Kahneman's *Thinking Fast and Slow* (2011) discusses cognitive errors in general and draws attention to many that bedevil physicians in particular. (See the Appendix for a list of such errors affecting medical diagnoses.) And Crosskerry's (e.g., 2003a, 2003b, 2004) work in developing curriculum units for the Dalhousie University Faculty of Medicine using cognitive error research findings to devise preventive strategies is an attempt to inculcate in medical students the habits of mind that are requisite for recognizing (among other things) when help is needed.

Crosskerry's "forcing strategies" are worth describing in a bit of detail. He calls them a "metacognitive approach" aimed at "practicing clinicians and those in training to inoculate them against making diagnostic errors" (2003b, p. 110)—in the emergency department, in this case. According to Crosskerry, a necessary condition of the whole process is to learn the technique of metacognition, that is, the technique of standing back from one's own thinking and analyzing it. Next, one needs to be familiar with a wide range of specific cognitive errors to which people are prone in one's particular area of medical practice. This means learning a vocabulary (not unlike learning a list of fallacies), and being able to apply it to examples (see Crosskerry 2003a for list of cognitive errors that may lead to diagnostic error). Then one identifies scenarios in one's area in which error is likely to occur. For instance, in radiology, there's a tendency to stop checking closely once a positive finding has been made on the radiograph, or in the emergency department there's a tendency to fail to look for any more foreign bodies in the patient after one has been found and to fail to look for medical problems once a psychiatric diagnosis has been made. (These three are examples of the error called "search satisfying"—the tendency to call off a search once one explanation has been found.) One can develop "generic cognitive forcing strategies," such as the strategy of always conducting a secondary search after a positive finding has been made. Lastly, one develops a "particular cognitive forcing strategy" to avoid particular cognitive errors. For instance, since "one of the most significant errors of omission for a patient who has had an animal bite wound lies in failing to elicit a history of immunocompromise where it exists" the appropriate cognitive forcing strategy in the case of an animal bite wound is to make oneself work through a checklist of immunocompromise indicators in eliciting a past medical history (pp. 116–117).

My hypothesis is that a physician who has internalized habits of avoiding cognitive errors by acquiring habits of metacognition and habitually employing generic and particular cognitive forcing strategies will be much more likely than one who hasn't to recognize a situation in which the diagnosis is not clear or the case has complications beyond his or her competence. As Groopman says, "Studies show that most physicians are unaware of their cognitive errors" (2009, p. 147—citing Graber et al., 2005; Ghandi et al., 2006; Crosskerry, 2004; Redelmeier, 2005). If you don't know you might be making a mistake, you can't know that you need to seek help to avoid it. Cognitive error awareness and sensitivity would seem to be a necessary causal condition of routine appropriate further opinion consultation.

7. CONCLUSION

This investigation began with a rather naïve assumption that there are or should be guidelines to advise physicians when they should seek a further medical opinion. As it progressed, it became clear that while criteria for such a decision have been fairly extensively worked out by the medical profession, what are needed are not guidelines to help meet those criteria but rather training in the ability and disposition to recognize, in particular situations, that one needs to get advice. The closing hypothesis of the paper is that training in metacognition, cognitive error theory, and cognitive forcing strategies is likely to be a useful way of inculcating that ability and disposition.

REFERENCES

American Medical Association Code of Medical Ethics. (2011, 17 October). Retrieved from
http://www.ama.assn.org/ama/pub/physician-resources/medical-ethics/code-medical-ethics/opinion304.page?
British Medical Association. (2011, 17 October). Patients requesting a second opinion. Retrieved from
http://www.bma.org.uk/ethics/doctor_relationships/SecondOpinion.jsp
Canadian Medical Association Code of Ethics. (2011, 17 October). Retreived from
http://policybase.cma.ca/PolicyPDF/PD04-06.pdf
Crosskerry, P. (2003a). The importance of cognitive errors in diagnosis and strategies to minimize them. *Academic Medicine, 78*(8): 775–780. Retrieved from
http://www.cchil.org/hospitalmedicine/images/resources/020912-090533pm-CED.pdf
Crosskerry, P. (2003b). Cognitive forcing strategies in clinical decision making. *Annals of Emergency Medicine, 41*(1), 110–120.
Crosskerry, P. (2004). *Cognitive errors in clinical decision making: A cognitive autopsy* [Powerpoint slides]. Ottawa: Quality Healthcare Network, Spring Forum. Retrieved from
http://www.qhn.ca/pdfs/croskerry2004sf.pdf
Gandhi, T. K. et al. (2006). Missed and delayed diagnoses in the ambulatory setting: A study of malpractice claims. *Annals of Internal Medicine, 145*, 488–496.
Graber, M. L. et al. (2005). Diagnostic error in internal medicine. *Archives of Internal Medicine, 165*, 1493–1499.
Groopman, J. (2007). *How Doctors Think*. Boston/New York, MA/NY: Houghton Mifflin Co.
Kahneman, D. (2011). *Thinking Fast and Slow*. Toronto: Doubleday Canada.
Redelmeier, D. A. (2005). The cognitive psychology of missed diagnoses. *Annals of Internal Medicine, 142*, 115–120.
Royal College of Physicians and Surgeons of Canada. (2012, March 9). About CanMEDS. Retrieved from
http://www.royalcollege.ca/public/resources/aboutcanmeds.

APPENDIX

Selected and abbreviated from "Cognitive Dispositions to Respond (CDRs) That May Lead to Diagnostic Error" (Crosskerry 2003a, 777–778)

Aggregate bias: when physicians believe that aggregated data, such as those used to develop clinical practice guidelines, do not apply to individual patients (especially their own). ...

Anchoring: the tendency to perceptually lock onto salient features in the patient's initial presentation too early in the diagnostic process, and failing to adjust this initial impression in the light of later information. ...

Ascertainment bias: occurs when a physician's thinking is shaped by prior expectation. ...

Availability: the disposition to judge things as being more likely, or frequently occurring, if they readily come to mind. ...

Base-rate neglect: the tendency to ignore the true prevalence of a disease, either inflating or reducing its base-rate, and distorting Bayesian reasoning. ...

Commission bias: results from the obligation toward beneficence .... It is the tendency toward action rather than inaction. ...

Confirmation bias: the tendency to look for confirming evidence to support a diagnosis rather than look for disconfirming evidence to refute it, despite the latter often being more persuasive and definitive.

Diagnosis momentum: once diagnostic labels are attached to patients ..., what might have started as a possibility gathers increasing momentum until it becomes definite, and all other possibilities are excluded.

Framing effect: how diagnosticians see things may be strongly influenced by the way in which the problem is framed, e.g., physicians' perceptions of risk to the patient may be strongly influenced by whether the outcome is expressed in terms of the possibility that the patient might die or might live. ...

Fundamental attribution error: the tendency to be judgmental and blame patients for their illnesses (dispositional causes) rather than examine the circumstances (situational factors) that might have been responsible. ...

Gambler's fallacy: ... the belief that if a coin is tossed ten times and is heads each time, the 11th toss has a greater chance of being tails (even though a fair coin has no memory). ... the pretest probability that a patient will have a particular diagnosis might be influenced by preceding but independent events.

Gender bias: the tendency to believe that gender is a determining factor in the probability of diagnosis of a particular disease when no such pathophysiological basis exists. ...

Hindsight bias: knowing the outcome may profoundly influence the perception of past events and prevent a realistic appraisal of what actually occurred. ...

Multiple alternatives bias: a multiplicity of options on a differential diagnosis may lead to significant conflict and uncertainty. The process may be simplified by reverting to a smaller subset with which the physician is familiar but may result in inadequate consideration of other possibilities. ...

Omission bias: the tendency toward inaction and rooted in the principle of nonmaleficence. ...

Order effects: information transfer is a U-function: we tend to remember the beginning part (primacy effect) or the end (recency effect). ...

Outcome bias: the tendency to opt for diagnostic decisions that will lead to good outcomes, rather than those associated with bad outcomes, thereby avoiding chagrin associated with the latter. ...

Playing the odds: (also known as frequency gambling) is the tendency in equivocal or ambiguous presentations to opt for a benign diagnosis on the basis that it is significantly more likely than a serious one. ...

Posterior probability error: occurs when a physician's estimate for the likelihood of disease is unduly influenced by what has gone on before for a particular patient. ...

Premature closure: ... the tendency to apply premature closure to the decision-making process, accepting a diagnosis before it has been fully verified. ...

Representativeness restraint: the representativeness heuristic drives the diagnostician toward looking for prototypical manifestations of disease: ... Yet restraining decision-making along these pattern-recognition lines leads to atypical variants being missed.

Search satisfying: ... the universal tendency to call off a search once something is found. ...

Sutton's slip: ... The diagnostic strategy of going for the obvious is referred to as Sutton's law. The slip occurs when possibilities other than the obvious are not given sufficient consideration.

Sunk costs: the more clinicians invest in a particular diagnosis, the less likely they may be to release it and consider alternatives. ...

Triage cueing: ... triage is a ... process that results in patients being sent in particular directions, which cues their subsequent management. Many CDRs are initiated at triage, leading to the maxim: ''Geography is destiny.''

Unpacking principle: failure to elicit all relevant information (unpacking) in establishing a differential diagnosis may result in significant possibilities being missed. ...

Vertical line failure: routine, repetitive tasks often lead to thinking in silos—predictable, orthodox styles that emphasize economy, efficacy, and utility. ... the approach carries the inherent penalty of inflexibility.

Visceral bias: the influence of affective sources of error on decision-making ....

The Cycle of Deliberative Inquiry: Re-conceptualizing the Work of Public Deliberation

MARTÍN CARCASSON

Department of Communication Studies
Colorado State University
Campus Delivery 1783
Fort Collins, CO 80523-1783
United States
mcarcas@colostate.edu

ABSTRACT: As the deliberative democracy movement continues to gain momentum, the theories and practices that underlie that momentum must continue to evolve, particularly in terms of the connections between deliberative processes and policy expertise. This essay introduces "deliberative inquiry" as a way of re-conceptualizing deliberative practice as a distinct mode of inquiry which produces unique research products that can significantly impact the quality of public discourse and improve community problem-solving.

KEYWORDS: public deliberation, deliberative inquiry, facilitation, polarization, community problem-solving.

1. INTRODUCTION

The Center for Public Deliberation (CPD) at Colorado State University (www.cpd.colostate.edu), established in the fall of 2006, was developed to serve as an impartial resource to the northern Colorado community. Its mission is to *enhance local democracy through improved public communication and community problem-solving* by providing independent policy analysis, process design, facilitation, and reporting services. It serves as a hub for what I have termed "passionate impartiality" (Carcasson, 2010). The CPD is based on the belief that a diverse democracy requires high-quality communication, a requirement that unfortunately is rarely met in our current political environment. The theory behind the CPD initially drew from the academic fields of argumentation (Crosswhite, 1996; Goodnight, 1982; Perelman & Olbrechts-Tyteca, 1969; Toulmin, 1958), rhetorical criticism (Booth, 2004; Condit, 1993; Zarefsky, 2010), and post-empirical public policy analysis (Fischer, 2009; Hajer & Wagenaar, 2003; Lindblom, 1990; Stone, 2002) and was then complemented by the interdisciplinary work in deliberative democracy and collaborative problem-solving (Briand, 1999; Mathews, 1999; Gastil & Levine, 2005), as well as the challenges to deliberative democracy from critical theorists (Sanders, 1997; Fraser, 1992; Young, 2001). The work of the CPD thus reconceptualises public deliberation by bringing together interdisciplinary theory and practice in a way that hopefully significantly enhances both and strengthens relationships between scholars and practitioners.

This essay provides an overview of the concept of deliberative inquiry that has been developed at the CPD over the seven years of completing projects in the northern Colorado area. I argue that high-quality deliberative practice should be considered as a specific type of research or inquiry entitled "deliberative inquiry" (DI). Overall, DI is focused on helping a

Carcasson, M. (2012). The cycle of deliberative inquiry: Re-conceptualizing the work of public deliberation. In J. Goodwin (Ed.), *Between scientists & citizens: Proceedings of a conference at Iowa State University, June 1-2, 2012* (pp. 85-97). Ames, IA: Great Plains Society for the Study of Argumentation. Copyright © 2012 the author(s).

community or organization make better decisions and solve problems more effectively, collaboratively, and sustainably. As the cycle of DI shows (Figure 1), DI combines issue analysis with getting people together across perspectives to talk in innovative ways about the issue, and thus combines research-based inquiry (in terms of examining texts) with engaged, interactive inquiry that brings people together in particular, purposeful ways and draws insights from those interactions. These interactions are critical for multiple

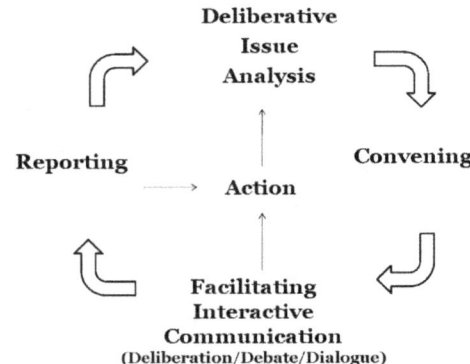

Figure 1: The Cycle of Deliberative Inquiry

reasons. They help create shared understanding across perspectives, and thus support the necessary building of community while working against the misunderstandings, cynicism, and polarization that are too often caused by the prevalence of low-quality political communication. The interactions also support the production of new public knowledge, the honing of key democratic skills and attitudes, and the development of a broader sense of ownership and legitimacy of actions, all of which are essential to community problem-solving (Carcasson, 2009a).

This short essay is a summary of a much broader ongoing project that will hopefully result in a book manuscript by the end of the summer of 2012. I begin the essay by briefly reviewing a typology of methods of inquiry in order to distinguish DI from its more dominant cousins, expert and strategic inquiry. I will then review the four key products of DI that further clarify its particular value, before closing with a review of the phases of the cycle from Figure 1.

2. BASIC FEATURES OF SCIENTIFIC, STRATEGIC, AND DELIBERATIVE INQUIRY

In order to quickly situate DI, table 1 identifies many of the key features of three broad forms of inquiry. Each form is not mutually exclusive, but nonetheless the categories help highlight important distinctions. I argue that most current inquiry on public issues is either scientific or strategic. *Scientific inquiry* is a particular sort of inquiry that focuses on rigorously discovering valid information, typically about empirical (i.e. observable and generally quantifiable) issues. It is very useful, but limited in important ways because it tends to avoid values and emotions, since such things are not susceptible to scientific analysis. *Strategic inquiry* is inquiry that focuses on developing evidence or arguments for a particular pre-set point of view, and thus politicizes the inquiry process. At its best, strategic inquiry informs a vibrant marketplace of ideas that supports high-quality decision-making, but unfortunately strategic inquiry often leads to situations where the marketplace is dominated by simplistic, manipulative appeals that undermine the ability for communities to address difficult problems well. *DI* seeks to avoid the problems and limitations of these other forms, while bringing out their best features. Each form of inquiry has particular strengths and weaknesses, thus my argument is not to abandon scientific and strategic inquiry, but rather that communities need to have capacity in all three forms in order to function well. Current capacity in DI is often very low, but steadily developing.

Table 1: The Three Forms of Inquiry

	Scientific Inquiry	Strategic Inquiry	Deliberative Inquiry
Overall goal	Discovery of valid information	Supporting particular points of view, winning arguments	Improving public decision-making and problem-solving, clarifying choices and their consequences
Primary Question	What is?	What evidence is available for my point of view? (or against the other side)	What should we do? What are our choices and their implications?
Primary method	Scientific observation	Strategic research or rhetorical invention	Open ended impartial research, issue mapping, and facilitation of interactive communication
Facts and fact questions	Focus of the work, seeking consensus	Often utilized as ammunition in the broader debate	Used as a common base to start from, but focus is often more on values
Tough choices and tradeoffs	Often bracketed and avoided due to unscientific nature	Often avoided or framed strategically	Often the focus of the research (to uncover and assist communities to work through them)
Common ground	Scientifically valid facts are common ground	Utilized if useful, often ignored, misrepresented, or manufactured	Issues are framed to start at a common point, and process seeks to build additional broad support
Primary audience	Narrow, specific expertise is required to be a part of the conversation, at times targeted to government officials, rarely to the public	Strategic, audience often limited in terms of those that already agree or target audience in the middle, rarely seriously address opposing views	Broad, seeks to connect public, government, and expert sources in the conversation
Scope of involved stakeholders in the analysis and solutions	More descriptive than proscriptive, so may avoid specific suggestions for solutions. When offered, solutions may be narrowly defined in terms of changes privileging governmental solutions.	Often limited by strategic goals and use of blame game or "magic bullet" solutions. Often specifically seek to exclude particular audiences that are opposed.	Broad, based on the notion of democratic governance and inclusion, considers all sort of potential actors (individuals, nonprofits, businesses, groups, and governments at all levels)

A key theoretical point concerning the need to further develop the capacity for DI is the growing prevalence of "wicked problems" (Rittel & Webber, 1973). Wicked problems are a class of complex, systemic, and interrelated problems that inherently involve competing underlying values and tradeoffs that cannot be avoided. They call for high-quality communication and collaboration to address well. Both strategic and scientific inquiry struggle with addressing wicked problems to the point they can be counterproductive, while DI is designed to address those problems in particular.

3. THE FOUR KEY PRODUCTS OF DELIBERATIVE INQUIRY

In general, DI focuses on much of the same material that scientific and strategic inquiry may focus on, such as defining the problem and its impacts, identifying a range of causes and potential solutions to the problem, establishing criteria for judging the value of solutions, and weighing the positive and negative consequences among solutions. DI, however, has four particular and distinct products that warrant specific attention. By "product" I mean particularly tangible items of information that represent the useful and distinct output of DI, and, as I will show as I walk through the stages in the cycle, are of interest to deliberative practitioners throughout each stage.

3.1 Product #1: The identification and attempted resolution of key obstacles to collaborative problem-solving

This first product of DI is primarily focused on addressing a wide range of troublesome issues, many of which are the result of adversarial politics and strategic inquiry. These barriers function similar to fallacies in argumentation that draw attention away from more important aspects of the issue and tend to make it more difficult for people to communicate productively. So the first step to improve the conversation is to undo the damage done.

For example, one of the big problems with public discourse is that issues are greatly simplified, often as a strategic ploy. People inherently tend to rely on *wishful thinking* when faced with complex decisions (Yankelovich, 1991), and many of the strategies used in adversarial politics tend to take advantage of this natural impulse. We want things to be easy, and they make them seem so. In many ways, deliberation involves complicating issues, but then providing genuine opportunities for participants to work through those difficulties (Kaner, 2007). Some of the particular tactics to address include *magic bullets* (assuming there is one solution to complex problems), *devil figures/scapegoats* (assuming the problem is caused by one individual or entity), or *paradox splitting* (Bryan, 2004) (attempts to resolve a difficult issue by focusing one side of a paradox and ignoring or dismissing the other).

Additional barriers that warrant attention involve the misrepresentation of motives of opposing groups. Indeed, much public discussion involves each side attacking positions that no one actually holds. Deliberative inquirers dig deeper to get past these assumptions, and events are often set up so that people from various perspectives can get a chance to explain their own motivations and learn from each other. Public discussions also often get derailed because opposing sides operate with a different set of facts. Unfortunately, without productive interaction, such separate assumptions tend to live on and are not resolved. DI again seeks out and tries to resolve such questions. Sometimes they can actually be resolved, or sometimes simply identifying the fact question as an open question can improve the discussion. At their best, deliberative practitioners can play important roles as "honest brokers of information" to rehabilitate the value of facts in our decision-making processes, while understanding their limitations.

In sum, this first key product is focused on helping communities get past the barriers that often arise due to an over-reliance on strategic inquiry and the misinformation it causes. It is clearly a very difficult task that can often test one's impartiality, but one that a local organization that has developed a strong reputation and a commitment to improved communication can take on and make significant, positive impacts.

3.2 Product #2: The identification and working through of tough choices or tradeoffs

The second key product of DI is perhaps the most important, at least in terms of a key aspect of democratic decision-making that is not adequately addressed by scientific or strategic inquiry. The work that has been completed in the "deliberative democracy movement" for the last 30 years in organizations such as the National Issues Forum, Study Circles, and Public Agenda already tends to focus on these concerns and the "choice work" they require. The Kettering Foundation's David Mathews (1998) and Public Agenda's Daniel Yankelovich (1991) have often written on the importance of surfacing and working through tradeoffs, and considering all the consequences, positive and negative, of our preferred actions.

Briefly, tough choices or tradeoffs are inherent to most public decisions, and simply involve judgments that must be made between various values. Authors have referred to them in a number of ways—including tough choices, tensions, tradeoffs, value dilemmas or conflicts, competing interests, policy paradoxes, etc.—I use these terms interchangeably while generally favoring the term "tough choices." Michael Briand captured the thinking behind the focus on tough choices when he wrote:

> Because the things human beings consider good are various and qualitatively distinct; because conflicts between such good things have no absolute, predetermined solution; and because to know what is best requires considering the views of others, we need to engage each other in the sort of exchange that will enable us to form sound personal and public judgments. This process of coming to a public judgment and choosing—together, as a public—is the essence of democratic politics. (1999, p. 42).

Work in argumentation concerning values is critical here as well, particularly Perelman's insights on value hierarchies. As argued in *The New Rhetoric*, "the simultaneous pursuit of these values leads to incompatibilities, [and] obliges one to make choices" (Perelman & Olbrechts-Tyteca, 1969, p. 82).

Unfortunately, scientific inquiry tends to avoid values and hence value dilemmas, and strategic inquiry tends to obscure or misrepresent them. DI is focused on uncovering them and helping audiences work through them, which, I argue, is a defining feature of deliberative practice. Diverse democracies will inherently be confronted with multiple value dilemmas, and they must develop the capacity to address them productively.

3.3 Product #3: The identification and building upon of common ground

The third key product of DI is the identification and building upon of common ground. Once again, DI is able to produce insights into this critical aspect of collaborative decision-making and community problem-solving much more than either scientific or strategic inquiry. Scientific inquiry tends not to focus on this issue—other than perhaps assuming that building a strong base of empirical, impartial information in itself represents critical common ground—

and strategic inquiry tends to make things more difficult by framing issues and opposing perspectives in ways that exaggerate certain similarities and differences, and thus tends to obscure common ground. DI, on the other hand, focuses to some degree in identifying and helping communities develop clarity concerning their common ground.

Common ground is surprisingly not difficult to identify and develop in communities. A significant amount of common ground exists based simply on geographical proximity and basics of human nature. Most citizens prefer excellent schools, good roads, low taxes, high paying jobs, vibrant economies, healthy environments, and minimal crime. Regardless of political ideology, there is always more that connects than divides a democratic citizenry. However, individuals that have 98% of their opinion in tandem will focus on the remaining 2% they disagree on and think they are worlds apart. Indeed, one of the typical tactics of political strategists is to exaggerate differences in order to gain votes and support, and unfortunately the media often follows suit as well.

In other words, whether citizens focus on their similarities or differences is a matter of framing and perspective. DI, therefore, attempts to cut through the chaff of false stereotypes and get to the heart of the issue. In important ways identifying common ground is closely related to the identification and working through of tough choices. Whereas the two features can be considered opposites—one focused on differences and one on similarities—the fact that many value dilemmas in our political culture are exaggerated results in a situation where helping people identify their actual differences and conflicts—rather than their perceived differences and conflicts—leads to them discovering they have more common ground than they previously realized. For example, opponents may at first see themselves with little if any common ground between them ("I care about national security, they don't"). Moving from that frame to a frame of "We both care about national security and individual liberty, but I rank national security higher, while they rank individual liberty higher" is a move that *both* clarifies tensions and identifies common ground.

3.4 Product #4: The identification of and development of support for action from a broad and inclusive range of community actors

The final key product of DI is based on the realization that the problems our communities face will require the involvement of a very broad range of actors from multiple sectors, across private, public, and non-profit lines, both in terms of engagement in the process of inquiry—to be able to understand the issues from multiple perspectives and allow voices to be heard—as well as in terms of action. Once again, scientific and strategic inquiry fall significantly short on both counts. Scientific inquiry tends to narrow its focus to particular actors, either individuals—see the volumes of research on the individual pathologies of poverty, for example—or the governmental, particularly the federal level. Said differently, scientific inquiry, when focused on public policy, tends not to imagine broad possibilities for action, primarily because breadth does not fit well with rigor and validity. Strategic inquiry will focus on whatever range of actors is most beneficial to their point of view, which is often rather narrow as well. Since strategic inquiry is not particularly focused on convincing opposing sides, it rarely involves consulting them for their viewpoints. Advocates may frame particular actors as simple solutions ("elect me and I'll solve the climate crisis"), or focus on the blame game, which also inherently narrows the scope of problem-solvers ("if we get rid of the evil oil companies, the climate crisis will be solved"). Blame-based solutions typically ask very little of most, because they frame problems as caused by a few (often either victims or victors).

DI, on the other hand, specifically seeks to engage broad audiences, particularly going beyond the usual suspects and empowering new audiences previously detached from "politics." DI begins with the notion that difficult problems will require a broad range of actors to understand and to address them. It connects to developing notions of democratic governance and public acting. As explained by Harry Boyte:

> Governance intimates a paradigm shift in the meaning of democracy and civic agency—that is, who is to address public problems and promote the general welfare? The shift involves a move from citizens as simply voters, volunteers, and consumers to citizens as problem solvers and cocreators of public goods; from public leaders, such as public affairs professionals and politicians, as providers of services and solutions to partners, educators, and organizers of citizen action; and from democracy as elections to democratic society. Such a shift has the potential to address public problems that cannot be solved without governments, but that governments alone cannot solve, and to cultivate an appreciation for the commonwealth. Effecting this shift requires politicizing governance in nonpartisan, democratizing ways and deepening the civic, horizontal, pluralist, and productive dimensions of politics. (2005, p. 536).

A focus on a broad range of potential actors, particularly citizens as problem-solvers and co-creators of public goods, develops somewhat naturally from a perspective that seeks to identify and work through tough choices and develop common ground. Such choices are not clarified unless a broad range is engaged, and then when citizens realize there are no easy solutions and that their "opponents" hold reasonable views, a shift to considering their collective role in solving problems often results.

3.5 Summary of the four key products of DI

These four key products just reviewed work together to support deliberative practice and work toward the ultimate goal of increasing the capacity of local communities to make better decisions about difficult issues. Another way to think about these products is as specific forms of "public knowledge" that can be created when citizens interact productively. One last key distinction between the three forms of inquiry merits comment here. The end product of DI is never a definitive opinion or recommendation (hence the cycle, not a linear process). The goal of DI is to clarify choices. The best a deliberative inquirer or any expert can do will be to more fairly lay out a set of options, each with their own distinct value set supporting it, and the ultimate decision will depend on the values in the community.

4. THE CYCLE OF DI: THE FOUR KEY TASKS

I return now to Figure 1 to walk through the four key tasks related to DI. As explained in the introduction, the cycle was developed specifically to bring together a set of academic traditions primarily connected to rhetorical studies, argumentation, and public policy and deliberative efforts that have been performed by a growing number of practitioners in recent years. The cycle represents an extension of my earlier work that focused on the need for deliberative practitioners to develop more focus and capacity on what occurs before and after deliberative events in order to increase their impact and to address key criticisms (Carcasson & Christopher, 2008; Carcasson, 2009b). Due to space constraints, a full explanation of these tasks cannot be provided here, but I hope to provide enough to give readers a good sense of how they work together.

4.1 Task #1: Deliberative issue analysis

The first task of deliberative practitioners is to analyze the issue and the situation in order to determine if the issue would benefit from DI and identify key features that will be critical to the remaining tasks of the cycle. At a basic issue analysis level, deliberative issue analysis involves researching issues, positions, and community voices in order to develop the best possible framework and process design for deliberation. Said differently, this analysis is focused on building a clear map of the issue and then identifying how best to frame the issue for deliberation, rather than persuasion (Friedman, 2007). At this stage, the analysis utilizes basic research techniques such as referencing books, articles, newspapers, web pages, message boards, and interviews with various stakeholders. At times, open-ended surveys may be utilized to gather specific perspectives on the issue from key stakeholders. Overall, deliberative issue analysis should include a wide variety of sources both in terms of perspective and in terms of form (such as expert information, activist information, public opinion, etc.). Analysts must be particularly careful to go beyond simply summarizing the dominant voices to help ensure broader inclusion, which is critical for supporting ongoing concerns for addressing power imbalances that should be addressed throughout the cycle.

Borrowing from Gastil's (2008) framework for the basic features of deliberation, deliberative issue analysis would focus on information relevant to the "analytic" aspects of deliberation: creating a solid information base, prioritizing the key values at stake, identifying a broad range of solutions, and weighing the pros, cons, and tradeoffs among solutions. Beyond all these issue analysis basics, however, DI would also focus in on the four key products of DI. When initially beginning a project on an issue, discovering these aspects begins to set the stage for the deliberative work to follow. All the major practitioner organizations already do various forms of deliberative issue analysis, particularly when developing backgrounders or issue guides to support deliberative forums, such as the process of "naming" and "framing" promoted by NIF. Deliberative issue analysis, however, could certainly be developed to a much greater extent, particularly in order to take more advantage of relevant scholarly traditions, situate itself more within the academic world, and to respond more effectively to theoretical criticisms of deliberative practice.

For example, a group of scholars in public policy and planning led by Frank Fischer (2009) and John Forester (1999) have applied argumentation theory to the work of deliberative democracy more directly. Their work combines rigorous empirical analysis with subtle understandings of narratives and normative values, all the while being cognizant of the unbalanced power relationships than inherently impact deliberative work. These authors discuss the "argumentative turn" in policy science, which led many of them away from a detached, scientific view of policy studies to one that realized policy will always involve judgment, and thus must always involve an inclusive public in important ways. A closer examination of the contributions of their work, as well as the scholarship in argumentation and rhetorical criticism, to deliberative practice is certainly warranted.

4.2 Task #2: Convening

The second task in the deliberative cycle is convening, which at the most basic level involves deciding who should be brought together to discuss the issue, and how to go about bringing them together. A wide variety of opinions and methods exist with the deliberation field concerning how to convene, though clearly more attention needs to be paid to this important step. Convening is a critical task for deliberative practitioners to live up to their ideals of inclusion and equality, and to address some of the criticisms of our work, particularly from diversity scholars and activists who believe democracy by discussion would inherently favor powerful voices and exclude those on the margins. The work of Jim Fishkin is important here, particularly his analysis of the difficulties of balancing the need for full participation, deliberation, and equality, which he terms the trilemma of democratic politics (2009). Practitioners too often focus on how many people attend, rather than who attends, which is a key limitation of much deliberative practice.

Similar to the situation with deliberative issue analysis, there is a great deal of related work being done in this area that would be useful for deliberative practitioners to be aware of and incorporate into their repertoire. I am much less connected to these areas, but have begun to examine work in community development, social movements, community organizing, and conflict management on stakeholder analyses. Much of the work done in these areas, however, is primarily from a strategic perspective—how to mobilize an audience toward your own point of view—rather than from a deliberative perspective. Borrowing key concepts from these literatures but then adapting the ideas to fit the deliberative mindset will be useful as we move forward.

4.3 Task #3: Facilitating interactive communication

The third task within the deliberative cycle is by far the most well known by practitioners, and simply involves process design and bringing people together to discuss issues in some specific way. Obviously, there are volumes of work on what deliberation is and how to facilitate deliberation and public engagement (for a review, see Gastil & Levine, 2005), and I will not rehash that work here in any degree. The main point is that improving the conversation and moving forward will almost always necessitate bringing people together with good process. High-quality deliberative issue analysis could perhaps improve the conversation on its own, but cannot replace real people engaging each other face to face (or, increasingly, online).

The primary "news" I present in this section is a broadening of the scope of deliberative work. I originally used "deliberating" as the label for the third task, but ultimately moved to "facilitating interactive communication" to indicate a focus both on the importance of facilitating and interaction. In a way, "interactive communication" may seem redundant, but in our polarized, mediated society, communication is often unilateral. Many have the opportunity to "express themselves," but the degree of listening, learning, and interacting is likely rather low. Interactive communication, therefore, emphasizes that the communication will involve participants actually engaging each together, particularly across perspectives.

In addition, the broader term allows me to include debate and dialogue alongside deliberation under the umbrella of interactive communication. Inspired by a presentation by Pete Bsumek and Kai Degner at the 2008 NCDD conference in Austin, I agree that deliberative practitioners are often overly "anti-debate." We should certainly be opposed to unproductive forms of debate, and unfortunately most forms of public debate are unproductive, but we need not throw the baby out with the bathwater. Deliberative practitioners, I would argue, need to rehabilitate, not simply reject, debate, and thus include debate within their community toolkit.

Overall, therefore, I see debate, deliberation, and dialogue as three primary interactive communication tools for communities to address problems, and each has their place in the discussion. The three are also not mutually exclusive, so we can talk about combinations such as deliberative dialogue and deliberative debate. The relationships between these three forms of interactive communication warrant more examination. Each in some ways could be framed as useful before or after the other. For example, if stakeholders are particularly polarized, dialogue would be useful before debate or deliberation. If the relevant facts are unclear and the public uninformed, a well-framed and moderated expert or activist debate may help clarify issues before a deliberation. If a deliberation exposes misunderstandings and distrust between participants—or if certain audiences refuse to even engage in deliberation due to fear or perceived disrespect—a dialogue may help move the conversation forward. Likewise, if a deliberation results in specific policy ideas of unclear merit, a debate between experts or activists concerning the outputs of the deliberation may be interesting. Overall, returning to the broad overall point of the cycle, the goal should be to improve the conversation in substantive ways each time people are brought together. Depending on the issue, the situation, and the participants—and the state of the four key products—various forms and combinations of debate, deliberation, and dialogue may be used. In all cases, having a passionately impartial entity dedicated to high-quality processes can be critical to community capacity for all three forms.

4.4 Task #4: Reporting

The final task within the deliberative cycle involves the analysis and reporting of what occurred during the first three tasks, particularly the third. Similar to convening, this is a task that many deliberative practitioners are familiar with, but we nonetheless lack clarity or overall understanding of the various techniques used. I believe further developing this task will significantly improve the quality of DI. Developing this skill is also closely connected to many of the academic traditions mentioned during the first task, as well additional traditions such as ethnography, discourse analysis, and small group communication. I would argue that much more is going on during forums than we realize, and the more we can understand how to capture, analyze, and then present all that is happening when citizens interact, the better. In particular, capturing and reporting what is unique about a deliberative forum should be a critical focus of deliberative practitioners (Carcasson, 2011).

There are many reasons why deliberative practitioners should take their reporting function seriously. Reports from deliberative forums provide a competing source of information to the products of strategic inquiry that dominate our political conversations. There are very few places citizens can go for information that is not purposefully biased, so the more deliberative practitioners can build up a record and establish reputations for useful, fair information, the better.

Returning once again to the four key products, reports are the culmination of their discovery and development. Key barriers that arise could be highlighted, and potentially even resolved before publication of the report if possible. Tough choices, common ground, and potential broad stakeholders could be discussed to expose a broader audience to the possibilities. True to the notion of the cycle, all this information would then be utilized to feed back into deliberative issue analysis once again as we begin to consider how next to move the conversation forward.

4.5 Summary of four key tasks of the deliberative cycle

As shown on Figure 1, the deliberative cycle involves the four key tasks of deliberative issue analysis, convening, facilitating interactive communication, and reporting. Ideally, the tasks should all flow together to continuously improve the quality of public communication, and thus hopefully the quality of community problem-solving. Figure 1 also shows "action" situated in the middle of the cycle, with arrows from both facilitating interactive communication and reporting pointing toward action, and an arrow from action pointing back up to deliberative issue analysis. The point here is that action is not technically part of DI, which again is focused on providing insights on the choices communities may make on important issues. Clearly the connection between deliberative talk and action is an essential issue for deliberative practice, and much of deliberative practice is designed specifically to spark productive, collaborative action. The degree to which deliberative practitioners are involved in the action varies considerably based on various processes and practitioner styles. Finally, the arrow going up from action to deliberative issue analysis is included to signify the fact that the cycle does not stop with action, but rather that action may change some of the dynamics of the problem. Borrowing from John Dewey's notion of democracy as a way of associated living (1916), the cycle assumes the conversation must always continue. Of course, communities improve when that conversation is of higher quality.

5. CONCLUSION

To conclude, I briefly shift focus to highlight the role of experts within DI. A key aspect of the cycle is the possibility for productively bringing together experts and the public on critical public issues, though not necessarily at the same time. Experts can play critical roles particularly during the first and fourth stages, but are also used in some processes during the third. During deliberative issue analysis, experts are used to help provide a clear map of the issue and improve understanding of the problem and the consequences of various actions to address it. Expert research is utilized, but experts can also be used to vet background information developed by deliberative practitioners. During the reporting stage, experts can again react to key themes derived from the public deliberation, and assist in moving forward on the key products. Depending on the issue, experts may also actually be the most useful stakeholders to convene for a particular pass around the cycle, particularly if the state of the research is murky or opposing perspectives tend to operate from incapable sets of facts.

In sum, the cycle of DI provides a model for the work of "passionately impartial" deliberative practitioners to impact the quality of communication and problem-solving in their community. It is important to note that the work of deliberative inquiry will always involve striving for an unreachable ideal. The hope for the cycle is that with practice, study, and reflection, deliberative practitioners can intervene positively on local issues, and with each project build their reputation and skills and thus increase their capacity to make significant impacts, moving our communities closer to a more perfect union.

ACKNOWLEDGEMENTS: I would like to thank the Kettering Foundation who supported my initial research into deliberative inquiry and has provided me the opportunity to have numerous vibrant conversations with theorists and practitioners over the past seven years.

REFERENCES

Booth, W. (2004). *The rhetoric of rhetoric: The quest for effective communication.* Malden, MA: Blackwell.

Briand, M. K. (1999). Practical politics: Five principles for a community that works. Chicago and Urbana: University of Illinois Press.

Boyte, H.C. (2005). Reframing democracy: Governance, civic agency, and politics. *Public Administration Review, 65,* 536–545.

Bryan, T.A. (2004). Tragedy averted: The promise of collaboration. *Society and Natural Resources, 17,* 881–896.

Carcasson, M. (2009a). Beginning with the end in mind: A call for goal driven deliberation. Occasional Paper No. 2, Center for the Advancement of Public Engagement at Public Agenda.

Carcasson, M. (2009b). The cycle of deliberative inquiry. Research report prepared for Kettering Foundation.

Carcasson, M. (2010). Facilitating community democracy from campus: Centers, faculty, and students as key resources of passionate impartiality. *Higher Education Exchange,* 15–26.

Carcasson, M. (2011). Reporting on deliberative forums: Current practices and future developments. Research report prepared for Kettering Foundation.

Carcasson, M., & Christopher, E. (2008). The goals & consequences of deliberation: Key findings and challenges for deliberative practitioners. Research report prepared for Kettering Foundation.

Condit, C.M. (1993). The critic as empath: Moving away from totalizing theory. *Western Journal of Communication, 57,* 178–190.

Dewey, J. (1916). *Democracy and education.* New York, NY: Free Press.

Fischer, F. (2009). *Democracy and expertise: Reorienting policy inquiry.* New York, NY: Oxford University Press.

Fishkin, J. S. (2009). *When the people speak: Deliberative democracy and public consultation.* New York, NY: Oxford University Press.

Forester, J. (1999). *The deliberative practitioner.* Cambridge, MA: MIT Press.

Fraser, N. (1992). Rethinking the public sphere: A contribution to the critique of actually existing democracy. In C. Calhoun (Ed.), *Habermas and the Public Sphere* (pp. 109–142). Cambridge, MA: MIT Press.

Friedman, W. (2007). Reframing framing. Occasional Paper #1, Center for Public Engagement at Public Agenda: New York.

Gastil, J.. (2008). *Political communication and deliberation.* Thousand Oaks, CA: Sage.

Gastil, J. & Levine, P. (Eds.). (2005). *The deliberative democracy handbook: Strategies for effective civic engagement in the 21st century.* New York, NY: Jossey-Bass.

Goodnight, T. G. (1982). The personal, technical, and public spheres of argument: A speculative inquiry into the art of public deliberation. *Journal of the American Forensic Association, 18,* 214–227.

Hajer, M. & Wagenaar, H. (Eds.) (2003). Deliberative *policy analysis: understanding governance in network society.* UK: Cambridge University Press.

Kaner, S. (2007). *Facilitator's guide to participatory decision-making* (2nd ed.). San Francisco, CA: John Wiley & Sons/Jossey-Bass.

Lindblom, C.E. (1990). Inquiry *and change: The troubled attempt to understand and shape society.* New Haven, CT: Yale University.

Mathews, D. (1998). *Politics for people: Finding a responsible public voice* (2nd ed.). Chicago, IL: U of Illinois Press.

Perelman, C., & Olbrechts-Tyteca, L. (1969). *The new rhetoric: A treatise on argumentation.* J. Wilkinson & P. Weaver, Trans.). Notre Dame, IN: University of Notre Dame.

Rittel, H., & Webber, M. (1973). Dilemmas in a general theory of planning. *Policy Sciences, 4,* 155–169.

Sanders, L. M. (1997). Against deliberation. *Political Theory, 25,* 347–376.

Stone, D. (2002). *Policy paradox: The art of political decision-making.* New York, NY: W.W. Norton.

Toulmin, S. (1958). *The uses of argument.* Cambridge: Cambridge University Press.

Yankelovich, D. (1991). *Coming to public judgment: Making democracy work in a complex world*. Syracuse, NY: Syracuse University Press.

Young, I. M. (2001). Activist challenges to deliberative democracy. *Political Theory*, *29*(5), 670–690.

Zarefsky, D. (2010). Plenary address: Reclaiming rhetoric's responsibilities. In M. Smith and B. Warnick (Eds.), *The responsibilities of rhetoric* (pp. 13–24). Long Grove, IL: Waveland Press.

Co-Evolving Expertise in Environmental Policy Debates: Rethinking Values and Participants through an Ecological Model of Rhetoric

PIPER CORP

Department of Communication
University of Pittsburgh
4200 Fifth Ave., 1117 Cathedral of Learning, Pittsburgh, PA 15260
USA
pwc6@pitt.edu

ABSTRACT: Environmental policy controversies reflect a struggle between "subjective" (human) and "objective" (scientific) knowledge, which a more rhetorical science could reconcile. I draw from actor network theory and rhetorical identification to suggest a model of ecological rhetoric, which I apply to two science policy projects: ecosystem service markets and adaptive management.

KEYWORDS: Latour, Haraway, ecology, adaptation, rhetoric, ecosystem services, identification, community, objectivity.

1. INTRODUCTION

On Earth Day in 2001, UN Secretary-General Kofi Annan launched the Millennium Ecosystem Assessment in response to the expressed need for international coordination of ecological knowledge and research.

The project synthesized the contributions of more than 1,360 "experts"—primarily but not exclusively scientists—to report on the state of the world's ecosystems. The basic diagnosis should sound familiar to those following the climate debate: there is resounding agreement among the international scientific community that human activities are harming the environment, and if we don't modify our behavior quickly and significantly, we could face unavoidable, irreversible, and in some cases devastating repercussions. But in spite of the urgency and cultural authority that such diagnoses ostensibly carry, the people and lawmakers of the United States rarely give them priority. Why?

The partial explanation with which this essay begins involves two related tensions that appear quite regularly in environmental policy debates: one between expert and non-expert knowledge, the other between environmental and human needs. Drawing from Latour, I will argue that both distinctions are caught up in the problematic binary of objective and subjective accounts of reality—understood as scientific expertise and embodied experience, respectively. Considering how we might reintegrate these two realms will be the primary focus of this essay.

Of course, *why* is a deceptively simple question, and the subject-object binary should not be seen as a complete answer. Indeed, one of the central assumptions in this essay is that *there are no complete answers* to decontextualized questions of causation. To suggest a cause or set of causes for a particular event, one must first define the context and the objects involved.

Corp, P. (2012). Co-evolving expertise in environmental policy debates: Rethinking values and participants through an ecological model of rhetoric. In J. Goodwin (Ed.), *Between scientists & citizens: Proceedings of a conference at Iowa State University, June 1-2, 2012* (pp. 99-108). Ames, IA: Great Plains Society for the Study of Argumentation. Copyright © 2012 the author(s).

For example, if I rear-end someone at a stoplight, I accept the blame, but the person I've bumped could easily think, "If I had just waited for rush hour to end, I would have never gotten hit." Both attributions of cause are partial but not inaccurate, yet they may lead to very different actions. The other driver may decide to travel at different times, while I may decide to pay better attention, and my insurance provider may decide to increase my premium. Partial causes help us predict and influence the future, but they do so only in a particular context.

Demarcating context is a kind of rhetorical selection integral to scientific inquiry. If, as Sarewitz (2000) puts it, "experiment serves to hold nature's complexity at abeyance," rhetoric is the means by which one decides what to study and what to fix. Sarewitz describes the political implications of such decisions in the climate change debate:

> To the climate modeller, a small, anthropogenic contribution to global temperature does not amount to climate "disruption," because the climate system is not fundamentally destabilized. To an ecologist, however, small temperature variations could stimulate significant changes in ecosystem function. The latter view might suggest the need for rapid policy action to control greenhouse gas emissions, even at high economic cost; the former might support a more cautious, less economically disruptive approach. (Sarewitz, 2000)

The question therefore shifts from why climate change happens to why climate change matters in the first place. If we are concerned about the stability of the climate system, we will take one approach; if we are concerned about sustaining the services provided by particular ecosystems, we will likely take another. And if there are additional, competing concerns involving economics, politics, and so forth, our chosen approach may shift dramatically as a result. In all cases, science is tremendously valuable in identifying a mechanism for change (e.g. reducing industrial emissions at a particular rate). But the mechanism(s) it identifies will always speak to particular investments and concerns.

In the section that follows, I discuss how the political function of "objective" science creates a rift between expert and non-expert knowledges, as well as natural and human needs. Noting the visible but muted role of science in current policy debates, I suggest that an explicitly rhetorical science, by bringing the objective and subjective realms into conversation, could *increase* the credibility and usability of scientific findings in political matters. Indeed, the rhetoricity of scientific facts, far from diminishing the contributions of science in policy, suggests an opportunity to more explicitly and thoroughly incorporate human values and experience into scientific knowledge production. The exigencies of our current moment suggest a constructive, *ecological* model of rhetoric that reflects the contingency and relationality of an era characterized by, among other things, globalization, environmental crises, and an emerging network culture.

To appreciate how such a model might contribute to current environmental policy discussions, I conclude by comparing two nascent efforts to integrate human and natural systems: ecosystem services markets and adaptive management programs. The former aims to put price tags on elements of the natural world that our economic system does not currently address—things like clean air and water, pollination and cultural significance. The goal is to fit "nature" into the abstract economic system that currently governs our affairs. Adaptive management, meanwhile, aims not to stabilize or change the "natural" world, but to change *with* it, as constituents of the ecosystem.

2. "OBJECTIVE" AND RHETORICAL SCIENCE IN POLICYMAKING

In science policy debates, we typically expect scientists to supply the facts and politicians to supply the values.

But if science tells us inconvenient or even ideologically impossible truths, the only "rational" choice is to accept them. Scientists can make their facts more interesting or accessible but they cannot make them otherwise. It is not surprising, then, that we are witnessing what science journalist Chris Mooney (2005) calls a "war on science": a backlash against scientific findings that appear to conflict with the beliefs and experiences of certain communities. Still, the *idea* of science has retained much of its legitimizing power (Weingart, 1999), as we see even in Mooney's example of the intelligent design debate. The Discovery Institute (a major proponent of intelligent design) provides an annotated bibliography of "Scientific Publications Supportive of Intelligent Design Published in Peer-Reviewed Scientific Journals, Conference Proceedings, or Academic Anthologies." The Institute is not rejecting science in principle; it is performing and evoking scientific authority to legitimate its own perspective. In a response to Mooney (Luskin, 2006), they quote Thomas Kuhn, accusing evolutionists of wearing the blinders of normal science—of lacking true scientific skepticism. Mooney's "war," it seems, is less between religion and science than between competing definitions of "objective" science.

Ironically, the studies that people deem objective are typically those consistent with their existing beliefs and values. Attempting to understand why the apparent scientific consensus has not convinced people that climate change is a serious threat, Kahan et al. (2011, p. 148) found that people rarely attack science itself, but instead reassign credibility to "experts" whose findings support their worldview.

Here, "expert" is a pragmatically valid designation for one who knows what works in a certain system. If literal readings of the Bible provide an epistemic base against which all other claims are tested, evolution is a no-go—as unfathomable as a neutrino outracing light. But whether we find knowledge in scripture, physics, or both, we maintain it through reference to *some* source of common ground. Objects, then, are phenomena of interest, fixed by some knowledge system, whether religious, empirical, or otherwise. Goodwin (1994) illustrates this point in his study of archeologists, where expertise comes from "socialization through language" (p. 13). The well-trained archeologist can identify with considerable nuance those elements in the landscape that other practitioners will also value. The basis for expertise is not universality or "objectivity," but the opposite: embededness and sociality. It is one's acceptance of a particular mode of selection that is, I would argue, rhetorical.

Appreciating the role of rhetoric here is critical, given the persistent technocratic tendencies at the science-policy interface, tendencies rooted in the belief that there are scientific solutions for political problems, given enough information (Weingart, 1999, p. 154). Technocratic discourse gives validity to calls for more research in policy debates (whether out of genuine concern or what Paroske (2009, p. 149) calls "epistemological filibustering"), and it sustains the fantasy that perfect knowledge will yield perfect solutions. But if such discourse appears to promote the full scientization of politics, it in fact does the opposite. Weingart (2009, p. 158) argues that since science became a source of legitimacy for policy decisions, politicians have raced to get "the latest, and therefore supposedly most compelling, scientific knowledge, driv[ing] the recruitment of expertise far beyond the realm of consensual knowledge right up to the research frontier where knowledge claims are uncertain, contested, and open to challenge." The paradox, as Weingart sees it, is that this tendency effectively delegitimizes scientific findings. When one can find (or hire) "expert" testimony to support virtually any position, science seems to give us very little.

The technocratic celebration of science as objective and apolitical obscures the rhetoricity of facts. The irony of objectivity, one might say, is that all objects are socially constructed. The *danger* of objectivity is that it takes for granted the ideology of those who constructed it. This is the central concern in Haraway's "Situated Knowledges" (1988, p. 580): "Science has been about a search for translation, convertability, mobility of meanings, and universality—which I call reductionism only when one language (guess whose?) must be enforced as the standard for all the translations and conversions." Whenever we speak of an object, we are evoking a context in which it is salient. Haraway's project, which this essay embraces, is to make these perspectives known, thereby "construct[ing] a usable, but not an innocent, doctrine of objectivity" (p. 582).

Construction figures importantly into both Haraway and Latour's work. Latour (1998, p. 81) writes, "To discover...is not a matter of revealing at last the 'true agent' *under* all the other, now 'false' ones ...To discover is not to lift the veil. It is to construct, to relate, and then to 'place under." To discover is to create a composite agent or entity—something that comprises a set of phenomena that act similarly in certain situations, like the microbe, greenhouse gas, or attention deficit disorder (ADD). But Kenneth Burke and many others remind us that such categories invariably run into trouble at their margins, always leaving something out or including something that doesn't belong. The designation of ADD, for example, likely encompasses a range of "disorders," which may have different causes and treatments. In some cases, ADD is an important designation (for example, it allows universities to identify students who need more time on exams); in other cases it may obscure disparate causes or lead to ineffective treatments. We could try to resolve the matter by parsing out the differences, but no two cases will ever be exactly the same. To study and treat any disease or disorder, we must be able to define an agent on the basis of relevant commonalities. But we must also be able to revise these definitions to address the needs of the situation. An explicitly rhetorical approach to science considers what different definitions do and revises them to reflect the exigence and community in question.

Here is the crucial distinction between rhetorical science and "objective" science: the former, as *techne*, constructs; the latter, as *episteme*, ostensibly uncovers. A rhetorical approach builds knowledge on the ground; an objective approach imposes it from above. Since modern science achieves its predictive power by controlling all variables but those being studied, Latour (1988, p. 89) argues that scientific theories hold only to the extent that practitioners are able to make the world mimic the lab so that the experimental results can be replicated outside. When being treated for a disease, medication is shorthand for the cure only if we take it as instructed. Recovery depends not upon the pill itself, but upon a range of physiological interactions, which may not occur if we ignore the directions attached to our prescription—directions that tell us how to behave like test subjects so we can expect a similar response.

We cannot effectively apply objective science to society without policies to control the world around us—a task that appears increasingly absurd. Sarewitz (2000) calls this understanding of science the "physics view". He proposes instead a "geological view" that "recognizes that nature, as experienced by humans and as recorded in the lithosphere and cryosphere, is the evolving product of innumerable complex and contingent processes and phenomena…" Whereas the physics view emphasizes "control and rigidity," this geological view, like rhetoric, privileges "adaptation and resilience."

Below I sketch out a framework for environmental decision-making that builds upon, but also complicates, Sarewitz's approach. The geological view does much to move science toward a more local, contingent, and dynamic understandings of the human-nature relationship. But if we take seriously the call to disrupt the divide between subjects and objects, and to make use of the contingency and radical relationality that Latour and Haraway espouse, it seems we must avoid even metaphorical evocations of bedrock. Co-evolution requires a transformation of both humans and nature. A rhetorically facilitated process of *critical* co-evolution—an ecological model of rhetoric—would help us take a more active role in this transformation, drawing from the predictive capacities of science and the context-building capacities of rhetoric to agree on and cooperatively pursue a common, if imperfect and impermanent, goal.

3. AN ECOLOGICAL MODEL OF RHETORIC

Millions of acres of the southeastern U.S. are draped in kudzu—an incredibly dense and fast-growing vine that smothers the trees and other vegetation beneath it.

Native to Japan, kudzu was introduced in the U.S. to control erosion, and for decades conservation groups and the federal government *encouraged* farmers to plant it. Unchecked by its new ecosystem, the vine flourished, rapidly crowding out the native vegetation. Government agencies have since put a great deal of time and money into developing an effective means of control, but kudzu remains unstoppable.

Hardin's First Law of Ecology states that "you cannot do only one thing." An ecological perspective recognizes that everything is interconnected—small changes may reverberate through the system, making it impossible to isolate and manipulate single parts without fundamentally changing the whole. Having in most cases co-evolved with the rest of the system, organisms cannot be meaningfully defined without reference to the others with which they interact. Antlers, camouflage, and mimicry are among the more obvious examples of how organisms embody their relationships. But while mechanistic attempts to "tweak" parts of the environment can and do have major and unexpected repercussions, we continue to try. Hence the growing interest in addressing climate change through "geoengineering"—that is, by initiating large-scale, technological transformations of the environment. Options include injecting aerosols into the atmosphere to block solar radiation and dumping iron into the ocean to boost plankton populations and therefore photosynthesis.

As these strategies suggest, we see ourselves as acting *on* rather than *in* ecosystems, making it difficult for us to grasp the degree to which our daily lives both impact and depend upon them. Latour connects this perceived divide with what he argues is the central dilemma in science policy debates—a false distinction between all things human (subjectivity, values, contingency, society) and all things nonhuman (objectivity, facts, rigidity, nature):

> If we concede too much to facts, the human element in its entirety tilts into objectivity, becomes a countable and calculable thing, a bottom line in terms of energy, one species among others. If we concede too much to values, all of nature tilts into the uncertainty of myth, into poetry or romanticism; everything becomes soul and spirit. If we mix facts and values, we come from bad to worse, for we are depriving ourselves of both autonomous knowledge and independent morality. (2004, p. 4)

A more useful mode of differentiation, Latour says, groups humans and nonhumans according to what they *do*, not what they "are" in some metaphysical sense—that is, according to habits rather than essences (2004, p. 86). After all, much human activity is functionally interchangeable with nonhuman activity (difficult to deny after Marx), just as nonhumans can often be said to argue just as loudly as humans (an accepted notion among many sub-fields in rhetoric—visual rhetoric, for example, or the rhetoric of technology). Human and "natural" phenomena are equally capable of action. Networked together, they are simultaneously objects *and* subjects, always influencing and being influenced. By removing these traditional distinctions, we can form communities or networks based on shared needs and associations, giving voice to both human and nonhuman participants.

To be clear, Latour is not engaged in a kind of nonhuman suffrage; nonhumans already speak all the time, whether we listen or not. If we ignore ecological needs, we will still have to face the consequences of our actions later on. Nonhuman voices, we could say, express what Burke calls recalcitrance. When materialists pound on tables to make their point, the tables talk back with equal and opposite force. Still, what we hear in their response is not a single voice, but a chorus that contains human voices as well—table makers and designers, lumber mill operators, and so forth. The story of any action can be thought of as a conversation between humans and nonhuman agents.

An ecological approach to environmental policy doesn't force us to face the facts, but it does ask us to listen. Environmental decision-making is a process of community formation. When we invite nonhumans to the table, everything, not just the supposedly human realm, is open to negotiation. And through these negotiations we can build a more sustainable world.

This was the vision that McKeon championed at the Wingspread Conference: rhetoric *as technology*—an architectonic art of world production (1971, p. 53). In an age where technology, not ideology, provided the foundation for collective action, McKeon argued for a blend of social constructivism and technological determinism that could allow us to participate in creating the structures that give order to our lives.

At the time McKeon was writing, the field of ecology was on the cusp of its own paradigm shift, articulated in Holling's (1973) seminal essay, "Resilience and Stability of Ecological Systems." Up to that point, ecosystems were treated as orderly machines; management practices sought to keep them stable and consistent. But as Holling (p. 17) made clear, if we approach ecosystems as engineers, tightly controlling them to minimize fluctuations, we may drastically reduce their *resilience* (their capacity to persist through major disturbances) by stifling the dynamic processes that would normally allow them to adapt. So if we, like McKeon, aim to create a space for cooperative action, perhaps we should build it as a resilient ecosystem rather than a stable technology. While the "physics" or "objective" approach to science policy pursues stability through universal knowledge and careful control, ecology and rhetoric offer tools for handling the complexities and uncertainties of our messy life on earth.

The remainder of this essay explores the potential value of an ecological perspective by applying it to two developing environmental policy strategies: ecosystem services markets and adaptive management programs.

4. ECOSYSTEM SERVICES AND ADAPTIVE MANAGEMENT

It's late winter, but strawberries are on sale: $1.99 for a twelve-ounce carton—a steal, considering the environmental toll of the pesticides, fertilizers and fuel needed to supply so many perfect (if tart) red berries out of season.

The abundance of the produce aisle is as miraculous as it is unsettling, offering strawberries when it's snowing while masking the profound, if untraceable, impacts of industrialized agriculture: toxic algae blooms, pollinator mortality, and nutrient leaching, to name a few.

Individuals troubled by these hidden costs can attempt to purchase more sustainably grown produce, though the price difference rules out this option for many. The relatively recent efforts by government agencies, nonprofit groups, and others to create ecosystem services markets (ESMs hereafter) aim to level the playing field between mass-produced foods and their more sustainable counterparts by factoring in the environmental costs of production.

ESMs, one might say, help nature speak in the universal language of (powerful) humans: money. Gretchen Daily and several other prominent ecological scientists argue that a large-scale turn to sustainability will only be possible only after institutions begin to "view...ecosystems as capital assets" (2009, p. 26). These assets would be revalued on the basis of scientific, economic, and political research, coupled with the contributions of various stakeholders (community groups, ecosystem managers, businesses and so forth). Ideally, ESMs would provide an incentive structure to promote more sustainable decision-making—companies would make the best choice for all parties involved simply by pursuing maximum profit.

Of course, efforts to create ESMs are often fraught with controversy, precisely because this ideal of a single best choice is unreachable. Moreover, translating ecosystems into "natural capital" necessarily foregrounds phenomena that lend themselves to quantification. For this reason, Robertson critiques concepts like "natural capital" and "social capital":

> Suggesting that cultural beliefs or ecosystems can be analyzed ... [as capital] amounts to a rhetorical dismissal of everything about culture and nature that cannot be reduced to an input to the production of a commodity fulfilling the utility function of the mythical *homo economicus*. And such rhetorical dismissals can be immensely powerful on the world stage, encouraging us all to assume that any value worth expressing can be expressed in price. (2009)

And as Robertson notes, rhetorical dismissals have material effects. A useful illustration of this relationship can be found in Klumpp's model of ecological argumentation, wherein "not only the speaker is at risk ... so are the beliefs, principles, and values invoked in the exchange" (2009, p. 188). Klumpp is evoking his concept of pragmatic risk—the principle that *if it doesn't work, it isn't true*. In ecological argumentation, beliefs and values are sustained through their constant evocation and reconsideration as parts of the argumentative network, and this continuous reconsideration serves as a selection mechanism. The survival of a belief or perspective depends on its utility. Argument constitutes and dissolves ideological commitments and their symbolic articulations in the objective realm. And without symbolic coherence, objects may lose material coherence as well. If the "national forest" designation were to vanish, for example, forests previously defined as such could be fragmented by roads and dispersed by logging

ESMs threaten to push out of sight and, indeed, existence, phenomena that cannot be standardized or quantified—in other words, objectified. So while they appear to let nature speak in human conversations, they do so through a process of objectification that codifies the subject-object divide. While an ecological rhetoric could provide opportunities for consubstantiality between humans and nonhumans through ongoing deliberation, the rationalizing and abstracting force of ESMs would force nonhumans into human systems, while foreclosing future discussion. Just as technocracy circumvents politics, such commodification circumvents rhetoric, making it impossible for humans and nature to transform and evolve together

Adaptive management programs, meanwhile, are based upon co-transformation. Regarding humans as part of the ecosystem, they use workshops, town meetings, and citizen science projects to engage a wide range of stakeholders, inviting input on how different people interact with and experience the environment in their daily lives. Still, these projects are typically dismissed as tokenistic or futile: time-intensive and politically complex, they frequently lose momentum and falter at impasses, ending in plans that are not discernibly different from the ones initially proposed (e.g. van Bommel et al., 2009).

Collins and Ison (2009a) argue that for adaptive management to succeed, we must rethink what we mean by "adaptation." The authors emphasize two key meanings: "adaptation as fitting into" and "adaptation as a good pair of shoes" (p. 354). In the first, adaptation is a matter of fitting together predetermined pieces to solve predetermined problems. In the second, it is a process of "co-evolution," wherein problems, participants, and the situation itself are deliberately (and deliberatively) constructed. This "social learning" model emphasizes, albeit not explicitly, rhetorical identification and construction, rather than the optimization of outcomes based on pre-established interests.

In contrast to the educational approach to public engagement, wherein experts share a non-technical version of their knowledge with stakeholders who then decide what to do, a social learning approach involves the on-site construction of both knowledge and "stakeholding" (2009b, p. 363). Rather than arrive with pre-assigned stakes in mind, participants develop a sense for what is important as they assess the situation and establish goals. This process arguably promotes the sort of situated self-awareness that Haraway pursues:

> As each stakeholder engages with situations from their different traditions of understanding [Russell and Ison, 2007], they begin to make sense of the issues from a partial perspective and different value judgments and so they construct their 'stakeholding.' (Collins & Ison, 2009b, p. 363)

The critical piece is not expert knowledge, taken to be universally applicable to all parties involved, but the ways in which different knowledges intersect and bump up against each other. Participants constitute new social identities and ways of knowing by reorienting themselves in relation to one another. While the role of scientists is somewhat vague in Collins and Ison's analysis, Latour's work makes it quite clear: scientists give voice to nonhumans.

But in spite of the improvements in this model, success remains elusive. Van Bommel et al. (2009) followed such a social learning initiative in the Netherlands, observing that because of unequal power relations, the knowledge produced reflected the perspective of those in power—social learning, they concluded, is "wishful thinking" (p. 410). They end on an optimistic note, however, reiterating Leeuwis's (2004) emphasis on 'instrumental/persuasive strategies to help create the pre-conditions for social learning and … enhance feelings of interdependence upon relevant stakeholders" (as cited in van Bommel et al., p. 410).

Still, the futility of current adaptation projects and the comparative success of certain market-based environmental protections suggest that some self-reflexivity may be in order. Having argued that knowledge is always situated and pragmatic, I should note that the position taken in this essay may not be viable outside of the sheltered epistemological environment of academia. Ecosystem services may, and probably should, appear untenable to many in our field, but that does not mean that they aren't among the best means available for sustaining human well-being—an aim that transcends most ideological divisions.

This is not to say that we should content ourselves to work within the status quo. In concluding their discussion of local knowledge systems, Watson-Verran and Turnbull write:

> The strength of social studies of science is its claim to show that what we accept as science and technology could be other than it is; its great weakness is the general failure to grasp the political nature of the enterprise and to work toward change. (2001, p. 138)

The need articulated here is for *re*constructive efforts to follow our deconstructive ones. Latour's work shows us how objects and relations are continually constructed and maintained; Haraway urges us to recognize our role in such constructions and to intervene with a critical, and self-consciously situated, eye. To the extent that rhetoric is concerned with making rather than uncovering and the contingent rather than the invariant, it seems that it has an important and unfilled role to play. An ecological rhetoric could operate in its constitutive, constructive, and persuasive capacities by facilitating identification and offering new models of deliberation that engage humans and nonhumans in a process of critical co-evolution.

REFERENCES

Collins, K. & Ison, I. (2009a). Living with environmental change: Adaptation as social learning. *Environmental Policy and Governance, 19*, 351–357.

Collins, K. & Ison, I. (2009b). Jumping off Arnstein's ladder: Social learning as a new policy paradigm for climate change adaptation. *Environmental Policy and Governance, 19*, 358–373.

Daily, G.C., Polasky, S., Goldstein, J., Kareiva, P.M., Mooney, H.A., Pejchar, L., Rickets, T.A., Salzman, J., & Shallenberger, R. (2009.) Ecosystem services in decision making: Time to deliver. *Frontiers in Ecology and the Environment, 7*(1), 21–28.

Goodwin, C. (1994). Professional vision. *American Anthropologist, 96*(3): 606–633.

Haraway, D. (1988). Situated knowledges: The science question in feminism and the privileged of partial perspective. *Feminist Studies, 14*(3), 575–599.

Holling, CS. (1973). Resilience and stability of ecological systems. *Annual Review of Ecology and Systematics, 4*, 1–23.

Kahan, D., Jenkins-Smith, H., & Braman, D. (2011). Cultural cognition of scientific consensus. *Journal of Risk Research, 14*(2), 147–174.

Klumpp, J. (2009). Argumentative ecology. *Argumentation and Advocacy, 45*(2), 183–197.

Latour, B. (1988). *The pasteurization of France.* (A. Sheridan & J. Law, Trans.). Cambridge, MA: Harvard University Press.

Latour, B. (2004). *Politics of nature.* Cambridge, MA: Harvard University Press.

Luskin, C. (2006, Sep. 22). Whose "war" is it, anyway? Exposing Chris Mooney's attack on intelligent design. *The Discovery Institute.* Retrieved from http://www.discovery.org/a/3739

McKeon, R. (1971). The uses of rhetoric in a technological age: Architectonic productive arts. In L. Bitzer & E. Black (Eds.), *The prospect of rhetoric* (pp. 44–63). New Jersey: Prentice Hall.

Millennium Ecosystem Assessment. (2005). Overview of the Milliennium Ecosystem Assessment. Retrieved from http://www.maweb.org/en/About.aspx

Mooney, C. (2005). *The Republican war on science.* Basic Books.

Paroske, M. (2009). Deliberating international science policy controversies: Uncertainty and AIDS in South Africa. *Quarterly Journal of Speech.* *95*(2), 148–170.

Remnick, D. (2006, April 24) Ozone man. *The New Yorker.* Retrieved from http://www.newyorker.com/archive/2006/04/24/060424ta_talk_remnick

Robertson, M. (2009, Jan. 4). Five hidden challenges to ecosystem markets. *The Katoomba Group's Ecosystem Marketplace.* Retrieved from http://www.ecosystemmarketplace.com/pages/dynamic/article.page.php?page_id=6415

Sarewitz, D. (1999). Science and environmental policy: An excess of objectivity. *Center for Science, Policy and Outcomes.* Retrieved from http://www.cspo.org/products/articles/excess.objectivity.html

van Bommel, S., Roling, N.,Aarts, & N. Turnhout, E. (2009). Social learning for solving complex problems: A promising solution or wishful thinking? A case study of multi-actor negotiation for the integrated management and sustainable use of the Drentsche Aa area in the Netherlands. *Environmental Policy and Governance, 19,* 400–412.

Watson-Verran, H. & Turnbull, D. (2001). Science and other indigenous knowledge systems. In S. Jasanoff, G.E. Markle, J.C. Petersen & T. Pinch (Eds.), *Handbook on Science and Technology Studies* (pp. 115–139). Thousand Oaks: SAGE.

Weingart, P. (1999). Scientific expertise and political accountability: Paradoxes of science in politics. *Science and Public Policy, 26*(3), 151–161.

Framing Science: The Influence of Expertise and Jargon in Media Coverage

DESERAI ANDERSON CROW

Environmental Studies
University of Colorado Boulder
1511 University Ave., 478 UCB
Boulder, CO 80309
USA
deserai.crow@colorado.edu

J. RICHARD STEVENS

Journalism and Mass Communication
University of Colorado Boulder
1511 University Ave., 478 UCB
Boulder, CO 80309
USA
rick.stevens@colorado.edu

ABSTRACT: The authors use an experimental pilot survey to examine the effects of message framing on a local environmental topic—urban development. Researchers focus on liberal respondents and find that scientists are considered the most credible sources of environmental information. However, some respondents appear less likely to agree that the inclusion of scientific information or jargon increases a news story's credibility or persuasiveness. The paper then discusses the significant potential implications that the findings have for public deliberation and the role of experts in public discourse.

KEYWORDS: environment, expertise, framing, jargon, science.

1. INTRODUCTION

Recent studies indicate that Americans are interested in but uninformed about science (National Science Board, 2010; Pew Research Center for the People and the Press, 2009) and that they get the majority of their information about science from the media (Pew Research Center for the People and the Press, 2002). Though the press plays a critical role in the public communication of science, scientists and media scholars often malign news media for an inability to cover science and complex environmental topics accurately (Pew Research Center for the People and the Press, 2009; Tankard & Ryan, 1974). These deficiencies are compounded by the economic issues imposed upon traditional news media over the past decade, and the resulting resource crisis that exists among American newspapers (Meyer, 2009). Even television news experienced an observable decline in dedicated coverage of science and environmental topics, despite relatively healthier profits (Brainard, 2008).

This paper draws from an online quasi-experimental survey designed to examine the effects of message framing (the use of technical and non-technical information) on a local environmental topic—urban air pollution.

Crow, D.A., & Stevens, J.R. (2012). Framing science: The influence of expertise and jargon in media coverage. In J. Goodwin (Ed.), *Between scientists & citizens: Proceedings of a conference at Iowa State University, June 1-2, 2012* (pp. 109-119). Ames, IA: Great Plains Society for the Study of Argumentation. Copyright © 2012 the author(s).

2. COMMUNICATING TECHNICAL INFORMATION AND POLICY IMPLICATIONS

As environmental problems grow increasingly difficult to solve (Vig & Kraft, 2003), examination of the role that science and environmental communication plays in the formation or change of opinions, knowledge, and policy becomes critical. In the deficit model of communication, scholars generally assume that citizens who possess more knowledge participate more fully in the policy process, policies are better informed, and democratic deliberation is better served. One hurdle to achieving the public awareness of environmental problems is the challenge journalists face in effectively communicating complex ideas to the public. Such efforts are increasingly important because "the public may be highly susceptible to influence by changes in media attention and media characterization" of scientific issues (Nisbet, 2004, p. 139).

One important issue concerns whether news media coverage of technical issues more effectively informs individuals through specific reporting of scientific facts or through more generalist narrative treatment of the issues. The increased complexity of a modern "knowledge society" (Giddens, 1990) and the increased centrality of science and technology (Lane, 1966) demands that journalists consult and interpret expertise in order to explain complex issues to the public.

2.1. Science Journalism in America

Americans are "knowledgeable about basic scientific facts that affect their health and their daily lives," topics widely covered by mainstream media (Pew Research Center for the People and the Press, 2009, p. 8), but are much less informed about more complex scientific topics (National Science Board, 2010). Despite high levels of science interest among Americans (National Science Board, 2010; Nunn, 1979; Pew Research Center for the People and the Press, 2009), news organizations tend to relegate science to a niche or beat subject, leading to uneven coverage by beat reporters, general assignment reporters, and wire stories (Friedman, 1986). The 1980s saw an increase in the number of science newspaper sections (Lewenstein, 1987), but the number of sections later declined from 95 sections in 1989 to 47 in 1992 (Jerome, 1992).

The Pew Research Center found in 2002 that Americans acquired 89 percent of their science and technology information from news media (Pew Research Center for the People and the Press, 2002). By 2010, "television and the Internet are the primary sources Americans use for science and technology information," with the Internet serving as the primary source for information about climate change (National Science Board, 2010, pp. 7–4). The Pew Center (Pew Center for the People and the Press, 2004) also found that 80 percent of the journalists surveyed reported media were not paying enough attention to complex stories, and half were pessimistic concerning the general state of journalism. While the Pew studies found television as one of the most important sources of science information, Sachsman, Simon and Valenti (2006) found that newspapers employed more than six times the environmental reporters. While nearly half of newspapers employed at least one environmental reporter, only 12.9 percent of television stations did.

Several past studies examined the accuracy of scientific news reports. Despite the public's positive attitude towards science, scientists overwhelmingly report dismay at public knowledge of science and blame the news for inadequately covering the issues (Pew Research

Center for the People and the Press, 2009). Tankard and Ryan (1974) found that scientists judged only 8.8 percent of science articles to be error-free, compared to 40 to 59 percent error-free in other types of stories. Studies repeatedly show scientists judging media reporting on science topics as inaccurate and distorted (Dunwoody & Scott, 1982; Tichenor, Olien, Harrison, & Donohue, 1970). Few science journalists possess scientific expertise (Palen, 1994), primarily because only three percent of journalists with college degrees major in mathematics or science areas, while most major in communication fields (Weaver & Wilhoit, 1996).

Beyond issues of whether scientific information is reported in media, and whether reportage is accurate, emerges the question of how reporters cover complex topics. Previous studies indicate the most compelling way in which reporters cover complex issues—whether they include science, economics, or other complexities—is through narrative use. Narratives feature compelling characters and typically possess a clear beginning, middle, and end, often with a villain, victim, and hero (Shanahan, Jones, & McBeth, 2011; Shanahan, McBeth, & Hathaway, 2011; Stone, 1997). Narrative structures have been reported the most effective in changing opinion, communicating information, or making people care about a topic (Golding, Krimsky, & Plough, 1992; Jones, 2010; Stone, 1997). Similar narrative strategies are also used by policy stakeholders to change public opinion regarding policy questions (McBeth & Shanahan, 2004; McBeth, Shanahan, Tigert, Hathaway, & Sampson, 2010).

Narrative strategies, however, can misinform the public with regard to broader policy problems, implications, causes, and consequences of policy action. By focusing on a single case (episodic framing), at the expense of the broader picture (thematic framing), media consumers often blame the victims for their plight instead of engaging broader societal trends and issues (Iyengar, 1990; Iyengar & Kinder, 1987). Thus, reporters can perform a disservice simply by attempting to tell a story using common journalistic norms of personification and localization of broad stories (Graber, 2006).

2.2 The Role of Expertise and Jargon in Science Communication

Expertise is an important factor in the communication of science, technology, and environmental events and issues. Expertise generally refers to skills or prowess, but is more often seen in communication research as a vehicle for creating authority or credibility (Hartelius, 2011). Journalists rely on scientific experts for context, legitimization, explication, and balance (Conrad, 1999).

Studies of expertise have reported expert sources in news stories had a strong positive impact on viewers' opinion change (Page, Shapiro, & Dempsey, 1987). Cozma (2006) found that stories containing a mix of expert and government sources were perceived more credibly than those including only government sources. Expertise uses are on the rise: a survey of newspaper stories from 1990 found twice as many experts quoted compared to 1978 (Soley, 1992).

One marker of expertise that often creates barriers to the public's understanding of science is the use of jargon. Jargon is essential for designating new entities for which the language has no name, producing an economy of effort and the accuracy and precision required in scientific research (Wilkinson, 1992). However, jargon does not always assist the communication process:

> Jargon has several meanings, one of which is neutral and the other negative. Neutrally defined, it refers to the "technical terminology or characteristic idiom of a special activity of group"; its negative definition refers to the inappropriate use of "obscure and often pretentious language marked by circumlocutions and long words." (Rowan, 1989, p. 171)

Jargon can also provide an air of technical or scientific authority while making the concepts referred to inaccessible to non-specialists. Such examples are said to be "mystificatory in aim and power-building in effect" (Fowler & Marshall, 1985, p. 3). Journalists appear to favor science sources that limit jargon and frame science in easily communicated terms (Conrad, 1999), even though they themselves often use jargon to demonstrate their scientific proficiency (Berglez, 2011). As for consumers, they cannot directly assess the technical knowledge of experts (Goodwin, 2011).

Science writers generally believe that defining scientific terminology is important to reducing public confusion about science and medical stories (Cooper & Yukimura, 2002), and particularly in environmental stories (Wigington, 2008), as well as risk-related stories (Jardine & Hrudey, 1997). But detailed scientific terminology is often perceived as a barrier to public understanding (De Boer, McCarthy, Brennan, Kelly, & Ritson, 2005).

Jargon often impedes public discourse about science or policy (Sandrelli, 2008) and reduces reader interest (Steinke, 1995). In policy research, the use of jargon often keeps the public confused, anxious about complexities, and uninvolved in the policy process (Schneider & Ingram, 1997).

2.3 Informed Citizens, Jargon, and Experts

Public opinion models indicate the presence of two separate publics—one with highly informed opinions and the other with less predictable and less-formed opinions (Zaller, 1992). The highly informed public is more likely to have access to information, education, and resources that influence their opinions. When such individuals are grouped, one expects higher education, greater employment, and more resources available (Brady, Verba, & Schlozman, 1995). Similarly, the individuals expected to most support environmental causes generally self-identify as liberal, educated, and having more resources (Dunlap & Mertig, 1992). Higher education levels suggest these individuals face lower comprehension barriers to complex scientific information than less-educated individuals. The question, then, becomes to what degree these highly educated individuals can synthesize technical information and whether technical information plays a role in opinion change.

Hartelius (2011) explained that "scientese" performed different roles for different audiences (p. 106). While internally jargon serves as an authenticator for scientific and professional communities, to outward audiences it can actually impede the persuasiveness of an expert's message.

Based upon the literature, the following research questions and hypotheses were developed for this study. Wondering whether Hartelius's observation was true only for the public at large, this study focused on higher-educated liberals.

- R1: Does the presence of jargon in media coverage of science topics inhibit or enhance understanding of the concepts presented?
  - o H1: The use of jargon, when presented to educated liberals, will not inhibit understanding of the concepts presented.

- R2: Does the use of jargon or lay language affect opinions related to the topics covered in media presentations of science? If so, what relationship can be observed?
  - H2: The use of jargon and data, when presented to educated liberals, will increase support for policies or positions relevant to the science presented.
- R3: Does the presence of jargon lead readers to assign higher levels of credibility to science or scientists compared to readers of less technical presentations?
  - H3: The presence of jargon, when presented to educated liberals, will lead to reader assignment of higher levels of credibility of science and scientists than non-technical presentations.

3. SURVEY RESEARCH METHODS

The study drew from an online quasi-experimental survey administered in July 2011. Subjects responded to social media advertising directed at Colorado residents. This paper focuses on research questions related to the liberal ideological and education demographics (N=108). While the response rate was low, the findings provide insight into barriers to the communication of expert information.

The survey design presented a series of opinion and knowledge questions focused on broad ideological positions as well as specific environmental and media-related questions. Subjects were randomly assigned to a group (two treatments and a control), each receiving a news article "treatment" (the treatments appear in the appendices below). The news article reported on the heat island effect on urban air pollution in Colorado. The control group received the survey questions but not the news article. Treatment 1 received a jargon-laden article, while treatment 2 received an article with a more lay presentation of the issues. The articles were of similar length (231 words in treatment 1 and 269 words in treatment 2). To avoid conflation with findings involving narrative language (Golding, et al., 1992; Jones, 2010; Stone, 1997), this study avoided narrative storytelling techniques to focus on the role of jargon and portrayals of expertise in media coverage of science and environmental topics.

4. RESEARCH FINDINGS

From the survey data, the respondents reported overwhelmingly (94.3%) that scientists were the most credible source for environmental stories. When asked for the second most credible sources, environmental activists (61.1%) were rated highly. When asked which sources were the least credible, the respondents reported industry representatives (36.8%), clergy (31.1%) and celebrities (29.2%) were least credible sources on environmental issues. Given the progressive ideological bias inherent in the selected sample, these results conformed to the researchers' expectations.

The researchers expected a more technical presentation of scientific research findings, in this case related to air pollution, would not present a barrier to comprehension for the sample surveyed.

- H1: The use of jargon, when presented to educated liberals, will not inhibit understanding of the concepts presented.

Indeed, as Table 1 illustrates, there was no correlation between treatment group and the measures of knowledge tested in the survey. Hypothesis one is therefore supported by the data.

Table 1: Correlations: Treatment Group and Knowledge Measures

	Knowledge	Article Knowledge	Non-attainment Knowledge	EPA Knowledge
Group	-.129	-.142	-.019	-.179

Second, it was expected that liberals would place high value on scientific information and therefore, the inclusion of jargon would sway their opinions to support pro-environmental policies and statements.

- H2: The use of jargon, when presented to educated liberals, will increase support for policies or positions relevant to the science presented.

Hypothesis two was not supported by the data, showing no correlation between the presence of scientific data and significant differences with the opinion measures.

Table 2: Correlations: Treatment Group and Pollution Opinions

	Agreement Urban Growth Limits	Agreement Gasoline Additives	Agreement Clean Air is Important	Agreement Regulate Business for Environment	Agreement Regulate Business for Health
Group	-.112	.003	.153	.029	.024

Third, it was expected that liberals would assign a high level of credibility to scientific sources and that the use of technical information to tell a news story would increase its credibility.

- H3: The presence of technical science information, when presented to educated liberals, will lead to reader assignment of higher levels of credibility of science and scientists than non-technical presentations.

Hypothesis three was not supported by the data. There was no significant correlation between treatment group and respondents' likelihood of assigning higher levels of credibility to scientific data generally, or to specific sources—both scientific and non-scientific. The presence of scientific data did not seem to influence opinions of credibility of science or news sources.

Table 3: Correlations: Treatment Group and Assigned Credibility of Scientific Data

	Scientific findings are generally credible	Data improve credibility of story
Group	-.003	.083

Table 4: Correlations: Treatment Group and Assigned Credibility of Sources

	Environmental Activists	Industry Representatives	Scientists	News Media
Group	.001	-.022	-.038	-.086

Finally, linear regression analysis was conducted to determine the influence of treatment group and demographic variables on the assignment of source and scientific credibility. Table 5 indicates that while the model helps explain a relatively low amount of the variance in credibility assignment (R2=.140), the treatment group only explains a small amount of this variance.

Table 5: Linear Regression Analysis for Influence of Treatment and Demographics on Credibility (N=108)

Variable	B	SE(B)	β	t	Sig. (p)
Treatment Group	.005	.301	.002	.017	.986
Gender	-.174	.299	-.070	-.582	.562
Age	-.025	.012	-.255	-2.14	.036
Education	-.199	.114	-.217	-1.75	.086
Income	.159	.065	.312	2.46	.017

Note: R2=.140.

The researchers therefore conclude that for the sample surveyed in this study, consumption of news information with or without data and jargon do not produce significantly different opinions regarding the credibility of science or scientific and non-scientific news sources. Similar tests conducted to determine the influence on knowledge and opinion resulted in even lower regression coefficients.

5. DISCUSSION

The results contain important implications for understandings of science communication and its effects on opinions, knowledge, and the trust or credibility of scientific sources. If news consumers show aversion to the language of scientific information and therefore do not receive the news content, this content cannot inform their opinions or political and personal decisions.

The presentation of technical information did not decrease understanding of the article content. However, the use of science in the news article did not affect opinion change, or the trust respondents reported for scientific sources of information. Previous studies (Page, Shapiro, & Dempsey, 1987; Cozma, 2006) consistently find that the presence of science sources in stories raise the claim credibility of scientific statements. The current study suggests this effect is more a judgment of status than performance, particularly since more detailed information does not increase message persuasiveness. Perhaps Goodwin's (2011) finding that consumers cannot directly assess technical knowledge themselves suggests the role of scientific expertise relies more on image than substantive claims.

Previous studies indicate that educated liberals place a higher degree of trust in science, sufficient to elicit a change of opinion. The failure to support H2 and H3 suggests the relationship of the technical parts of the message are more complicated than originally supposed, perhaps serving as more of a reinforcement of pre-existing ideological positions than a factor in opinion change.

Given that the use of jargon and data does not increase persuasiveness among those predisposed to a liberal political views, the tendency for those elements to impede discourse in the populace at large (Sandrelli, 2008) suggest their usage is not justified when the communication goal is to inform or persuade. Jargon itself may increase the credibility of the individual expert source through the "mystificatory" function (Fowler & Marshall, 1985), but such credibility does not appear to extend to the message or its content.

6. CONCLUSION

This study challenged earlier understandings of the role of expertise in the public communication of science. Specifically, by attempting to separate the role of the expert source from the expert language utilized, the study sought to glean insight into what elements of expert sourcing increase credibility and persuasiveness of a topic or issue.

The study also attempted to test the general findings from the literature regarding general populations against a sample predisposed to embrace science as authoritative. The results suggest that the politically liberal pro-science bias might be more superficial than substantive, as respondents did not appear motivated by detailed scientific information.

Due to sample limitations, the authors recommend the expansion of this inquiry to a national and representative sample in order to adequately understand whether these findings are generalizable to the population, or merely a function of preconceived views serving as a function of ideology. In particular, one would expect that the presentation of highly technical information would increase barriers to comprehension among the general public. One would also expect that among some demographics, a reliance on science as a primary news source may decrease the influence that a news article has on individual opinions or knowledge formation. It is important to next pursue studies that attempt to measure these effects within a broader sample.

Additionally, as stated above, it is evident from this study that among this presumably science-friendly audience, knowledge barriers were not evident. However, opinions were also not changed by the scientific data presented. Future research should combine analyses of technical information in news sources along with narrative framing studies to measure more discretely the most effective combination of the two approaches when attempting to shift opinion among media consumers. This information could provide both communicators and policymakers with a valuable source of information regarding teaching and persuading constituents about important, but highly technical issues.

REFERENCES

Berglez, P. (2011). Inside, outside, and beyond media logic: Journalistic creativity in climate reporting. *Media, Culture & Society, 33*(3), 449–465.

Brady, H., Verba, S., & Schlozman, K. L. (1995). Beyond SES: A resource model of political participation. *American Political Science Review, 89*(2), 271–294.

Brainard, C. (2008, December 4). CNN cuts entire Science, tech team. *Columbia Journalism Review: The Observatory*. Retrieved from http://www.cjr.org/the_observatory/cnn_cuts_entire_science_tech_t.php

Conrad, P. (1999). Uses of expertise: Sources, quotes, and voice in the reporting of genetics in the news. *Science, 8*(4), 285–302.

Cooper, C. P., & Yukimura, D. (2002). Science writers' reactions to a medical "breakthrough" story. *Social Science & Medicine, 54*(12), 1887–1896.

Cozma, R. (2006). Source diversity increases credibility of risk stories. *Newspaper Research Journal, 27*(3), 8–21.

De Boer, M., McCarthy, M., Brennan, M., Kelly, A. L., & Ritson, C. (2005). Public understanding of food risk issues and food risk messages on the island of Ireland: The views of food safety experts. *Journal of Food Safety, 25*(4), 241–265.

Dunlap, R. E., & Mertig, A. G. (1992). *American environmentalism: The U.S. environmental movement, 1970–1990*. Philadephia, PA: Taylor & Francis.

Dunwoody, S., & Scott, B. T. (1982). Scientists as mass media sources. *Journalism Quarterly, 59*(1), 52–59.

Fowler, R., & Marshall, T. (1985). The war against peacemongering: Language and ideology. In P. A. Chilton (Ed.), *Language and nuclear arms debate: NukespeaktToday* (pp. 3–22). London: Pinter.

Friedman, S. M. (1986). The journalist's world. In S. M. Friedman, S. Dunwoody & C. L. Rogers (Eds.), *Scientists and journalists: Reporting science as news* (pp. 17–41). New York, NY: Free Press.

Giddens A. (1990). *The Consequences of Modernity*. Palo Alto, CA: Stanford University Press.

Golding, D., Krimsky, S., & Plough, A. (1992). Evaluating risk communication: Narrative vs. technical presentations of information about radon. *Risk Analysis, 12*(1), 27–35.

Goodwin, J. (2011). Accounting for the appeal to the authority of experts. *Argumentation, 25*(3), 285–296.

Graber, D. A. (2006). *Mass media & American politics*. Washington, D.C.: CQ Press.

Hartelius, E. J. (2011). *The rhetoric of expertise*. Lanham, MD: Lexington.

Iyengar, S. (1990). Framing responsibility for political issues: The case of poverty. *Political Behavior, 12*(1), 19–40.

Iyengar, S., & Kinder, D. R. (1987). *News that matters*. Chicago, IL: University of Chicago Press.

Jardine, C. G., & Hrudey, S. E. (1997). Mixed messages in risk communication. *Risk Analysis, 17*(4), 489–498.

Jerome, F. (1992). For newspaper science sections: Hard times. *SIPIscope, 20*, 2–4.

Jones, M. D. (2010). *Heroes and villains: Cultural narratives, mass opinions, and climate change.* (Unpublished doctoral dissertation). University of Oklahoma, Normal, OK.

Lane. R. (1966). The decline of politics and ideology in a knowledgeable society. *American Sociological Review, 31*(5), 649–662.

Lewenstein, B. V. (1987). Was there really a popular science 'boom'? *Science, Technology & Human Values, 12*(2), 29–41.

McBeth, M. K., & Shanahan, E. A. (2004). Public opinion for sale: The role of policy marketers in Greater Yellowstone policy conflict. *Policy Sciences, 37*(3-4), 319–338.

McBeth, M. K., Shanahan, E. A., Tigert, L. E., Hathaway, P. L., & Sampson, L. J. (2010). Buffalo tales: Interest group policy stories and tactics in Greater Yellowstone. *Policy Sciences, 43*(4), 391–409.

Meyer, P. (2009). *The vanishing newspaper: Saving journalism in the Information Age*. Columbia, MO: University of Missouri Press.

National Science Board. (2010). "Science and technology: Public attitudes and understanding." In *Science and Engineering Indicators 2010* (7-1-7-49). (NSB Publication No. 10-01). Arlington, VA: National Science Foundation.

Nisbet, M. C. (2004). Public opinion about stem cell research and human cloning. *Public Opinion Quarterly, 68*(1), 131–154.

Nunn, C. Z. (1979). Readership and coverage of science and technology in newspapers. *Journalism Quarterly, 56*(1), 27–30.

Page, B. I., Shapiro, R. Y., & Dempsey, G. R. (1987). What moves public opinion. *American Political Science Review, 81*(1), 23–44.

Palen, J. A. (1994). A map for science reporters: Science, technology and society studies: Concepts in basic reporting and newswriting textbooks. *The Michigan Academician, 26*(3), 507–519.

Pew Research Center for the People and the Press. (2002). *The state of the news media 2002: An annual report on American journalism.* Washington, D.C.: Pew Research Center for the People and the Press.

Pew Research Center for the People and the Press. (2004). *The state of the news media: An annual report on American journalism.* Washington, D.C.: Pew Research Center for the People and the Press.

Pew Research Center for the People and the Press. (2009). *Scientific achievements less prominent than a decade ago: Public praises science; scientists fault public, media.* Washington, D.C.: Pew Research Center for the People and the Press.

Rowan, K. E. (1989). Moving beyond the what to the why: Differences in professional and popular science writing. *Journal of Technical Writing and Communication, 19*(2), 161–179.

Sachsman, D. B., Simon, J., & Valenti, J. M. (2006). Regional issues, national norms: A four-region analysis of U.S. environment reporters. *Science Communication, 28*(1), 93–121.

Sandrelli, S. (2008). A dialogue on hard sciences is possible. Is it useful too? *Journal of Science Communication, 7*(1), 1–4.

Schneider, A. L., & Ingram, H. (1997). *Policy design for democracy.* Lawrence, KS: University Press of Kansas.

Shanahan, E. A., Jones, M. D., & McBeth, M. K. (2011). Policy narratives and policy Processes. *Policy Studies Journal, 39*(3), 535–561.

Shanahan, E. A., McBeth, M. K., & Hathaway, P. L. (2011). Narrative policy framework: The influence of media policy narrative on public opinion. *Politics & Polity, 39*(3), 373–400.

Soley, L. C. (1992). *The news shapers.* New York, NY: Prager.

Steinke, J. (1995). Researching readers: Assessing readers' impressions of science news. *Science Communication, 16*(4), 432–453.

Stone, D. (1997). *Policy paradox: The art of political decision making.* New York, NY: Norton.

Tankard, J. W., & Ryan, M. (1974). News source perception of accuracy of science coverage. *Journalism Quarterly, 51,* 219–225.

Tichenor, P. J., Olien, C. N., Harrison, A., & Donohue, G. (1970). Mass communication systems and communication accuracy in science news reporting. *Journalism Quarterly, 47,* 673–683.

Vig, N. J., & Kraft, M. E. (2003). *Environmental policy: New directions for the twenty-first century* (5th ed.). Washington, D.C.: Congressional Quarterly Press.

Weaver, D., & Wilhoit, G. C. (1996). *The American journalist in the 1990s.* Mahwah, NJ: Lawrence Erlbaum.

Wigington, P. S. (2008). Clear messages for communication. *Environmental Health, 70*(10), 71–73.

Wilkinson, A. M. (1992). Jargon and the passive voice: Prescriptions and proscriptions for scientific writing. *Journal of Technical Writing and Communication, 22*(3), 319–325.

Zaller, J. R. (1992). *The nature and origins of mass opinion.* Cambridge, UK: Cambridge University Press.

APPENDIX A: NEWS ARTICLES USED IN QUASI-EXPERIMENTAL SURVEY DESIGN

Treatment 1: Technical science

Urban development linked to pollution concentration

By Steve Riggs

Widespread urban development results in a significant increase in ambient particulates accumulated during summer on paved surfaces, rather than being distributed across significant regions, says a new study.

The reason for this is that the proliferation of strip malls, subdivisions and other paved areas may cause a significant decrease in the evening breezes.

The international study, led by the National Centre for Atmospheric Research (NCAR) in the US, could have implications for the air quality of fast-growing coastal cities and other mid-latitude regions globally, the Journal of Geophysical Research-Atmospheres reports.

For instance, paved surfaces, a consequence of worldwide urbanization, keep the city warmer than more natural surfaces, according to an NCAR statement.

"The paved surfaces in metro Denver can trap as many as 19 extra joules of heat per square meter," said NCAR scientist Fei Chen, who led the study.

This difference lowers the contrast between high and low elevated area temperatures and causes an average 7 mph reduction in nighttime winds.

The stagnant conditions persist during the day because of larger-scale wind patterns.

"If the city continues to expand we're going to hit the 35 micrograms per cubic meter limit set by the Environmental Protection Agency around 2014," he added.

The research team combined extensive atmospheric measurements with computer simulations to examine the impact of pavements on the breezes in the area.

Treatment 2: Lay-science

Urban development linked to pollution concentration
By Steve Riggs
Widespread urban development alters weather patterns in a way that can help pollution accumulate during summer on paved surfaces, rather than being dispersed over large areas, says a new study.

The reason for this is that the proliferation of strip malls, subdivisions and other paved areas may interfere with the breeze needed to clear away smog and other pollution.

The international study, led by the National Center for Atmospheric Research (NCAR) in the US, could have implications for the air quality of fast-growing cities and other mid-latitude regions globally, the Journal of Geophysical Research-Atmospheres reports.

For instance, paved surfaces, a consequence of worldwide urbanization, keep the city warmer than more natural surfaces, according to an NCAR statement.

Researchers found that because pavements soak up heat and keep land areas relatively warm overnight, the contrast between temperatures at different elevations is less during the summer. This, in turn, causes a reduction in nighttime winds.

Consequently, overnight temperatures are often similar between the city and nearby low areas, which weakens summertime breezes and enables air pollution to build up.

The stagnant conditions also persist during the day because of larger-scale wind patterns. "The developed area of Denver has a major impact on local air pollution." said NCAR scientist Fei Chen, who led the study.

"If the city continues to expand, it's going to make the winds even weaker in the summertime, and that will make air pollution much worse," he added.

The research team combined extensive atmospheric measurements with computer simulations to examine the impact of pavements on the breezes in the area.

The Problem of Communicating Beyond Human Scale

MICHAEL DAHLSTROM

Greenlee School of Journalism and Communication
Iowa State University
215 Hamilton Hall
USA
mfd@iastate.edu

RAEANN RITLAND

Greenlee School of Journalism and Communication
Iowa State University
101 Hamilton Hall
USA
raeannr@iastate.edu

ABSTRACT: Human beings can only experience a thin ribbon of reality constrained by the biological limits of our perceptual systems. Yet, science routinely examines processes and phenomenon outside of human scale and science-related policy requires us to use our conceptions of these toward informed policy making. It often falls to experts to assist non-experts in constructing conceptions beyond human scale. This paper will organize relevant literature from varied fields to introduce the cognitive challenge of comprehending concepts beyond human scale and to suggest what communication techniques experts may find useful to help non-experts arrive at a perception more closely aligned with reality.

KEYWORDS: construal theory, grounded cognition, narrative, numeracy, psychology, psychophysics, risk communication, science communication, visual communication.

1. INTRODUCTION

Human beings can only experience a thin ribbon of reality constrained by the biological limits of our perceptual systems. While we can conceive of varied timeframes, sizes and dimensions, we cannot perceive them directly. Our conceptions of reality beyond human scale are therefore dependent upon the language used to describe it and will be built on analogues from our direct experience, both of which make an accurate conception of such reality unlikely.

Science routinely examines processes and phenomenon outside of human scale and science-related policy routinely asks us to assess our likely inaccurate conceptions of these processes and phenomenon toward informed policy making. While experts have the same perceptual limits as all humans, they often have a much richer conception of such constructs due to their training and experience that enables them to identify patterns that are not perceivable by non-experts (Ericsson & Charness, 1994; Ross, 2006). Therefore, it often falls to experts to explain these processes and phenomenon toward assisting the construction of non-expert perceptions. Since many political decisions depend on these non-expert perceptions, the expert communication of processes and phenomenon beyond human scale becomes crucial for informed decision making.

Dahlstrom, M., & Ritland, R. (2012). The problem of communicating beyond human scale. In J. Goodwin (Ed.), *Between scientists & citizens: Proceedings of a conference at Iowa State University, June 1-2, 2012* (pp. 121-130). Ames, IA: Great Plains Society for the Study of Argumentation. Copyright © 2012 the author(s).

To take an example, in 2010 the Deepwater Horizon oil spill expelled 4.9 million barrels of crude oil into the Gulf of Mexico (Restore the Gulf, 2010). To arrive at some conception of this magnitude, an individual must take something about which they have direct experience—something within human scale—and perceptually extrapolate to the desired magnitude outside of human scale. In this case, how does an individual extrapolate what they know about liquid volume, possibly thinking of a gallon of milk or the water-filled safety barrels surrounding highway onramps, to construct a perception of 4.9 million barrels? Whatever perception is formed will then determine the perceived severity of the event and opinions about what should be done. For some individuals, 4.9 million may be so meaningless that the risk is never cognitively engaged. To others, 4.9 million may fall within a categorically "large" conception similar to any other value considered "large," even if the actual values differ by many factors of magnitude. Either perception would likely result in different "appropriate" responses, neither of which may necessarily approach the best action to address the actual magnitude of effect.

Therefore, the primary questions of interest become (1) how do individuals extrapolate information taken from their direct experience in an attempt to comprehend processes and phenomenon beyond human scale and (2) how can experts best use communication to either assist or counter this extrapolation to arrive at a perception more closely aligned with reality? Unfortunately, the cognitive challenges of perceiving beyond human scale remain understudied. No single research area examines these questions. Rather, multiple strands of tangential research occasionally intersect within this scope of inquiry. Therefore, this paper will not attempt to answer these questions, but rather to define the scope of the problem by organizing relevant literature from varied fields linking perceptual limits and psychological extrapolation and suggesting research into communication techniques that may assist experts in improving the accuracy of non-expert conceptions beyond human scale.

2. DEFINING THE PROBLEM

2.1 Psychophysics

The field of psychophysics examines the relationship between physical stimuli and the sensations they produce (Snook, 1999) and attempts to develop mathematical functions that describe these relationships. These functions allow the calculation of the amount of sensory input needed to register a certain psychological perception at a certain magnitude.

One of the most cited of these functions is Weber's law, named after one of the founders of the discipline, Ernst Heinrich Weber. Weber's law states that as the magnitude of sensory input increases, it takes a greater increase in sensory input to cause the same increase in perceptions of that sensory input (Solomon, 2009). For instance, it would be easier to notice a difference between objects weighing 10 and 20 pounds than a difference between objects weighing 110 and 120 pounds. The term "Just Noticeable Differences" are used to describe such a unit change in perception (Bartoshuk, 2004; Moskowitz, 2003).

Psychophysics in general, and Weber's law in particular, can serve as a testable model for our first question of interest. Specifically, Weber's law provides a function predicting how perceptions can be extrapolated across different magnitudes of sensory input. Our analogous question seeks to develop a function describing how perceptions can be extrapolated across different magnitudes of imagined sensory input. Because it seems reasonable that imagining the difference between a 10 and 20 pound object is easier than imagining the difference between a 110 and 120 pound object, it is possible that this mental extrapolation of perceptions to different magnitudes may also follow Weber's law.

Current trends in psychophysics research do not address this possibility, instead applying the concept to enhance product development (Moskowitz, 2003) or in the context of visual processing (Murray, 2011). More relevant to our questions concerning human scale is the work of Bartoshuk (2004) who seeks to create a scale that equates people's perceptions across subjects. For example, people can understand the statement, "It was a large mouse that run up the trunk of a small elephant" (Bartoshuk, 2004, p. 17), but it is only after applying the terms within the context of the sentence that people know exactly what "large" and "small" mean. Considering this, Bartoshuk raises the question, "Are individual differences in experience similar? The 'Strongest pain experienced' will obviously denote a more intense pain to a woman who has experienced a particularly painful childbirth than to someone whose most intense pain to date is a stubbed toe" (2004, p. 17). Likewise, "We can compare very weak and very strong rose odors and we can compare very weak and very strong pains, but a very strong pain will be much stronger than a very strong rose odor for most individuals" (Bartoshuk, 2004, p. 17). More research is needed to understand this variability in perceptions based on previous experience and how to best standardize perceptual differences across topics.

2.2 Numerosity and Numeracy

An obvious factor underlying differences in perceived magnitude is that of how individuals perceive numbers. Numerosity refers to how individuals judge how numerous something is, or "how many." Researchers often employ spatial reasoning tasks, done over very short periods of time, involving visual tests of different groupings, arrangements and "connections" of dots (Allik & Tuulmets, 1991; Anobile, Turi, & Burr, 2010; Franconeri, Bemis, & Alvarez, 2009; Frith & Frith, 1972; Gordon, 2004; He, Zhang, Zhou, & Chen, 2009). For instance, a researcher may show two pictures for a split second, each containing varied dot patterns, one with lines connecting dots and one without, and then ask which picture contained more dots.

Results indicate that people use different cognitive processes in number estimation depending on the size of what is to be counted (Anobile et al., 2010; Dehaene & Changeux, 1993; Trick & Pylyshyn, 1994). When faced with small numbers of objects, individuals use subitizing, which is an automatic and accurate process based on visual recognition, for instance noting the difference between two and three coins at a glance. For larger numbers, such as if a coin jar spilled, individuals must resort to counting, an effortful and more inaccurate process (Anobile et al., 2010; Trick & Pylyshyn, 1994). Results also find that people group objects and estimate amounts according to proximity and similarity, but often increase errors with doing so (Allik & Tuulmets, 1991; Frith & Frith, 1972; He et al., 2009). For instance, when objects are clustered, people tend to underestimate their numbers because clustered objects take up less space and appear less numerous. Conversely, when objects are far apart, people tend to overestimate their numbers (Franconeri et al., 2009; He et al., 2009).

Such number estimating techniques may also be culturally dependent. In America, people perceive numbers between 1 and 10 as linear, but logarithmically for larger numbers (Dehaene, Izard, Spelke, & Pica, 2008) whereas an indigenous Amazonian community showed logarithmic perceptions across all values. Likewise, Gordon (2004) found the indigenous Amazonian community follows a "one, two, many" counting system where the individuals could not accurately represent exact quantities for medium-sized values, such as four or five (Gordon, 2004). When asked to perform matching exercises, the individuals successfully matched groups 0–3, success dropped to a near 0% for numbers 3–6 and then returned to near perfect for groups of 7–10. Gordon (2004) suggests this is perhaps due to chunking, grouping into smaller, more manageable, groups of 2–3.

Numeracy differs from numerosity in that it refers to the representation of numbers rather than amount of objects. One aspect of numeracy involves individual differences of mathematical capabilities based on experience or innate ability and suggests that larger numbers are often more difficult to understand than smaller numbers. Yet beyond individual differences, numeracy also suggests the representation of numbers can influence perceptions (Dehaene et al., 2008; Garcia-Retamero & Galesic, 2011; Gigerenzer & Edwards, 2003; Reyna & Brainerd, 2008). Identical values presented as a frequency (1 out of 100) versus a probability (1%) resulted in different perceptions with the frequency seen as representing greater risk (Gigerenzer & Edwards, 2003; Reyna & Brainerd, 2008). Likewise, the denominator used to present a fraction also influences perceptions. Using smaller denominators that are closer to "plausible" group sizes in human society (x / 125) allow easier risk determinations that are less influenced by message framing than when the denominator represents a larger value beyond normal human group sizes (x / 100,000) (Garcia-Retamero & Galesic, 2011; Wang, 1996).

Both numerosity and numeracy suggest that how the mind estimates general amounts and comprehends specific number representations can influence changing magnitudes of concepts and may play a role in generating perceptions beyond human scale.

2.3 Grounded Cognition

Grounded cognition is a field of psychology that claims that functions of the mind are limited by biology and heavily influenced by our physical bodies. Unlike traditional theories of cognition that assume the mind and body to be separate, grounded cognition can help explain why there may be a bias for easier comprehension of concepts at the human scale and more difficult comprehension when asked to go beyond—namely that the mind evolved to account for sensory input that directly led to bodily survival. Involved in this process are affordances, which are relationships between action and perception, or possibilities for interaction (Kaschak & Maner, 2009). Simulations of these affordances "allow us to consider, and have our behavior controlled by, knowledge and goals that are not directly signaled by the current environment" (Kaschak & Maner, 2009, p. 1241). Thus, within the frame of grounded cognition is the belief that thoughts are not words or symbols but visual and motor images that are driven by everyday goals (Pecher, Boot, & Van Dantzig, 2011).

Beyond biological limitations, another aspect of our first question of interest that can be addressed through grounded cognition is that of abstraction. Up until this point, we have defined concepts beyond human scale as being of a different magnitude. Yet, much of science is outside of human scale not because of magnitude, but due to being more intangible or abstract than what can be experienced directly. Abstract transfer within grounded cognition can serve to explain how people use grounding to understand abstract ideas. For example, telling a story is an abstract concept, while handing an object to someone is a concrete action. Yet, studies find that both acts of "giving" are represented by a hand movement away from the body (Pecher et al., 2011). Abstract concepts with an analogous action are therefore easy to "ground" in direct experience. However, concepts that become so abstract that there are no analogous actions in which to ground the experience may result in a categorically different challenge to create perceptions beyond human scale.

2.4 Construal Theory

Abstraction may not just be specific to the process or phenomenon beyond human scale, but also relative to the individual doing the perceiving. Construal theory states that the perception of an event is dependent upon the psychological distance between the event and the perceiver (Kanten, 2011). Four kinds of psychological distance exist: temporal, spatial, social, and probability (Jia, Hirt, & Karpen, 2009; Trope, Liberman, & Wakslak, 2007). In other words, decisions and judgments depend on how "close" an event is to an individual. If an event is farther away, a person's construal level is higher, leading to a perception that is more abstract, structured and less detailed. In contrast, an event that is closer would cause a person's construal level to be lower and lead to one that is more concrete, unstructured and more detailed. Events with a high construal level usually answer the "why" of an event, and low construal thoughts answer the "how" (Garcia, 2011).

Results show that all four forms of psychological distance can affect people's perceptions of events. For example, temporal distance affects descriptions of life satisfaction. People predicted the near future in concrete terms, involving mixed affect, but they predicted the distant future more abstractly, with more positive affect (Heller, Stephan, Kifer, & Sedikides, 2011). Likewise, even subtle changes in spatial distance can influence the creativity in problem solving (Polman & Emich, 2011), and the extremity of affective evaluations is larger when abstract and smaller when concrete (McCarthy & Skowronski, 2011). Visual and verbal representations have also been found to differ in construal levels. Predictably, text should require higher construal because it is more abstract, whereas photographs, which are concrete representations, should require lower construal. Results support these predictions, finding that responses were faster when the medium fit its construal level—photos of domestic objects and names of foreign objects (Trope et al., 2007).

These results suggest that while the human mind may have physiological limits bounded by scale, individual reference points may also play a role in determining what is processed within or beyond human scale for an individual at a particular point in time.

3. POSSIBLE SOLUTIONS

The previous section aimed to summarize literature that could better define the scope of our problem, specifically the challenges involved when individuals extrapolate information from their direct experience in an attempt to comprehend processes and phenomenon beyond human scale. In this section, we begin to suggest areas where a partial solution may be found, specifically areas of communication research that may provide experts with a practical toolkit to assist non-experts in arriving at a perception more closely aligned with reality. We suggest three fields of research that may hold particular promise: metaphor, narrative and visual communication.

3.1 Metaphor

Metaphors are linguistic devices that link two concepts together through some shared trait to emphasize an aspect of the original concept or make it observable in a new light. The concept of metaphor consists of two distinct positions. The first, termed constructivism, views metaphor as an intrinsic quality of all language and the process by which humans create new knowledge. The second, termed non-constructivism, views metaphor as a literary embellishment that allows for a novel interpretation of reality based on emphasizing certain features of the paired concepts (Baake, 2003). The second position is more relevant to the current study.

Science in particular is known for its reliance on metaphor, not just in communication with the public, but also between scientists at the research development stage to formulate hypotheses and interpret data (Baake, 2003). By linking two concepts, metaphors can offer the non-expert public avenues to understand complex scientific phenomenon in terms of something more familiar. However, scientific metaphors can also backfire when they blur the concept they are trying to clarify or introduce unintended associations, and caution is recommended (Weigmann, 2004).

Metaphors may help experts communicate beyond human scale by both removing the need of potentially confusing values and by substituting something concrete in place of something abstract. Returning to our previous example, the 4.9 million barrels of crude oil released in the Gulf of Mexico could be described using a metaphor of a newly constructed toxic 51st state of the union. Such a metaphor describes the size of the event, associates it geographically and emphasizes what about it demands attention, namely the toxic properties, without needing to extrapolate specific values of any unit.

3.2 Narrative

A narrative, or story, represents a specific and temporally structured form of communication where characters and their desires hold a string of events together through a cause and effect relationship. Narratives are thought to be processed differently than non-narrative communication (Fisher, 1984) and are considered by some to represent the fundamental constituents of human cognition (Schank & Abelson, 1995). Supporting this view is research that finds narrative communication is read twice as fast and remembered twice as easily as non-narrative communication (Graesser, Olde, & Klettke, 2002) and can more easily persuade otherwise resistant audiences (Moyer-Guse & Nabi, 2010).

Processing a narrative essentially involves engaging in a mental simulation of some aspect of reality from a particular human point of view (Oatley, 1999). Narratives are therefore a method of packaging an event into a particular viewpoint of human scale. Narratives may help experts communicate beyond human scale by providing a structure that juxtaposes how something beyond human scale looks within human scale.

Returning again to our previous example, the 4.9 million barrels of crude oil released in the Gulf of Mexico may be better communicated as a story or collection of stories told from the viewpoint of individuals nearby or within the event (for example, a narrative of a fisherman who has lost his livelihood or a regular vacationer who was shocked to see changes in the recreational beach or even a fictional crab struggling to escape the suffocating death of the spreading oil). These narratives could portray the magnitude of the event through a mental simulation that provides an even richer and more complex interplay of context without the need to extrapolate specific values.

3.3 Visual Communication

While communication is often thought of in spoken or written form, visuals represent an additional form of communication that may offer benefits through pattern recognition. In particular, visual communication has been found to improve comprehension of risks, particularly in health contexts (Ancker, Senathirajah, Kukafka, & Starren, 2006; Garcia-Retamero & Galesic, 2010; Lipkus & Hollands, 1999; Zikmund-Fisher et al., 2008). Different forms of visuals are even recommended depending on the type of comprehension desired: tables are better at communicating verbatim information whereas pictographs are better when the overall gist of the information is more important (Hawley et al., 2008).

However, it is important to avoid "hiding" pieces of information when constructing visuals for the communication of risk. Important information can be lost due to the foreground/background salience effect that states that in graphical situations, people are drawn to foreground information and pay less attention to what is in the background. This can skew the risk perception and lead people to incorrect perceptions, especially when the number of people at risk is emphasized at the expense of all those at risk of harm (Stone et al., 2003).

Again returning to our example, rather than attempt to portray the magnitude of 4.9 million barrels of crude oil in words, a visual comparing the volume of oil released in the Deepwater Horizon oil spill, the amount of oil used daily in the U.S. and the amount of oil necessary to cause damage to certain types of ecosystems could assist in constructing a perception of the magnitude of the event that may align more closely with reality, again without the need to extrapolate values beyond human scale.

4. CONCLUSION

The challenge of communicating beyond human scale is an understudied problem within science communication of which experts are often expected to be able to address. The successful communication of these processes and phenomenon outside of human scale is crucial for informed decision making within a democratic society, yet most experts are not equipped to address these challenges and unfortunately very little theoretical work exists to assist them.

Therefore, the purpose of this article was not to propose a solution but to better define the scope of the problem by summarizing literature than begins to explore the challenges of communicating beyond human scale and identifying areas of communication literature where a partial solution might be found.

Specifically, we suggest that the field of psychophysics offers a testable model in Weber's law to examine how perceptions are extrapolated across different magnitudes of imagined sensory input. Likewise the areas of numerosity, numeracy, grounded cognition and construal theory offer theories and empirical evidence that suggest perceiving concepts beyond human scale becomes increasingly difficult and likely divorced from reality. We also suggest that the areas of metaphor, narrative and visual communication represent promising areas of communication research that may offer practical methods for experts to better address these challenge when communicating to non-experts. Specifically, metaphor addresses the bias of magnitudes and abstractness, narrative represents a structure that packages abstract concepts into human scale and visual communication can leverage the comprehension benefits of pattern recognition over language-based description.

With the problem defined and relevant literature identified and summarized, this article represents a call for future research to explore these areas to better understand the boundaries of human cognition on an axis of scale, the processing strategies used to address these boundaries, the inaccuracies or biases inherent in these processing strategies and finally how experts can use communication to help correct for these biases. An additional question not addressed in this paper is that even when the answers to these questions are better understood, it is likely that experts will not be adequately trained in using the successful communication techniques. Therefore, additional research needs to examine the organizational structures that may best equip and reward experts for their ability to assist non-experts in the construction of more accurate perceptions beyond human scale.

REFERENCES

Allik, J., & Tuulmets, T. (1991). Occupancy model of percieved numerosity. Perception & Psychophysics, 49(4), 303–314. doi: 10.3758/bf03205986

Ancker, J. S., Senathirajah, Y., Kukafka, R., & Starren, J. B. (2006). Design features of graphs in health risk communication: A systematic review. Journal of the American Medical Informatics Association, 13(6), 608–618. doi: 10.1197/jamia.M2115

Anobile, G., Turi, M., & Burr, D. C. (2010). Subitizing but not estimation of numerosity requires attentional resources. Perception, 39, 80–80.

Baake, K. (2003). Metaphor and knowledge: The challenges of writing science. New York, NY: State University of New York Press.

Bartoshuk, L. A. (2004). Psychophysics: a journey from the laboratory to the clinic. Appetite, 43(1), 15–18. doi: 10.1016/j.appet.2004.02.005

Dehaene, S., & Changeux, J. P. (1993). Development of elementary numerical abilities—A neuronal model. Journal of Cognitive Neuroscience, 5(4), 390–407. doi: 10.1162/jocn.1993.5.4.390

Dehaene, S., Izard, V., Spelke, E., & Pica, P. (2008). Log or linear? Distinct intuitions of the number scale in Western and Amazonian Indigene cultures. Science, 320(5880), 1217–1220. doi: 10.1126/ science.1156540

Ericsson, K. A., & Charness, N. (1994). Expert performance: Its structure and acquisition. American Psychologist, 1–23.

Fisher, W. R. (1984). Narration as a human-communication paradigm: The case of public moral argument. Communication Monographs, 51(1), 1–22.

Franconeri, S. L., Bemis, D. K., & Alvarez, G. A. (2009). Number estimation relies on a set of segmented objects. Cognition, 113(1), 1–13. doi: 10.1016/j.cognition.2009.07.002

Frith, C. D., & Frith, U. (1972). Solitaire illusion—Illusion of numerosity. Perception & Psychophysics, 11(6), 409-10. doi: 10.3758/bf03206279

Garcia, D. (2011). Happy today, happy tomorrow: The (non-)effect of temporal distance on judgments of Life Satisfaction. Personality and Individual Differences, 51(8), 1048–1051. doi: 10.1016/j.paid.2011.07.031

Garcia-Retamero, R., & Galesic, M. (2010). Who proficts from visual aids: Overcoming challenges in people's understanding of risks. Social Science & Medicine, 70(7), 1019–1025. doi: 10.1016/j.socscimed.2009.11.031

Garcia-Retamero, R., & Galesic, M. (2011). Using plausible group sizes to communicate information about medical risks. Patient Education and Counseling, 84(2), 245–250. doi: 10.1016/j.pec.2010.07.027

Gigerenzer, G., & Edwards, A. (2003). Simple tools for understanding risks: From innumeracy to insight. British Medical Journal, 327(7417), 741–744. doi: 10.1136/bmj.327.7417.741

Gordon, P. (2004). Numerical cognition without words: Evidence from Amazonia. Science, 306(5695), 496–499. doi: 10.1126/science.1094492

Graesser, A. C., Olde, B., & Klettke, B. (2002). How does the mind construct and represent stories? In M. C. Green, J. J. Strange & T. C. Brock (Eds.), Narrative impact: Social and cognitive foundations (pp. 229–262). Mahwah, NJ: Lawrence Erlbaum.

Hawley, S. T., Zikmund-Fisher, B., Ubel, P., Jancovic, A., Lucas, T., & Fagerlin, A. (2008). The impact of the format of graphical presentation on health-related knowledge and treatment choices. Patient Education and Counseling, 73(3), 448–455. doi: 10.1016/j.pec.2008.07.023

He, L., Zhang, J., Zhou, T., & Chen, L. (2009). Connectedness affects dot numerosity judgment: Implications for configural processing. Psychonomic Bulletin & Review, 16(3), 509–517. doi: 10.3758/pbr.16.3.509

Heller, D., Stephan, E., Kifer, Y., & Sedikides, C. (2011). What will I be? The role of temporal perspective in predictions of affect, traits, and self-narratives. Journal of Experimental Social Psychology, 47(3), 610–615. doi: 10.1016/j.jesp.2011.01.010

Jia, L., Hirt, E. R., & Karpen, S. C. (2009). Lessons from a Faraway land: The effect of spatial distance on creative cognition. Journal of Experimental Social Psychology, 45(5), 1127–1131. doi: 10.1016/j.jesp.2009.05.015

Kanten, A. B. (2011). The effect of construal level on predictions of task duration. Journal of Experimental Social Psychology, 47(6), 1037–1047. doi: 10.1016/j.jesp.2011.04.005

Kaschak, M. P., & Maner, J. K. (2009). Embodiment, evolution, and social cognition: An integrative framework. European Journal of Social Psychology, 39(7), 1236–1244. doi: 10.1002/ejsp.664

Lipkus, I. M., & Hollands, J. G. (1999). The visual communication of risk. Journal of the National Cancer Institute Monographs, 1999 (25), 149–163.

McCarthy, R. J., & Skowronski, J. J. (2011). You're getting warmer: Level of construal affects the impact of central traits on impression formation. Journal of Experimental Social Psychology, 47(6), 1304–1307. doi: 10.1016/j.jesp.2011.05.017

Moskowitz, H. R. (2003). The intertwining of psychophysics and sensory analysis: Historical perspectives and future opportunities—A personal view. Food Quality and Preference, 14(2), 87–98. doi: 10.1016/s0950-3293(02)00072-1

Moyer-Guse, E., & Nabi, R. L. (2010). Explaining the effects of narrative in an entertainment television program: Overcoming resistance to persuasion. Human Communication Research, 36(1), 26–52. doi: 10.1111/j.1468-2958.2009.01367.x

Murray, R. F. (2011). Classification images: A review. Journal of Vision, 11(5). doi: 10.1167/11.5.2

Oatley, K. (1999). Why fiction may be twice as true as fact: Fiction as cognitive and emotional simulation. Review of General Psychology, 3(2), 101–117.

Pecher, D., Boot, I., & Van Dantzig, S. (2011). Abstract concepts: Sensory-motor grounding, metaphors, and beyond. Psychology of Learning and Motivation: Advances in Research and Theory, 54, 217–248.

Polman, E., & Emich, K. J. (2011). Decisions for others are more creative than decisions for the self. Personality and Social Psychology Bulletin, 37(4), 492–501. doi: 10.1177/0146167211398362

Restore the Gulf. (2010). U.S. scientific teams refine estimates of oil flow from BPs well prior to capping. Retrieved from http://www.restorethegulf.gov/release/2010/08/02/us-scientific-teams-refine-estimates-oil-flow-bps-well-prior-capping

Reyna, V. F., & Brainerd, C. J. (2008). Numeracy, ratio bias, and denominator neglect in judgments of risk and probability. Learning and Individual Differences, 18(1), 89–107. doi: 10.1016/j.lindif.2007.03.011

Ross, P. E. (2006). The expert mind. Scientific American, 295(2), 64–71.

Schank, R. C., & Abelson, R. (1995). Knowledge and memory: The real story. In R.S. Wyer, Jr. (Ed.), Knowledge and memory: The real story (pp. 1-87). Hilldale, NJ: Lawrence Erlbaum Associates.

Snook, S. H. (1999). Future directions of psychophysical studies. Scandinavian Journal of Work Environment & Health, 25, 13–18.

Solomon, J. A. (2009). The history of dipper functions. Attention Perception & Psychophysics, 71(3), 435–443. doi: 10.3758/app.71.3.435

Stone, E. R., Sieck, W. R., Bull, B. E., Yates, J. F., Parks, S. C., & Rush, C. J. (2003). Foreground : Background salience: Explaining the effects of graphical displays on risk avoidance. Organizational Behavior and Human Decision Processes, 90(1), 19–36. doi: 10.1016/s0749-5978(03)00003-7

Trick, L. M., & Pylyshyn, Z. W. (1994). Why are small and large numbers enumerated differently—A limited capacity preattentive stage in vision. Psychological Review, 101(1), 80–102. doi: 10.1037//0033-295x.101.1.80

Trope, Y., Liberman, N., & Wakslak, C. (2007). Construal levels and psychological distance: Effects on representation, prediction, evaluation, and behavior. Journal of Consumer Psychology, 17(2), 83–95. doi: 10.1016/s1057-7408(07)70013-x

Wang, X. T. (1996). Domain-specific rationality in human choices: Violations of utility axioms and social contexts. Cognition, 60(1), 31–63. doi: 10.1016/0010-0277(95)00700-8

Weigmann, K. (2004). The code, the text and the language of God—When explaining science and its implications to the lay public, metaphors come in handy. But their indiscriminate use could also easily backfire. Embo Reports, 5(2), 116–118. doi: 10.1038/sj.embor.7400069

Zikmund-Fisher, B. J., et al. (2008). Communicating side effect risks in a tamoxifen prophylaxis decision aid: The debiasing influence of pictographs. Patient Education and Counseling, 73(2), 209–214. doi: 10.1016/j.pec.2008.05.010

Testimony Traces in Appellate Review: Expertise Extension in Cases of Domestic Abuse and Eyewitness Identification

PER FJELSTAD

Department of Communication
University of New Hampshire
10 Academic Way
Durham, NH 03824
U.S.A.
Per.Fjelstad@unh.edu

ABSTRACT: Appellate rulings contribute to policy deliberations on uses of and parameters for expert testimony. As courts perform gatekeeping and evaluative roles, opinions highlight investigative independence, calculations of probability, and consistency across studies. Even so, the elements from expert testimony most commonly extended into precedent stand out for their summative concision and figurative cogency.

KEYWORDS: expert testimony, expertise extension, figurative cogency, battered women's syndrome, eyewitness reliability.

1. INTRODUCTION

Testimony in court by victims of domestic abuse often is supplemented by expert testimony that modifies and supersedes a victim's voice (Hamilton, 2010). Expert testimony also serves as an interpretive filter and evaluative caution when a jury evaluates the reliability of eye-witness testimony (Terrance, Thayer, & Kehn, 2006; Cutler, Dexter, & Penrod, 1989). In both instances, the testimony introduces and explains criteria that a jury and judge use to understand, and at times reject, other testimony at trial. In some circumstances, attorneys also strategically use expert testimony to convey a defendant's point of view while avoiding cross-examination (Miller, 2003). Given these uses of external expertise in trials, often to filter and qualify factual accounts, the conditions under which such testimony is admitted, presented, and interpreted is a matter of significant interest.

Several factors influence procedural decisions on whether a court admits the testimony of an external expert on a given subject in a particular trial: policy preferences, role of requesting party, criteria used for evidentiary review, legal sub-domain, type of expertise, degree of consensus about a social problem and the sufficiency of existing legal solutions (Harris, 2008; Buchman, 2007). Less has been written, however, about the process in particular legal domains by which that expertise incrementally is naturalized. This essay uses the ideas of transmutation and adoption to describe a process by which knowledge and recommendations derived from a particular expertise are actively imported and incrementally reified. The analysis suggests that domain elaboration in legal sub-fields is solidified in a process of transmutation that distills elements with figurative cogency from expert narratives.

Fjelstad, P. (2012). Testimony traces in appellate review: Expertise extension in cases of domestic abuse and eyewitness identification. In J. Goodwin (Ed.), *Between scientists & citizens: Proceedings of a conference at Iowa State University, June 1-2, 2012* (pp. 131-139). Ames, IA: Great Plains Society for the Study of Argumentation. Copyright © 2012 the author(s).

2. EXPERTISE EXTENSION IN DYNAMIC LEGAL FIELDS

The case opinions analyzed in this study all evaluated uses of expert testimony to frame understandings of fact. Some cases cited expert testimony explicitly to revise a legal standard (*State v. Henderson*, 2011). In others, the original and declared purpose for the testimony was to make factual determinations in the case (*State v. Haines*, 2006; *People v. Midyette*, 2011). Nonetheless, as this analysis shows, even in those cases, the use of that testimony subtly developed elaborations and applications of legal doctrine. The expert testimony was interpreted and adopted in ways that developed law.

The sample consists primarily of two types of cases: (1) ones that involved allegations of domestic abuse, a subject matter concerning which courts and some legislatures increasingly have enshrined a right at trial to expert psychological testimony, and (2) cases in which eyewitness observations may have been contaminated by inappropriately suggestive line-up procedures or other reliability-decreasing factors. These two types of cases give the sample a limited and symmetrical diversity, particularly since the common exclusion at trial of expert testimony concerning suggestive line-up techniques typically benefits the prosecution, while the increasingly guaranteed right to call expert psychological testimony in cases of alleged battering benefits the defense, especially for the particular charge that generated public support for a statutory right to expert testimony, where a battery victim is charged with a crime against her abuser. Not only do these two types of cases reflect on the one hand a restrictive, and on the other a lighter, admission threshold for expert testimony, they also reflect tendencies in which the testimony carries different advantages for the prosecution and defense. As a whole, the sample shows distinctive proclivities in how expert testimony has been adopted into and transformed for legal understandings.

The study uses primarily three cases in which appellate courts ruled on matters of expert testimony. One modified the legal standard by which factors that reduce eyewitness reliability can be identified and mitigated (*State v. Henderson*, 2011). The original trial court convicted Larry Henderson of reckless manslaughter and aggravated assault, based primarily on his identification by James Womble, an eyewitness. The trial court allowed Womble's testimony, based on a pre-trial hearing in which it found the police officers had not been "impermissibly suggestive," even though Womble said in that hearing that the investigators had strongly "nudged" him to make a choice from a photo line-up of possible suspects. An appellate court reversed that ruling and, showing some uncertainty, requested certification of that decision from the state supreme court. The high court heard arguments from third parties that "raised questions about possible shortcomings in the [legal standard used]." As a result, the state supreme court appointed a Special Master to "evaluate the scientific and other evidence about eyewitness identifications" (p. 2). This testimony, summarized in a separate report and recounted at length in the court's subsequent opinion, focused not on matters of fact decided by the trial court, but instead on the validity of the legal standard used to determine the reliability of eye-witness testimony. Based on the scientific evidence and expert testimony, the court recommended a revision of the law governing eye witness identifications.

In *People v. Midyette* (2011), the defense called Dr. Lenore Walker, a widely-cited social scientist whose work first identified and documented Battered Women's Syndrome (BWS), for the purpose of showing that the defendant suffered from the syndrome, which may have contributed to her ineffective defense at trial. The expert testimony thus was not presented at the original trial, but instead on appeal at the district court, where the defendant argued that the testimony might effectively have been presented earlier, but for her emotional and cognitive incompetence at that time. In this case, the district court allowed and heard the expert testimony, which it eventually found unpersuasive. The opinion included an explanation, offered by the district court, as to why the expert testimony was not persuasive.

In the third case, *Ohio v. Haines* (2006), the Ohio Supreme Court reversed convictions for some domestic assault charges, while leaving other assault convictions intact. The court let stand the trial court's decision to allow expert testimony on BWS, but under the condition that the testimony address only the issue of social framework. The court found irreparable prejudice introduced when the expert presented at trial also a diagnostic opinion that the abuse victim suffered from BWS. According to the court, that diagnosis implied the alleged crimes had been committed and intruded on the jury's fact-finding responsibility.

Finally, the study draws on multiple-cases analyses by Melissa Hamilton, Julie Stubbs, and Julia Tolmie, which place the use of expert testimony on domestic abuse within an evolving and dynamic context. Stubbs and Tolmie (1999) focused narrowly on the use of BWS testimony on behalf of defendants accused of murder or assault. In other words, they did not consider the testimony as used in prosecutions against the men perpetrating the abuse. Even in that limited context, however, they saw a developmental trajectory in Canadian and U.S. law, though it had not yet developed that way in Australian courts. In particular, they noted that expert testimony increasingly was used not just as evidence of a qualifying social framework, but also as a means for assessing responses by the abused women as plausible and "reasonable" defensive actions.

Hamilton focused on a presumably later stage of development in that general domain. She analyzed judicial uses of expert testimony in sixty-two appellate opinions presented in California between 1996 and 2004, particularly those in which the expert testimony was admitted "under a special evidentiary statute in a prosecution of a male abusing his female partner" (Hamilton, 2009, p. 61). California was unusual at that time for this type of testimony since in 1990 its legislature broadly guaranteed admission of such testimony in all domestic violence cases. While legislatures in other states authorized such testimony only as part of a defense by a woman accused of murdering her abuser—and appellate courts in other states used case law to allow testimony through case law (e.g., *State v. Koss*, 1990)—California was the first state to change evidence law to allow such expert testimony for all cases in which domestic abuse was alleged. Hamilton's data sample of such testimony use documents the evolving relationship between the testimony and changing legal standards.

3. PRESENTATION AND EXAMINATION OF TRANSMUTED EXPERTISE

In order to give effectiveness to knowledge, people possessing specialized expertise testify in courts of law about scientific theories, patterns of behavior, and occasionally to justify medical diagnoses. In doing this, designated experts endeavor to make a kind of translation. Things that may be understood and discussed in one way among other experts of the same domain, perhaps more exactly, perhaps more technically, are presented in a court of law in such a way that ordinary non-specialists, primarily judges, lawyers, and members of a jury can understand and accept the primary claims. This act of translation requires all parties involved, the experts and non-experts, to find a middle ground, where some shared standards of validity can be applied to the matters in question. The parties must speak in an intermodal language that realizes, as Hans-Georg Gadamer put it, a "fusion" of linguistic and evaluative horizons (1997, p. 302). As the following analysis shows, the dimensions of expert testimony that broadly explain the acceptance and uses of expertise in the sample opinions are allusions to professional consensus, selective indices of methodological validity, and illustrations of political and figurative cogency.

3.1 Criterion of General Acceptance

A common measure for admitting expert testimony is the prevalence of consensus among specialists. Insofar as a party or witness can show unanimity of opinion within a scientific or technical community, then the court generally may accede to the presentation of that evidence, that is, if the testimony also meets other criteria for admissibility, such as usefulness to the jury and relevance to one of the legal questions being considered. In one sense, this focus on possible scientific consensus is a legacy of the so-called *Frye* (1923) or "general acceptance" standard. It also may reflect a developmental model of scientific inquiry, alluded to by Harry Collins and Robert Evans (2007) in the distinction between disputed and consensual science (pp. 20–21). Thomas Kuhn's well-known account of paradigm shifts in the development of science similarly reinforces this notion that divisions created by a revolutionary paradigm shift are eventually subsumed again in a newly stable comprehensive theory (1996). The criterion of unanimity serves the court as a sign of scientific validity and maturity.

The bane of a disputed science also has been used to disqualify marginal knowledge claims at the point of accepting or rejecting possible expert testimony. In the original *Frye* case, the District of Columbia Court of Appeals upheld the trial court's exclusion of testimony by which an expert offered to explain how a blood pressure test could be used to detect truth and lies. The criterion used to exclude that testimony in *Frye* (1923), and increasingly cited for that purpose since the 1970s, was that the scientific principle or discovery used as a basis for a deduction should be "sufficiently established to have gained general acceptance in the particular field in which it belongs" (as cited in Lyons, 1997, para. 4).

Following that prevailing *Frye* standard for admitting expert testimony, the Ohio Supreme Court first had ruled "no general acceptance of the expert's methodology [on BSW] had been established" (*State v. Thomas*, 1981, p. 521). Nine years later, however, the same court explained that "since 1981, several books and articles have been written on this subject" (State v. Koss, 1990, p. 214). The opinion further explained that "[i]n jurisdictions which have been confronted with this issue, most have allowed expert testimony on the battered woman syndrome (p. 214). The court's analysis suggested that over time a tipping point had been achieved, at which point the theory, together with its investigative methodology, was no longer characterized as disputed but instead as established science.

State v. Henderson (2011) too acknowledged an emergent scientific consensus about factors that may undermine the reliability of eyewitness identifications. Over the course of a ten-day remand hearing, a Special Master heard testimony from seven experts in the field and reviewed 360 exhibits, including more than 200 scientific studies on human memory and eyewitness identification. The eventual opinion considered particularly important the results of meta-analyses, which looked for statistical trends in all available results for certain types of studies. These meta-analyses, of which there were more than twenty-five, allowed a look at the degree of consensus across studies. The advantage of this kind of study, according to the court, was that "[t]he more consistent the conclusions from aggregated data, the greater confidence one can have in those conclusions" (p. 29). Furthermore, in two sections near the end of the opinion, the court described the degree of consensus the Special Master found among scientists, expert witnesses who testified in person at the hearing, and people beyond the scientific community who were engaged in related law enforcement and reform efforts (pp. 43–44).

3.2 Fidelity to Ideal Models of Scientific or Diagnostic Methodology

The evaluative standard of consensus, however, breaks down under conditions of shared mistake and habitual neglect. As the U.S. Supreme Court eventually noted in *Daubert v. Merrell* (1992), the "general acceptance" test for scientific knowledge does not independently justify exclusion of opinions and studies that, while not yet widely accepted, still represent valid findings. Sending the case back to a lower court for rehearing, the Court charged judges to evaluate scientific evidence based on specific criteria for scientific validity. For example, the opinion instructed judges to evaluate the question whether a study tested a theory that was in fact falsifiable, as well as whether it reported or otherwise quantified the margin of error. While it does not appear that courts as a whole have excluded or included more expert testimony since that decision (Buchman, 2007), many have hailed the decision as a procedural turning point, one that might embolden judges to be more active and articulate gatekeepers in qualifying and disqualifying testimony by experts (Gatowski et al., 2001).

Melissa Hamilton (2009) showed how external expertise was employed by trial and appellate courts at a time of self-conscious procedural transformation. She surveyed a broad set of appellate opinions in the State of California since the law had been changed to allow experts on domestic abuse to testify in trials, not only, as the original law had allowed, when the abuse victim defended herself against a charge of murder or assault, but also when the abuse victim spoke in court as a witness in a case against her alleged perpetrator of that abuse. Hamilton examined the differences that emerged in the uses of expert testimony in those cases, as well as the criteria applied in evaluating the validity and relevance of that testimony.

The opinions Hamilton reviewed frequently noted calculations of probability that experts reported. These statistical figures represented a tangible measure, an epistemic product that the testimony in given cases yielded. In separate instances, the judges affirmed that a particular claim had been shown by a named expert to have a given percentage of probable truth. These figures thus were implicit quantifications of likely error. Whether or not this probability of truth had any relevant relationship to levels of certitude required for different charges (e.g., "beyond a reasonable doubt" or a "preponderance of evidence") was not possible to discern from the opinions. Still, the regularity with which the opinions alluded to indices of probability suggested that the figures served an epistemic function in representing factual validity.

Nonetheless, when these opinions are read in aggregate, the seeming crispness of the probability calculations was fuzzier than in individual reports. Some testimony focused on the likelihood that battered women recanted earlier accusations they had made. Estimates ranged from "around 50" to 85 percent. In two instances, the very same experts testified to different rates of recantation in different trials. Although Hamilton granted there may have been some operational difference between "refusal to cooperate" and "becoming uncooperative," she still was troubled by the variation of percentages, as well as the lack of background information retained in the record to help explain the discrepancies (2009, pp. 102–103). Hamilton also found significant discrepancies in the reports on the number of times women typically tried to leave a relationship before successfully doing so. Some opinions reported that number to be five to seven times, one as three to five times, and another as an average of five times (2009, p. 111).

Particularly striking to Hamilton in her review of these opinions was the lack of attention given to issues of validity and methodology in assessments of this expert testimony. She wrote:

> Many courts report the expert testifying about the prevalence of recantation, denial, and minimization (e.g., commonly, frequently) without providing any research support for their characterizations, even when citing specific statistics (e.g., 71% of battered women recant). . . . As a sociologist, I clamor for more details regarding the source material, such as the date any study was done, the research methodology used, the identity of the primary investigators, or any other information that could reveal bias or overgeneralization. (Hamilton, 2009, p. 110)

From the perspective of an outsider looking in, the appellate opinions seemed unreflective at best and even cavalier in how minimally they summarized and evaluated scientific knowledge about psychological phenomena.

Recently, a district court in Colorado used performance criteria to the evaluate the testimony of a renowned expert on BWS, Dr. Lenore Walker (*State v. Midyette*, 2011). This case differed from the set of opinions Hamilton reviewed since the appellate court itself heard the expert testimony and evaluated its validity. The case also was different in that the expert offered an opinion that the claimant suffered from the syndrome. In other words, the expert presented a particular diagnosis, rather than just a description of a more general phenomenon or scientific theory. In evaluating the validity of that testimony, the court cited a competing expert, Dr. William Hansen, who had been introduced by the prosecution. According to Hansen, a forensic interview and evaluation of the type Dr. Walker had conducted required "more than a client's self report" (*State v. Midyette*, 2011, p. 8). Citing guidelines described by Hansen, the court noted:

> Evaluators ask attorneys for information, but it is the evaluator's responsibility to get whatever is missing, particularly prior mental health history. He noted that Dr. Walker did not have Defendant's therapist's notes and was not even aware that Defendant had been in counseling. (*State v. Midyette*, 2011, p. 8)

In rejecting the expert's diagnosis, the court noted the limited observations Walker had made of the claimant, the mixed answers she gave about the method for scoring personality tests, and the extent to which she had allowed the claimant's lawyer to edit and shape her report.

3.3 Figurative Cogency and Political Knowledge

Figurative cogency refers to a form of cultural accretion in which a symbolic construction works in a communicative locale as a vehicle for summing up a perception or judgment. For example, in describing a "geographics of identity," Susan Stanford Friedman characterized a type of experience in which elements of "cultural hybridity" attain a "material reality[,] political urgency[, and] figurative cogency" (2000). Such an identity-forging construction also describes the process by which external expertise, once admitted to the court room and transmuted into actual and prospective meaning, enters the language of the court and subsequent appellate guidance.

In her analysis of appellate opinions in California on domestic abuse cases, Hamilton (2009) noted the regularity and consistency with which the opinions adopted metaphoric language from the expert testimony, as well as the ease with which metaphoric language came to represent factual reality in descriptions of experience and patterns of behavior. The most common metaphors were accounts of "power and control," the "cycle of violence," a "honeymoon period, and "window of opportunity" (p. 112). Often the court summarized how certain actions or patterns of behavior by a man reinforced an expectation for a special "male privilege" and a corresponding female responsibility. For example, one cited expert explained how abusive behaviors served a man's "need to feel like the 'king of the castle'" (p. 114). Another testified that the man's "masculinity is based on the extent of power and control over his female partner, whom he sees as his property" (p. 114). These illustrations of the drive for power and control were often associated with the linked metaphor of a cycle of violence, which characterized the repetitive pattern that conditioned and propelled the parties' behavior and responses.

Striking about Hamilton's finding is the degree to which these essentially metaphoric resources stood out in the legal record as higher profile traces of the testimony's legacy in the law than any assessment of the science or diagnostic accuracy of that testimony. In the set of opinions Hamilton studied, "the court opinions were far more likely to embrace a definition [of BWS] originating in a legal precedent . . . [than by invoking Lenore] Walker or other authority external to the law" (2009, pp. 78–81). In a similar and parallel way, the opinions generally avoided describing the situations in the cases as "psychological conditions" or by using "clinical terminology" (2009, p. 81).

Hamilton (2009) offered two explanations for this preference for testimony elements that either cited precedent or represented figurative cogency. The first was that they provided a mechanism for understanding the possible reasonableness of the beliefs held by and actions taken by the involved parties. She noted that "judicial writings utilize these particular phrases and often expound upon them to create judicial knowledge about the dynamics of abusive relationships and, more particularly, to account for the women's seemingly vulnerable emotional states and inexplicable behaviors" (p. 117). The metaphoric resources cited from the expert testimony thus serve as a kind of bridge between worlds. It helps judges map, as Friedman (2000) might say, an otherwise unfamiliar and hybrid terrain.

In addition to this function of epistemic mapping, the prominence of these interlinked metaphors also served to represent for the court a context for possible legal or other remedial action. Experts used the term "window of opportunity" to refer to a period of time that recurred within the "cycle of violence," during which a battered woman could extricate herself from the cycle and from her relationship with the abuser (Hamilton, 2009, p. 117). The significance for the court in this "window" was that, according to the experts' recounted narratives, the woman's ability to use that period of time to escape the cycle often depended on encouragement and support of others who could assist the woman in breaking the cycle (pp. 117–118). The metaphor thus served as a framing device to underscore the possible importance of external actions, as a verdict might be, that could hold a perpetrator accountable.

Extensions of expert testimony, particularly in cases concerning BWS, can also substantially transform the character and meaning of underlying law. Robert Mosteller (1996) observed "political influence" in how courts used and built on social framework or "group character" evidence. Of course, in several cases, political bodies explicitly approved statutes in evidence law to guarantee BWS testimony in specific kinds of cases. At the same time, though, other state courts expanded this right to plaintiffs and witnesses for defenses other than coercion, and in ways quite distinct from the more restrictive rulings courts have made in other legal sub-domains, such as the potentially parallel issue of potentially unreliable eyewitness identifications (1996, pp. 485–491).

The plausibly political influence in how expert testimony has been admitted and used of course may not necessarily be a bad thing. Mosteller (1996) characterized this influence not only as a form of pressure and advocacy by interest groups, but also as a modality in which a "moral component" is "integrated and incorporated into the law" (pp. 465-466, note 15.) Mosteller wrote:

> The broad political consensus is both that social reality of the battering relationship is badly imbalanced and that the legal process has not appropriately responded to self-help violence by women. As a result, and despite scientific uncertainty about the existence of a true syndrome, the judgment is that jurors should nevertheless receive such evidence to help redress the imbalance. (1996, pp. 490-1)

The selective use of expert testimony about group character thus appears to institute a substantive and normative change in the processes and standards applied. It represents a change in the scope and force of law, primarily to solve a widely acknowledged social and adjudicatory problem.

4. CONCLUSION

The appellate opinions reviewed here show both openness and caution about the uses of external expertise by courts to frame and construe factual assessments. Concerning the matter of domestic abuse and its possible long-term effects on victims, the opinions readily adopted metaphoric representations of the violent relationships and consistently relied on definitions and explanations of the phenomenon that reinforced through cited precedent to stand as provisional legal categories. In the matter of potentially prejudicial eyewitness testimony, the cases studies showed an opposite tendency: extreme caution, even aversion, to any blanket presumption for expert testimony at trial when such concerns are raised. Instead, in the one specific case reviewed here (*State v. Hendersen*, 2011), the New Jersey Supreme Court held a specialized hearing that re-evaluated scientific research on the subject and, finding inadequacies in the prevailing legal standard, recommended procedural adjustments in pre-trial hearings and new directives for model jury charges.

The opinions showed significant deference for generalizations and causal theories presented by the subject matter experts. In part, this caution may have been due to the exacting "abuse of discretion" standard used for reversals of trial court decisions on controversial testimony. Still, the opinions also accepted many elements of the testimony at face value. This was particularly true of metaphoric renderings of reality that were consistent with familiar precedent, previous expert testimony, and political consensus about legal problems and possible solutions. While the opinions referred to various standards of scientific reliability, for the most part they did not delve meaningfully into matters of methodology or research design.

Even so, in spite of this tendency to grant scientific expertise a general and autonomous credibility, the opinions also fiercely guarded other knowledge prerogatives for judges and triers of fact. This reluctance to cede adjudicative authority was most striking in *State v. Haines* (2006) and *State v. Midyette* (2011). Yet it also animated the strictly procedural refashioning of the legal standard in *State v. Hendersen* (2011). While that decision left trial courts the discretion in cases where witness reliability was challenged to allow expert testimony at the request of parties, the decision did not suggest case-specific expert testimony as a possible or preferred course of action. When the question was particular, as to whether a given witness was reliable or a particular defendant a recipient of abuse, the courts appeared to tighten the standard of admissibility for expert opinion. Overall, the court was most willing to hear and admit expert testimony on scientific questions if it could extract and retain figuratively cogent traces of the testimony, in the form of precedential legal accretion, while guarding against practices in the testimony that could potentially displace or usurp the voices of actual parties and witnesses of fact.

REFERENCES

Buchman, J. (2007). The effects of ideology on federal trial judges' decisions to admit scientific expert testimony. *American Politics Research, 35*, 670–693. doi: 10.1177/1532673X07302339.

Collins, H. & Evans. R. (2007). *Rethinking Expertise.* Chicago, IL: University of Chicago Press.

Cutler, B. L., Dexter, H. R., & Penrod, S. D. (1989). Expert testimony and jury decision making: An empirical analysis. *Behavioral Sciences & The Law, 7*(2), 215–225.

Friedman, S. S. (2000, September). Locational feminism: Gender, cultural geographies, and geopolitical literacy. Paper presented at 4th European Feminist Research Conference, Bologna, Italy. Retrieved from http://www.women.it/quarta/workshops/literatures7/sstanford.htm

Frye v. United States of America. (1923). 54 App. D.C. 46, 293 F.1013.

Gadamer, H. (2004). *Truth and method* (2nd rev. ed.). (J. Weinsheimer & D. G. Marshall, Trans.). New York, NY: Crossroad.

Gatowski, S. I., Dobbin, S. A., Richardson, J. T., Ginsburg, G. P., Merlino, M. L., & Dahir, V. (2001). Asking the gatekeepers: A national survey on judging expert evidence in a post-*Daubert* world. *Law and Human Behavior, 25(*5), 433–458. Retrieved from JSTOR database.

Hamilton, M. (2009). *Expert testimony on domestic violence.* El Paso, TX: LFB Scholarly Publishing.

Hamilton, M. (2010). Judicial discourses on women's agency in violent relationships: Cases from California. *Women's Studies International Forum, 33*(6), 570–578. doi:10.1016/j.wsif.2010.09.007

Harris, R. C. (2008). *Black robes, white coats: The puzzle of judicial policymaking and scientific evidence.* New Brunswick, NJ: Rutgers University Press.

Kuhn, T. S. (1996). *The structure of scientific revolution* (3rd ed.). Chicago, IL: University of Chicago Press.

Lyons, T. (1997). Frye, Daubert, and where do we go from here? *Rhode Island Bar Journal, 45*, 5–12. Retrieved from Lexis-Nexis Academic Universe database.

Miller, R. D. (2003). Testimony by proxy: The use of expert testimony to provide defendant testimony without cross-examination. *Journal of Psychiatry and the Law, 31*, 21–41.

Mosteller, R. P. (1996). Syndromes and politics in criminal trials and evidence law. *Duke Law Journal, 46*(3), 461–516.

People v. Midyette. (2011). Retrieved from http://www.thedenverchannel.com/download/2011/1122/29828345.pdf

State v. Haines. (2006). 112 Ohio St.3d 393.

State v. Hendersen. (2011). Unapproved syllabus. Ohio Supreme Court, Office of the Clerk. Retrieved from http://pdfserver.amlaw.com/nj/Henderson-A8-08.pdf

State v. Koss. (1990). 49 Ohio St. 3d 213.

State v. Thomas. (1981). 66 Ohio St. 2d 518.

Stubbs, J. & Tolmie, J. (1999). Falling short of the challenge? A comparative assessment of the Australian use of expert evidence on the battered woman syndrome. *Melbourne University Law Review. 23,* 709–747.

Terrance, C., Thayer, A., & Kehn, A. (2006). Undermining eyewitness confidence inflation: Effecting change through expert testimony. *Journal Of Forensic Psychology Practice, 6*(1), 73–82. doi:10.1300/J158v06n01_05

Reason, Values and Evidence: Rational Dissent from Scientific Authority

BRUCE GLYMOUR

Department of Philosophy
Kansas State University
201 Dickens Hall
Manhattan, KS, 66506
USA
glymour@ksu.edu

SCOTT TANONA

Department of Philosophy
Kansas State University
201 Dickens Hall
Manhattan, KS, 66506
USA
stanona@ksu.edu

ABSTRACT: We argue that value-based dissent from scientific consensus need not be irrational, as is often supposed. Instead it may commonly be a rational response to information which, if accepted, induces a conflict in core values. We briefly survey normative theories of rationality, drawing specific attention to the role values play in those theories. We then characterize the conditions under which it is rational simply to reject the contextual facts generating conflict among values. We close with some observations about the values to which science communicators may effectively appeal without relinquishing scientific authority.

KEYWORDS: auxiliary hypotheses, framing, means-ends rationality, methodological naturalism, public understanding of science, rational choice, rational dissent, science and values, science denial, science communication.

1. INTRODUCTION

In this paper we briefly survey the relation between values, evidence and rational belief revision. Though incomplete, even a brief survey warrants some conclusions, among them this: what counts as evidence, and what counts as rational inference from the evidence, depend essentially on prior value commitments. That conclusion, in turn, has implications for science communication. First, it illuminates the basic logic of the role of values in science communication. Second, even scientists and journalists who recognize the importance of values in effective communication often frame that importance by appeal to the 'irrational' aspects of 'science denial.' But in fact there need be nothing irrational about such dissent when the relevant science induces a conflict between sufficiently important values. Finally, as a consequence, such values should not be used to frame science communication.

Glymour, B., & Tanona, S. (2012). Reason, values and evidence: Rational dissent from scientific authority. In J. Goodwin (Ed.), *Between scientists & citizens: Proceedings of a conference at Iowa State University, June 1-2, 2012* (pp. 141-150). Ames, IA: Great Plains Society for the Study of Argumentation. Copyright © 2012 the author(s).

2. REASON: A SHORT HISTORY

Philosophy has always been intimately concerned with rationality. There are roughly three kinds of theory about good reasoning in philosophy: theories of logic, theories of scientific method, and theories of rational choice. Values play an ineliminable role in all three kinds of theories. Unfortunately, the role of values in logic and in theories of scientific method is often obscured. To uncover the way in which values inform both logic and science, it will be useful to attend to their more explicit role in Rational Choice Theory.

2.1 Values in Rational Choice

Values matter in rational choice theory twice over. A theory of rational choice is, nominally, a theory about which *actions* are rational and which are not.[1] To reach such a determination a rational choice theory must stipulate a decision rule—a rule for determining of each possible action whether it is or is not rational. Examples include such rules as *maximin*—the rational action is the one with the best worst possible payoff, *maximax*—the rational action is the action with the best best possible payoff, and *maximize expected utility*—the rational action is the action which has the highest expected utility. But any application of such a decision rule requires input; loosely, part of that input must be a set of valuations for each action. The valuations arise as a function of the utility of the outcomes each action makes possible. The utility of the outcomes is in turn derived from a preference ordering over all possible outcomes. This preference ordering is an expression of the values of the agent whose actions are to be modeled. Because utilities express values—preferences, desires, aims—their role in rational choice theory is both explicit and ineliminable.

But values enter in a second way too, namely in the choice of decision rule. So, for example, in some contexts (decision under uncertainty with no dominant strategy), it may intuitively make sense to maximin *or* to maximax, depending on how risk-averse the decision maker is. If some actions have possible outcomes that are regarded as entirely unacceptable, while on other actions those unacceptable outcomes are impossible, it may make sense to play it safe, i.e., to avoid risk by using the maximin decision procedure. This is essentially the idea behind the so-called 'precautionary principle' (see, e.g., O'Riordan & Cameron, 1994). On the other hand, suppose in a particular decision context, no action can produce outcomes that are truly disastrous, but all of the actions that produce the very best possible outcomes also risk the worst outcomes. In such contexts, it may intuitively make sense to maximax—to risk a (relatively) bad result in order to have a chance at getting the overall best possible result. As it turns out, there is *no* theory of rationality that can decide when and where one should be risk averse (i.e., choose using maximin reasoning) or risk seeking (i.e., choose using maximax reasoning) for decisions under ignorance (i.e., when the probabilities of each outcome conditional on taking a given action are not known) (Luce, 1959/2005).

The choice between risk-aversion and risk-tolerance, between pessimism and optimism, is a choice of *values*, and as such theories of rational action cannot be brought to bear on it; rather, theories of rational action require such choices as input. Said another way: it only makes sense to talk about the rationality of an action relative to a whole host of prior

---

[1]   Within the decision theoretic framework, decisions about what to believe are themselves actions (see e.g.,, Levi, 1991).

decisions concerning what one cares about—without knowing what your aim in acting is, how you value different outcomes, and how much risk you are willing to bear, one cannot say very much of anything about the rationality, or irrationality, of your actions. As a philosophic aphorism: values are *prior* to rationality.

2.2 Values in Logic

Curiously, values matter essentially in logic as well. To see why, a little bit of logical terminology is useful. A logical theory is a theory about what inferences are and are not 'valid,' i.e., to be countenanced as good. Generally the standard for goodness in logic is 'truth preserving': a logical theory should count as valid only arguments such that if their premises are in fact true, it is not possible that their conclusions should be false. In order to classify arguments as valid or invalid, logical theories have to be formalized. Technically, then, a theory or system of logic is a triple of things: a formal language, a model theory or semantics for the language, and a proof theory, or set of rules (inference rules and axioms) defining what does and does not count as a valid (i.e., good) inference in the logic.

Because there are lots of possible languages, semantics and proof theories, there are lots of different logics. A standard first course in logic introduces students to two theories in logic, namely Propositional Logic and Predicate (or Quantifier) logic. As it turns out, Propositional Logic can be embedded in Predicate logic; consequently it is not necessarily obvious that the theories are really all that different. But these two theories do not exhaust the logics that have been developed, championed, and put to good use by logicians, philosophers, computer scientists and mathematicians.

So for example, Propositional and Predicate logics cannot be used to express modal claims—claims about what is possible and/or necessary. If one wants to know which inferences about possibilities and necessities are valid, one requires a Modal Logic. For example, inferences from the fact that one person has a right, say a property right to a manuscript, to the fact that some other person is thereby obliged not to act in particular ways, e.g.,, not to copy the manuscript without permission, can be shown valid only using a Modal Logic. There are many Modal Logics, and they are not simply extensions of one another. For example, two famous modal logics are the systems D and S5; each is useful but they are not compatible. And in fact the variety of modal logics are but a tithe of the available logics, each useful for its own purposes, but not for others.

The implication is straightforward. There are lots of logics that have, provably, the qualities desired of a logic (they are complete, sound and consistent), but which are not reconcilable—they differ about which arguments are and which are not rational inferences. Many of them are useful—for some chores Predicate Logic is a good choice; for others a Modal Logic such as D, for yet others S5, and for yet others an intuitionistic logic, a quantum logic, a fuzzy logic, or some even more esoteric logic is required. No one logic is 'right,' so far as we know—while each logic defines how to judge 'good reasoning' within the logic, such standards apply only after one has chosen the logic, and no logic determines standards for how to choose which of those logics to adopt in the first place. Instead, the choice between alternative logics is a matter of pragmatics—it depends on what you want to do with the logic; i.e., it depends on prior commitments about what you value. Again, values are prior to reason.

2.3 Values in 'the' Scientific Method

Empirical inferences, inferences from data to theory in science, are also grounded in values in much the same way that a choice among logics or decision rules are grounded in values. So for example the statistical methods that it makes sense to use when building a model of some phenomenon—population size in a managed species, the influence of television on aggressive behavior, the causes of recessions, or what have you—depend on the way in which the model will be used. One might simply want an efficient representation of the data, and in this case it might make sense to use factor analysis or regression methods to construct the model. Differently, one might want to use a model to predict what will happen in an undisturbed system. In this case maximal likelihood methods such as AIC inference recommend themselves. Neither sort of method is particularly good at recovering causal dependencies, however, and so if one wants to use the model to choose policies that will control some outcome—population size, aggression, or the unemployment rate—yet other methods are required (Pearl, 2000; Spirtes, Glymour & Scheines, 2000). Inference methods are claims about reasoned belief formation, and so again values are prior to reason.

But there is another way in which scientific inferences are grounded in values. Science is aimed to produce various useful products, the most central of which are theories. Theories can be understood in a variety of ways, but whatever else it may be, a scientific theory is a story about how the world works. Some of us, scientists and consumers of science alike, demand truth of theories because theories sustain scientific explanations, and we are interested in finding true explanations of patterns in the observable world. For like-minded scientists and their consumers, the one overriding aim of science is the reliable discovery of true theories from observations of the natural world. Note that this aim, like any other, is a value commitment in disguise—to adopt an aim is to do endorse certain sacrifices as warranted by the goal, i.e., as costs worth bearing in order to achieve the aim. To adopt an aim as overriding is to endorse all possible sacrifices as costs worth bearing to achieve the aim.

Science is not the only human endeavor to produce stories about how the world works, but science claims for itself a special place. The familiar justification for this privileged status is that the methods of science are the most reliable methods for producing true stories about the machinery that generates observable phenomena. This reliability is in turn justified on the ground that science, uniquely, tests its theories against empirical observation. That is true, but also incomplete. The basic problem is that the number of alternative theories that might possibly account for any domain of phenomena (e.g., motion or heat or adaptation) is literally infinite. Assuming that at most one of these theories is true, science cannot hope to discover this true theory if it proceeds by testing theories one by one. A better image is that of a filter or sieve: scientists use empirical tests to filter infinitely large classes of theories, winnowing good theories from bad and retaining only the contenders that are consistent with the available data. Typically, these remaining contenders will disagree about many fundamental things. But often enough they agree about a subset of claims. What we learn from the data using scientific methods are those claims on which all surviving theories agree.

The filtering process requires more than just data and alternative theories. Philosophers of science call these extra ingredients 'auxiliary hypotheses.' There are broadly two kinds of auxiliary. One type of auxiliary is used to connect theory to data in the following way. The theory will say that under certain conditions (e.g., that a particular trait is a heritable cause of survival and reproductive success) a certain outcome (an increase in the frequency of that trait) is to be expected. But generally one cannot directly observe whether or not the specified conditions are satisfied (e.g., whether a trait is heritable or whether it causes reproductive success); instead, in any particular study one must assume that the conditions are satisfied. Often these assumptions are themselves testable, using different data and yet further auxiliary assumptions. But not always: there are some auxiliaries that cannot be tested. These auxiliaries are instead justified pragmatically.

To see the pattern and force of a pragmatic justification, consider Glymour's youngest son Johannes, who is bent on becoming a world-class soccer player. From one perspective, Johannes has chosen an unwise career goal. The odds against him are long, and in any serious attempt to succeed he will likely compromise his abilities to pursue other, more easily achieved, goals. Johannes, however, is willfully blind to the long odds he faces: he simply refuses to recognize that it might be impossible for him to become a professional soccer player.

Curiously, he is in so doing behaving in a completely rational way. For whatever chance he has, that chance depends essentially on lots and lots of practice playing soccer. And it would be psychologically impossible for him to devote the necessary time and effort to practice if he believed he had no chance of success. Whatever the truth about his chances now, Johannes would be doomed to failure if he took seriously the fact that he almost certainly doesn't have the genetic endowment for a career as a professional athlete. Given that soccer is Johannes' heart's desire, his overriding aim, it is perfectly rational for him to ignore the facts, and proceed to practice.

Such pragmatic justifications are important, because they are the only kind of justification that can be given for a second class of auxiliary assumption. These assumptions are needed because among the infinity of alternative theories through which a scientist must sort, there are some that are 'sticky.' If one puts them into the scientific sieve, they turn out to be untestable. And what is very much worse, they make it impossible to test any other theory as well. These theories gum up the works.

If the central, most fundamental aim of science is reliable discovery of true theories from observational data, then there is a pragmatic justification for ruling out such sticky hypotheses, more commonly known as 'skeptical hypotheses.' The most well known of the anti-skeptical auxiliaries is 'methodological naturalism.' Methodological naturalism rules out, from the very beginning, any theory about the machinery producing observed patterns that appeals to supernatural causes. The assumption is made for the very good reason that if it is not made, one cannot learn from the data. This is because any theory can be made consistent with any pattern in the data, provided one assumes an auxiliary appealing to the local action of a non-natural cause (e.g., the interventions of supernatural beings). If we do not rule out such alternatives from the very beginning, if we allow appeal to supernatural causes, *every* theory can be made consistent with *any* data whatsoever, and it becomes impossible to learn from the data. Importantly, this is true even if the skeptical hypotheses are in fact false—they do not have to be true to gum up the works. Just as Johannes' only hope of becoming a pro soccer player depends both on his actually having enough talent *and* on his assuming that he does (because only then will he practice enough to develop that talent), our only hope for reliable discovery in any given domain of observation is that first, no supernatural causes influence the observations, and second, that we assume this is true (because only given that assumption will our sieve work). So if reliable discovery of the truth from observational data is our overriding aim, then methodological naturalism is justified, but only by appeal to that overriding value. Again, values are prior to reason.

What does it mean to say that discovery is an overriding value? In the context of scientific inference, we can specify that priority quite precisely. First, this value determines which claims stand in need of *evidential*, empirical, warrant, and which do not. Second, this value determines which *inferential methods* are applicable and which are not. Third, because different methods will countenance different subsets of the data as evidence, this value determines which data are and which are not *evidence*. And finally, because different methods applied to different evidence will yield different conclusions, this value determines which conclusions, which *beliefs about how the world works*, are and are not rational (cf. Kitcher, 1995).

Science yields explanatory stories about how the world works, about which human activities are changing the climate or causing extinctions and so on, because science can tell us the truth about what is happening, why it is happening and the likely consequences of various changes in social behavior. But rational belief in what science says depends on accepting the reliable discovery of the truth as an essential aim. There is a corollary: *to the extent that we value something, anything, more than we value reliable discovery of the truth, it may be perfectly rational to reject scientific truths.* (cf. Quine, 1951) In effect, this is what Johannes does when he persists in his beliefs about his future as a professional soccer player. And he is not irrational in such persistence. This sort of rational justification of science denial is available more broadly, but to see why we need a more careful discussion of theories of value.

3. THEORIES OF VALUE

As a working hypothesis, stipulate that for each one of us there is at each time a set of facts about what we care about—the outcomes such that we think, all else being equal, it would be good if they occurred, and the sacrifices we would be willing to make to secure those outcomes. A descriptive theory of the values held by a person at a time is a systematic description of those facts. In the worst case, such a theory is no more than a long conjunction of such facts: Boris would prefer better public health and more single malt, he is willing to pay higher taxes, even on single malt, to secure the first aim, and is willing to pay up to but not more than his disposable income for a bottle of single malt to secure the second aim, and so on. But potentially there are simpler descriptions in terms of principles that systematize those facts: Boris prefers better public health to single malt, and single malt to dollars. Unfortunately, the multiplicity of things we care about—public health, single malt, and dollars, but also world peace and eliminating poverty and saving endangered species and so on—raises a problem in systematizing any description of what we care about. In particular, the relative priority of these many aims is often problematic.

There are three ways in which a pair of values may be related to one another. One possibility is that the values are subject to some common measure, i.e., are commensurable, and hence an exact trade-off can be specified. If Boris is willing to pay as much as but no more than $8 for a six-pack of pale ale and $100 for a bottle of single malt, then he prefer bottles of single malt to six-packs of pale ale by roughly a factor of 12.5. A second possibility is that one value might be incommensurable with another—in effect Boris values public health infinitely more than single malt, because he'd be willing to give up *any* amount of single malt to secure better public health. Philosophers speak here of a *lexical* preference for the first good. Were all goods or all goods but one commensurable, we would be on safe ground. But sometimes more than one good is incommensurable with a third. Boris may care infinitely more about world peace and public health than about single malt and so have lexical preferences for fewer wars and for lower rates of infant mortality over increases in the single malt supply. But how many wars will Boris countenance to secure a 20% decrease in infant mortality rates? Here there may be no clear answer. And that is the third way in which values may be prioritized—we may simply not have ranked them against one another, so that their relative priority is, for us, undetermined.

Values with undetermined priority are familiar ground in ethics, as they provide the stuff of moral tragedy: Is it more important to obey a just law or to honor your family? Is it more important to save lives or not to kill? If one's value commitments are sufficiently clear cut that each of the values one cares about is ranked, either by degree on some common measure or by lexical preference, then these values will not conflict with one another in any circumstance. Those so fortunate as to have completely worked out and prioritized values are said by ethicists to be in 'reflective equilibrium' (Rawls, 1971). But most of us are not in reflective equilibrium—do you have a clear answer to Sophie's choice, to Antigone's dilemma, or to trolley cases in which one must kill some to save many? If not, if you find it difficult to resolve such moral dilemmas, then you are not in reflective equilibrium.

Reflective equilibrium is rare because humans are inordinately good at adopting some values in one domain, and quite other values in different domains, and simply hoping that the domains never overlap in ways that induce conflict. Of course conflicts often do arise, and this is what underpins the possibility of moral tragedy. When such conflicts appear, one solution is to actually prioritize (or reprioritize) the values one has. But this is psychologically quite difficult—which really is more important, the obligation not to kill or the obligation to save lives? If the answer is context dependent, as it must be if sometimes killing is better and sometimes failing to save lives is better, then in which contexts does each value take precedence, and why? It is much, much easier to simply deny that the values conflict. Just so, the initial response of students in introductory ethics classes to any hard case of moral tragedy is, quite commonly, to look for a course of action that realizes, or anyway respects, both values—is there really no other way to save the 5 railway workers except to throw the fat man onto the tracks, derailing the trolley? Preserving the values in this way turns on denying the conflict between them, which in turn depends on finding an 'out,' i.e., changing or expanding the specified facts so that some available action will respect both of the nominally conflicting values.

The need to find an out will be especially pressing in certain circumstances. If the relative priority of two values has not been resolved, we may hold a third value, namely holding both the first two values without actually prioritizing them. Let us call such a third value, i.e., a desire either to avoid prioritizing two other values or to leave their relative priority undetermined, an *equal-importance* commitment. Resolving a conflict between the two initial values, say between not killing and saving lives, often requires that we either reject an equal-importance commitment, or reject as illusory the apparent conflict. When one gives the equal-importance commitment a priority at least as high as that of the conflicting values, the second option will be preferable. This is one way of explaining the fact that many of us are willing to devote a good deal of cognitive labor to finding outs when confronted by a moral dilemma. If one can respect the dead without violating the laws, then the conflict is resolved by showing that the initial specification of the context was false, and the equal-importance commitment is thereby preserved. Importantly, this way of resolving a conflict between values is just a way of denying the (purported) facts about the context in which the conflict arises.

Suppose now that one prioritizes three values over truth: two values that in fact conflict, and a third equal-importance value with respect to them. That is, one cares more about holding onto those two core values, as currently unprioritized, than about believing the truth, e.g., about the specific context in which those values are supposed to conflict. In such circumstances some versions of rational choice cannot be applied, and on them we simply have no grounds on which to judge irrational a decision to reject the facts. Other versions of rational choice theory, arguably, can be applied and yield the result that the rational choice is to reject the facts, to simply deny that the circumstances in which the first two values conflict is actual. On those views of rational choice, science denial is sometimes plainly and fully rational.

4. VALUES, INFERENCE AND RATIONAL DISSENT

Scientists and science journalists have recently begun to be more aware of the importance of values in science communication. If nothing else, there is growing awareness of various 'cognitive biases' such as confirmation bias, assimilation bias and attitude polarization. Even so, it is still common for scientists and journalists to see the operation of such biases as species of irrationality (see, for example, vanden Huevel, 2011; Giberson & Stephens, 2011; Borthwick, 2011). To the extent that science denial is regarded as essentially irrational, the point of appealing to values in science communication will be to remediate this irrationality. But if we are right, this is to mistake the very nature of the beast. Science denial is not (or anyway need not be) irrational—it can arise from equally prioritized, fundamentally important values that are brought into conflict by the scientific facts. If those values are more important than truth itself, then the rational response in such cases is simply to reject the facts.

By way of illustration, there are various conditions under which it will be rational to actively reject Evolutionary Theory (ET) as false. The obvious, and least interesting, case is that of literalism about the collection of writings known as the 'Old Testament.' If one's highest, most central, aim is preserve in oneself a belief that those writings are literally true in every respect, then it will be perfectly rational to reject ET, and indeed a great deal of the rest of science. Just as Johannes is justified in rejecting the facts in order to pursue his most central aim of becoming a soccer player, so too a literalist is justified in rejecting ET. In general, rejection of ET is rational if two conditions are satisfied. First, acceptance of ET must be incompatible with particular aims which one privileges above that of discovering true theories. And second, those aims must take precedence over all other aims, success with respect to which would require believing (or anyway using) the true theory about the origin of biodiversity. If the first condition is not satisfied, one has no reason to reject ET. If the second condition is not satisfied, one has reason to reject ET, but this reason is outweighed, overridden, by even more powerful reasons to accept ET. But if both conditions are satisfied, it is rational to reject ET.

If one is trying to decide for oneself whether it makes sense to accept or to reject a bit of science, one must confront some hard questions. But these questions are not about science, or even about truth at all. They are about values, and in particular, they are about the values one wishes to endorse, to voice, to live by, to embody. Some sets of values are, in fact, inconsistent with acceptance of certain bits of science, while others are not. The choice between these values is not a matter of reason, it is simply a choice. Suppose that for Svetlana, say, the most important, but as yet unranked or equally ranked, values include literalism about the Bible and the physical well-being of her children. Call such values 'core values.' As it turns out, Svetlana's core values are in conflict, and because they are, there is a third possible core value that Svetlana may or may not hold: that of holding onto both literalism and the well-being of her children as equally important (or unranked) core values. If she does not hold this third core value, then she will respond to facts, say about the role of ET in controlling disease, by prioritizing her two core values, thereby removing one from her core. But if she holds the equal-importance commitment as a core value, then the rational thing for her to do is simply reject the facts that generate the conflict—i.e., to deny that ET is true, and hence that it plays any essential role in public health or medical science. Of course, in so doing she is in fact undermining her ability to preserve the well-being of her children. But if she were to accept the facts, either this value or the value of literalism would have to be reprioritized out of the core, and in either case so too the equal-importance value would have to be rejected. The three values can only be co-satisfied if the facts are wrong—and so it is rational for her not only to hope they are wrong, but moreover to proceed with the full belief that they are.

And that is bad news for the science, because in order to accept ET she really does have to accept the fact that ET influences public health and medical science. For the only possible rational motivation for removing a value from the core, i.e., prioritizing some core values over others, is that there is a conflict among the core values, here between the well-being of Svetlana's children and literalism. In this case, if Svetlana is to rationally accept ET, her core values must first change, which means she must first recognize the conflict. But the conflict can be recognized at all only if the relevant facts are recognized! And if the facts are introduced in ways that explicitly or implicitly evoke the conflict, the rational choice for Svetlana will simply be to deny the facts.

How then to generate acceptance of ET, i.e., of conflict-inducing facts? A good first step is to introduce them in contexts where the relevant values are consistent with those facts. That is, by communicating the science using frames that entrain core values which are themselves jointly consistent with the facts being communicated. Even better if all the values implicit in the frame are such that effective pursuit of them demands knowledge of the truth, whatever it may be, because then the goal of reliable discovery of the truth from observation becomes instrumentally co-equal with the core values whose pursuit requires knowledge of the truth. In that context, science speaks with authority, and scientific inference is rational inference.

5. CONCLUSION

In lieu of a summary, we offer some lessons which follow if the above remarks are broadly correct.

(1) To the extent that the facts about a particular decision context force one to reject core values, the rational response is to deny the facts.
(2) Such contexts occur when scientific results force a conflict between core unprioritized values.
(3) Those who deny science for these reasons do so rationally.
(4) Such science deniers will not be swayed by more data or by efforts to improve their reasoning: they are in fact reasoning rationally, and will, rationally, reject new data.
(5) Effective science communication really is a matter of values, not of facts.

(6) Science speaks with authority in contexts in which knowledge of the truth is prioritized, but has no special standing regarding the rationality of core value choices or prioritizations.

(7) To speak both productively and with authority, scientists should use frames that appeal to core values, the effective pursuit of which requires knowledge of the truths being communicated.

(8) To prevent those truths from being rejected, scientists should avoid forcing a choice between science and core values, or among core values.

ACKNOWLEDGEMENTS: This paper has benefited from much discussion with two colleagues, Amy Lara and Jon Mahoney.

REFERENCES

Borthwick, L. (2011, July 1). Communicating rationality to irrational beings (or why climate change communication is so darn hard). OnEarth blog. Retrieved from http://www.onearth.org/blog /communicating-rationality-to-irrational-beings

Giberson, K. W., & Stephens, R. J. (2011, October 17). The evangelical rejection of reason. *New York Times*. Retrieved from http://www.nytimes.com/2011/10/18/opinion/the-evangelical-rejection-of-reason.html

Kitcher, P. (1995). *The advancement of science*. Oxford: Oxford University Press.

Levi, I. (1991). *The fixation of belief and its undoing*. Cambridge: Cambridge University Press.

Luce, R. D. (2005). *Individual choice behavior: A theoretical analysis*. Mineola, NY: Dover Publications. (Original work published 1959).

O'Riordan T., & Cameron, J. J. (Eds.). (1994). *Interpreting the precautionary principle*. London: Earthscan.

Pearl, J. (2000). *Causality: Models, reasoning, and inference*. New York, NY: Cambridge University Press.

Quine, W. V. O. (1951). Two dogmas of empiricism. *Philosophical Review, 60*, 20–43.

Rawls, J. (1971). *A theory of justice*. Cambridge, MA: Harvard University Press.

Spirtes, P., Glymour, C., & Scheines, R. (2000). *Causation, prediction, and search* (2nd ed.). Cambridge, MA: MIT Press.

vanden Heuvel, K. (2011, Oct 26). The Republicans' war on science and reason. *Washington Post*. Retrieved from http://www.washingtonpost.com/opinions/the-republicans-war-on-science-and-reason/2011/10/24 /gIQALl3BEM_story.html

What is "Responsible Advocacy" in Science? Good Advice.

JEAN GOODWIN

Department of English
Iowa State University
310 Carver Hall
Ames, IA 50011
goodwin@iastate.edu

ABSTRACT: Debates over scientists' appropriate contributions to policy-making are prominent in a variety of natural resources fields. The issue is often presented as one of "responsible advocacy." But this framing locks us into a paradox: Scientists who advocate aim to be effective in the policy arena, but by advocating lose their credibility. In this preliminary review of the issue, I argue that we can avoid the paradox by acknowledging a wider range of speech acts structuring scientists' obligations in the policy process. Scientists can advocate–but they can also report, give their assessments, make recommendations, and especially, offer good advice.

KEYWORDS: science-policy interface, scientists, advocacy, advice, credibility, trustworthiness.

1. INTRODUCTION

We need scientists to contribute to public deliberations over wicked problems: wicked problems as local as development in the floodplain here in Ames and as global as the wickedest problem of them all, climate change. Managing the interface between science and democratic policy-making is itself a wicked problem, however, as suggested by the variety of ways that have been proposed to describe what is going wrong. Is the problem that we have too much or too little public participation in decisions that have a technical aspect (Collins & Evans, 2007; Wynne, 2003)? Is it that science is being politicized, and/or that politics is being scientized (Weingart, 2002)? In this paper, I want to focus on yet another approach to conceptualizing what is going wrong between scientists and citizens: an approach that takes advocacy by scientists as either a key problem in—or a leading solution to—getting the nation's wealth of science into the policy process. Scientists, it is argued, should or should not advocate in the policy realm; or if they should advocate, they should only do it responsibly, in accordance with some set of guidelines.

This framing of the problem at the boundary between science and policy is of interest for at least two reasons. First, talk of "(responsible) advocacy" directs our attention to the specifically *communicative* conduct of scientists on particular occasions. This raises the hope that communication scholarship such as that in evidence at this conference may have something useful to say, by way of refining—or challenging—conceptions of advocacy.

Second, "(responsible) advocacy" is a conception of communication at the science/policy interface that is being advanced within scientific communities. Some of the conversation within biology fields will be reviewed below; for now, it may be enough to point to the series of events on "advocacy in science" that have been organized at the national level since 2006 (American Association for the Advancement of Science, 2006, 2008, 2011). The invocation of "(responsible) advocacy" by scientists gives the conception a certain validity; it deserves respect as an attempt by skilled practitioners to articulate the ideals which regulate their communication practice (Craig & Tracy, 1995). Further, the debate over "(responsible) advocacy" opens opportunities for interdisciplinary dialogue. While many scientists may resist learning about theories of science in society put forward by humanities or social science scholars who study science, they may welcome outsiders who can offer increased clarity for conceptions they are already deploying.

In the next section of this paper I sketch the debate as it has occurred in one scientific community, identifying the main reasons advanced for and against scientists' obligation to advocate and the normative standards that have been proposed in order to mitigate some of advocacy's undesirable consequences. I next review what we know about the normative structure of the ordinary communicative activity of advocating. Audiences have strong expectations about what a good advocate will do. In light of these expectations, what is called "responsible" advocacy by scientists will either be taken as bad advocacy, or simply ignored. Does this relieve scientists of the obligation to participate in policy-making? I close by arguing a strong "no." There are many alternative communicative activities through which scientists can contribute to deliberations. Among these, the act of advising stands out as achieving many of the high purposes identified by proponents of responsible advocacy. I conclude that when scientists talk of "responsible advocacy", what they really mean is "good advice."

2. ADVOCACY BY SCIENTISTS: THE STATE OF THE DEBATE IN ONE DISCIPLINARY COMMUNITY

Discussions of advocacy by scientists have occurred in a variety of scientific fields as well as in scholarship on science. In this preliminary survey of the issue, I will focus almost exclusively on discussions within biology fields related to natural resources: ecology, conservation biology, invasion biology, marine biology, wildlife management and forestry, for example. Commitments to—and concerns about—involvement in policy-making have deep historical roots in these fields (Nelkin, 1977), and continue to provoke deep disagreements (Young & Larson, 2011). Their love of the natural world pushes these scientists into policy arenas, especially when they perceive the biodiversity and ecosystems they cherish under imminent threat of annihilation (Barry & Oelschlaeger, 1996; Myers, 1999). They also experience a pull into policy-making from environmental advocacy groups (Kaiser, 2000; Lindeman, 2007) and natural resource managers and regulators (Mills & Clark, 2001; Steel, List, Lach, & Shindler, 2004) who welcome their expertise. The nature and ethics of advocacy has thus been the subject of discussion in dedicated fora at conferences and in journals (e.g., *Conservation Biology* 10.3, 1996; 21.1, 2007; *Human Dimensions of Wildlife* 6.1, 2001; *BioScience* 51.6, 2001). In addition, a series of empirical studies have surveyed attitudes towards and conceptions of advocacy among natural resource scientists and other stakeholders (Gray & Campbell, 2009; Kinchy & Kleinman, 2003; Lach, List, Steel, & Shindler, 2003; J. Scott et al., 2007; J. M. Scott & Rachlow, 2011; Steel et al., 2004; Young & Larson, 2011)

Within this literature, "policy advocacy" is commonly and I believe correctly defined as activity aimed to support a policy proposal (Ehrlich, 2001; Gill, 2001; Lach et al., 2003; Lackey, 2007; J. M. Scott & Rachlow, 2011). It "involves advancing the most convincingly reasoned suggestions for change, informed by defensible, rigorous evidence" (Foote, Krogman, & Spence, 2009).

Those defending the legitimacy of advocacy argue:

- *Pro 1*: that their science is inherently value-laden, and thus that pursuing objective knowledge cannot be separated from advocacy for valued outcomes; pretending anything else will only lead to value commitments being covert and unexamined (Barry & Oelschlaeger, 1996; Noss, 2007; Shrader-Frechette, 1996; for a trenchant defense of a version of this view, see also Sarewitz, 2004, 2012)
- *Pro 2*: that all citizens, including scientists, are required to serve the public good and participate in democratic policy-making (Blockstein, 2002; Kaiser, 2000; Karr, 2006)
- *Pro 3*: that scientists' special knowledge gives them a special obligation to contribute to the common good (Karr, 2006; Lovejoy, 1989; Nelson & Vucetich, 2009)
- *Pro 4*: that scientists as recipients of public support are obligated to contribute back their knowledge to help solve public problems (Foote et al., 2009; Karr, 2006)
- *Pro 5*: that if scientists don't advocate on policy issues, the vacuum will be filled by misinformation from interested stakeholders; further, failure to advocate for change is equivalent to advocating for the status quo (Foote et al., 2009; Karr, 2006; Nelson & Vucetich, 2009)

Those questioning the legitimacy of advocacy, and perhaps even calling for its complete avoidance, argue:

- *Con 1*: that scientific objectivity requires scientists to aspire to a value-free stance (or a stance committing them only to epistemic values like objectivity) which is incompatible with advocacy (Tracy & Brussard, 1996)
- *Con 2*: that scientists who advocate will experience negative consequences, including time lost from research, lowered reputation among peers, and personal attacks by political opponents (Foote et al., 2009; Karr, 2006)
- *Con 3*: that advocacy will tend to corrupt the scientific process, e.g. illegitimately influencing the interpretation of data, either because advocacy will tend to distort the scientist's own reasoning process, or because the scientist will be forced to "keep up appearances" once committed to a specific policy (Aron, Burke, & Freeman, 2002; Kaiser, 2000; Lackey, 2007; Nielsen, 2001; Wiens, 1997)
- *Con 4*: that independent of the actual integrity of their science, scientist-advocates will be perceived as being motivated by personal interests, with the result that their credibility or trustworthiness as scientists will come into question, and indirectly, the credibility/trustworthiness of their field and of the scientific enterprise as a whole (Gill, 2001; Lackey, 2007; Mills & Clark, 2001; Rykiel Jr, 2001; J. Scott et al., 2007)
- *Con 5*: that scientists are poorly prepared to be policy advocates, and are better off leaving that task to professionals (Aron et al., 2002)

The clash between Pro 1 and Con 1, while significant, raises epistemological issues beyond the scope of this paper. Further, it seems to me that scientists and ordinary citizens should be able to figure out what communicative activity is appropriate on a particular occasion, without waiting for philosophers to definitively solve the puzzle of values in the scientific process. In this, as in many other cases (e.g., Goodwin, 2002), skilled communicators must work out practical solutions to (or work-arounds for) theoretical problems. Objection Con 2, as has been pointed out by Nelson and Vucetich (2009) does not have much relevance to a debate over scientists' obligation to advocate; advocacy could still be owed, even if it is hard and painful. Objection Con 3 raises a significant psychological point; but in this paper I will take scientists' avowals of their own integrity at face value. Pro/Con 5, finally, raise interesting empirical questions about who is to blame for the present dismal state of science communication; these are both beyond the scope of this paper, and also tend to advocacy themselves.

It is consideration Con 4 (credibility with the broader public) that has proven to be the most compelling, judging by the number of replies it has attracted. Where Con 3 focuses on the trustworthiness of the science produced by the scientist-advocate, Con 4 focuses on the trustworthiness of the scientist-advocate herself (for this distinction, see Goodwin, 2011)—and in particular, on the trustworthiness of the scientist-advocate manifest to her audience of citizens and policy-makers. One response has been to declare this concern negligible. Scientists are not to blame (it is argued) if the public wrongly perceives them as being biased when in fact they are promoting the public good (Nelson & Vucetich, 2009). But blaming only citizens for the breakdown of trust between scientists and citizens seems—well, a little too convenient a response from the scientific community. More commonly, proponents of advocacy respond to Con 4 by imposing limits on advocacy, so that advocacy when correctly pursued will not in fact threaten scientists' manifest trustworthiness. Principles of "responsible" (Foote et al., 2009) or "honest" (Noss, 2007) advocacy that have been proposed include:

- *RA 1*: scientist-advocates should be fully open about their value commitments, interests, funding sources, etc. (Foote et al., 2009; Meyer, Frumhoff, Hamburg, & de la Rosa, 2010; Nielsen, 2001)
- *RA 2*: scientist-advocates should change their public positions when the evidence demands it (Meyer et al., 2010; Nelson & Vucetich, 2009; Noss, 2007)
- *RA 3*: scientist-advocates should advocate only on topics within their areas of expertise, and/or should openly indicate where they are going beyond their expertise (Foote et al., 2009; Meyer et al., 2010)
- *RA 4*: scientist-advocates should not (like "sophists"—Nelson, 2009) use the most effective arguments for their policy positions; they should use only the best available, peer-reviewed, data-supported science to make their cases (Blockstein, 2002; Foote et al., 2009; Meyer et al., 2010; Nelson & Vucetich, 2009)
- *RA 5*: scientist-advocates should be fully open about uncertainties, margins of error, and caveats (Blockstein, 2002; Meyer et al., 2010)
- *RA 6*: scientist-advocates should bring forward counter-considerations that weigh against the policies for which they advocate (Foote et al., 2009; Nielsen, 2001)
- *RA 7*: scientist-advocates should avoid hyperbole (Blockstein, 2002; Meyer et al., 2010)

Many of these normative principles are attractive. I do wish scientists would heed them, or at least some of them, when engaging with me and other citizens on policy issues. However, we don't expect advocates to follow them, whether scientists or not; and for good reason. The ordinary communicative activity of advocating already has norms. These aren't them.

3. THE NORMS OF THE ORDINARY COMMUNICATIVE ACTIVITY OF ADVOCATING

As consumers in a capitalist economy, members of a litigious society, and citizens in an adversarial democracy, we are quite familiar with advocacy. Our well-being depends in part on recognizing advocacy when we're subject to it, and we are capable of judging advocacy as good or bad. What are the norms we use in making such judgments?[1]

A full analysis of the communicative activity of advocating has not yet been accomplished. In the meantime, a good place to start is with the central normative principle articulated for legal advocates: to represent their clients "zealously, within the bounds of the law" (American Bar Association [ABA], 1983, Canon 7; or, "as advocate, a lawyer zealously asserts the client's position under the rules of the adversary system," ABA, 2004). When we say that an attorney was a "good advocate," we don't necessarily mean that she won her suit. We do mean that she did everything she could, given the facts, laws and procedures, to urge the judge or jury to reach a decision in her client's favor—everything short of outright dishonesties like presenting evidence that she knew was fake or bribing a juror. A zealous advocate is responsible for making the strongest case possible on behalf of the position her client has taken.

The norm of zeal is not imposed on legal proceedings from the outside; it is invoked by the participants in legal proceedings themselves, as I have shown in a study of the closing arguments of the OJ Simpson criminal trial (Goodwin, 2001). The excellent (or at least expensive) advocates there excused the length of the trial by explaining that they were bound to take as much time as necessary to defend their clients, for example. They also argued that when their opponents had not produced evidence in support of one of their points, it must be because there was no such evidence, since their opponents were bound to support their position as strongly as they could. In these and similar arguments, advocates were encouraging the jury to recall and apply the basic presumption that advocates ought to be zealous, attempting to make the best case possible.

We don't necessarily like to be subjected to another's zeal. Zealous advocacy obviously is good for who- or whatever the advocate is speaking for. But when we're the audience of advocacy, why is it good for us? Often it isn't; so it's not surprising that a common response to advocacy is to ignore or resist it. Indeed, a large part of instruction in "critical thinking" consists in helping students recognize and withstand advocacy directed at them.

Why then do we keep advocacy around as a social practice? What use could it be? In some cases, good advocacy allows us to make prudent use of our own scarce cognitive resources. We can, for example, presume that the evidence or arguments are no stronger than the zealous advocate has made them out to be (Klonoff & Colby, 2007). If the best evidence that an advocate can produce is ambiguous, or the best arguments weak, we may be able to dismiss her position efficiently, without ourselves investing in a search for evidence or an

---

[1] I employ here without defending a normative conception of communicative (speech) acts initially put forward by Paul Grice, elaborated by Dennis Stampe and brought into argumentation studies by Fred Kauffeld (Kauffeld, 2009). In this view, in any given speech act the speaker undertakes a specific set of responsibilities to the auditor—undertakings which give the auditor a good reason to respond in the desired way. Every instance of a speech act can thus be seen as creating a local "normative terrain" between speaker and auditor. The speaker's conspicuous undertaking and fulfillment of the local normative requirements is what allows the speech act to be effective; echoing Hegel, for a speech act, "the practical is the normative."

elaboration of our own reasoning. In fact, we can save even more time by making sure that we receive advocacy from both sides. That way we can not only benefit from the advocates making the best cases for us, we can also count on them to make obvious to us the weaknesses in each other's arguments. Advocacy, in sum, outsources some of the work of reasoning.

Still, despite their occasional usefulness, we have mixed feelings about the ethical status of advocates. On one hand, the advocate's norm of zeal is in some ways stronger than the obligation of veracity binding any speaker who says something seriously; as when I tell you that the food at The Café is good, I am not obliged to defend that statement by every means necessary. From this perspective, dedicating oneself to "speaking for" a cause has a certain nobility. But in other ways, the norm of zeal is more limited than the norms of ordinary sayings. In particular, we understand that in advocating a position zealously, an advocate may not believe everything she is saying. As long as she is not saying something that she knows is untrue, she may be putting forward colorable claims that she would not personally endorse. The case she makes has to be good; it does not have to be her own. Indeed, court rules traditionally prohibit advocates from "vouching"—making known their personal opinions on the case—and advocates at the Simpson trial actively encouraged jurors not to believe their say-so (but see Audi, 1995; Goodwin, 2001). From this perspective, "speaking for" a cause has a certain baseness. An advocate is capable of saying for another things that the other cannot truthfully say for him or herself. The accused cannot make claims such as "I wasn't there at the time, and if I was, he hit me first," since he knows which (if either) is true; the advocate not only can, but must—if it would support the plea.

It should be apparent from this brief discussion that the proposed norms of responsible advocacy by scientists are incompatible with the actual norms of the ordinary speech activity of advocating.

RA 1–3 require scientists-advocates to be open about their personal values, to draw only on their personal knowledge, and to defend only positions that they themselves hold after full consideration. Once a speaker has undertaken to be an advocate, however, the advocate's personal values, knowledge and position are irrelevant. She will be presumed to be committed to defending her position, whatever her personal values. She will be expected to seek out all the evidence that will support her position, whatever her personal expertise. And she will be expected to continue to defend her position even when it becomes apparent that other positions have something to say for themselves.

RA 4–7 restrict the kinds of rhetorical techniques scientist-advocates can deploy. While sometimes the best science may also be the grounds for the strongest argument for her position, the advocate is committed to drawing from it not because it is best, but because it helps her make her case. Similarly, an advocate may judge that revealing weaknesses, uncertainties and opposing considerations may help her defend her position (the jury is out on this, in communication theory); if so, she ought to be open about them, but only because it is the zealous thing to do. And as for hyperbole—where would an advocate be without a little of that?

In short, following the proposed rules for responsible advocacy may frequently render the scientist-advocate a bad advocate. Now, that might be a necessary sacrifice, if it helped preserve her credibility and the credibility of her science—that is, if following the rules would reduce the concerns about advocacy by scientists expressed in Con 4. Unfortunately, even open and explicit commitments to the proposed rules will be unlikely to preserve the scientist's manifest trustworthiness. As soon as an audience understands that they are listening to an advocate, they will presume that everything she says is in the service of zealously making her case. Avowals of personal commitment, personal knowledge, personal expertise, of reliance on the best evidence and of full transparency about weaknesses—all will be discounted as attempts to bolster her case. The audience of a scientist-advocate will allow her to be a bad advocate, but won't permit her to be a responsible one.

4. BEYOND ADVOCACY

It would be unfortunate if we were faced with a choice between advocacy by scientists and their silence, since neither serves to get reliable knowledge into the policy process. Luckily, we have other options. "There are many ways" for scientists "to express and act upon values" within the policy process, as one article has put it (Meine & Meffe, 1996); "the notion that a scientist is either an advocate or does nothing at all to shape policy is a false dichotomy that has muddied the debate about science and advocacy" (J. M. Scott & Rachlow, 2011).

Indeed, within the debate in natural resource fields we find already named a variety of other communicative activities scientists could undertake. The empirical work has largely relied upon a five part categorization developed by Steel and his colleagues based on interviews with ecologists (Lach et al., 2003): reporting, interpreting, integrating, advocating, and deciding. Blockstein (2002) identifies interpretation (which he also calls reporting), advice and counsel as alternatives to advocacy; Minnis and McPeake (2001) distinguish education and promotion from advocacy. Even those defending advocacy tend to refer to other communicative activities when discussing the details of what scientists ought to do: informing (Brussard & Tull, 2007), assessing (Meyer et al., 2010; Nelson & Vucetich, 2009), recommending (Meyer et al., 2010; Noss, 2007) and advising (Meyer et al., 2010).

We can appreciate how different these communicative activities are from each other, and from advocacy, by considering a decision context more familiar than the realm of policy-making. Imagine you have been diagnosed with a serious illness. You might look to your doctor to report to you what is known about the success rate of the different treatment options; to offer her assessment of the different options; to recommend the option that in her judgment has proven most successful; to advise you to choose a specific treatment; or to advocate for one treatment. Each of these establishes its own normative expectations–each of them establishes a specific, normatively charged relationship between your doctor and you. Your doctor is undertaking different obligations to you when reporting than when recommending, for example; in reporting, you expect her to be accurate and thorough (and you would criticize her if she weren't), while in recommending, you expect her to employ her best judgment.

In general, we don't want our doctors to advocate; when my dentist did, I began to think he was more interested in getting me to pay for an expensive bit of equipment than in alleviating my pain, and switched practitioners. In ordinary parlance, what we often seek is a doctor's advice. Similarly, when policy-makers seek scientists' help, they often organize them into "advisory committees" and ask them to produce "advisory reports." Let us look briefly then at the normative standards underlying the ordinary communicative act of advising, as developed in the work of Fred Kauffeld (esp. 1999).

Kauffeld started from the presumption that we all should be minding our own business. Every individual is autonomous, with the right to make his or her own decisions on matters of concern to him or her. Occasionally, however, situations arise where someone else may actually know better. As Kauffeld has noted, in these situations "talking to another about that person's concerns is a delicate matter for both parties." The auditor may be legitimately cautious about why the speaker is going out of her way to share her knowledge: is it for the speaker's own good, or for the auditor's? As Kauffeld has reminded us,

> We are all, I think, familiar with and resent the prospect that what others want to tell us about our business will amount to little more than meddlesome interference which complicates the task of taking care of our concerns but provides little assistance, because the interference issues from the other's perspective and is not based on an understanding our situation and responsibilities.

The speaker may be equally concerned to avoid the appearance of intruding into the auditor's business. How then can the speaker's knowledge get communicated?

The communicative act of advising is a practical solution to this "delicate" problem. In Kauffeld's analysis,

> Where a speaker gives advice, (i) she tells the advisee something which she at least purports to believe he needs to know, and (ii) she openly takes responsibility for trying assist him in determining what to do about his concerns.

The speaker's second undertaking—her acceptance of responsibility for addressing the auditor's concerns—opens the speaker to resentment, criticism and even perhaps reprisal should it become apparent that she isn't out to serve him. The auditor can thus reasonably presume that the speaker would not undertake such a risk unless she was indeed proceeding with his concerns in mind. The open undertaking thus alleviates the concern both speaker and auditor might have about meddling, and opens the way for the auditor to consider the speaker's advice in making up his mind.

Applied to the case of interest in this paper, Kauffeld's account of advising suggests that what citizens are looking for from scientist-advisors is not a value-free *disinterestedness*, but instead a dedication to *their* interests. Citizens expect the scientist-advisor to take responsibility for helping them make decisions that will further their own concerns. From this perspective, the proposed principles of responsible advocacy are not norms for scientific advice; a scientist can give good advice without, for example, providing all the considerations pro and con. Instead, the principles appear to be useful methods for the speaker to provide extra reassurances to her audience that she is indeed speaking with their concerns in mind. Such supererogatory efforts may be necessary to bolster audiences' trust in purported scientist-advisors during policy controversies where the stakes are high, the interests diverse, and the conflicts of interest apparent. Consider:

The scientist-advisor's openness about her own interests (RA 1) positions the audience to better assess whether she is honestly trying to assist them, or is really out for herself. Her willingness to reveal what is normally private information also demonstrates the depth of her concern for them.

The scientist-advisor's openness about the limits of her knowledge (RA 3), about uncertainties (RA 5) and counter-considerations (RA 6) again serve to put the audience in a better position to judge the quality of her advice. Providing this additional information over and above the core of what the audience "needs to know" (point (i) in Kauffeld's analysis above) reinforces the audience's ability to make an autonomous decision, and confirms that the scientist-advisor is not meddling.

Finally, the principle that the scientist-advisor be willing to change her advice when her assessment of the science changes (RA 4) also bolsters the trustworthiness of what she says. Unless the concerns of the advisor and advisee are manifestly different, it seems odd if an advisor won't take her own advice. And that oddity raises reasonable suspicions that the advice is not well-intended.

In sum, while RA 1–7 are either irrelevant to or bad for scientist-*advocates*, they may often be good working principles for scientist-*advisors* in policy contexts. So when the natural resources scientists talk of "responsible advocacy," what they really seem to mean is "good advice."

4. CONCLUSION

The arguments Pro 2–4 (and possibly Pro 1 as well) all give strong reasons—reasons for scientists to contribute to the policy process *somehow*. But, as I have argued, there are many communicative activities through which that contribution can be made.

Advocacy is one. Scientists, like all citizens, have the right to advocate zealously for policies they believe will serve the public good. In fact, scientists can make quite good advocates, since they start with a deep knowledge of the issue and are well-positioned to select and develop the strongest appeals.

But inevitably, such advocacy will have the consequence of reducing the scientists' credibility to zero. Advocates undertake to make a case, not to convey their best judgment. Audiences who even suspect the presence of advocacy will presume that what they hear is the strongest defense of a position, not the best science relevant to it.

When scientists and citizens want scientists' knowledge—not their arguments—respected, scientists must refrain from advocating positions however dear to their hearts and, in their view, well-justified by the best scientific results. This looks like a paradox: the more people know, the less politically effective we allow them to be. As Gill (2001) put it:

> When professionals decide to use the power of their expert knowledge to control policy outcomes, the public image of professionalism subtly metamorphoses. It transforms the professional's role from reliable expert into competing interest, and credibility erodes. The erosion of credibility reduces political power and a paradoxical futile cycle ensues. The paradox lies in the fact that the political power of professionals can be retained only if it is not exercised.

This paradox, however, is exactly what we should expect in a democratic polity: authority turns out to be self-limiting. When epistemic authority is exercised, both scientists and citizens have something at risk if their communicative transaction goes awry: scientists, their public repute; citizens, their sound decision-making. In the context of policy controversies, it is not surprising that it is difficult or impossible to get authority to work.

There are alternatives to advocacy—indeed, many alternatives. "It is time," as Scott and Racklow have put it, "to shift the question from whether conservation professionals should be advocates to how the expertise of scientists and professional societies can be given greater weight in ongoing discussions regarding policies and management actions that affect biological diversity" (J. M. Scott & Rachlow, 2011). In advising and other communicative activities, by laying aside their personal values and views in favor of serving the deliberative process and the interests of the citizens they are addressing, scientists can make a contribution and preserve their manifest trustworthiness, both.

ACKNOWLEDGEMENTS: This work was supported in part by a grant from Iowa State University's Center for Excellence in the Arts and Humanities. Thanks also to Mark Frankel and the other participants for a stimulating AAAS workshop on Advocacy in Science.

REFERENCES

American Association for the Advancement of Science. (2006). AAAS convenes experts to review role of advocacy in science. Retrieved from http://www.aaas.org/news/releases/2006/1011advocacy.shtml

American Association for the Advancement of Science. (2008). Panel says scientific advocacy can help resolve important issues, warns of its misuse. Retrieved from http://www.aaas.org/ news/releases/ 2008/ 0519stpf_advocacy.shtml

American Association for the Advancement of Science. (2011). Workshop on advocacy in science. Retrieved from http://srhrl.aaas.org/projects/advocacy/workshop/index.shtml

American Bar Association. (1983). Model code of professional responsibility. Retrieved from http://www.law.cornell.edu/ethics/aba/mcpr/MCPR.HTM

American Bar Association. (2004). *Model rules of professional conduct*. Retrieved from http://www.law.cornell.edu/ethics/aba/

Aron, W., Burke, W., & Freeman, M. (2002). Scientists versus whaling: Science, advocacy, and errors of judgment. *BioScience, 52*(12), 1137.

Audi, R. (1995). The ethics of advocacy. *Legal Theory, 1*, 251–281.

Barry, D., & Oelschlaeger, M. (1996). A science for survival: values and conservation biology. *Conservation Biology, 10*(3), 905–911.

Blockstein, D. (2002). How to lose your political virginity while keeping your scientific credibility. *BioScience, 52*(1), 91–96.

Brussard, P., & Tull, J. (2007). Conservation biology and four types of advocacy. *Conservation Biology, 21*(1), 21–24.

Collins, H., & Evans, R. (2007). *Rethinking expertise*. Chicago, IL: University of Chicago Press.

Craig, R., & Tracy, K. (1995). Grounded practical theory: The case of intellectual discussion. *Communication Theory, 5*(3), 248–272.

Ehrlich, P. R. (2001). Intervening in evolution: ethics and actions. *Proceedings of the National Academy of Sciences, 98*(10), 5477.

Foote, L., Krogman, N., & Spence, J. (2009). Should academics advocate on environmental issues? *Society and Natural Resources, 22*(6), 579–589. doi: 10.1080/08941920802653257

Gill, B. R. (2001). Professionalism, advocacy,and credibility: A futile cycle? *Human Dimensions of Wildlife, 6*(1), 21–32. doi: 10.1080/10871200152668661

Goodwin, J. (2001). Cicero's authority. *Philosophy & Rhetoric, 34*, 38–60.

Goodwin, J. (2001). The noncooperative pragmatics of arguing. In E. T. Nemeth (Ed.), *Pragmatics in 2000: Selected papers from the 7th International Pragmatics Conference* (Vol. 2, pp. 263–277). Antwerp: International Pragmatics Association.

Goodwin, J. (2002). Designing issues. In F. H. van Eemeren & P. Houtlosser (Eds.), *Dialectic and rhetoric: The warp and woof of argumentation analysis* (pp. 81–96). Dordrecht: Kluwer.

Goodwin, J. (2011). Accounting for the appeal to the authority of experts. *Argumentation, 25*, 285–296.

Gray, N., & Campbell, L. (2009). Science, policy advocacy, and marine protected areas. *Conservation Biology, 23*(2), 460–468. doi: 10.1111/j.1523-1739.2008.01093.x

Kaiser, J. (2000). Ecologists on a mission to save the world. *Science, 287*(5456), 1188.

Karr, J. (2006). When government ignores science, scientists should speak up. *BioScience, 56*(4), 287–288.

Kauffeld, F. (1999). Arguments on the dialectical tier as structured by proposing and advising. In C. W. Tindale, H. V. Hansen & E. Sveda (Eds.), *Argumentation at the century's turn: Proceedings of the Third OSSA Conference*. St. Catharines, ON: OSSA.

Kauffeld, F. (2009). What are we learning about the arguers' probative obligations. In S. Jacobs (Ed.), *Concerning argument* (pp. 1–31). Washington, D.C.: National Communication Association.

Kinchy, A., & Kleinman, D. (2003). Organizing credibility. *Social Studies of Science, 33*(6), 869.

Klonoff, R. H., & Colby, P. L. (2007). *Winning jury trials: Trial tactics and sponsorship strategy* (3rd ed.). Louisville, CO: National Institute for Trial Advocacy.

Lach, D., List, P., Steel, B., & Shindler, B. (2003). Advocacy and credibility of ecological scientists in resource decisionmaking: a regional study. *BioScience, 53*(2), 170–178.

Lackey, R. (2007). Science, scientists, and policy advocacy. *Conservation Biology, 21*(1), 12–17.

Lindeman, N. (2007). Creating knowledge for advocacy: The discourse of research at a conservation organization. *Technical Communication Quarterly, 16*(4), 431–451.

Lovejoy, T. (1989). The obligations of a biologist. *Conservation Biology, 3*(4), 329–330.

Meine, C., & Meffe, G. K. (1996). Conservation values, conservation science: A healthy tension. *Conservation Biology, 10*(3), 916–917.

Meyer, J., Frumhoff, P., Hamburg, S., & de la Rosa, C. (2010). Above the din but in the fray: Environmental scientists as effective advocates. *Frontiers in Ecology and the Environment, 8*(6), 299–305.

Mills, T., & Clark, R. (2001). Roles of research scientists in natural resource decision-making. *Forest Ecology and Management, 153*, 189–198.

Minnis, D., & Stout McPeake, R. (2001). An analysis of advocacy within the wildlife profession. *Human Dimensions of Wildlife, 6*(1), 1–10. doi: 10.1080/10871200152668643

Myers, N. (1999). Environmental scientists: Advocates as well? *Environmental Conservation, 26*(03), 163–165.

Nelkin, D. (1977). Scientists and professional responsibility: The experience of American ecologists. *Social Studies of Science, 7*(1), 75–95.

Nelson, M., & Vucetich, J. (2009). On advocacy by environmental scientists: What, whether, why, and how. *Conservation Biology, 23*(5), 1090–1101. doi: 10.1111/j.1523-1739.2009.01250.x

Nielsen, L. (2001). Science and advocacy are different–and we need to keep them that way. *Human Dimensions of Wildlife, 6*(1), 39–47. doi: 10.1080/10871200152668689

Noss, R. (2007). Values are a good thing in conservation biology. *Conservation Biology, 21*(1), 18–20.

Rykiel Jr, E. (2001). Scientific objectivity, value systems, and policymaking. *BioScience, 51*(6), 433–436.

Sarewitz, D. (2004). How science makes environmental controversies worse. *Environmental Science & Policy, 7*, 385–403.

Sarewitz, D. (2012). *Science advocacy is an institutional issue, not an individual one*. American Association for the Advancement of Science. Washington, DC. Retrieved from http://shr.aaas.org/projects/advocacy/workshop/Sarewitz.pdf

Scott, J., & Rachlow, J. L. (2011). Refocusing the debate about advocacy. Conservation Biology, 25(1), 1–3. doi: 10.1111/j.1523-1739.2010.01629.x

Scott, J., Rachlow, J., Lackey, R., Pidgorna, A., Aycrigg, J., Feldman, G., . . . Steinhorst, R. (2007). Policy advocacy in science: prevalence, perspectives, and implications for conservation biologists. *Conservation Biology, 21*(1), 29–35.

Shrader-Frechette, K. (1996). Throwing out the bathwater of positivism, keeping the baby of objectivity: Relativism and advocacy in conservation biology. *Conservation Biology, 10*(3), 912–914.

Steel, B., List, P., Lach, D., & Shindler, B. (2004). The role of scientists in the environmental policy process: A case study from the American west. *Environmental Science and Policy, 7*(1), 1–13.

Tracy, C. R., & Brussard, P. F. (1996). The importance of science in conservation biology. *Conservation Biology, 10*(3), 918–919.

Weingart, P. (2002). The moment of truth for science: The consequences of the 'knowledge society' for society and science. *European Molecular Biology Organization Reports, 3*(8), 703–706.

Wiens, J. A. (1997). Scientific responsibility and responsible ecology. *Conservation Ecology, 1*(1), 16.

Wynne, B. (2003). Seasick on the Third Wave? Subverting the hegemony of propositionalism: Response to Collins & Evans (2002). *Social Studies of Science, 33*, 401–417.

Young, A. M., & Larson, B. M. (2011). Clarifying debates in invasion biology: A survey of invasion biologists. *Environmental Research, 111*(7), 893–898. doi: 10.1016/j.envres.2011.06.006

Using Delphi to Track Shifts in Meanings of Scientific Concepts in a Long-term, Expert-lay Collaboration on Sustainable Agriculture Research in the Midwest

NANCY GRUDENS-SCHUCK

Department of Agricultural Education and Studies
217 Curtiss Hall
Iowa State University
Ames, IA 50011 USA
Email: ngs@iastate.edu

GL DRAKE LARSEN

Department of Natural Resources Ecology and Management
Iowa State University
Ames, IA 50011 USA
Email: dlarsen@iastate.edu

ABSTRACT: Dilemmas for ongoing expert-lay collaborations for agricultural conservation include divergence of meaning of new scientific concepts. Delphi method documented stakeholders' understandings of a new term, "ecosystem services." Flood mitigation and pest management benefits as ecosystem service attributes were ranked lower by stakeholders than anticipated by scientists. Recreation and aesthetics, and food production, were ranked higher.

KEYWORDS: conservation, ecosystem services, erosion, land-grant university, National Wildlife Refuge, participatory, prairie, water quality

1. INTRODUCTION

Participation of citizens and stakeholders in the development of agricultural technologies and in land management is common turf for expert-lay collaborations, as well as expert-lay conflict. When successful and authentic, participation contributes to greater democratization of science and technology (Peters, Jordan, Adamek, & Altern, 2005; Sclove, 1995; Williamson & Smoak, 2005). Academics and practitioners have studied mainly the start-up phase of collaborations. Scott Peters and associates (2005) dwell on the legacy of participation of citizens with land-grant researchers and its extension and outreach functions – key responsibilities for many of the researchers discussed herein. The authors bemoan the demise of vigorous cooperative ventures that provide a foil for transfer-of-technology approaches, and other processes in which experts dominate. MacKenzie (1996) details a multi-year, multi-institutional process for improving water quality of the Great Lakes. She lauds the production of new scientific knowledge, but speaks to its limited influence on outcomes: "Despite additional scientific knowledge about the nature of water resources and their interactions within the ecosystem, incremental and fragmented management has limited the capacity of institutions to respond meaningfully" (MacKenzie, 1996, p. 5). Academics and professionals studying the phenomenon of multi-stakeholder collaborations worry, like MacKenzie, about the difficulties associated with convening disparate groups, and keeping science in the mix.

After convening, a next step would be for members of collaboration to "do something" together. If exemplary actions are taken, a collaboration may continue to involve both lay and experts in a process of contributing to, and creating knowledge for, solving intractable conservation and environmental problems (see also Carolan, 2006; de Groot, Alkemade, Braat, Hein, & Willemen, 2010; Hein, 2006; Röling & Wagemakers, 1996). The literature also relays cases of partial, disrupted, co-opted, or otherwise failed collaborations, with an undercurrent that suggests that participatory projects drain time and resources; and may pose risks to one's personal, professional, or institutional status. The literature on participation theory and practice has blossomed since the 1980s, and studies of "How To" are now common. The stance of the literature has described goals that have remained, for the most part, in service to democratic or other social change agendas (Sclove, 1995). Sclove delineates the types of gains from collaborations that contribute to democratization, yet recognizes uneven progress:

> Insofar as participatory design results in technologies or services that go on to help constitute the agenda for a democratic politics of technology, then RD&D can also embody that special dignity... However, RD&D processes can also be organized to be crimped and growth-impairing. (1995, p. 182)

2. NEW SCIENCE TERMS

This paper focuses on a different phase of collaboration, that of continuance and renewal of collaborations, dilemmas that typically appear in the middle of hard-won partnerships. It is our presumption that if the challenges are not solved using democratic processes similar to the ones that created them, collaborative ventures risk, (a) falling apart, (b) becoming co-opted by a dominant faction, and (c) doing nothing worthwhile.

In the study, empirical data were collected on a dilemma experienced by an already-established collaboration whose members have worked together long enough to (a) witness the emergence of significant findings ("hard results") from a project, and to (b) be introduced to new science terms used to describe their work. The "dilemma of the middle" challenges us to learn how established multi-stakeholder groups successfully maintain integrity while weathering turn-over in membership, fluctuating financial resources, and reconciliation of an individual's participation in the face of revolving-door strategic plans of home institutions. Our paper focuses on a change in scientific terminology associated with the fields of ecology and environmental economics. The term, "ecological services," was missing from most early (around 2002) conversations regarding the research, substantiated by verbatim field notes by Grudens-Schuck of early discussion of the stakeholder group and research team. By 2010, however, "ecosystem services" was employed in titles and project statements of most grant proposals, publications, and other deliverables (video, websites, press releases) associated with the project. The timing of arrival of the new term with respect to social development of the collaboration allowed theoretical investigations into dynamics of complex, multi-player, participatory environmental and agricultural research and development.

The Delphi study was conducted as part of research for a master's thesis of co-author Larsen. Grudens-Schuck is a university collaborator working with the research team, and also served as a member of Larsen's Graduate Committee. The "STRIPS at Neal Smith National

Wildlife Refuge" project within which this research report is nestled is much larger than the Delphi study; consequently, not wholly described herein (see Jarchow et al.., 2012; Home page http://www.nrem.iastate.edu/research/STRIPs/). STRIPS is an acronym for *Science-based Trials of Row Crops Integrated with Prairies*. "Neal Smith" names a central partner for the biophysical research: the *Neal Smith National Wildlife Refuge*, dedicated to prairie, including animals (such as American bison, elk, other native mammals, birds, and insects). An aspect that sets the larger project apart from more commonly conducted field-trial type research is (a) the project is a long-term ecological research project, and (b) the project is conducted on a large scale—that of three "whole" watersheds. In the main, the research team sought to test, and then introduce (where appropriate), "prairie strips" for use by farmers on row crop fields (corn and soybean) to lessen erosion and improve water quality. Funding was provided internally and externally, including by government agencies. Stakeholders spanned farm commodity groups, environmental nonprofits, and government staff. Day-long meetings, at least annually since 2003, also included walking the watershed-sized experimental "plots." Frequently, there was standing room only, with enthusiastic graduate students lining the room of the large educational conference room at the Neal Smith Wildlife Refuge, Prairie Learning Center (Home page: http://www.fws.gov/midwest/nealsmith/). A brief schematic follows which illustrates the experimental configuration.

Experimental Setup

The systems being studied include a range of percentage and placement of perennial vegetation ("prairie strips") in Figure 1. The project is being conducted (replicated) on fourteen small watersheds in and around the Neal Smith National Wildlife Refuge ranging in size from 1.2 acres to 13 acres.

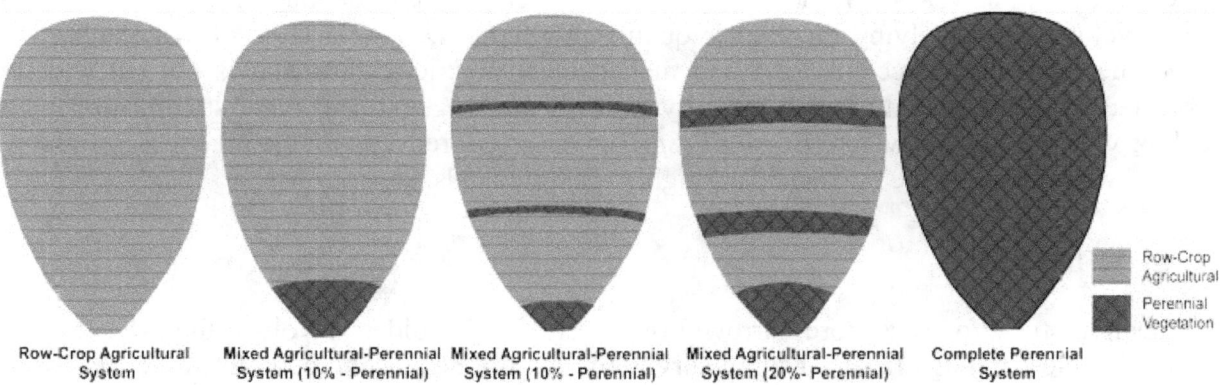

Fig. 1. Conceptual design of watersheds. Copyright 2012 STRIPS Research Team.

Stakeholders—up to 40 participating at a time in some meetings—have been involved at different stages of research planning, data collection, interpretation of annual data; and in discussions about securing external funding, and the impact funding would have on the core activities. What is referred to as the "research team" is the smaller group, composed of scientists from university and government; program planners associated with biophysical, engineering, or social science; and those enrolled in a the land-grant MS or PhD degree.

2.1. Why "Prairie Strips"?

- Improved water cycling, including filtration that results in clean water.
- Nutrient cycling, which may reduce the amount of fertilizer needed
- Carbon cycling, which may address the need to retain carbon in other places than in the atmosphere, where it contributes to global climate change (global warming).
- Biodiversity, which seeks to retain species that contribute in ways that are known (fox eat rabbits; birds eat mosquitoes) and unknown.

These benefits to society may potentially be made easier, and more affordably delivered if the prairie strip idea works. So far, the data support the use of prairie and prairie strips in managing water at a level that is the same or better than established conservation measures (see also Atwell, Schulte, & Westphal, 2010; Jarchow & Liebman, 2010; Larsen, 2011; URL: http://www.nrem.iastate.edu/research/STRIPs/research/index.php). Findings have been breathtaking in their success thus far.

The research topic was considered key to the mission of several of the participating researchers and stakeholders. Surface and ground water contamination by agricultural run-off in the Midwest continues to make the news; especially in states such as Iowa—in the "breadbasket" of the U.S. For example, this state, a prime producer of corn, soybeans, and livestock, recently requested "more time" of the Environmental Protection Agency (EPA) to remediate its many "impaired waterways." Iowa, the site of the research, could not meet the first deadline, one of many such delays since creation of the Clean Water Act. Efforts to keep soil on the farm and out of ground and surface water predated the Act and creation of the EPA in this same area of the country. Intractable yet hard to ignore, land-grant colleges and universities, and newer government and non-profit agencies, have devoted decades of research and development to solving the "water quality problem." Despite a suite of affordable, and sometimes subsidized, technologies and management practices, farmland is not sufficiently protected from high-level erosion rates by riparian plantings, stream bank stabilization with rock or willow or crown vetch (*Coronilla sp.*), cover crops, reduced tillage practices, terracing, grassed waterways, or the "set aside" of highly erodible lands.

3. METHODS

The Delphi study focused more narrowly on ideas of stakeholders involved the project. This paper provides findings from a Delphi survey of twenty-one stakeholders, from the Master of Science thesis of Larsen (2011). Delphi is an individual, private, and multi-step survey process with an established history of success in predicting a "likely future" (Angus, Hodge, McNally, & Sutton, 2003). Delphi—as a method alone—is not presumed to be a participatory process itself. Typically, face-to-face dialogue, the more intense the better, is put forward as ideal in participatory and collaborative endeavors (Cuppen, 2012; Gutmann & Thompson, 1996; Oels, 2000; Röling & Wagemakers, 1998; Williamson & Smoak, 2005). However, there are differences of opinion regarding the warrant for employing one social science method over another, even within participatory endeavors. For the STRIPS stakeholder group, face-to-face and phone contact were frequent. Individuals crossed paths professionally in associations such as the Iowa Prairie Network, and in farm conferences and equipment shows. What was novel was the opportunity that Delphi afforded: a chance to complete a questionnaire that (a) was close to their hearts, (b) asked about a topic stakeholders knew a lot about, and (c) was conducted absent the din of normal meeting-room chatter that does not always permit reflective thinking. Delphi seemed to fit also because the question was direct, and because there was not prior consensus about the term "ecosystem services." It was known among the team and stakeholders that the terms were not yet an easy fit. Classic Delphi technique requires a purposefully selected group of experts. This study fulfilled this criterion. In this study, a census sample was conducted; in other words, all stakeholders were provided opportunities to contribute to the confidential, 3-round Delphi process. Delphi's power comes from its ability to predict the most likely future, mainly because the people surveyed are intimately involved in the issue, and are able to express both conscious and unconscious preferences and restrictions.

3.1 Where did "ecosystem services" come from?

The strong wave that thrust the research team into the semantic arena of "ecosystem services" as well as "multifunctional agriculture," according to retained ethnographic field notes and team meetings, field days, and stakeholder meetings [by Grudens-Schuck and others], was the demand from external funders—mainly federal. It would be too much to list the funding sources large and small, and all important, in this paper. However, one must speak plainly: the emphasis on conjoining ecosystem services with the project goals and rhetoric was funder-driven. Funders, as far as we can tell, adopted the term from the ecological sciences and environmental economics. In a 2010 editorial, Rudolf de Groot, editor of a fairly new peer-review journal called *The International Journal of Biodiversity Science and Management*, declared the new title, *The International Journal of Biodiversity Science, **Ecosystem Services**, and Management* [authors' emphasis]. De Groot explained that the term migrated from science to policy making, and emphasized the human element—a potential recipient of benefits from ecological management. Heretofore, biodiversity and ecology were broadly viewed to benefit wildlife and habitat, sometimes holding out the possibility that scientific and medical advances could be identified by a yet-unknown species. Most descriptions in the publications of the team, and in talks given to each other and stakeholders, emphasize the benefits of landscapes to provide much more than food. In this way, the "ecosystems services" phrase made visible the formerly externalized gains (i.e., ecological services) from landscapes; benefits that were not sufficiently realized or valued. In sum, ecosystem services were talked about by scientists and other experts as distinct from production capabilities of landscapes. Conceptually, ecosystem services were additive, not integral to, farm and ranching.

4. FINDINGS

The Delphi study produced both confirming and disconfirming ideas. A few will be highlighted to illustrate how research functioned within a multi-stakeholder group. Next, it will be necessary to explain the gaps or differences in understanding of scientific terms in terms of the dialectic of agreement and disagreement embedded in the concept of democracy (Bills & Gross, 2005; Cuppen, 2012; Gutmann & Thompson, 1996).

(A) Flood mitigation and pest management were ranked lower by stakeholders than by most members of the research team.

(B) Recreation and aesthetics were ranked higher by stakeholders as key objectives than by members of the research team.

(C) Production of food was ranked highly as a feature of ecosystem services—even though scientists' talk indicated that ecosystem services consisted of benefits above and beyond commodity food production (corn and soy).

All three findings stand out for reasons related to the history of the group, its funding, and the changing interests of the institutions that support stakeholders' role in the group.

4.1 Flooding and Water Quantity

The surprise regarding the lower ranking for flood mitigation potential of the STRIPS project was palpable. In four years (2008-2012), large portions of the state of Iowa received flash floods, devastating rising waters over levies of rivers near key cities in Iowa. Distressingly, each event was "not supposed to happen" statistically or historically. Neither team members nor stakeholders were spared from effects of the floods. Moreover, team members had for many years been involved in shared goals for reducing soil eroded from Iowa farm fields contaminating the major rivers into which the watersheds (i.e., Mississippi, Missouri) delivered their mix of sediment, as well as trash, and chemicals and biological contaminants, mixed with more of the same from other sectors.

4.2 Recreation and Aesthetics

When working on big projects with big goals, congruence with larger missions of the major institutions is usually present. Moreover, it is a grapevine phenomenon that farmers and their experts risk being taken less seriously if recreation and open-space views are a major thrust of an agro-ecological project. Consequently, it came as a surprise that these elements were ranked so highly.

4.3 Food Production

Food production was, most of the time, talked about by scientists as a given. What were needed sorely were the ecosystem services that would help to clear impaired waters; reduce soil erosion; and slow the movement of water, which routinely de-stabilized river banks and collapsed bridges and dams, overwhelming the ability of local governments to manage infrastructure repairs.

4.4 Other Findings

Stakeholders' initial list of ecosystems services terms contained sixty items describing an array of services and processes across a wide range of scales from field to global. Similar to flooding, wildlife and biodiversity were surprisingly ranked lower than expected in light of emphasis given in research talks and publications.

5. DISCUSSION

The three findings of discrepancies (broadly speaking) between research team members and stakeholders would provide sufficient fodder for long meetings and fervent emails. With respect to development of the wherewithal of stakeholders and citizens to participate wisely in democratic decision making about complex landscape topics, is it good or bad that the survey arrived at these findings? Moreover, would it be wise or foolish to place the discord squarely in front of both stakeholders and team members—perhaps while in the same room? Answers to these questions can move in opposite directions. In a command-and-control social environment, "negative data" are seen as an element to be hidden or glossed over. In a democratic environment, authentic differences are valued for their potential for learning laterally; and for allowing for stakeholders to make recommendations for change to the project in line with their understandings and rankings (Röling & Wagemakers, 1998; Ruhl, Kraft, & Lant, 2007). This claim falls in line with Gutmann and Thompson's (1996) distinctions between procedural democracy, which is more rule-based; constitutional democracy, which demands a just outcome; and deliberative democracy, which emphasizes ongoing discussion, with attention to power issues within the group. They write:

> Deliberative democracy rejects [this] dichotomy. It sees deliberation as an outcome-oriented process; citizens deliberate with the aim of justifying their collective decisions to one another as best they can. (Gutmann & Thompson, 1996, p. 27)

These discussions have begun, and will take months to impact decision making. A deliberative democratic thrust would suggest allowing—even facilitating—exploration and disagreement regarding the term "ecosystem services." It is anticipated that the road of deliberative democracy will allow members of the research team (old and new) and the stakeholder group (old and new) to vigorously support the research while remaining ambivalent or in disagreement about the scope or extent of adoption of prairie strips. There are other benefits to a deliberative approach. Taking care of smaller disagreements renews trust and builds understanding at a human pace. On the other hand, learning at a later time about data that were "hidden" undermines collaborations (Hein, 2007). Moreover, there is always the wonderful event when collaborators learn that stakeholders are correct and scientists wrong. In fact, one of the findings fits this last phenomenon. It is understandable, given the degree to which prairie has been destroyed for crop and livestock production, that scientists emphasize the "other than" quality of ecosystem benefits with respect to returning prairie to the state with the largest historical swatch of Tall Grass prairie. However, it is incorrect per the emerging definition of the "ecosystem services" term. For example, de Groot (2010), in his editorial justifying the addition of "ecosystem services" to the journal discussed earlier, provides a list of elements of ecosystem services—and it includes the production of food.

6. CONCLUSION AND REFLECTION

It is inaccurate to state that the goal of the Delphi study was to stimulate argumentation, reflective thinking, and the exercise of deliberative democracy. The goal was to clarify, in a short period of time, a term that has become central to the narratives of the research team. Moreover, stakeholders and team now have sufficient data to move into an "adoption" phase. This phase would demonstrate and explain prairie strips, and the research behind them, to farmers and land owners. The overall goal—integral to a land-grant mission—is to facilitate good decision making. The hope is to ready Iowa and its watersheds to provide more functions than it has in the last 150 years (e.g., food and fiber). Members of the research team badly want a lot of farmers to use the prairie strip technology; but it cannot be a "sales job" as long as the wherewithal of farmers to make some decisions about their land is respected. A practical question for the collaboration might be: should the term "ecosystem services" be featured in "technology transfer"? By how many, or to what extent, do the members of the collaboration need to be in agreement for the term to be featured? These questions are useful; but in the end, Gutmann and Thompson (1996) would suggest that disagreement and the need "to discuss" will not go away, and should not go away.

ACKNOWLEDGEMENTS: Appreciation is extended to all STRIPS team members and stakeholders, whose expertise and good questions added immeasurably to our small effort herein. Lisa Schulte-Moore and John Tyndall were co-chairs of Drake Larsen's graduate committee. Special thanks also to an anonymous reviewer of our abstract, and to the kind soul who read our paper and provided special comments at the end of the presentation. This research was provided funding from many sources, including the Experiment Station of the Iowa State University of Science and Technology; and the College of Agriculture and Life Sciences.

REFERENCES

Angus, A. J., Hodge, I. D., McNally, S., & Sutton, M. A. (2003). The setting of standards for agricultural nitrogen emissions: A case study of the Delphi technique. *Journal of Environmental Management, 69*, 323–337.

Atwell, R.C., Schulte, L.A., & Westphal, L.M. (2010). How to build multifunctional agricultural landscapes in the U.S. Corn Belt: Add perennials and partnerships. *Land Use Policy, 27*, 1082–1090.

Bills, N., & Gross, B. (2005). Sustaining multifunctional agricultural landscapes: Comparing stakeholder perspectives in New York (US) and England (UK). *Land Use Policy, 22*, 313–21.

Carolan, M. S. (2006). Science, expertise, and the democratization of the decision-making process. *Society and Natural Resources, 19*, 661–668.

Cuppen, E. (2012). Diversity and constructive conflict in stakeholder dialogue: Considerations for design and methods. *Policy Science, 45*, 23–46.

de Groot, R. S. (2010). Editorial. *International Journal of Biodiversity Science, Ecosystem Services & Management, 6*, 1-2.

de Groot, R. S., Alkemade, A., Braat, L., Hein, L., & Willemen, L. (2010). Challenges in integrating the concept of ecosystem services and values in landscape planning, management and decision making. *Ecological Complexity, 7*, 260–272.

Gutmann, A., & Thompson, D. (1996). *Democracy and disagreement: Why moral conflict cannot be avoided in politics, and what should be done about it.* Cambridge, MA: Harvard University Press.

Hein, L., van Koppen, K., de Groot, R. D., & van Ierland, L. C. (2006). Spatial scales, stakeholders and the valuation of ecosystem services. *Ecological Economics, 57*, 209–228.

Jarchow, M.E., Kubiszewski, I., Larsen, GL D., Zdorkowski, G., Costanza, R., Gailans, S.R., . . . Liebman, M. (2012). The future of agriculture and society in Iowa: Four scenarios. *International Journal of Agricultural Sustainability, 10*(1), 76–92.

Jarchow, M.E., & Liebman, M. (2010). *Incorporating prairies into multifunctional landscapes: Establishing and managing prairies for enhanced environmental quality, livestock grazing and hay production, bioenergy production, and carbon sequestration.* PMR 1007. Ames, IA: Iowa State University Extension.

Larsen, GL D. 2011. *Farming for ecosystem services: A case study of multifunctional agriculture in Iowa, USA.* (Unpublished MS thesis). Iowa State University, Ames, IA.

MacKenzie, S. H. (1996). *Integrated resource planning and management: The ecosystem approach in the Great Lakes Basin.* Washington, DC: Island Press.

Oels. A. (2000). *Evaluating stakeholder participation in the transition to sustainable development. Methodologies, case studies, policy implications.* (Doctoral dissertation.) Piscataway, NJ: Rutgers University, Transaction Publishers.

Peters, S. J., Jordan, N. R., Adamek, M., & Altern, T. A. (Eds.). (2005). *Engaging campus and community: the practice of public scholarship in the state and land-grant university system.* Dayton, OH: Kettering Foundation Press.

Röling, N. G., & Wagemakers, M. A. E. (Eds.). (1998). *Facilitating sustainable agriculture: Participatory learning and adaptive management in times of environmental uncertainty.* Cambridge, UK: Cambridge University Press.

Ruhl, J., B., Kraft, E. C., & Lant, L. C. (2007). *The law and policy of ecosystem services.* Washington, DC: Island Press.

Williamson, R., & Smoak, E. (2005). Engaging campus and community to improve science education. In S.J. Peters, N.R. Jordan, M. Adamek, & T.R. Alter (Eds.), *The practice of public scholarship in the state and land-grant university system* (pp. 265–308). Dayton, OH: The Kettering Foundation.

Examples, Illustrations, Inductions, Anecdotes, Analogies, Precedents, Narratives, and Personal Testimonies: Are They Essentially Different?

DALE HAMPLE

Department of Communication
University of Maryland
College Park MD
USA
dhample@umd.edu

ABSTRACT: This essay addresses the question of whether these argument schemes—example, illustration, induction, anecdote, analogy, precedent, narrative, and personal testimony—are distinct from one another. Each of them is essentially based on a single case (although the cases can be multiplied, perhaps converging into an informal induction). "Example" is the prototypical scheme. The critical questions for "example" apply to the other argument schemes as well.

KEYWORDS: analogy, anecdote, example, illustration, induction, narrative, personal testimony, precedent.

1. INTRODUCTION

Public health and information campaigns often make use of personal testimonies (sometimes authentic and sometimes conveyed by actors). These might offer an example of the benefits of better health practices and so are designed to guide the public to healthier and more accurate understandings of drinking, cigarette smoking, safe sex practices, and other such issues. The basis for thinking that the messages contain information that is in fact healthier and more accurate than uninformed opinion is scientific, but often the science is missing from the public campaigns. Material from the technical sphere is suppressed and easier information is supplied to people to use in making decisions in the personal sphere. Instead of scientific reports or summaries, recipients are sometimes given something that we variously term an example, a story, or a self-disclosure.

Here I take up the question of whether a number of argument schemes—example, illustration, induction, anecdote, analogy, precedent, narrative, and personal testimony—are genuinely distinct from one another. In one sense there is no suspense about whether these schemes can all be reduced to a common form. We can take any group of argument types and notice, say, that they all have a conclusion and one or more premises, and announce a successful reduction. The question is whether the reduction is a useful one—whether it advances our thinking, our research, or our teaching. I will give reasons to suppose that the reduction I offer is clarifying and useful.

A considerable amount of progress on my project has already been made by van Eemeren and Grootendorst (1992, pp. 94–102). Their proposed classification of argument schemes already combines example and analogy, and I take their work as a key contribution. My plan is to examine these various kinds of argument to see in what ways they are similar and different.

Hample, D. (2012). Examples, illustrations, inductions, anecdotes, analogies, precedents, narratives, and personal testimonies: Are they essentially different? In J. Goodwin (Ed.), *Between scientists & citizens: Proceedings of a conference at Iowa State University, June 1-2, 2012* (pp. 173-182). Ames, IA: Great Plains Society for the Study of Argumentation. Copyright © 2012 the author(s).

2. DISTINGUISHABLE ARGUMENT SCHEMES

In this section I identify and analyze a number of different argument schemes. I consider these all to be distinguishable. This does not mean that the distinctions are necessarily important. It only means that the argument schemes have different labels, and that analysts can tell when to apply one label and not another. I restrict myself to those argument schemes that are of immediate interest. I have chosen these particular argument schemes precisely because I believe that I can make a case for their essential similarity.

2.1 Example

The first scheme to consider is example, and I will eventually suggest that all the others be reduced to this one. We can begin the analysis with the Walton, Reed, and Macagno (2008, p. 314) skeleton of the scheme. They specify a single premise and a conclusion:

> Premise: In this particular case, the individual a has property F and also property G.
> Conclusion: Therefore, generally if x has property F, then it also has property G.

This scheme permits us to reason directly from the instance a to the instance or category x. The conclusion's phrasing, "generally if x," is usefully ambiguous since it might be construed to be a general statement to the effect that anything having F will have G, or it might be restricted to the particular instance x.

I want to work with the same concrete material for as long as I can, so let me introduce it here:

> Munich and Taliban. In hopes of preventing a wider war, in September 1938 British Prime Minister Neville Chamberlain negotiated an agreement with Germany's Adolph Hitler. Called the Munich Agreement, this treaty confirmed the legitimacy of Germany's annexation of Czechosolovakia's Sudetenland. In return, Hitler agreed that he would not engage in any further territorial aggression. The outcome of the agreement, however, was not only that Germany kept the Sudetenland; Germany also continued its policies of expansion. This is just what will happen if the United States tries to appease the Taliban.

We can fit this into the Walton et al. (2008) scheme in this way:

> In this particular case, the Munich agreement had certain properties (diplomacy, an aggressive diplomatic partner, a paper promise, appeasement) and also had the property of failure;
> Therefore, since any agreement with the Taliban would have those same properties it will also have the property of failure.

Here, an example has been used to draw a specific conclusion, one about the particular case of the Taliban rather than about all cases that resemble Munich. The question of whether an argument from example results in a general or specific conclusion has been awkward from the first (cf. Benoit, 1980). Aristotle described examples as rhetorical inductions, suggesting the appropriateness of a general conclusion (1356b). However, he also explained, "When two statements are of the same order, but one is more familiar than the other, the former is an example" (1357b). The second remark legitimizes a specific conclusion. I am willing to entertain either sort of conclusion.

2.2 Illustration

Perelman and Olbrechts-Tyteca (1969, pp. 350–362) insisted on clear distinctions between example and illustration.

They saw argument from example as being intended to establish a generalization, rather as an induction would (p. 350). They begin their discussion by treating a legal precedent as an example and showing how the decision in an earlier case (the example) supplies a rule for the present case and all similar ones.

Perelman and Olbrechts-Tyteca would describe the Munich material as argument "from the particular to the particular" (1969, p. 352). They insist that the passage from one particular to another is mediated by an implicit rule: "It is by their relation to a given rule that phenomena become interchangeable" (p. 353). That rule would seem to be a general conclusion drawn from the example in the premise. So the Belgians might prefer this reconstruction of the Munich material (I have italicized the implicit rule):

> In this particular case, the Munich agreement had certain properties (diplomacy, an aggressive diplomatic partner, a paper promise, appeasement) and also had the property of failure;
> *Any instance having those certain properties will also have the property of failure;*
> Therefore, since any agreement with the Taliban would have those same properties it will also have the property of failure.

This summary of the Belgians' understanding of the relationship between example and rule allows us to appreciate their distinction between example and illustration:

> Whereas an example is designed to establish a rule, the role of illustration is to strengthen adherence to a known and accepted rule, by providing particular instances which clarify the general statement, show the import of this statement by calling attention to its various possible applications, and increase its presence to the consciousness. (Perelman & Olbrechts-Tyteca, 1969, p. 357)

Walton et al. (2008, p. 315) accept this distinction and offer a schematic representation of argument from illustration:

> Premise 1: Usually, if x has property F (belongs to class F), x has property G.
> Premise 2: In this case, k has property F and property G.
> Conclusion: The rule is valid.

Premise 1 expresses the rule (expressed in terms of x, which here seems to like the X in algebra so that it means any instance rather than a particular one), premise 2 supplies the illustration (the specific instance k), and the conclusion re-asserts the rule.

I have some hesitation about all this. For one thing, I am not sure that illustrating is a kind of proving, and so I am reluctant to allow that the last thing in the scheme should actually be called "concluding." On my reading of the *New Rhetoric*, illustration seems to be essentially a rhetorical device that amplifies the rule. An illustration can make a rule more attractive, more memorable, more comprehensible, or more present, but I do not see how these rhetorical functions amount to proving (as contrasted to persuading).

Let us examine the two schematic skeletons a little more closely. Setting aside the differences in how the argument elements are labeled, some clear similarities appear. The single premise in *example* is the same as the second premise in *illustration*: both identify the specific instance as having two properties. The conclusion to *example* is essentially the same as the first premise in *illustration*: both express the generalization that any other case that has the first set of properties will also have the second set. So there are two differences in the schemes: the materials are in a different order, and *illustration* contains the conclusion that the rule is valid. We have seen that some believe that 'the rule is valid' is actually an implicit premise in arguing from one instance to another. If so, this means that the only difference between the two schemes is order. In argument from example, if we include the implicit premise we have:

> The first case has certain properties F and also another property G;
> Any case that has properties F will have property G;
> So a second case, having properties F, will have property G.

In argument from illustration, we have:

> Any case that has properties F will have property G;
> The first (only) case has properties F and G;
> So, any case that has properties F will have property G.

Several things are apparent. Only the argument from example has the capacity to move from one instance to saying a new thing about a second one (when multiple instances are used as illustrations, they just amplify one another, more or less). Second, the conclusion of the illustration schema is simply a restatement of the first premise. Third, the first two premises of the example schema are the last two of the illustration schema. If the second premise of the example schema is understood as being distinct from the first premise— that is, the general statement is externally established by some other proof—then the two sequences (*example* P1-P2 and *illustration* P2-C) are substantively different. This is possible. However, if we join Aristotle and Perelman in supposing that an example can used to prove (however weakly) a generalization, then the two sequences are identical.

So the only reliable differences are these. First, argument from example can move from one specific case to another and argument from illustration does not. But this is only because the example argument has additional content; if we wanted to add another premise about a second specific case to the illustration scheme, we would have the same thing. Second, argument from illustration has a certain rhetorical emphasis built into it, with the triumphant restatement of one of its premises. This is not a probative difference.

The essential distinction between example and illustration, then, has to do with rhetorical efficacy or strategic maneuvering (van Eemeren & Houtlosser, 1999; see van Eemeren, 2010) not with proving. When we look for probative movement, we see the same relations in either sequence. Consequently I regard these as the same sort of argument.

2.3 Induction

Induction is a multiplication of examples. The conclusion of an induction is always a generalization and is never a statement about a particular case. This is by definition. Inductions can of course supply premises that lead to conclusions about particulars, but those will be arguments in which the induction is subordinate to some other argument scheme. To display an induction, the Munich-Taliban material would need to have several other examples of failed appeasements added to it, and the conclusion would have to be restated to refer to a general case rather than to the Taliban.

Structurally, an induction can be understood in two different ways.

First, it can be seen as a conductive (or convergent) argument (Henkemans, 2000; Wellman, 1971). On this understanding, we have a set of reasons for the conclusion. Each reason consists of a particular instance combined with whatever linking premise is required to move probatively from example to conclusion. The reasons have no relationship to one another beyond simple companionship: that is, none implies another, none is a consequence of another, none supplies a premise for another, and so forth. Each reason is independent and points directly to the conclusion. Presumably, each reason is individually insufficient to support the conclusion. That is why there are several instances rather than one. The arguer implicitly proposes that the examples are *collectively* sufficient to support the conclusion. Because it does not matter in principle what order the premises are in, this argument scheme has a rhetorical flavor of nonlinearity and merely-narrative-discipline to it.

The second way of understanding induction's structure makes induction into a more linear sequence, one that requires a premise that comments on the other premises. Italicizing the commenting premise, this scheme might be:

> The first case has certain properties F and also another property G;
> The second case has certain properties F and also another property G;
> The third (up to *N*) case has certain properties F and also another property G;
> *These three (N) cases are sufficient to justify a general conclusion;*
> Therefore, any case that has properties F will have property G.

This formulation is immediately attractive to those of us who have studied induction in the guise of sampling and social science, because we are taught to think explicitly about issues such as representativeness, sampling frames, sample size, and so forth. As a practical matter, the italicized premise simply locates the fissure point for several of the critical questions used to evaluate an induction. The same critical questions apply equally whether this or the first structural description is used.

I suggest that the example be regarded as the base schema. A good argument can be made for induction being the prototype, however. Considering only the forms discussed so far—example, illustration, and induction—induction is most encompassing. An example can be thought of as a very simple induction, a sort of degenerative case that requires no additional apparatus in comparison to induction (provided that we draw a general conclusion from the example rather than a particular conclusion). In contrast, an induction does require an element not present in arguments from example. This additional element is most clearly expressed by the commenting premise (if one wishes to grant that there is one), or is otherwise apparent in the idea of conduction. A simple argument from example does not involve convergence or joint sufficiency. So it might make sense to identify induction as the overarching idea. I prefer to say that example is the key idea, because I think it is the essential component and the base probative relationship that appears in each of these argument schemes.

2.4 Analogy

Perelman and Olbrechts-Tyteca (1969, pp. 371–398) distinguish analogy from example. They consider the prototypical form of analogy to be "A is to B as C is to D." The key distinction between example and analogy is that argument from example asserts that two things (the instances or category) directly resemble each other, but in analogy it is the two *relationships* (among four things) that resemble each other (pp. 372–373). This is clear enough, and the difference seems unbridgeable. But perhaps the clarity is deceptive. Consider: negotiating with the Taliban is analogous to negotiating with the Nazis. Or more elaborately: Taliban negotiations are to oppression as the Munich negotiations were to territorial conquest. Wasn't this an example?

Perhaps we had better look closely at the schematic summary of analogy from Walton et al. (2008, p. 315):

> Similarity Premise: Generally, case C_1 is similar to case C_2.
> Base Premise: A is true (false) in case C_1.
> Conclusion: A is true (false) in case C_2.

This does not immediately join up with the schema in the *New Rhetoric* (i.e., A is to B as C is to D), so we need to expend a moment on explication.

Consider a simple argument: A writer's pencil is like a carpenter's hammer. Or more elaborately: A pencil is to a writer as a hammer is to a carpenter. As written, the analogy proposes that carpenters and hammers are better known than writers and pencils, so that the hearer is drawing a conclusion about pencils and writers (case C_2) from information about hammers and carpenters (case C_1). What is the A in the Walton et al. (2008) schema? It is the relationship in each case: the relationship between writers and pencils is the same as the relationship between carpenters and hammers. So A might be expressed as "the defining tool" or something similar, leaving us with "a pencil is A, the defining tool, for a writer, just as a hammer is A, the defining tool, for a carpenter." In other words, we have one example of a defining tool (hammer) that we propose as resembling another instance (pencil), which is also thereby offered as a defining tool.

So in examining the Walton et al. (2008) skeleton, we can now perceive that the A for C_1 is an example that resembles the A for C_2. We are now in a position to see how this can be covered by the Walton et al. description of example:

> Premise: In this particular case, case C_1 has various properties F and also property A.
> Conclusion: Therefore, since case C_2 also has properties F, then it also has property A.

This explains how carpenters' hammers are examples that resemble writers' pencils; how hammers and pencils are analogous; how Munich's outcomes are examples of expected outcomes for Taliban negotiations; and how Munich and Taliban negotiations are analogous.

In both the tool and negotiation materials, I have chosen argument parts that are elements of roughly the same conceptual domain. Perelman and Olbrechts-Tyteca (1969) make a point of noticing that often the theme and phoros of an analogy come from different domains, as for instance when some description of light is made analogous to spiritual illumination. I think my analysis still works in such cases, but the argument itself (i.e., the analogy) is likely not to be as tight. Perhaps it is this domain-crossing capability that results in analogy being "viewed with distrust when used as a means of proof" (Perelman & Olbrechts-Tyteca, 1969, p. 372). Not all examples are good ones.

2.5 Precedents, Anecdotes, Narratives, and Personal Testimonies

My title promises consideration of four other distinguishable sorts of argument. Precedent has already been mentioned. Perelman and Olbrechts-Tyteca (1969) see a precedent as a judgment in an earlier legal case that is used as a forcible example in deciding a current case. The court's reasoning in the old case is held to be an example of how the current court should reason about the instant case.

An anecdote is a narrative, sometimes humorous and usually brief. A two sentence story might be called an anecdote and ten pages a narrative. Varying lengths of arguments do not immediately justify different analytical schemata. Stories told probatively make a point and they do so by encasing the point in plot and/or character so that the conclusion emerges. The Munich story yields the point that appeasement fails. A story, short or long, is therefore an example.

Sometimes arguers tell stories about themselves. We call these self-disclosures or personal testimonies. The Munich materials might be more striking if I had summarized them as Neville Chamberlain's own reflections on his actions. A first person story might well be more present and involving than a third person story such as the one about Munich, but it is still essentially a story and therefore an example.

This is a convenient place to make a general point: in distinguishing all these sorts of argument our basis may have been rhetorical rather than structural. The same schema can be used for different presentations that can be more or less persuasive. Well-told stories, for instance, will probably make more impression than recital of boringly similar instances; first person narratives might elicit more identification and emotional resonance than dry objective accounts; analogies might seem more creative and intellectually exciting than a simple enumeration of parallel instances; and so forth. It might be important for rhetorical scholars to insist on the distinctions that I am setting aside here. But in labeling argument types, I propose we stick to structure. The base structure for all these things seems to be argument by example.

3. CRITICAL QUESTIONS

My argument to this point can be evaluated directly, based on what I have said. But there is also an indirect way to test it. If I am right, then the critical questions for all these argument variants ought to be comparable, if not essentially the same.

Let us begin with Walton et al.'s (2008, p. 314) critical questions for argument from example. Recall that their scheme for the argument involved one premise about the example and drew an immediate conclusion about a general class of instances, which might be taken to be a statement about a second instance.

> CQ1: Is the proposition claimed in the premise in fact true?
> CQ2: Does the example cited support the generalization it is supposed to be an instance of?
> CQ3: Is the example typical of the kinds of cases the generalization covers?
> CQ4: How strong is the generalization?
> CQ5: Do special circumstances of the example impair its generalizability?

The first question would seem applicable to any argument with a premise. The study of narrative reminds us, however, that we can give further specification of what we might mean by "true" in this question. Fisher (1987) says that there are two considerations in a narrative's fidelity. One is whether the material is internally consistent: that is, whether characters act the same way in different parts of the story, whether acts have the same consequences throughout, etc. The other matter is external: whether people in the story act the way people in the world act, whether consequences play out in the story the way they do in the wider world, etc. These specifications are only pertinent in a well-developed example (they could be asked about the

Munich and Taliban negotiations, but probably not about the pencil-owning writer). Where relevant, however, they develop the idea of truth beyond simple factual verification.

Question 2 is specific about the relation between an instance and its generalization: is the example relevant and sufficient? Whether we consider that the example schema has an "implicit premise" or not, it is the implicit premise that is at issue here. It is the bridge to a particular conclusion about another instance, or it is essentially the same as the conclusion in the Walton et al. (2008) schema. The same movement from instance to generalization appears in argument from illustration, as we saw. For induction, the movement to generalization is complicated by the fact of conductive argumentation, but we need only to make "example" plural in the critical question to accommodate this fact. Since I was able to offer a schema for analogy that matched the Walton et al. skeleton for example, we can see that the same sort of "implicit premise" is operative there. In analogy the issue is whether the case's features justify adding A to what we believe about the case, and this is the parallel generalization. Perelman and Olbrechts-Tyteca's (1969) skepticism about the power of analogies is probably aimed at this critical question, since analogies will often only be suggestive of the generalization about A rather than giving the sounder grounds we expect in argument from example or induction. Since the other argument types—narratives, personal testimonies, and the like—only seem to be renaming of "example" the critical question is equally pertinent to them.

Question 3 concerns the example's typicality and its capacity to generalize. Walton et al. (2008) take for granted that this and the other questions apply straightforwardly to illustration, and that seems to be so. The matters in question 3 are worked out with precision in scientific treatments of induction, where we learn details about random sampling, likely error tolerances and sample size, and similar ideas. When the examples are numerous (or should be) these matters can be used to detail the third question. For analogy, this question translates into a concern for whether the theme and phoros are properly matched. Walton et al. (2008, p. 315) offer a critical question for analogy ("Are there differences between C_1 and C_2 that would tend to undermine the force of similarity cited?") that seems to be asking in part whether C_2 is typical of C_1 in important respects.

The fourth question asks about the strength of generalization. This has to do with the degree to which the premises support the conclusion. Like the first critical question, this one seems to apply to any argument scheme. Certainly it is reasonable to ask whether an illustration, induction, analogy, or story is over- or under-claiming its point.

The last question inquires about special characteristics of the premise's example that interfere with its generalizability. In most respects, this repeats the concern of question 3 about typicality. But particularly in the case of examples, it has a special relevance. No two cases of any general category are likely to be identical, especially when the instances are empirical (i.e., historical, observational, etc.). The fact that the examples are different in the first place is a guarantee that some differences can be found. The theme and phoros of an analogy, for instance, are supposed to be different so that a lesson from one pairing can be applied to a second one. Especially when theme and phoros are taken from different intellectual domains, substantial differences between them are going to be immediately apparent. For analogy, this question transforms into a more particular concern that the element A that is moved from C_2 to C_1 is unique to a feature of C_2 that does not appear in C_1. This is a very good question.

For analogy, Walton et al. (2008, p. 315) have a critical question that does not appear to be contemplated in the list of questions for example. Perhaps it should be. The question is, "Is there some other case C_3 that is also similar to C_1, but in which A is false (true)?" The parallel question for example would be, "Is there another example that has property F but not property G?" Whether applied to analogy or example, this is the opportunity to inquire about counter-instances or competing stories. Possibly Walton et al. regard this as the sort of thing that might be offered to support a criticism on the grounds of question 4, but it seems to me that it is a sufficiently important matter that it ought to be elevated to be its own critical question. The current critical questions all have to do with the internal workings of the argument by example, but the importance of an instance external to the constructive argument seems to be an advance that ought to be transported from analogy to example in general (cf. Johnson, 2000, on the dialectical tier).

So I hold that the same set of critical questions, elaborated when circumstances allow (e.g., by inquiring about narrative coherence or sampling frames), serves any of the sorts of arguments I have been examining. This indirect test of my thinking seems not to offer any objection to it.

4. CONCLUSION

As I remarked earlier, van Eemeren and Grootendorst (1992, pp. 96–97) proposed a system of argument schemes that this paper has some consistency with. They say that there are three general kinds of argument: symptomatic arguments (these rely on a relation of concomitance between premise and conclusion; argument from sign is an example), similarity arguments (these rely on a relation of analogy; this category is the present paper's topic), and instrumental arguments (these rely on a relation of causality; means-end arguments are an example). In developing the similarity scheme, they supply a useful list of ways of specifying the sort of relationship we might find between two instances (p. 99). The first instance might have these relations to a second one: it might be *comparable* to the second; it might *congrue* with the second; it might *remind* one of the second; it might *be the same as* the second; it might be *analogous* to the second; it might be *related to* the second; it might *correspond* to the second in a crucial way; it might be *defined along the same major lines* as the second; or it might be *just like* the second. Van Eemeren and Grootendorst present these as characteristic expressions rather than an exhaustive list. Still, this seems to be a nicely detailed set of ways that instances can be related to one another. Since we are also concerned with the possibility that one or more instances can support a generalization, it is worth noticing that these are also ways of connecting a particular case to a category of cases as well.

Although example, illustration, induction, analogy, precedent, anecdote, narrative, and personal testimony have important differences in their rhetorical character, they all seem to have the same essential structure. They move from one instance (or a set of them) to a conclusion about either another instance or a category of instances. In this, they work as examples. Therefore I believe that they should all be seen as instances of the same argument scheme, answerable to the same critical questions.

REFERENCES

Aristotle (1984). *Rhetoric*. (W. R. Roberts, Trans.). In J. Barnes (Ed.), *The complete works of Aristotle* (Vols. 1–2). Princeton, NJ: Princeton University Press.
Benoit, W. (1980). Aristotle's example: The rhetorical induction. *Quarterly Journal of Speech, 66*, 182–192. doi:10.1080/00335638009383514

Fisher, W. R. (1987). *Human communication as narration: Toward a philosophy of reason, value, and action.* Columbia, SC: University of South Carolina Press.

Henkemans, A. F. S. (2000). State-of-the-art: The structure of argumentation. *Argumentation, 14*, 447–473. doi:10.1023/A:1007800305762

Johnson, R. H. (2000). *Manifest rationality.* Mahwah, NJ: Lawrence Erlbaum Associates.

Perelman, Ch., & Olbrechts-Tyteca, L. (1969). *The new rhetoric: A treatise on argumentation.* (J. Wilkinson & P. Weaver, Trans.). Notre Dame, IN: University of Notre Dame Press.

van Eemeren, F. H. (Ed.) (2010). *Strategic maneuvering in argumentative discourse.* Amsterdam: John Benjamins.

van Eemeren, F. H., & Grootendorst, R. (1992). *Argumentation, communication, and fallacies: A pragma-dialectical approach.* Hillsdale, NJ: Lawrence Erlbaum Associates.

van Eemeren, F. H., & Houtlosser, P. (1999). Strategic manoeuvring in argumentative discourse. *Discourse Studies, 1*, 479–497.

Walton, D., Reed, C., & Macagno, F. (2008). *Argumentation schemes.* Cambridge: Cambridge University Press.

Wellman, C. (1971). *Challenge and response.* Carbondale and Edwardsville, IL: Southern Illinois University Press.

Analysis of Arguments Favoring Vaccine Resistance

JESSICA M. HAMPLE

Department of Communication
Western Illinois University
Macomb, IL
USA
jm-hample@wiu.edu

ABSTRACT: This study uses data collected from an internet message board to analyze arguments favoring vaccine resistance. The results replicate previous research into vaccine-resistant groups and identify three themes in vaccine-resistance discussion. The themes identified were: first, feelings of persecution and conspiracy theories; second, feelings of guilt; and third, community-building strategies.

KEYWORDS: vaccine resistance, message boards, persecution, conspiracy, guilt, community.

1. INTRODUCTION

Vaccination is one of the most important and successful public health innovations in medical history. The word vaccine derives from Edward Jenner's work using the *vaccinia*, or cowpox, virus to induce immunity against smallpox. Jenner's work in creating the smallpox vaccine allowed the complete eradication of the disease in 1979. Recently, vaccine-resistance has been a topic of debate. However, this resistance is not a new phenomenon. Resistance to compulsory public vaccination has existed since the first public vaccination program was begun in the United Kingdom.

The Vaccination Act of 1840 provided free vaccinations for the poor while the Vaccination Acts of 1853 and 1857 made smallpox vaccination compulsory for all children under the age of 14. The Vaccination Act of 1853 sparked violent public protests and led to the formation of the Anti-Vaccination League. Resistance to compulsory vaccination increased within the United Kingdom and eventually, the Vaccination Act of 1898 allowed for the first certificates of exemption (Sharpe & Wolfe, 2002). Public vaccination programs also incited protest in the United States. The Anti-Vaccination Society of America was founded in 1879 and anti-vaccination activists were able to repeal compulsory vaccination laws in a number of states (Sharpe & Wolfe, 2002).

While most parents in the United States now choose to follow the recommended childhood vaccination schedule, there remains a community of parents who question the safety of these vaccines. This vaccine-critical community rose to prominence once again as the result of a 1998 study by Andrew Wakefield, et al. Wakefield and twelve other authors published an article in the British medical journal *The Lancet* suggesting a possible link between the MMR vaccination, gastrointestinal disease, and the onset of autism in children (Wakefield, et al., 1998). The assertion of a causal link between the MMR vaccine and autism was retracted in 2004 by eleven of Wakefield's twelve coauthors and the entire article was retracted by the editors of *The Lancet* in 2010 (Murch et al., 2004; The Editors of the Lancet, 2010). However, the phenomenon of vaccine-resistance continues today.

Hample, J.M. (2012). Analysis of arguments favoring vaccine resistance. In J. Goodwin (Ed.), *Between scientists & citizens: Proceedings of a conference at Iowa State University, June 1-2, 2012* (pp. 183-193). Ames, IA: Great Plains Society for the Study of Argumentation. Copyright © 2012 the author(s).

In this study, online conversations regarding vaccine-resistance were collected and analyzed for overarching themes. Findings from previous literature were supported by the data in this study and three further themes were identified: first, feelings of persecution and conspiracy arguments; second, feelings of guilt; and third, community-building.

2. PARENTS' DECISION TO VACCINATE OR NOT

In the United States, vaccinations begin at birth and continue throughout childhood. Very young children receive a large number of vaccinations during their early doctor's visits. For some parents, the practice of injecting a young child with such a large number of vaccines all at once is frightening. They may question the safety of the vaccines, the wisdom of administering many vaccines at once, or the necessity of vaccines to protect against relatively uncommon diseases. Ultimately, each parent must decide whether or not to adhere to the recommended vaccination schedule and understanding the ways parents think about vaccination, the factors that influence their decisions, and their evaluation of information sources should illuminate that decision.

Parents have been shown to display attentional biases to risk information regarding vaccines (Gardner, Davies, McAteer, & Mitchie, 2010). A 2003 British study conducted of parents whose children had not received the complete recommended course of immunizations revealed that, although most (82%) associated vaccines with disease prevention, only 6% reported having no concerns about the safety of vaccination and 34% reported a belief that some vaccines were more risky than the disease they protect against (Smailbegovic, Laing, & Bedford, 2003). Petts and Niemeyer (2004) analyzed group discussions of parents in England's West Midlands area who either had already immunized or intended to immunize their children with the MMR vaccine. They reported that mothers in particular felt a responsibility to make the right medical decisions for their children and that these mothers attempted to balance the risks of disease and vaccination (Petts & Niemeyer, 2004). Vaccination in general presents a problematic decision for parents because they are acting on behalf of their children. The need to make the right choice is paramount and competing messages regarding vaccination risks makes that choice extremely difficult.

Gellatly, McVittie, and Tiliopoulos (2005) surveyed parents in Edinburgh, UK in order to identify factors that would predict the decision to vaccinate or not. All parents listed the same factors as important, regardless of vaccination status, but rated the factors' relative importance differently. Factors that predicted a decision not to vaccinate were perceived adverse reactions of the vaccine (including autism and bowel disease) and current research (Gellatly, McVittie, & Tiliopoulos, 2005). This suggests that parents may not be convinced of their doctors' expertise regarding MMR and erroneously believe that current research still suggests the possibility of an MMR-autism link. Thus, parents who choose not to vaccinate rate the perceived risks of vaccination as more important than the perceived risks of not vaccinating. Factors that predicted a decision to administer the MMR vaccine were leaflets and information packs provided by health professionals and information regarding the risks of contracting rubella. In other words, parents who chose to vaccinate did so because of the perceived risk of not vaccinating.

Petts and Niemeyer (2004) reported that information about possible side effects of measles, mumps, and rubella was new to many parents and was generally regarded as very persuasive information. Combined with the Gellatly et al. (2005) finding that rubella information predicted immunization, it appears that parents who choose to vaccinate do so because the perceived risk of not vaccinating is higher than the perceived risk of vaccinating.

2.1. Information Seeking Behavior

Acquiring and understanding information regarding vaccination is a primary goal for parents. Parents report using multiple sources when gathering information, with most parents reporting that health professionals are their primary source (Smailbegovic et al., 2003). However, one common complaint from parents wondering about vaccine safety is the feeling that their children's doctors do not provide adequate information. Some report that they feel uncomfortable asking doctors for help out of a fear of wasting the doctor's time (Petts & Niemeyer, 2004). Others feel that the doctors are too busy to provide the necessary information and that there was no opportunity to talk to health professionals about safety concerns (Smailbegovic et al., 2003; Gardner et al., 2010). Despite a willingness and desire to discuss concerns with or seek information from health professionals, many parents are still unable to obtain the information that they need to make a confident decision.

While health professionals are the most common source of information, other sources do exist. Parents report using the internet to find information, but also report treating such information with suspicion (Gardner et al., 2010). A further source of information is advice from other parents. Many suggest that information received from other parents is extremely trustworthy as these parents are considered to be honest and unbiased.

2.2. Evaluation of Trustworthiness.

Most parents in the UK do rate health professionals as the most helpful sources of information, but those who do not often cite concerns that these professionals are biased by government policy and may withhold information about vaccine risks (Smailbegovic et al., 2003)—a concern perpetuated by vaccine-critical groups and anti-vaccination arguments. In interviews with parents who had their children immunized but who still expressed concerns with the MMR vaccine, information provided by health professionals was rated as being of poor quality. This poor rating was due to a lack of information about vaccine testing and about research concerning negative side effects of the MMR vaccine (Smailbegovic et al., 2003). Gardner et al. (2010) also found that parents considered more "balanced" information regarding MMR to be more trustworthy. A balanced MMR message was considered to be one that included information both for and against the administration of the vaccine, despite the preponderance of research supporting the vaccine's safety (Gardner et al., 2010). This shows that parents are not content with a simple reinforcement of one message or the other; rather they attempt to evaluate the risks of both pro- and anti-vaccination appeals.

In addition to governmental policy, parents may perceive officials to have financial incentives to vaccinate their children. In the United Kingdom, government sources are especially distrusted for this reason (Petts & Niemeyer, 2004; Gardner et al., 2010). Occasionally this distrust is extended to doctors and other health officials (Gardner et al., 2010). Thus, even when information is made available to parents, it is sometimes not trusted.

3. CHARACTERISTICS OF VACCINE-CRITICAL DISCUSSIONS

A 2007 study by Pru Hobson-West identified a number of vaccine-critical groups in the United Kingdom, categorized them, and characterized their shared and differentiating features. Vaccine-critical groups tend to be small and geographically diverse. They do not necessarily meet in person but use websites usually run by one or a small number of parents. Their discussion of risk and trust is of particular interest when investigating the anti-vaccination fear appeals.

3.1. Talk About Risk

Vaccine-critical groups tend to reframe the concept of risk and portray it as unknown (Hobson-West, 2007). They portray risk information provided by health professionals as being strategic rather than objective. Thus, group members can easily discount it as untrue or incomplete. The ability of vaccinations to actually prevent disease is sometimes questioned while vaccines are simultaneously portrayed as introducing new health risks.

Vaccine-critical groups question the sufficiency of safety trials. For example, experiments testing new vaccines against old vaccines instead of against placebo are said to be irrelevant because they merely prove that the new vaccine has the same side-effects as the previous option. Members of vaccine-critical groups also cite the length of safety trials as being insufficient; side-effects that do not manifest until much later in life would be missed by most experimental designs. As a result of the perceived strategic nature of official risk statistics and the insufficient nature of safety trials, vaccine-critical groups portray officially provided risk information to be largely inaccurate or irrelevant (Hobson-West, 2007).

3.2. Talk About Trust

Trusting healthcare officials and complying with recommendations to vaccinate is considered the easiest option for busy parents, but is also portrayed as being dangerous to their children (Hobson-West, 2007). Parents are encouraged to be "free-thinkers" and to make their own choices. Vaccine-critical groups push members to become experts about vaccination and to make decisions specific to their own child. Educating oneself is positioned as the most important action a parent can take, more even than making any specific decision.

4. METHOD

This study sought to identify and analyze persistent themes in discussions among vaccine-critical individuals. A vaccine-critical message board from the BabyCenter Community forums provided the data for this study. BabyCenter.com is a website devoted to pregnancy and parenting and has an extensive and active message forum community. The None/Select/Delayed Vaccinations board has 3187 members, 6097 threads, and 48445 comments as of the time of this study (BabyCenter, L.L.C, 2012). Data consisted of every discussion thread active during a randomly selected week of February 2012. Conversation threads were excluded from the final analysis only if both the initial topic and the ensuing discussion were unrelated to vaccination or alternatives to vaccination. Thirteen discussion threads were excluded, leaving 66 threads consisting of 597 comments for analysis.

Data were analyzed using the grounded theory method (Corbin & Strauss, 2008). First data were open coded into separate concept units. Through a constant comparative method of rereading and reinterpreting the lists and descriptions of concepts, some of the initial topics were deleted from further consideration because they were not central enough to the corpus of material. Through applying axial coding to other initial topics, the relationships among individually coded concepts were grouped into themes, which are described in the next section.

5. RESULTS

The discussions included in this study supported previous research findings. The risks of vaccines were characterized as unknown and understudied. Parents were often encouraged to research vaccines and not to blindly trust medical advice. Government and pharmaceutical organizations were clearly considered untrustworthy and many parents expressed displeasure concerning their interactions with their doctors. This study will not discuss data that expressly replicates previous findings, however this data set expands the previous research by identifying the same themes in a primarily USA-based message board (most previous research had been conducted in the UK) and by replicating these findings in discussions not based primarily on MMR.

In addition to these expected findings, three new themes were identified: feelings of persecution and conspiracy arguments, feelings of guilt, and community-building. Feelings of persecution and conspiracy arguments work together to cast the vaccine-resistant community as the victim or the underdog. Feelings of guilt regarding previous decisions to vaccinate are addressed explicitly by the community while feelings of guilt regarding the choice not to vaccinate are carefully managed with discussions of vaccine alternatives. Finally, community-building is accomplished through both unifying and distancing language.

5.1 Feelings of Persecution and Conspiracy Arguments

The first and most striking theme to emerge from the discussions was a sense of persecution by pro-vaccination individuals and organizations. Group members often wrote of instances in which they felt attacked by pediatricians or family members. Members created discussion threads asking for advice in finding doctors in their area who were accepting of delaying or refusing vaccinations:

> Any of you live in NYC? I am looking to switch pediatricians, and trying to find a good ped in NY (preferably Brooklyn) who is at least tolerant of no vaxing and is receptive to alternative medicine.

Other threads were devoted to stories of exceptionally negative or positive experiences with medical professionals:

> So today we had our well check apt that I dred going to cause our dr pressures me to vaccinate and I say no every time. There are 3 dr in our office. I go to our apt today and our original dr had something come up and couldn't see us. So we went to one of the other drs. I llloooovvvveeee her. She didn't pressure me at all about not vaccinating and said that was fine with her. She didn't even make me sign the form that says I am putting my child at risk by not vaccinating. She even encourages her patients to try natural things instead of jumping to prescriptions. We switched to her and she is wonderful!!!!!!!!

Occasionally, a poster would ask advice regarding how to talk to a pediatrician or to a family member who disagreed with her decision to avoid vaccinations. Often, the advice given was to refuse to discuss the decision or even to lie about the reasons for the decision. In response to a question about refusing the Hepatitis B vaccine, one poster replied:

> We didn't do it and I basically just lied and said our ped starts it at the 2 month visit. Which isn't entirely a lie. They do, we just don't get it.

Other suggestions were made to "unfriend" a Facebook friend who argued in support of vaccination or to challenge family to research vaccine-safety with the reasoning that the family member would not do the research and that the challenge would effectively end the conversation.

In addition to resistance from individuals, many group members alluded to persecution from organizations. Many group members accused the government of infringing too far into their private lives. Posters commented that "people are violating our rights as parents" and "our rights are eroding every day." References to parental rights generally appeared in discussions concerning school vaccination requirements or governmental programs, such as WIC (a government program that helps pregnant women and parents of young children buy healthy foods), that require up-to-date vaccinations for children. Other references to parental rights appeared in discussions surrounding parents' groups devoted to lobbying state governments to allow philosophical exemptions to vaccination.

Some members of the discussion group also leveled accusations of outright conspiracy against the government, medical community, and the media. Posters wrote of frightening encounters with CDC or public health officials paying visits to their homes:

> Hi there-
> Today I receive a letter today:
> [letter text]
> Clearly, I won't be participating or "helping" in anyway, BUT this does incite a bit of paranoia in me. Has anyone else received this letter?

Responses to these stories again suggested that parents lie to the officials or simply refuse to answer any questions. It was even suggested that children might be taken by Child Protective Services if the parents admitted to choosing not to vaccinate. In response to the above story about a letter from the CDC/NIS, a poster wrote the following:

> On Monday afternoon I caught the tail end of a radio program discussing this very issue. I have no idea who the man being interviewed was but he said he is normally very verbal about his non-vaxing and fights for our rights but that to the CDC he will outright lie. He suggested going as far as having a list of the recommended vaccines next to your phone and to even have the corresponding dates written of when your LO would have had each vaccine. So, I didn't catch whether he knew this for a fact or if it was just a fear but he said those who responded to the survey that they did not vax would be reported to CPS.

Along with the government, the healthcare and pharmaceutical industries were also subject to accusations of conspiracy. One commenter wrote:

> I think vaccines are used for population control too, and to shorten our lifespans/make us sick with lifelong illness (because that makes the pharm. and healthcare industry money).

Another comment suggested that doctors vaccinate other people's children because:

> They must know the world is a competitive place and want their kids to get ahead of everyone else's by destroying them while collecting fees for their 'services'

While previous research has suggested that vaccine-critical parents often question the trustworthiness of the pharmaceutical and healthcare industries, the level of malice attributed to them in the above examples is worthy of note.

A third target for conspiracy accusations is the popular media. One discussion thread concerning a television company inviting vaccine-critical parents to comment on an upcoming program revealed the community's belief that the media cannot be trusted to present their arguments accurately. In advising group members not to appear on the program, one poster wrote:

> They will just portray you as a nutjob in any media piece for not vaccinating—they will quote something you said out of context and contrast it to something Paul Offit said to make you look awfully stupid and dangerous. The media has no interest in decently reporting about vaccine issues or even remotely reporting anything that faintly might criticize vaccines or even mention that there might be bad side effects. Stay away people.

While these conspiracy theories are by no means universal within the community, they do appear in a range of discussion threads. Comments following accusations of conspiracy sometimes ignore the assertion but no comment in the data set directly challenges the accusation. Often, as in the case of the popular media discussion and the comment suggesting that vaccines are a form of population control, subsequent posters expressly agree with the accusation.

5.2 Feelings of Guilt

The second general theme identified in the data set was parents' feelings of guilt regarding their decisions to vaccinate or not vaccinate. The most explicit feelings of guilt were expressed by parents who had already allowed their children to be partially vaccinated. However, many discussions served to alleviate or prevent guilt that might be felt by parents who chose not to vaccinate.

While the majority of group members appear to be expecting mothers or mothers to infants, many have only begun to question vaccinations after fully or partially vaccinating older children. As a result, they often write that they feel guilty for not researching vaccinations earlier. One member commented that her daughter "got hep b and I still regret it" while another began a discussion thread specifically asking if others also felt guilty over prior vaccinations. One poster responded:

> yes, I do feel guilty that I vaxed DS [darling son], (although on a very selective and delayed schedule), my 1st child and that my other children are not vaxed. Even he told me a few times that it's so great that I found out about vaccines and that his sisters did not get any. :) It's kind of bitter-sweet because he first asked me why we gave him vaccines whey they are bad. :(It sure made me feel guilty.

Other group members respond to this and other expressions of guilt by reassuring the original poster that they are making the right choice now and that they should consider themselves good parents compared to others who continue to vaccinate. One commenter even declared that a feeling of guilt was the mark of a good parent.

While commenters rarely discussed feelings of guilt over not vaccinating, many discussions seem to serve the function of alleviating such guilt. Mothers were commonly advised that it is better to delay vaccinations because the decision not to vaccinate could be reversed but that "You canNOT undo vaccine damage!!!" Other commenters reassured posters that "kids get sick, its part of life." One of the most common arguments against vaccination centered on the beliefs that naturally-acquired immunity is superior to vaccine-acquired immunity, that children are supposed to get sick in order to strengthen their immune systems, and that if a child does catch a disease (like chickenpox or whooping cough), that they will recover safely. One poster asked for stories of others' experiences with whooping cough as a way to reassure her that the disease was not overly dangerous.

Many discussion threads serve to relieve guilt by seeking and providing alternatives to vaccination. Most of these conversations centered on the benefits of breastfeeding as well as normal hygiene practices. Instead of the DTaP (Diphtheria, Tetanus, and acellular Pertussis) vaccine, one poster recommended "[Breastfeeding], good hygiene, [and] lots of vitamin D." Another declared that it is "Amazing what nutrition, indoor plumbing/handwashing and education can do!" Others suggested chiropractic, homeopathy, and herbal remedies. One person wrote that "What works best with any virus is homeopathy" while another directed a fellow commenter to a natural remedies discussion board with the suggestion that the people there "might suggest some anti-virals that are natural." Multiple posters mentioned seeing homeopathic pediatricians and one stated that she and her husband began to question vaccine safety after attending a presentation from their chiropractor. In fact, as much discussion was devoted to vaccine alternatives and to natural remedies for vaccine-preventable-diseases as was devoted to the actual choice of whether or not to vaccinate.

5.3 Community-Building

A consequence of the popularity of the internet is the opportunity for individuals to form communities without regard to geographic constraints. While vaccine-resistance is still relatively rare, message boards such as the None/Select/Delayed Vaccinations board allows vaccine-resistant parents to gather in fairly large numbers. It is clear that in this group, at least, there is a strong sense of community among the group members. This community provides advice and emotional support to its members and likely helps to strengthen their resolve in the face of persecution as well as helping to assuage any feelings of guilt they might experience. The community is formed in two ways: first, by unifying group members, and second, by distancing the group from outsiders.

5.3.1 Unifying group members

A feature of the BabyCenter message boards is the ability to give "hugs" to individual posts. The number of hugs given to each post (most do not receive any) is displayed underneath the text of the message. In this way, group members can express affection for each other without necessarily needing to post a new response. The posts with the largest amount of hugs generally related personal stories about purported vaccine-reactions. For example, one poster received five hugs (the most any post received was six) for a story about her little sister's developmental difficulties. Other posts received hugs for stories about a poster's own children. In addition to the hugs, many of the responses to these stories expressed sympathy, hope for the child, or the promise to pray for both the child and the poster.

Other comments that received multiple hugs served a less emotional community-building function. One member announced that she had been on the news as part of a group lobbying their state government for the right to file philosophical (rather than religious) exemptions to school vaccine requirements. This post received multiple hugs as well as comments supporting the original poster's cause and thanking her for her work. In this instance, one group member was thanked for publicly pursuing the community's goals.

Another discussion centered around group members' wish to have a similar vaccine-resistant parents' group outside of the internet. Many of the commenters in this thread wrote that they did not know anyone or only knew of one or two people outside of the internet who held similar views. More hugs were given to messages in a discussion asking how old group members were. This thread served the function of the "introduce yourself" threads found in many message boards but not in the None/Select/Delayed Vaccinations board.

5.3.2 Distancing from outsiders

The second community-building activity evident in the message boards is the distancing of the community from outsiders. Group members distance themselves from family, health professionals, and organizations through the persecution and conspiracy language outlined above. It is worth noting that the discussion of the media conspiracy was awarded multiple hugs. Stories of negative interactions with family members also received hugs and generally earned encouraging or sympathetic responses.

Group members also make efforts to distance themselves from parents who choose to vaccinate their children. Some posters declare that they feel sorry for parents who are tricked or pressured into vaccinating their children. Others express disbelief that any parent would choose to vaccinate. As one group member says:

> People continue to inject their babies with poison vaccines and just follow what the doctors say like good little sheeple. It blows my mind!

References to "sheeple" or to "drinking the Kool-Aid" appear regularly. In expressing sympathy or disgust for pro-vaccination parents, the members of the vaccine-resistant group declare that they are not only different from but also superior to those parents who choose to vaccinate.

Finally, the feelings of persecution occasionally serve to differentiate the vaccine-resistant group from the pro-vaccine parents. Posters refer to "rabid provaxxers," caution each other against engaging them in debate, and recount particularly offensive comments made by the "provaxxers." Much like the belief that the media will never listen to a vaccine-resistant argument, group members declare that the pro-vaccination community refuses to listen to vaccine-resistant arguments. Once again, group members advise each other to avoid the conversation entirely.

6. CONCLUSIONS

The data collected from the None/Select/Delayed Vaccinations message board corroborated previous findings regarding vaccine-critical groups. These conversations, however, were not focused on a single objection to vaccination (like the MMR-Autism controversy) but existed independently of any particular safety question. Thus, previous findings primarily regarding MMR resistance in the United Kingdom have been replicated in a general vaccine-critical United States sample, some time after the Andrew Wakefield controversy has faded from public view.

Furthermore, three new themes of vaccine-critical groups have been identified. The first, feelings of persecution and conspiracy theories illuminate the community's sense of victimization. Second, feelings of guilt due to both vaccinating and not vaccinating are evident, however the first is explicitly addressed while the second is addressed by exchanging vaccine alternatives to lessen the severity or susceptibility of the disease. Finally, community-building language serves both to unify the community members (providing support for each individual's decision not to vaccinate) and to distance the community from outsiders. This distancing language is tied to the persecution and conspiracy discussions and allows the community to jointly develop arguments against the outsiders (e.g., the "rabid provaxxers" cannot be reasoned with or the government cannot be trusted to tell the truth.)

7. IMPLICATIONS

Current scientific research overwhelmingly supports vaccination as a "best practice" in public health. While this study primarily sought to identify themes within vaccine-resistant arguments, the findings do have implications for future attempts to counter vaccine-resistance.

First, the pervasive feeling that the government and healthcare officials are violating parents' rights should not be ignored. The question of individual rights versus the good of society is perhaps the most objectively legitimate argument made by vaccine-resistant groups. Future attempts to address vaccine-resistance are unlikely to succeed if the parents feel "forced" or "bullied" into compliance.

Second, it is clear from this text that alternatives to vaccines are just as much a part of the vaccine-resistance argument as the safety of the vaccine or the danger of the disease. The practice of comparing the vaccine and the disease may be insufficient. Future messages might focus on the efficacy of the vaccine versus naturally-acquired immunity, or versus breastfeeding, or versus simple handwashing.

Finally, the use of anecdotes in the vaccine-resistance argument should be more closely studied. While anecdotal evidence serves a persuasive purpose, it also serves a community-building function, allowing community members to forge closer emotional bonds with each other. Care should be taken to position healthcare officials as caring and part of the parent's community if they are to be persuasive.

REFERENCES

BabyCenter, L.L.C. (2012). None/Select/Delayed Vaccinations. Retrieved from http://community.babycenter.com/groups/a233655/noneselectdelayed_vaccinations
Corbin, J., & Strauss, A. (2008). *Basics of Qualitative Research* (3rd ed.). Los Angeles, CA: Sage Publications.

Gardner, B., Davies, A., McAteer, J., & Michie, S. (2010). Beliefs underlying UK parents' views towards MMR promotion interventions: A qualitative study. *Psychology, Health & Medicine, 15*(2), 220–230.

Gellatly, J., McVittie, C., & Tiliopoulos, N. (2005). Predicting parents' decisions on MMR immunisation: A mixed method investigation. *Family Practice, 22*(1), 658–662.

Hobson-West, P. (2007). Trusting blindly can be the biggest risk of all: Organised resistance to childhood vaccination in the UK. *Sociology of Health & Illness, 29*(2), 198–215.

Murch, S. H., Anthony, A., Casson, D. H., Malik, M., Berelowitz, M., Dhillon, A. P.,...& Walker-Smith, J. A. (2004). Retraction of an interpretation. *The Lancet, 363*(9411), 750.

Petts, J., & Niemeyer, S. (2004). Health risk communication and amplification: Learning from the MMR vaccination controversy. *Health, Risk & Society, 6*(1), 7–23.

Sharp, L. K, & Wolfe, R. M. (200, August 24) Anti-vaccinationists past and present. (Education and Debate). *British Medical Journal, 325*(7361), 430. *Expanded Academic ASAP*. Retrieved from. http://go.galegroup.com/ps/i.do?id=GALE%7CA91560835&v=2.1&u=westerniul&it=r&p=EAIM&sw=w

Smailbegovic, M. S., Laing, G. J., & Bedford, H. (2003). Why do parents decide against immunization? The effect of health beliefs and health professionals. *Child: Care, Health & Development, 29*(4), 303–311.

The Editors of the Lancet. (2010). Retraction—Ileal-lymphoid-nodular hyperplasia, non-specific colitis, and pervasive developmental disorder in children. *The Lancet, 375*(9713), 445.

Wakefield, A. J., Murch, S. H., Linnell, J., Casson, D. M., Malik, M., Berelowitz,...& Walker-Smith, J. A. 1998). Retracted: Ileal-lymphoid-nodular hyperplasia, non-specific colitis, and pervasive developmental disorder in children. *The Lancet, 351*(9103), 637–641.

"Accommodating Science": A New Way of Thinking about Rhetorical Dynamics

THIERRY HERMAN

School of French as a Foreign Language
University of Lausanne
Anthropôle
1015 Dorigny
CH – Switzerland
Thierry.herman@unil.ch

CAMILLIA SALAS

French Literature Department
Module "Writing and argumentation"
University of Neuchâtel
Espace Louis Agassiz 1
2000 Neuchâtel
CH – Switzerland
Camillia.salas@unine.ch.

ABSTRACT: By analyzing three case studies (neutrinos, victimization survey and quality of mass media), our present issue is to figure out if underlying successive accommodations to new rhetorical situations will have an impact on the respective importance of logos, ethos and pathos. We would like to pinpoint the stakes of science's public dimensions considering the scientists' image, their expertise, and also the given results' implication. We will especially take into account scientific papers that may be or are potentially controversial in the political, media and civic spheres.

KEYWORDS: accommodations, ethos, logos, media sphere, pathos, rhetorical dynamics, scientific sphere, socialization of science.

1. INTRODUCTION

For a long time, the genre of scientific texts was considered *closed*—one of its goals being the production and the transmission of knowledge among members of the scientific community (Maingueneau & Charaudeau, 2002). But this is a double illusion since on the one hand, the scientific sphere is tightly linked to others—such as for instance the political, civic or media spheres. This *process of socialization of science* (Beacco, 2000) gets also much more attention from the media, when issues that are being discussed are involved with society. On the other hand, knowledge passes from one sphere to another through reformulations between primary and secondary discourses. "Studying such discourses is part of a contrastive perspective; a guideline that has only rarely been adopted by those interested in scientific popularization" (Jacobi, 1999, p. 150). Therefore, the main goal of this study is to see how such textual disseminations—involving many types of adaptations and reformulations—become readjustments to a rhetorical situation (Bitzer, 1968), which is different each time.

Herman, T., & Salas, C. (2012). "Accommodating science": A new way of thinking about rhetorical dynamics. In J. Goodwin (Ed.), *Between scientists & citizens: Proceedings of a conference at Iowa State University, June 1-2, 2012* (pp. 195-208). Ames, IA: Great Plains Society for the Study of Argumentation. Copyright © 2012 the author(s).

Our theoretical framework is based on two approaches. First of all, considering that plural differential comparative levels between two texts allow a hermeneutic scan of rhetorical and argumentative dynamics, we use comparative discourse analysis (Adam, 2005; Heidmann, 2005a, 2005b). This approach helps us to identify the rhetorical and argumentative[1] dynamics. Thus we will focus on three major pillars of the Aristotelian rhetoric, i.e., matters of ethos, logos and pathos. Indeed, following Aristotle's idea and his successors like Chaïm Perelman (1977) or Ruth Amossy (2010), every public discourse that aims to convince an audience, is necessarily under the interdependent influence of these means of persuasion. The second approach belongs to the recent tradition of media analysis. Indeed, Herbert Gans stated in 1979 that "media specialists have been too neglectful" (chap. 4) of studying the sources. In the same way, the publication of Schlesinger's article (1992) accuses the sociology of mass media of "media-centrism," because it encourages the study of journalism by looking at the position of information in society and a decentralization of the subject—all too often focused on the media themselves, isolated from their sources. Reacting to this article, several studies point out the impact that sources have on media accommodation. Following this idea we want to study all possible discursive changes of scientific reports: those external to the media field (primary discourse and press releases), as well as those internal to the media field (wire story of the Swiss telegraphic agency, newspapers articles).

Our purpose is to figure out if underlying successive accommodations (Fahnestock, 1985) to new rhetorical situations will have an impact on the respective importance of the three different appeals to persuasion (logos, ethos, pathos). By focusing our analysis on these categories, we would like to pinpoint the stakes of science's public dimensions considering the scientists' image, their expertise, and also the given results' implication, in the civic sphere. We will especially take into account scientific papers that may be or are potentially controversial in the political, media and the civic spheres. We also want to investigate if a series of adaptations may have escaped the control of the original texts' authors, or even ended up contradicting the primary discourse. In other words, it is interesting to study ethos when evaluating the assertion and the maintenance of scientists' credibility in controversial issues. Logos will help the investigation of the evolution of accuracy and caution enhanced by scientists from both the scientific and the media sphere. Finally, pathos will examine if the implication of results in the civic sphere will evolve according to discursive changes.

Also, in order to avoid getting trapped by the particularities of each study, we chose to build a corpus that offers as much diversity as possible. In the present case, our discursive sources are related to Physics (first study about neutrinos), Forensic Science (second study about victimization), and the Media and Communication Studies (third study about quality of mass media), respectively. According to Alice Toma (2005), the diffusion of science can be divided into three discursive frameworks: popularized discourse—informative activity (press release); didactic discourse—explicative activity (study 2&3); research discourse—discovery activity (study 1).

---

[1] Firstly as rhetoric of science "that emphasizes the interactions at work through texts and the way the scientific discourse, far from being focused only on logos, has also to convince its addressees of the legitimacy of what is previously announced" (Rinck, 2010, p. 432). Secondly, we also consider that rhetoric and argumentation are fields that work regarding a dynamic process where argumentativity is inherent in discourse and it is appearing in such different levels. Consequently, an analysis of a global discursive working is possible in which verbal means related to logos, ethos and pathos are implemented in order to act on an audience. In any way, to this approach, there is no need to consider rhetoric and argumentation as distinctive fields (Amossy & Koren, 2009).

Before starting the comparative analysis itself, we would like to point out that the first part of our project had to be modified. At first, we wanted to examine if moving from technical to civic sphere would transform arguments regardless of the scientific procedures used by scientists. We especially wanted to identify how the media cover qualitative, as opposed to quantitative procedures. We were however surprised to notice that the Swiss mass media we analyzed offered hardly any qualitative studies. Therefore the comparison would have been hard to establish. This observation also raises the question of the representativeness of a serious study, which a priori, would be more significant by using statistical rather than interpretative tools.

2. STUDY 1: NEUTRINOS CASE

An experience called OPERA, led between European Organization of Nuclear Research and an Italian laboratory in Gran Sasso, aroused an unexpected interest for physics of particles in news media when it was announced, in September 2011, that some elementary particles of light, called neutrinos, have been measured with a velocity faster than speed of light in vacuum. Since the velocity of light is supposed to never be exceeded, according to Einstein's theory of relativity, this announcement provoked astonishment and scepticism among physicists and made headlines in the news.

The rhetorical dynamics that we will study are based on four stages of wording and rewording (Authier-Revuz, 1982):

(1) The conclusion of the scientific paper published on September 23rd.
(2) The press release made by French National Center of Scientific Research the day before.
(3) The wire story based on the press release written by the Swiss Telegraphic Agency published after the press was put to bed, on September 22nd.
(4) Five news articles published in French that announce the results of measurements.

As mentioned earlier, our study will examine the strength of the three rhetorical poles (ethos, logos and pathos) in each rhetorical situation. The stakes in the scientific article of physics are indeed essentially connected to logos—how to justify this astonishing announcement—and to ethos—showing the credibility of an incredible speed. Such a persuasion strategy addressed to peers doesn't need pathos—probably because of the discursive genre, but also because scientific ethos of seriousness is precisely built on the rejection of emotion, above all when measurements are so exceptional and unexpected.

Let us look at how the article was reworded. For the press release, the stakes are different than for the original article: the goal is to increase the value of the results found by the scholars and catch the attention of the media—so, we can suspect that logos will be less important than ethos and pathos. Indeed, the focus won't be placed on the physicists' effort to eliminate hypotheses, but on the results of the experience; ethos will be crucial because the French Center needs their scholars to appear trustworthy; lastly, since the impact of such measurements could increase the CNRS' prestige, the temptation of writing a more pathemic press release could be expected.

Obviously, such a temptation might be increased by media adaptations of these discourses to the detriment of logos and ethos. We can indeed expect a loss of accuracy in information when a text is adapted for a wider readership, as well as a polarization of news coverage somewhere in between total acceptance or skepticism of the scientists' work. Let's have a look at texts that examine these hypotheses.

2.1 Logos

First of all, let's say that we have a fairly loose view of logos: what French linguists call argumentative orientation of language (Ducrot & Ancombres, 1983), argumentative dimension of discourse (Amossy, 2010) or schematization (Grize, 1996) are all included here: a selection of data for example "isn't yet argumentation, but it's already a persuasive strategy" asserts Marc Angenot (2008, p. 149). We'll focus on two angles here: communication of data and linguistic designation of the results. The only example, among others, that we'll show here is about the measured speed.

In the scientific article, speed of neutrinos is worded in a mathematic formula ($dt = (57.8 \pm 7.8$ (stat.) $- 5.9+8.3$ (sys)) ns.), whereas the press release states "60 nanoseconds faster (than speed of light photons)" and makes an analogy: "in other words, in a long-distance race of 730 kilometers, neutrinos cross the finishing line with an advance of 20 meters compared to hypothetical photons that have covered the same distance." The wire story asserts "a speed of 300,006 km/second, that is to say 6 km/s faster than speed of light" and quotes the analogy in the press release. Two examples from the news: "60 billionth seconds earlier" in the French newspaper *Le Monde* and "10 (sic) nanoseconds earlier, meaning an advance of 20 meters on a 730 km path" in the Swiss newspaper *24H*.

This example illustrates some well-known aspects of scientific popularization (Jeanneret, 1994; Jacobi, 1999; Jacobi & Schiele, 1988). Since rewording requires simplification, abstract unities like nanoseconds are redefined, adapted to a more familiar system (speed is stated in km/sec) or explained by an analogy of a race between photons and neutrinos. Measures are also rounded up leading to a loss of accuracy founded on approximations (e.g., speed of neutrinos at 300,006 km/sec) or even mistakes (an advance of 10 nanoseconds is the margin of error, not the measured speed).

Considering this from the point of view of linguistic designation, it was the "result" in the scientific article that was communicated. This result was also described as an "observed anomaly." News coverage emphasizes the necessary caution with this result. But the wire story already underlines a "regularly measured achievement." Three newspapers mention the word "discovery," which authors of the scientific paper advise journalists not to use in a press conference. Besides the words "results," "observation," and "information," we also find more colored wording like "something that will make Einstein turn around in his grave," "a possible revolution," a "bomb," if true, for physics. Although the neutral wording of "results" is quantitatively more present, the wording of "anomaly" at the end of the scientific article is never picked up again by news coverage.

2.2 Ethos

The impact of the result led physicists to ensure their credibility: first of all, they systematically rejected objections against their measurements; second, they chose to publish their lack of understanding instead of a result or, worse, of a discovery. Then the scientific paper is signed by 189 researchers: this great number of signatories builds a kind of credibility founded on a classical ad populum topos: the greater the number of researchers who work on an issue, the higher the chances for calculation mistakes or biases to survive criticism.

The last paragraph of the scientific article sheds an interesting light on the authors' belief in soundness of their results combined with an ethos made of caution and modesty:

> In conclusion, despite the large significance of the measurement reported here and the robustness of the analysis, the potentially great impact of the result motivates the continuation of our studies in order to investigate possible still unknown systematic effects that could explain the observed anomaly. We deliberately do not attempt any theoretical or phenomenological interpretation of the result.

On the one hand, the team of researchers underlines that they are not boasting: "large significance," "robustness," etc., but, on the other hand, they hold their certainty back, imposing "deliberately" further investigation on themselves rather than interpretation.

CNRS' press release, which may be worried first and foremost about their ethos of credibility as a respectable institution, will pinpoint this problem and work on it with considerable caution. Firstly, the result itself is called into question, since it is said that velocity faster than speed of light is "what seem to indicate carried out measurements." Now, the scientific article mentions an effectively obtained result. Epistemic modality of the verb "to seem" adds a layer of uncertainty on measurement that denotes extreme caution by CNRS on the result.

Other signs help the team of researchers from being condemned for eagerness to glory and haste to publish an uncertain and astonishing result. Press release says indeed: "After three years of very highly precise measurements and complex analysis, OPERA experience states a completely unexpected result." The left dislocation of a time adverbial that stresses the lapse of time, the mastery of complexity and the high value of measurements is a good sign of strengthening team's ethos. They did not hurry; they had a leading-edge technology and had made complex analyses. Quotation of the team in the press release helps also to show their caution and modesty. The aim is clearly to dismantle a hypothesis of a kind of burst of enthusiasm in front of the probable controversy that will be triggered by this result.

Is this ethos of caution and respectability retained by the media? The adopted persuasive strategy seems to have hit the bull's eye. The wire story quotes numerous verifying tasks made by the team and finishes with an intensive form: "Even continental drift and the devastating earthquake from L'Aquila have been considered." It also mentions authority ("Verified by great independent experts") and emphasizes the staunch position taken. Seriousness of the researchers' team is not questionable. Newspapers stress also caution from the physicists who do not believe measurements: "everything is passed over," states *Liberation* just before making a list of analyses made by the team, which is followed by a statement about the "extreme caution" of physicists. In short, no media calls into question the astonishing result or derides the scientific community for getting excited by discoveries. With remarkable consistency, they praise researchers' seriousness and caution, even with a sardonic hint hidden in the hyperbole from the word "extreme."

2.3 Pathos

Analysis of logos and ethos shows a form of relative control from original communication to mass-media: desires of researchers and National Center seem to be, more or less, respected.

We have nevertheless seen that temptation to describe the result as a discovery was so strong that some parts of the media gave in. The "spot" where control of information will escape from the scientific sphere is on the level of pathos. Emotional tone, which is missing in the scientific article, even if it surfaces with "potentially great impact," finds two essential ways starting with the press release: the physicists' own emotions and emotions provoked by an astounding result.

The physicists' emotion is present starting from the press release, either denoted (Micheli, 2008)—"astonishing result" for example—or connoted (Micheli, 2008)—"completely unexpected result" is said twice. The wire story, based on the press release, emphasized in its lead's section (known as a crucial place of information (Ross, 2005)) this emotion: it quotes precisely "completely unexpected" and "astonishing" from the press release and then adds "physicists didn't believe their instruments." The point of discovery is now depicted as a story: the media sphere tries to imagine physicists' incredulous reaction. This will snowball: emotions we picked up are excitation (physicists are in turmoil), fear of mistake, and, above all, "surprise." We find four newspaper articles with this word, sometimes going with attributive adjective "total." In *Liberation*, qualifying the event as surprising is a proof "to have an acute sense of understatement." In fact, this newspaper shows hyperbole and familiar tone to emphasize breaking news: "in March, Dario Autiero watches closely measurements and is flabbergasted" (French literal translation is "falling on his ass").

Our first analysis shows that expressed emotion in front of a possible discovery is interesting for media: the press release mentions emotions only indirectly, describing the result as astonishing and unexpected; the news source puts in the front page that the physicist is frozen by surprise and excited by the impact of the measurements.

The desire in the article to deliberately ignore all interpretations and calls for new experiences only resists slightly the media adaptation. The press release urges the need for further independent measurements for verification. But the wire story is clearly excited by the implications of result: it says "hence challenging Einstein's relativity theory!" with a very rare exclamation mark for this media genre. This way of emphasizing impact is put at the forefront in the newspapers; several titles mention Einstein: "Eintein outdistanced," "Einstein soon relative?" "Faster than light and farer than Einstein," "A fabulous speeding is threatening Einstein's theory." Every newspaper article envisions consequences of the measurements and finishes with a series of hypotheses, each with an explanation: new dimension, non-trivial space geometry, quantic mousse and so forth.

In brief, our review shows a kind of "heating of the minds" by the media sphere related to consequences of measurements. That creates new rhetorical dynamics, emphasizing pathos, which was precisely the dynamics that physicists try to prevent in their communications (article and press release).

3. STUDY 2: CRIME RATE CASE

The issue of rise in crime in Switzerland that communicates the 2011 national victimization survey creates many reactions among the civil society (particularly questions about sense of security, self-confidence in officers' work, Quid of the crime rate that reaches the European one). This survey was financed by the Conference of the Cantonal Heads of the Departments of Police and Justice (CCDJP) and supervised by the Police of Bern and the Institute of Criminology (University of Zürich).

The analysis of the rhetorical dynamics are based on the French version of the full report published on the 26th of August 2011 (focusing essentially on the results section), the French version of the CCDJP press release, the wire story of the Swiss Telegraphic Agency and 5 newspapers articles (French-speaking Switzerland) published the following day of the press release.

Much more attention needs to be paid to the particular aspect of this study that reflects the relationship between the scientific community (UniZh/Institute of Criminology)—considered as an "expert"—and the public authorities (CCDJP) who asked for the creation of the report. This specific context highlights a didactic feature that we can notice by the non-common use of explanation statistical terms, concrete examples, and the underlining of statistical standards such as, for example, the definition of the weighting factor.

3.1 Logos

As in the neutrinos' case, the logos undergoes a loss of accuracy at the beginning of the press release. This reduction from the original report to the press release reveals itself over 3 modes:

- Selection of the disclosed information (7 offenses treated against 13 in the report).
- Communication of not all the statistical results—which are indicated above a table in the primary discourse and changed in textual approximations on the accommodated one (e.g., "after a long period of stability," without the exact indication of the period).
- Mutation of the statistical information by using frequency adverbs ("The last 178 offenses occurred in the following places: [listing]. In 23 cases, a weapon was used, including 11 knives" – (report) vs. "recorded offenses especially occur in the street and do not concern the domestic violence area. Those tend to be more serious rather than by the past," "Threat increased in a sensitive way – (press release)).

These changes reflect the well-known effects of the press release: a pre-formulated text for the mass media (Jacobs, 1999), encouraging an easier readability (Pander Maat, 2007, 2008). However, our analyses show that the press release is not the only source allowing the diffusion of the results to the media. Information was not exclusively channeled by this one.

Loss of accuracy is emphasized in the wire story's discursive rewrite. This version seems to be inspired from the press release. Thus, we can identify a shift in meaning (offenses' consequences more serious (wire story) vs. offenses more serious (press release)), a temporal ellipsis generalizing the discourse (reported offenses' rates are very stable (wire story) vs. rates are the same in 2000 and 2005), or a change in the modality (the rise of crime is due to (wire) vs. this increase can be explained).

In the same way, media accommodations still compromise the accuracy; consequently, there is confusion between types of burglary and crime rate (*Le Matin*), or the percentage approximation between the burglary and burglary attempt.

3.2 Ethos

In the report, there is a credibility ethos, reflected by a kind of know-how—legitimating expert status. The report mentions scientists' skills, perhaps even those from Switzerland in the survey practice. Ex: "this survey illustrates a certain 'tradition' of Swiss studies that are accustomed to interest in the phenomenon of a national and also regional perspective."

Moreover, the report's structure plays a role in the credibility of the scientific community (graphics, overdeveloped method, detailed tables, sources indicated and appendix available).

Furthermore, we can identify the features of a quantitative approach (exclusive use of statistical tools) that encourage an ethos of certainty, since "figures speak for themselves" and work as a material evidence. This can be underlined by the use of constative verbs: "study (…) showed," "we observe."

Other kinds of discursive changes convey also an ethos of certainty. Indeed, in the press release, results are not questioned but are indicated in an assertive way: "New rise in the crime rate—Switzerland reaches the European mean." We found again constative verbs that stimulate the idea of evidence: "results show," "the study demonstrates."

Finally, in the coverage media, results are not contested, since they are privileged by the given credence to scientific authority: "Myth of Switzerland as 'the safest country in the world' is done: that's claimed by the Pr. Killias on the basis of a study" (*Le Matin*). This example illustrates the crystallization of the communicated result as unequivocal. On the one hand, we have the appeal of authority through the scientific aspect of the study, representativeness of the figured results, the extent of this survey, and the use of the social category "professor." On the other hand, modal verbs in the present tense allow the finalization of the crystallization's process. Several assertive formulations show the given credence to the scientific authority: "Martin Killias is categorical" (*La liberté*).

Furthermore, even if this certainty can be contested—"theory of the rise in crime isn't approved unanimously"—the newspaper strengthens this by taking a stand: "Martin Killias counterattacks vigorously" (*La liberté*). We notice that the media believes in scientific words—insofar Killias is a good media speaker and the scientific community shows consideration for him—regardless of the partnership.

3.3 Pathos

The pathos is also reactive to the rhetorical dynamics, as the matter of security is a recurrent political item in Switzerland. Prof. Killias's words related to security (i.e, "Myth of Switzerland as 'the safest country in the world' is done") work as a catalyst of discursive change. We identify in the media designations that call to mind the ending of one of the Swiss constitutive values: "This is the death of a myth: Switzerland is no more an island of security in Europe" (*24H*), or "Switzerland is no more an island against crime" (*La liberté*), or "Switzerland is no longer safe" (*Le Matin*). However, our analysis must be refined in several ways. Over the press attention-grabbing titles, pathos does not really spread itself in the different rewrites of scientific information. The media's dramatization in the diffusion of the information seems very weak to us, aside from the examples mentioned above.

Nevertheless, if pathos is massively found in opinion pages (editorials, global opinion) it gives also the impression that scientific results are brought forth for political purposes. For example, on the matter of security, we found a great deal of positive or negative terms, evaluative verbs, metaphor, or satirical tone: "But the political community is still ridiculing the legitimate anxiety of the population. The left refuses to face the truth" (*24H*). "The facts speak for themselves. The sense of security felt by the population is from now on approved by a serious study, conducted with victims and . . ." (*24H*). Thus, media work as a source for controversy, since they take for granted the scientific credence of the study, believe in Prof. Killias' words accentuating then the vision showed by the study itself.

On the contrary, the press release is more careful on political matters, highlighting on the officers' positive image. We notice that through the use of selective information (no indication of the assessment of officers' work quality), the press release makes claims that the majority is satisfied by the police or rephrasing over evaluated data (88.4% people interviewed consider officers' work as good or very good vs. very good or good enough). Consequently, disclosing a positive image of the police will contribute to the diffusion of a positive pathos towards policemen.

If we are talking about property crime, results of the report are neither minimized nor exaggerated in the press release. On the contrary, if we are talking about crime against life or personal freedom, we clearly observe the results are minimized: e.g., the press release brings out attacks on youth population who are less concerned about this kind of attacks. Moreover, places and seriousness of the offenses are "erased": there is no indication of public spaces, but the press release uses the term "street" in order to create an emphasis on the comparison with domestic violence term. This process can give the impression that we feel more secure at home.

It seems that the political dimension is more highlighted than the scientific one. The press release becomes aware, in a way, of the social controversy relating to the matter of crime. One of the effects is to limit the impact of the study's major results. This careful diffusion of the press release does not stop the disclosure of strong reaction through the coverage media.

Thus, what is encouraging in the case of this study in controversy is not the communication of the results but their implications in the different spheres of concern (political, economical or civic one). As an indirect consequence, the scientific research is useful as a powerful argument for texts for political purpose: expert's credibility is not questioned. Furthermore, we also notice that the issue of crime received widespread media coverage in Switzerland. So, in the case of media adaptation, the strong visibility about crime matter, already existing (priming effect) will make this information more interesting.

4. STUDY 3: THE CASE OF MASS MEDIA QUALITY

In August 2010, a yearly study edited for the first time indicates the decline in standards of Swiss mass media. Conducted by the Forschungsbereich Öffentlichkeit und Gesellschaft/ University of Zurich ("fög"), the major public dimension of this research underlines a rhetorical dynamics that are radically different from the two above studies.

As study 1 and 2, we will consider, briefly, the same documents: the French version of the annals "Quality of mass media: major findings," focusing essentially on the 1st chapter that presents major results (primary discourse)—French version of the fög's press release, wire story of the Swiss telegraphic agency, and 4 newspapers articles.

As in study 2, we have identified a didactic discourse, but there is also a popularized one. The rhetorical situation is different in this case, since the research is sponsored by private and publics funds, especially the fög, a non-profit foundation. Beyond the explicative reasoning through the didactic discourse, there is also an argumentative stake: diffusion of the results and publicizing of these annals. It seems that this process failed and its reason may be related to the rhetorical dynamics of the discursive changes.

4.1 Logos

Scientific precaution and exhibited precision noticed in the other studies do not really exist in the present results. We are surprised by the excessive use of a simplified vocabulary, assertive phrasing (present gnomic) that may tend to exaggerated generalizations, but also negative terms (to reduce, to disappear, to damage, to melt, survival, tragic, to suffer, to put pressure on, to be going downhill):

Ex: "new media, free dailies and news websites, but also major paid-for newspapers *reduce information to wire stories*." (Bayraktar et al., p. 7)

Ex: "*central task of journalism*, i.e. integrating events into their context—based on deep investigations, *is no longer* practiced nowadays." (Bayraktar et al., p. 8)

As many strong opinions as possible, not belonging to the scientific socio-discursive practice (Adam, 2011), forces us to question the reasons of these overdeveloped assertions: does scientific aura of the research permit changing opinions to valuable interpretation?

The assertive strength of the results in the study is nevertheless diminished in the press release. We will cite one of several examples: "the establishment of a free culture online and offline" becomes in the press release "Use of the media has shifted to an online and offline free culture."

Consequently, our analyses make it apparent that rephrasing differences occur when it is time to pass judgments on the quality of the media. These distinctions tend to minimize the original phrasing of the annals by the addition of toning down forms and by the rephrasing of negative words, even pejorative ones. Terms as "to disappear," "to be at its height," "at the expense of," "to weaken" are modified in the press release, even deleted. The media criticism via the research is finally minimized in the press release.

In the same way, the wire story, which is much more inspired from the press conference than the press release, is also going to tone down the wording of the release: "the *loss* in standards of media has an impact on democracy" vs. "*the decline* in standards of the media presses on democracy."

4.2 Ethos

As the logos expected of it, we can identify an ethos of certainty in the annals: "we are forced to admit that the significance of genres and types of news providers, who already participate in the decline of standards, will continue to increase." This ethos of certainty, expected in the scientific discourse, is related to an uncommon ethos—a sign maybe of an interpenetration of civic and the scientific sphere. Indeed, we can notice that scientists seem to be "entrusted on a mission": their goal is "*to make aware* of the *need* to have media of good quality in Switzerland" (Bayraktar et al., p. 4) Acting like judges of the media quality, they oppose journalism of low quality and investigative journalism. As they were very critical of new media, by asserting that the success of free journals are prejudicial to the quality of journalism, scientists show clearly their purpose of making a criticism of the quality of media. This attitude may question the scientific character of their reasoning.

However, the rhetorical dynamics change again in the press release: this ethos of certainty and the evaluative markers will not be communicated. Consequently, some of the familiar original phrasings that could spoil credibility of the experts (as "to be at its height") are avoided. Furthermore, the press release emphasizes seriousness of the study by highlighting, for example, institutional acronyms.

On the contrary, this ethos of credibility doesn't really appear in the wire story. The vague linguistic designation of the scientists could indicate their lack of credibility: "academic," "researchers," "authors' paper" or "a dozen of researchers of the University of Zurich."

Nevertheless, it is interesting to point out that, while the scientific discourse is circulated from one sphere to another, additional information is mentioned with regards to the original study. This process is more likely done not to "reinvent" information, but on the contrary to "diversify" it. We can imagine that the media works as a source for controversy: indeed the wire story, on its own initiative, contacted other actors who may express themselves on the quality of the media in Switzerland, delineating that the unilateral version of this report worked as cutting words towards the media.

4.3 Pathos

Concerning the evolution of the pathos, we notice its major apparition on the annals. The more rephrasing there is, the more the pathos is minimized. For example, in the wire story most of the elements of the pathos are in fact those of the scientists themselves or exposed by attributive verbs as "to deplore" or "to warn."

On the contrary, in the original report, we can identify the scientists' enthusiasm for their research, but also the results' dramatization:

Ex: "advertising that assures the major part of media receipts—according to private-sector principles—has *melted* in a *dramatic way* (table.1)."

Ex: "this reduction affects essentially the paid-for newspapers . . . in its *survivability*."

Consequently the rhetorical dynamics in this case are the opposite of the previous studies. Indeed, rephrasing seems to erase the discrepancies regarding expectations of the scientific genre. This lack of scientific standards in the annals may explain why there is scepticism from the media. That is why they, firstly, only take for granted what is on the wire story and secondly they start the controversy of the quality of the research itself. Indeed, a few months after the communication of the results were published, there appeared several protests of scientists' results (*Le Nouvelliste, le Temps*) or even reactions by the media sphere itself (*Le Courrier*).

5. CONCLUSION

From now on, we can draw some temporary conclusions regarding our questioning and a stronger one regarding our methodology.

First of all, it seems that scientific markers (communication of accurate figures, exclusive wordings to the scientific community, precaution in the primary discourse, and no trace of pathos) are considered by the media as a sign of study's strength: there will not be any controversy. Considerations for the scientist are always present (e.g., "serious study"). On the contrary, in the case of the study on the media, we identified that a hybrid scientific discourse (didactic and popularization aspects) with more pathos and less scientific markers, could lead the media not to consider, as it is supposed to be, a scientific research. In this case, there was a controversy on the quality of the scientists' conclusions.

Furthermore, the analysis reveals, regarding rhetorical dynamics changes, the importance of the press release in the adaptation process. At that point, there is some kind of "deal" between the scientific sphere, the media sphere and the socio-political one. But, that "deal," i.e., an arrangement of exchange, seems to be more visible on this document when the controversy is anticipated: calling into question the experiment results for the neutrinos, political hijacking concerning security's figures and the media fighting back in the third study. The science adaptation process is related to the consideration of the addresses' rhetorical situation and consequently it adapts itself by anticipating reactions of the media and civil society so as not to call into question the credibility of the scientists.

We also notice the rhetorical dynamics do not automatically imply the adaptation of a scientific discourse through all the means of persuasion (ethos, logos and pathos). The media seems to enhance pathos when ethos of credibility is predominant. The rise of ethos of credibility is related to news promoters, who are the authors of the press release.

Many of the exposed conclusions could be confirmed by other studies, but we come to a double certainty: on the one hand, it is fundamental to consider comparative discourse analysis of all the discursive changes (rather than a primary discourse and its popularization) and on the other hand it is important to consider it within rhetorical dynamics, as we define it earlier.

At that point in the communication, we can imagine several explicative concluding remarks: 1. Press release wants to protect itself from criticism by giving away to the media a report which is like a "serious study"; 2. Press release wants to defuse potential controversy with addressees who are also targeted by the report: the media. In any case, this defusing seems to constitute a relevant sign of the adaptation of a scientific study in the civic or political sphere.

REFERENCES

Adam, J.-M. (2005). Les sciences de l'établissement du texte et la question de la variation. *Etudes de Lettres, 1-2*, 69–96.

Adam, J-M. (2011). *La linguistique textuelle : Introduction à l'analyse textuelle des discours* (3rd ed.). Paris: Armand Colin.

Amossy, R. (2010). *L'argumentation dans le discours* (3rd ed.). Paris: Armand Colin.

Amossy, R., & Koren R. (2009). Rhétorique et argumentation : Approches croisées. *Argumentation et Analyse du Discours, 2*. Retrieved from http://aad.revues.org/561disciplinaires

Angenot M. (2008). *Dialogues de sourds. Traité de rhétorique antilogique*. Paris: Éditions Mille et Une Nuits.

Authier-Revuz, J. (1982). La mise en scène de la communication dans les discours de vulgarisation scientifique. *Langue française, 53*, 34–47.

Beacco, J-C. (2000). Écritures de la science dans les médias. *Les Carnets du Cediscor, 6*. Retrieved from http://cediscor.revues.org/319

Bitzer, L. (1968). The rhetorical situation. *Philosophy & Rhetoric, 1*,1–14.

Ducrot, O., Anscombres J.Cl. (1983). *L'argumentation dans la langue*. Brussels: Éditions Mardaga.

Fahnestock, J. (1986). Accommodating science: The rhetorical life of scientific facts. *Written Communication 3, 3*, 275–296.

Gans, H. (1979). *Deciding what's news*. New York, NY: Pantheon Books.

Grize, J-Bl. (1996). *Logique naturelle et communication*. Paris : Presses universitaires de France.

Heidmann, U. (2005a). Comparatisme et analyse de discours. La comparaison différentielle comme méthode, in : Adam, J-M., & Heidmann, U. (dir.), *Sciences du texte et analyse de discours*, Slatkine, Genève, pp.

Heidmann, U. (2005b). Epistémologie et pratique de la comparaison différentielle. L'exemple des (Ré)écritures du mythe de Médée. *Etudes de Lettres , 4*,141–159.

Jacobi, D. (1999). *La communication scientifique*. Grenoble: Presses universitaire de Grenoble.

Jacobi, D., & Schiele B. (Eds.). (1988). *Vulgariser la science, le procès de l'ignorance*. Seyssel: Collections Champ Vallon.

Jacobs, G. (1999). *Preformulating the news: An analysis of the metapragmatics of press releases*. Amsterdam: John Benjamins.

Jeanneret, Y. (1994). *Ecrire la science: Formes et enjeux de la vulgarisation*. Paris: Presses universitaires de France.

Maingueneau, D., & Charaudeau, P. (2002). *Dictionnaire de l'analyse de discours*. Paris: Edition du Seuil.

Micheli, R. (2008). L'analyse argumentative en diachronie : Le pathos dans les débats parlementaires sur l'abolition de la peine de mort. *Argumentation et Analyse du Discours, 1*. Retrieved from http://aad.revues.org/482

Pander Maat, H. (2007). "How promotional language in press releases is dealt with by journalists: Genre mixing or genre conflict." *Journal of Business Communication, 44*(1), 59–95.

Pander Maat, H. (2008). Editing and genre conflict: How newspapers journalists clarify and neutralize press release copy. *Pragmatics, 18*(1), 87–113.

Perelman, C. (1977). *L'empire rhétorique : rhétorique et argumentation*. Paris :Edition Vrin.

Rinck, F. (2010). L'analyse linguistique des enjeux de connaissance dans le discours scientifique. *Revue D'anthropologie des Connaissances*, 427–450. Retrieved from www.cairn.info/revue-anthropologie-des-connaissances-2010-3-page-427.htm

Ross, L. (2005). *L'écriture de presse*. Montreal: Éditions Gaëtan Morin.

Schlesinger, P. (1992). Repenser la sociologie du journalisme, les stratégies de la source et les limites du média-centrisme. *Réseaux, 51*, 75–98.

Toma, A. C. (2005). L'organisation informationnelle de vulgarisation scientifique. Revue Marge Linguistique, 9, 176–194. Retrieved from http://www.revue-texto.net/Parutions/Marges/00_ml092005.pdf

SOURCES

Adam, T., Agafonova, N., Aleksandrov, A., Altinok, O., Alvarez Sanchez, P., Anokhina, A.,...Zhigiche, A. (2011). Measurement of the neutrino velocity with the OPERA detector in the CNGS beam. *Journal of High Energy Physics* (preprint), 1–32. arXiv:1109.4897v2 [hep-ex]

Ats. (Wire story). (2010, August 14). La baisse de la qualité des médias pèse sur la démocratie suisse. *24Heures*, p. 6.

Ats. (Wire story). (2010, August 14). La baisse de la qualité des médias pèse sur la démocratie suisse. *Le Nouvelliste*, p. 6.

Aubert, L. (2011, August 31). En matière de criminalité, la Suisse a rejoint l'Europe. *24Heures*, p. 1.

Aubert, L. (2011, August 31). La criminalité est la face noire de la société des loisirs. *24Heures*, p. 3.

Bach, P. (2010, August 14). Démocratie et médias dans le même bateau. *Le Courrier*, p. 5.

Bayraktar, S., Bürgis, P., Eisnegger, M., Ettinger, P., Imhof, K., Kamber, E.,...Udris, L. (2010). *Annales 2010 : Qualité des médias: Principaux constats* (pp. 1-26). Basel : Editions Schwabe.

Brouet, A.-M. (2011, September 24). Les neutrinos, auteurs d'un irrésistible excès de vitesse. *24 Heures*, p. 3.

Donzé, V. (2011, August 31). La Suisse n'est plus sûre. *Le Matin*, p. 9.

Gumy, S. (2011, August 31). La Suisse n'est plus une île face au crime. *La liberté*, p. 6.

Huet, S. (2011, September 23). Einstein distancé. *Libération*, p. 30.

Imhof, K (2010, November 16). Pourquoi la presse décline et la démocratie en souffre. *Le Temps*. Retrieved from Schweizer Mediendatenbank.

Killias, M., Staubli, S., Biberstein,, L., Bänziger, M., & Iadanza, S. (2011). *Sondage au sujet des expériences et opinions sur la criminalité en Suisse: Analyses dans le cadre du sondage national de victimisation 2011*. Zürich: Université de Zürich.

Larcusserie, D. (2011, September 25). Excès de vitesse des neutrinos. *Le Monde*, p. 3.

Press release (CCDJP). (2011, August 30). Nouvelle hausse du taux de criminalité—La Suisse a atteint la moyenne européenne. CCDJP, pp. 1–4. Retrieved from http://www.kkjpd.ch/images/upload/ Communiqu%C3%A9%20de%20presse%20Sondage%20national%20de%20victimisation.pdf.

Press release CNRS. (2011, September 22). Plus vite que la lumière? CNRS. Retrieved from http://www2.cnrs.fr/presse/communique/2289.htm

Press release Forschungsbereich Öffentlichkeit und Gesellschaft & University de Zürich. (2010, August 13). La perte de la qualité des médias se répercute sur la démocratie. Forschungsbereich Öffentlichkeit und Gesellschaft & University de Zürich, pp.1–3. Retrieved from http://jahrbuch.foeg.uzh.ch/Broschren/ Medienmitteilungen_2010_08_13/fran%C3%A7ais.pdf

Vanlerberghe, C. (2011, September 24). La théorie d'Einstein bientôt relativisée?. *Le Figaro, L'Express, L'Impartial*, p. 21.

Vos, A. (2011, September 24). Plus vite que la lumière et plus loin qu'Einstein. *Le Temps*. Retrieved from Schweizer Mediendatenbank.

Windisch, Uli. (2010, September 14). Pluralisme médiatique romand : Oui, il est insuffisant, je tape sur le clou. *Le Nouvelliste*, p. 2.

Wire story. (2010, August 13). Etude sur les médias —La baisse de qualité des médias pèse sur la démocratie suisse.

Wire story. (2011, August 30). Criminalité—Le taux continue à augmenter et se rapproche de celui de l'Europe.

Wire story. (2011, September 22). Sciences-physique—La vitesse de la lumière et Einstein dépassés par une particule?

Authority in an Age of Expertise

CATHERINE E. HUNDLEBY

Department of Philosophy and Women's Studies Program
University of Windsor
401 Sunset Avenue, Windsor, Ontario
Canada N9B 3P4
hundleby@uwindsor.ca

ABSTRACT: The potential for experts to exploit their positions of authority requires attention to the role of epistemic work as part of the social division of labour. Expertise has not become so distant from social hierarchy as we sometimes fancy, and evaluating expertise requires political analysis.

KEYWORDS: argument, authority, bias, command, expertise, motivation, reasoning, social status, testimony, trust.

1. INTRODUCTION

Authority and trust have always raised questions for feminist scholars: Why do women and other socially marginalized people suffer disproportionately under supposedly just authorities? Why do we trust authorities that deny our moral and rational status, even refusing to count us as normal human beings for clinical trials? Understandings and procedures that count as expertise, such as the practice of giving women medicine tested only on men, contribute to women's suffering and marginalization.

Among the forms of expertise that have done damage to women, including legal expertise, the rise of medical expertise has been most cruelly ironic, Lorraine Code argues. The rise of medical science took over by various means the midwifery and lay-healing performed primarily by working-class women and thus subjected all women to the authority of white upper-class men. Likewise child-rearing and housework became matters in which women are no longer trusted but taken to need instruction and monitoring. So women have come to lack authority in the very domains that define their own gender and that submissiveness has come to define the gender. "When possibilities of being a 'good enough' woman and mother depend on relinquishing trust in their own skills in favor of a more distinguished expertise, it is not surprising that women would do what was expected of them." (Code, 1991, p. 207). The expectation that women will be *trusting* makes women especially vulnerable to self-doubt and inclined to defer to experts. The authority of experts draws further sustenance from and meaningfully contrasts with the cultural conflation of femininity with pathology (Code, 1991, pp. 203–219).

For a few years I've been struggling to engage the feminist critiques of science, medicine, and technology with accounts of how expertise operates in argumentation. However, the current analysis of *ad verecundiam arguments* ignores forms of authority that intersect with expertise, and argumentation theorists in general assume equal social status among interlocutors and thus seem oblivious to the principles of feminist analysis. The contemporary focus on expertise in accounts of the fallacy of appeal to authority suggests that other forms of authority are no longer important or problematic. Fortunately, authority receives more complex

Hundleby, C.E. (2012). Authority in an age of expertise. In J. Goodwin (Ed.), *Between scientists & citizens: Proceedings of a conference at Iowa State University, June 1-2, 2012* (pp. 209-217). Ames, IA: Great Plains Society for the Study of Argumentation.

analysis from the rhetoricians in the field. Jean Goodwin has opened up the discussion of authority to provide greater attention to the role of social status, looking back to its significance for the Roman orator Cicero (2001), and noting its centrality to John Locke's introduction of the term *ad verecundiam*.

Expertise has become increasingly possible as our society develops more forms of specialized labour and forms of knowledge to support that diversification. The transition from hierarchies of religion and aristocracy, such as in ancient Rome, to hierarchies influenced more by specialized knowledge has many implications for individual reasoning and argumentation. The intertwining of social hierarchies plays only a limited role in Goodwin's account of expertise as a principle-agent problem but her model opens up many social considerations. The trust that she recognizes to underpin expertise is subject to exploitation in a way that demands attention to hierarchical intersections, including gender, race, and class hierarchies.

2. EXPERTISE AND THE DIVISION OF LABOR

In contemporary scholarly contexts, the term "authority" often refers almost exclusively to expertise, also referred to as "epistemic" and sometimes "cognitive" authority, leaving behind the authorities of dignity and command recognized by Locke and Richard Whately, and the ordinary language of "calling in the authorities" to indicate the need for policing and regulation. Arguments that appeal to expertise have extensive company among forms of defeasible or presumptive reasoning. Other "schemes" for presumptive argumentation include *ad verecundiam*'s Lockean sisters *ad hominem* and *ad ignorantiam* but extend far beyond. A good deal if not all of reasoning and argumentation can be understood in this fashion, as dependent on defeasible inference patterns. However, expertise and appeals to it—good and bad—also depend on contingent social structures. Expertise functions in much the same way as other forms of authority, and it rests on the same foundations as the authorities of command and dignity, on social structures and status within them. Social diversification provides the need for expertise and resources for the education that provides the knowledge.

Recognizing expertise as a specialized form of authority does not reduce it in any simple fashion to status, but indicates how status makes expertise possible. Sometimes our expertise may account only for artefacts of social structures, such as when directory assistance provides phone numbers or information staff in an airport or shopping mall provide directions for the space. Experts also guide us regarding non-social phenomena, but we need not get into debates over realism to say that social structures make possible the *types* of engagement in the physical world that allow a person to become, say, a physicist or an engineer. Developing these specializations depends on technologies, schools, and professional associations, as well as on teaching and mentoring.

Reliance on expertise has increased in recent centuries as the division of epistemic labour brings with it the division of epistemic authority (Anderson, 1995, p. 59; Code, 1995, p. 175). We have no more polymaths; we cannot 'have it all' epistemically regardless of wealth, and much of our knowledge has become esoteric.

> A layperson defers to the authority of experts, not because in so doing one is guaranteed the truth-of-the-matter, but because one lacks the means to determine the issue oneself. In deferring to experts, one is not deferring simply to particular knowledge claims, but to a process for making those claims. (Pierson, 1995, p. 402)

In democracies, especially those based on representation, processes and documents as much as people hold and convey authority, moving away from the personal authority of aristocracy, which was more important in Rome than any other authority.

> Despite the Roman pride in their legal expertise, despite their pride in their military discipline, auctoritas was their preferred method of social control. Thus it is notorious that in Republican times there were no police within the city limits of Rome; order was maintained instead by an often symbolic display of personal dignity. (Goodwin, 1998, p. 277).

As sources of authority can become disembodied, moving beyond individual humans, they can rematerialize in such things as timetables (Walton, 1997), academic disciplines and methods (Hundleby, 2010), and even web sites. The division of cognitive labour and the proliferation of forms of expertise give expertise a systematic nature.

Specialization and correspondingly accreditation have proliferated with the growth of science and technology, in turn encouraging the development of education as an industry. Training has long been a prerequisite for the authority of command, but the demand for education has increased such that we have training for pretty much every job or vocation. While experience may sometimes substitute for education and accreditation in seeking a position of authority, training has become the gold standard because it guarantees a degree of expertise; we even have a distinct sector to the workforce described as "information technology." Education has become an industry in its own right, supported not only by governments and private endowments but by profit.

3. FORMS OF AUTHORITY

To distinguish among different types of authority Goodwin (1998) looks at the forms of corresponding recourse for not following authority. Refusal to defer to dignity by exhibiting reverence amounts to impudence and invites shame; refusal of command is disobedient and risks loss of position or punishment; while refusal of expertise is imprudent and risks loss of resources, including the willingness of others to cooperate.[1] These forms of authority intertwine in ways that aid the functioning of expertise. They connect to each other by social networks more general than and underpinning the systems that provide dignity. Such prestige is merely an especially high form of social status, I suggest, that in specific forms also includes the authority of command.

Goodwin argues that the types can be clearly distinguished despite how they compound and complement each other: "Authorities, like other goods, are often distributed on the principle that to those who have shall be given more" (1998, p. 273). So, those who are learned are given the authority of command and those who are able to wield command are granted status to support it. Goodwin maintains nonetheless that refusal to any authority reflects clear delineation among the three forms. So, for example, "the student history paper which is not typed as required is disobedient, the student history paper contradicting the instructor on a date, imprudent." (1998, p. 273)

That infringements of authority can sometimes be specifically diagnosed does not entail that forms of authority often operate independently. In particular, a degree of expertise

---

[1] The taxonomy is summarized in Hansen (2006), p. 323.

belongs to all forms of authority. A senator knows how to behave like senator, which many do not, and any officer has technical understanding about protocol.

The diversification of labour gives expertise such an important role in our society that we accord experts an elevated status—titles, good salaries, and general influence—that makes their authority complex, involving other forms. A person may be esteemed and granted significant prestige because of his knowledge or for that same expertise he may receive an administrative position and the power of command. An elevated position of some kind helps to mark expertise, and can also aid in its operation. Among experts, we prefer a plumber who can also order the materials for her recommended options, and a midwife to whom physicians will listen. Their authorities of command and dignity, respectively, are not specifically epistemic aspects of their power or not intrinsically matters of expertise but nonetheless contribute to their status and function as experts. Complementary forms of authority increase the relevance of an expert's knowledge.

3. DIGNITY AND BIAS

Dignity may seem to be old-fashioned, mostly obsolete. Even if it once influenced other forms of authority, many of us like to think we can accord someone dignity without believing what the dignitary says. However, recent work in the psychology of bias suggests differently: prestige can have a profound effect on people's evaluation and proper comportment can influence behaviours, decisions, and beliefs that seem independent of courtesy. Perceived social status is basic and remains standard, even generic to our engagement with other people. It certainly affects discursive authority in the sense of who gets to speak and who is heard; these elements of testimony are important for argumentation and public reasoning, and hence for expertise.

Dignity like command does not operate generically but provides exceptional status, say belonging to a religious figure or hero. Yet both connect from broader forms of authority and they are related to each other as forms of high regard. The status of dignity operates more loosely than the status of command and also has a more subtle influence. Fulfilling the obligations of etiquette often is more subject to interpretation than following an order. The authority of command is treated together with other social authorities by Douglas Walton (1997). His terminology of "administrative authority" might encompass dignity too, but it assumes a formality to social authority less common and less acceptable in democracies than in aristocracies.

Hierarchical distinctions that are important to reasoning in discursive contexts can be quite broad, operating generically; they need not grant exceptional status in the manner of the prestige providing for command or traditional dignity. The commonplace social ranks follow lines of race, class, and gender. Biases against people with lower social status and for people with higher social status are part of *status quo bias* which outweighs in-group preference. Thus white people tend to be perceived as generally good by the overwhelming majority of North Americans taking the implicit association test or IAT (https://implicit.harvard.edu/implicit/demo/), regardless of the tester's own racial identity and political views on race. No matter how opposed to racism one is one may show these unconscious biases (Jost, Mahzarin, & Nosek, 2004). Likewise, my decades of work on feminist research and identification as a "career woman" do not prevent me from testing with a moderate bias against associating women with careers. Bias against women in careers and in science is common across cultures and identities. More specific social evidence of these biases has been available for decades from studies of job and tenancy applications, for instance, revolutionizing some interview practices such as for orchestras. They now use screens to blind auditions and as a result employ many more women than before. Likewise, to be fair to our students we must view their work as anonymously as possible when evaluating it.

Despite our romanticism about the equalitarian nature of democracy, broad contemporary social categories relate closely to traditional dignity. We accord people with something-like-dignity in ways that we often do not recognize but that can have profound effects. Ordinary social divisions cumulatively may accord a level of status reminiscent of a Roman senator's dignity. Admittedly, white, middle-class, straight men are not elevated so much as they are taken to be standard, but that is the ideal position in a democracy, analogous to classical dignity. What is typical (in some sense) and viewed as being usual is also considered ideal. The rest of us have relatively compromised status (although some white, middle-class, straight young men face other forms of marginalization).

Trust operates implicitly in democratically tolerated forms of oppression. We adhere more strictly to gendered and raced behaviour in strange environments because maximizing our predictability helps us work with others. People become extremely confused and even unable to think and act in the face of indeterminate gender and sometimes for indeterminate race and age. We can better anticipate the behaviour of individuals whom we can locate along these axes (Ridgeway, 2011). We trust our social presumptions and that other people will conform to them; others do conform because they are trying to negotiate a world structured by the same roles. Consider that women and people of colour smile more than white men; it's typical of subordinates, stereotypical, and can be a reliable way to minimize social hindrances but it is a burden and creates an unequal discursive environment.

The social status that adheres when one is on the beneficiary side of lines of oppression thus can be understood as a generic form of authority. However modest this dignity it can sometimes amounts to the authority of command via its testimonial effects. Statements from white men carry weight; statements made by people of colour carry less, even among those sharing their ethnicity; women tend to be interrupted more than men even by other women. A man can be proud to be a man, in a way that people can rarely be proud of anything else except other dominant social statuses. Although it may be unacceptable in many contexts to appeal directly to one's masculinity, standards for politeness are gender-specific, demanding assertiveness from men and supporting roles from women and people of colour. The dynamics of compound privileges can also make the elevation to one's status consistent and resilient. The prestige is merely less explicit than the traditional authority of dignity, and people are less conscious of its operation.

4. POWER, TRUST AND EXPERTISE

Violence and other forms of power lack the trust characteristic of authority (Goodwin, 2001, p. 51). "The kind of authority that is everywhere and always open to challenge is . . . no authority at all" (Hanrahan & Anthony, 2005, p. 64) and so authority is the place where explanation, at least provisionally, stops. While good authority is in principle subject to scrutiny, in practice I must trust you to a degree if I have you take on any task for me, whatever that task might be: shovelling snow or diagnosing my headaches. Our trust in experts may be engendered and secured in various ways: I can oversee and note the impact of your work or advice, but the trustworthiness of experts is distinctly difficult to assess. The power of expertise can be evaluated and challenged, as feminist critiques of medicine show, and there may be increasing need to seek out such challenges.

What qualifies experts is that they have epistemic skills we lack, abilities to examine and understand. Experts hold power not only by way of complementary authorities for command and dignity; they make certain courses of action possible by giving novices instruction and confidence, opening up the recommended paths and making other choices more awkward (Goodwin, 2001, p. 50).

Expertise is notoriously difficult to assess. Like many of us, I cannot judge well the quality of work done by a plumber in the way I can the mowing of a lawn. Or to bring back the medical case, I can judge skill in bandaging but not in delivering a baby. Those physicians who specialize sometimes do so because of their inept bedside manner, so that my attempt to judge a physician's expertise based on his apparent ease may be systematically mistaken. On the other hand experts cannot judge other true experts if their skills are so close that they overlap, Alvin Goldman (2011) argues.

Because the ability of the putative expert to serve our needs cannot be adequately scrutinized and evaluated, it is difficult to justify our trust. The more we can oversee the less we need to trust. The problem is akin to the *Euthyphro* dilemma: Do we trust experts because they have cognitive authority or do they have cognitive authority because we trust them? In causal sequence our trust gives people cognitive authority in the sense of licensing them to investigate and advise; but ideally we come to trust people because they have cognitive authority in the sense of understanding, demonstrated already to some degree.

In the cases of medicine usurping midwifery and trials based on male subjects only, the putative experts did not deserve trust. Trust seems to have derived from social privilege, especially the gender and class status, an analysis that shows how other privileges of any kind involve trust that can be exploited to acquire expert status. The opacity of expertise may make it especially vulnerable to confusion with other forms of authority. Expertise may be especially compromised in technological democracies given the insidious nature of social status and rendering expertise especially in need of critique.

5. TRUST

An analysis of the ways that expertise goes wrong and trust is violated requires attention beyond the individuals in the exchange. It takes us back to the division of labour in which expertise is just one form of authority, one among several ways in which we entrust people. The more official this trust, the more likely that the trust is recognized to depend on authority, yet I have suggested that dignity operates in subtle but powerful ways. Gender and race have the company of more official institutions that have more transparent guarantees of expert services, and more trustworthy sources for the authority of individual experts. Just as official institutions encourage us to trust some people as experts, social institutions likewise affect who receives expert credibility, authorizing and deauthorizing both individuals and whole traditions.

To solve the apparent dilemma of expertise and provide a basis for trust Goodwin suggests a model of individual exchange. "Someone—the "principal"—needs to retain someone else—an "agent"—to do something she cannot or does not want to do for herself"

(Goodwin, 2010, p. 136) An individual expert *stakes her dignity* on the correctness of her claims.

> In appealing to authority, the speaker offers her dignity as a hostage for her judgment, wagers it on her judgment, or, to use another analogy, posts it as a bond guaranteeing the correctness of her judgment. (Goodwin 2001, p. 51)

Because of this stake, trust is not just given but earned, Goodwin argues (2001). The principle becomes empowered by expert advice insofar as being granted with the power to *challenge the status* of the expert. A physician can be shown to have provided a misdiagnosis, a teacher a false set of information, and such risk accrues to all offers of expertise. The expert has been authorized to perform a certain sort of labour and that authorization can be retracted, by the principle or the monitoring body, she suggests. The risk taken by experts depends on their self-presentation and the risk of appearing fraudulent and thus provides a reputational bond. Credentials are guarantees of expertise only insofar as they guarantee good behaviour: the licensing body may respond and there are other systems of accountability. Likewise, contributors on webpages may be trusted not because we know who they are (and sometimes especially because of their anonymity which can encourage honesty). Whether we trust them depends on the structure of the web site and the standards it follows. Online personas are sometimes carefully cultivated in ways well deserving of trust (Goodwin, 2011).

Goodwin's principle-agent model thus demands we are not alone in our relationships with experts, yet monitoring bodies are not adequate support for trust among individuals. Neither the risk of adverse selection or the hazard of individual immorality or irresponsibility can account for the exploitation of women by the medical industry, how a whole expert profession can perpetrate an injustice and falsehood. As Code explains:

> The rhetoric of voluntary agreement cannot account for the politics of trusting . . . Trust-based relationships lend themselves readily to the forms of exploitation to which women of all classes, races, ages, and persuasions have long been subject. (1991, pp. 184–185)[2]

A political account of trust can be found in social contract theories from Plato through Hobbes, Locke, and Rawls. Recently Miranda Fricker (1998) uses it to indicate the fundamental nature of testimonial authority to our social means of understanding, and Charles Mills (1997) shows how exploitative it can be, arguing that the social order of Western democracies rests on the exploitation of people of colour.

Goodwin suggests that we scholars need to attend to the specific techniques used to develop trust, a piece of advice that directs us to social contracts operating in the background:

> Where [the principle] may not have experience with this physician, for example, she does have experience with the medical system as a whole; and if her experience is good, she has some reason for confidence in the judgments of the Board of Medical Examiners and other professional organizations. (Goodwin, 2010, p. 141)

---

[2] Code suggests the general epistemological model of friendship, which is as individualist as Goodwin's. I am not sure how Code has developed her account of trust in her most recent work (2006) but a word search of it suggests that she has left behind "friendship" as the model.

The larger structures of authority that set the stage for novices to engage expert authorities include structures that give rise to the authorities of dignity and command; we must beware their effects, both beneficial and pernicious.

6. CONCLUSION

Our ability to judge trustworthiness itself needs to become specialized to fit our age of expertise. The opacity of expertise renders it especially vulnerable to confusion with other forms of authority, and thus to exploitation in the service of other forms of power, such as gender, race, and class privilege. Goodwin argues that we escape the dilemma of judging the authority of experts by way of our ability to judge general trustworthiness (2010, p. 141). Her strategy is on the right track but stops short. Some of our ability to judge is pervaded by systems of social authority. We have distinct reasons to trust experts that do not hold for other authorities, but these do not ease the problem.

The motivations people have for becoming experts will include those for becoming authorities of any kind, such as an ambitious disposition ('I climbed it because it was there'), the desire for power or personal gain, the intention to do good, the invigoration of competition, and the desire to contribute to society. More specific purposes will attract people to expertise as a form of authority over others, such as the love of learning, the desire to understand something in particular, and the intention to contribute to social knowledge. These more specialized motivations seem less vulnerable to corruption than those driving authority in general. Certainly expertise can make one more able to be a trickster and a con artist, but those do not rank highly among long-term personal ambitions. Should one have expertise in psychology or rhetoric and apply it to merchandising, there obtains a sense that one has "sold out" relative to other experts.

Nevertheless, expertise will attract those seeking obscure power; it is intrinsically ripe for corruption; and the implicit authority granted according to the status quo may mask poor excuses for expertise. Institutions that govern professions and practices may not be sufficient deterrent, and themselves engrained in the existing systems of power. Status quo thinking operates as the default. To scrutinize the politics of expertise requires special effort and attention to liberatory analysis and critiques.

REFERENCES

Anderson, E. (1995, August). Feminist epistemology: An interpretation and a defense. *Hypatia, 10*(3), 50–84.
Code, L. (1991). *What can she know? Feminist theory and the construction of knowledge*. Ithaca, NY: Cornell University Press.
Code, L. (1995). *Rhetorical spaces: Essays on gendered locations*. New York, NY: Routledge.
Code, L. (2006). *Ecological thinking: The politics of epistemic location*. New York, NY: Oxford University Press.
Daston, L. (1992, November) Objectivity and the escape from perspective. *Social Studies of Science, 22*(4), 597–618.
Fricker, M. (1998). Rational authority and social power: Towards a truly social epistemology. *Proceedings of the Aristotelian Society, 98*(2), 159–77.
Goldman, A. (2011). Experts: Which ones should you trust? In A. Goldman & D. Whitcombe (Eds.), *Social epistemology: Essential readings*. New York, NY: Oxford University Press.
Goodwin, J. (1988). Forms of authority and the real *Ad Verecundiam*. *Argumentation, 12*, 267–280.
Goodwin, J. (2001). Cicero's authority. *Philosophy & Rhetoric, 34*(1), 38–60.

Goodwin, J. (2010). Trust in experts as a principle agent problem. In C. Reed & C. W. Tindale (Eds.), *Dialectics, dialogue and argumentation: An examination of Douglas Walton's theories of reasoning and argument* (pp. 133–143). London: College Publications.

Goodwin, J. (2011). Accounting for the force of the appeal to authority. *Argumentation: Cognition and Community.* Proceedings of the 9th biennial conference of the Ontario Society for the Study of Argumentation. Windsor: Centre for Research in Reasoning, Argumentation and Rhetoric.

Hanrahan, R., & Anthony, L. (2005, November). Because I said so: Toward a feminist theory of authority. *Hypatia,* (4): 59–79.

Hansen, H. V. (2006). Whately on arguments involving authority. *Informal Logic, 26*(3), 319–340.

Hansen, H. V., & Pinto, R. C. (Eds.). (1995). *Fallacies: Classical and contemporary readings.* University Park, PA: Pennsylvania State University Press.

Hundleby, C. E. (2010). The authority of the fallacies approach to argument evaluation. *Informal Logic, 30*(3), 279-308.

Jost, J.T., Mahzarin, R.B., & Nosek, B.A. (2004). A decade of system justification theory: Accumulated evidence of conscious and unconscious bolstering of the status quo. *Political Psychology, 25*(6), 881-919.

Pierson, R. (1994). The epistemic authority of expertise. *PSA: Proceedings of the Biennial Meeting of the Philosophy of Science Association, 1994*(1), 398–405.

Ridgeway, C.L. (2011). *Framed by gender: How gender inequality persists in the modern world.* New York, NY: Oxford University Press.

Tirrell, L. (1993). Definition and Power: Toward authority without privilege. *Hypatia, 8*(4), 1-34.

Tirrell, L. (2012). Authority and gender: Flipping the F-switch. Unpublished manuscript. University of Massachusetts - Boston.

Walton, D. (1997). *Appeal to expert opinion: Arguments from authority.* University Park, PA: Penn State University Press.

Deliberative Systems View of Efforts to Democratize Energy in Arizona

TRAVIS JOHNSON

Consortium for Science, Policy & Outcomes
Arizona State University
PO Box 875603
Tempe, AZ 85287-5603
USA
tajohn1@asu.edu

ABSTRACT: Science often struggles to find answers in a world of complexity and uncertainty and deliberative democracy has been used as a way to bring in public values to help guide the scientific process. This study looks at Arizona's energy system through a theoretical lens and three separate efforts to democratize dialogue around energy in Arizona. The Arizona Town Hall, Emerge and the Solar Summit are all efforts to engage a more diverse range of stakeholders and keep dialogue progressing.

KEYWORDS: deliberative democracy, energy policy, dialogue, public engagement, democratization.

1. INTRODUCTION & FRAMEWORK

This study uses Dryzek's deliberative systems view as a framework to analyze three events designed to alter the deliberative system around energy in Arizona: the Arizona Town Hall; Emerge and the Solar Summit. First, the paper outlines Dryzek's deliberative systems view and then Arizona's energy system is explored in the context of the current political climate to give a general landscape of who is involved in dialogue, how they deliberate and how the system is networked. Information on the AZ Town Hall, Emerge and the Solar Summit, as it relates to deliberative systems, was gathered through first hand experiences and interviews with organizers and participants and is presented in section three. In section four, these accounts are analyzed through the deliberative systems framework developed from Dryzek and implications of the analysis is discussed relating to future actions and changes that could positively affect efforts to alter Arizona's energy system.

Science often struggles to find answers in a world of complexity and uncertainty (Cartwright, 1999; Dupre, 1993; Popper, 1968; Wilson, 1998). Deliberative democracy is a popular framework to help incorporate public participation and values into the scientific process (Abelson et al., 2003; Kleinman, Powell, Grice, Adrian, & Lobes, 2007 & 2009; Powell & Colin, 2008, 2009; Weeks, 2000). It has been very successful in countries around the world in dealing with complex science and technology policy issues (Anderson & Jaeger, 1999), and there are many forms of deliberative democracy, such as citizen juries, consensus conferences, and scenario workshops with many studies showing successes and failures of these different methods (Button & Mattson, 1999; Pelletier, 1999; Hagendijk & Irwin, 2006; Hendriks, Dryzek, & Hunold, 2007). The literature is full of varying claims about the efficacy of the various deliberative democracy methods but the deliberative systems' view outlined by Dryzek provides a useful framework with which to analyze varying deliberative efforts (2010).

Dryzek's deliberative system view can be summarized as a network of spaces where deliberation of different kinds happen and are linked together by various communication activities that make the network larger than the sum of the parts (C. Miller, personal communication, October 27, 2011*)*. While it sounds like a simple and straightforward idea, there are many details and concepts that must be elucidated to truly understand and apply it as a conceptual framework. The spaces where deliberation occurs within a deliberative system can be a formal network such as a government or an informal network like advocacy groups or even conversations in people's homes. Thus, defining a group of spaces as a deliberative system falls to its capacity as a system and not to the make-up of the individual actors within the system. For Dryzek, the *deliberative capacity* of a system is based on whether deliberation is authentic, inclusive and consequential (Dryzek, 2010, p. 10). *Authentic deliberation* is non-coercive in nature and uses language that all deliberators can understand and respect. *Inclusive deliberation* predicates the opportunity and ability of all affected stakeholders or their representatives to participate. *Consequential deliberation* must somehow make a difference in determining or influencing collective outcomes. Obviously, these criteria of deliberative capacity are not binary variables, there are many degrees to which a deliberative system can be authentic, inclusive and consequential; so when we look to analyze deliberative systems in the real world there will be only degrees to which a given system fits those criteria. Evaluation of a deliberative system requires some form of demarcation along these variables, so for this study a yes/no classification is assigned along with discussion and examples of how that classification was reached as well as possible contrary evidence.

Once a deliberative system's capacity is identified, we need a conceptual way to look at the system. Dryzek outlines six categories of a deliberative system; public space, empowered space, transmission, accountability, meta-deliberation and decisiveness (Dryzek, 2020, p. 11). *Public spaces* would be informal deliberations open to anyone who wishes to participate including lay citizens, media, advocacy groups or politicians and could happen over the internet, in a café or in a public square. *Empowered spaces* are the more formal institutions charged with decision making like the legislature or courts but could also include informal networks that produce collective outcomes. *Transmission* is a means through which deliberation in a public space can influence that in an empowered space and might include social movements or advocacy. *Accountability* is how the empowered space answers to the public space. A formal example would be elections or informally it could be decision makers explaining why they made a particular decision. *Meta-deliberation* is discussing how the deliberative system itself should be organized and *decisiveness* is the degree that the previous five categories effect the collective decisions. These six categories not only help give a clearer picture of what a deliberative system looks like but also offer a coherent framework to begin analyzing efficacy of the system itself. Dryzek states that a well-functioning deliberative system will have authentic deliberation in categories 1–5, be inclusive in 1, 2 and 5 and will be decisive in its collective outcomes. *Democratic legitimacy* is another fundamental theme in this deliberative systems view which simply means that people affected by a decision or action should be allowed to participate directly or through representation in a consequential deliberation about said decision (Dryzek, 2010, p. 3). There is no deliberative system that has all of the characteristics outlined above and sometimes improving one quality takes away from another but that is always the challenge inherent when applying a conceptual framework to real-world examples.

2. ARIZONA'S ENERGY SYSTEM

Arizona's deliberative system around energy is fairly unique in the U.S., in that the Arizona Corporation Commission is constitutionally charged with regulating the utilities and setting energy policy within the state. Commissioners are appointed to a four-year term via general election. The ACC makes up the principal body in the *empowered space* of the deliberative system. Also in the empowered space, are the utilities, power cooperatives and the Governor's Energy Policy Office. The legislature is currently trying to insert itself into this empowered space as another sphere of influence in the energy policy-making arena. The *public space* consists mainly of university groups that focus on energy issues and public advocacy groups on certain energy issues, most notably new project siting. *Transmission* and *accountability* both come from 'public meetings' held by the ACC, power providers and the Governor's Energy Policy Office. The availability of public meeting information does lead to questions of how effective they really are at providing accountability of the decision making bodies to the public as well as providing transmission opportunities from the public to the empowered spaces. The ACC and the state legislature are really only accountable to the public through the election process. University and public advocacy help transmit dialogue from the public to the empowered spaces, albeit limited. Intra-space communication is also fairly constrained. The empowered space is primarily connected through formal channels of regulation and policy creation with little deliberation within. The public space has more informal connections but seem to be more diverse and numerous, connecting through various institutions and across several groups.

Meta-deliberation does not appear anywhere in Arizona's energy system except in the very abstract at high level university discussions. With the lack of robustness in the first five categories, *decisiveness* is a quality absent within Arizona's energy system. Public and empowered spaces have very little transmission and accountability between them or even within themselves and meta-deliberation only happens at the very abstract level among a select few stakeholders. *Authenticity* of deliberations is often compromised with technical jargon or partisan political views and *inclusiveness* is low due to poorly advertised 'public meetings.' As a result of poor inclusiveness, *democratic legitimacy* is also sacrificed because there is not representation of all the relevant stakeholders affected by a given issue.

Table 1: Arizona's Energy Network: A Deliberative System Perspective

	Authentic	*Inclusive*	*Consequential*	*Legitimate*
Public Space	Yes	No	Yes	No
Empowered Space	No	No	Yes	No
Transmission	No		No	No
Accountability	Voting only		Voting only	Voting only
Meta-Deliberation	No	No	No	No

*Grayed boxes indicate where a deliberative capacity trait is not required within a certain part of the deliberative system (Dryzek, 2010).

Energy policy is an important topic for public input because we are in a unique situation historically to shape our energy future and the decisions that are made regarding our energy future will deeply affect our society as a whole for decades. Advocating for alternative energy and bringing together citizens to discuss topics is nothing new but understanding the deliberative system and the related activities that attempt to alter the dialogue process is a useful way to inform planning, structure, execution and outcomes of future deliberative system activities.

Our energy production methods as a civilization must change by definition; using a finite resource mandates an end to that supply. With future change a certainty, we are then left with the questions of how, when, where and with whom the change will happen. This is a unique opportunity in human history to actively and intentionally shape our energy future, how it will look, when it will happen, where it will take shape and most importantly *who* will decide what gets done. To take full advantage of this opportunity for collective change, we must understand the energy system and actions designed to intervene in it from a deliberative democracy point of view.

3. DELIBERATIVE EVENTS

The Arizona Town Hall was held on November 6th–9th, 2011 at the Grand Canyon, AZ (AZ Town Hall). It focused on Arizona's energy future, looking at technical, social, political and ethical dimensions around various energy issues. The Town Hall hosted 85 participants representing stakeholders from across the energy industry, local utilities and academia. It was noted that there were few lay citizens and mostly energy insiders at the event along with only one ACC commissioner and no state legislators. This event fit into a larger network of deliberations around energy by interfacing primarily with the empowered space. There were few non-experts in attendance and the ones that were may have been unable to understand the technical nature of the discussions. The Town Hall was structured into four panels of approximately 20 people with a moderated discussion around questions/issues derived from an in depth background report. Each panel discussed the same topics and strove for consensus around each with a moderator facilitating the discussion and a recorder present to record the opinions and recommendations of the panel.

On the second day, a large plenary session was held and the recommendations of each panel was read and discussed as one large group. Moderators fielded questions and concerns and the recommendations were modified in language and content by the recorder to reflect language that the group could all agree on. If debate persisted among a few participants, they were sent to the back where they discussed their opinions until they could reach common ground. This process resulted in a final document of consensus recommendations based on topics from the original background document. Insiders were seen to dominate the conversations and lay citizens and students did not contribute frequently, most likely due to the over-representation of stakeholders. The order in which the topics were discussed both in the panels and as a single group was important in that the first topics got more attention and debate leaving less time for the subsequent discussion topics. This format of debate and consensus building was thought to be empowering and non-conflicting where everyone had the opportunity to contribute to the conversation. Despite the majority of expert stakeholder representation there was a wide variety of issues brought up and a seemingly in-depth understanding of social, political and ethical concerns.

There are several communication and outreach activities in place to connect this event with other networks around energy. Mini-town halls are held statewide and are designed to involve lay citizens that may be underrepresented at the main, large event. These mini-town halls discuss the recommendations of the primary event and allow attendees to voice opinions and educate themselves on topics relating to Arizona's energy future. The recommendations report of the Town Hall is sent to all state legislators in an effort to inform them of the consensus opinions across these topics. In addition to the recommendations report, there is thought to be grassroots action that springs from the Town Hall resulting from enthusiasm and empowerment of participants. It is unclear, however, if there is a productive outlet for that enthusiasm. The Town Hall can be seen as democratically legitimate in so far as the outcomes (the recommendations report) were a consensus of all the participants.

Emerge was a three day event, March 1st–3rd, 2012 held at Arizona State University (Emerge). The event brought together artists, scientists, engineers and story-tellers to participate in workshops, festivals and lectures all looking at the future of the human species and the environments that we share. The specific workshop discussed here was titled "Humanist Narratives for Energy" focusing on exploring the future of energy in Arizona to 2030. It strove to move beyond the technical issues and delve into the social, political and economic drivers that play important roles in energy. Two main variables were decided on during discussions and put on corresponding x and y axis; a decentralized, high competition (many new players) vs. centralized, low competition (legacy players) x axis and a high investment vs. low investment y axis. These two axis subsequently made a four-quadrant grid with four corresponding scenarios (high competition/high investment, high competition/low investment, low competition/high investment, low competition/low investment). Individual scenarios were then brainstormed for each set of variables using freedom, innovation, social will and state of the environment as important drivers in each scenario.

Stakeholder representation at the event was primarily academic faculty and students with some government participants, most all of whom had energy backgrounds to varying degrees. The deliberation/discussions that went on during the scenario planning resembled high-level, expert discussions with point-counterpoints and clarifications. The main themes discussed were economic concerns and how those affect the future of energy, political aspects specific to Arizona and some human and social dimensions of energy. This type of scenario planning is very conducive to producing a myriad of differing views. Contrary to the Town Hall, Emerge encourages plurality of views; consensus need only be that ideas are logical and plausible. Currently, the only communication activity in place to connect Emerge with other networks around energy is an art exhibit conveying depictions of the futures scenarios designed at the event. Given that this is the first event of its kind and that it happened very recently, it can only be hoped that the next event will have more robust methods of communication and outreach following the event. The primary outcomes from Emerge are the scenarios themselves, the embedded ethnographers' notes and the art exhibit. Participants generally had no problem perceiving each other's ideas as legitimate in their particular scenario.

The Arizona Solar Summit was held on March 26th & 27th, 2012 at the Arizona Biltmore. This was the second Solar Summit. Stemming from the first Solar Summit in 2011 were four working groups (Supply Chain/Workforce Development, Applied Research Collaborations/Pilot Projects, Policy/Finance & Building and Strengthening the Narrative) that continued the dialogue and initiative between the Summits and reported back on progress made during the previous year at the most recent Solar Summit. Over 200 people attended the Solar Summit in March, and in addition to working group reports there were several expert panel discussions on relevant issues to solar energy in Arizona and the Southwest region. The format was a panel discussion of the topic followed by audience Q&A and panel recommendations for action on that particular topic. In the context of a stakeholder forum or a policy-making forum, the Solar Summit would offer little in the way of representation or legitimacy to issues discussed but viewed through the lens of a conference the Solar Summit fairs more positively. The working groups are made up of participants and audience Q&A is incorporated into action items from each panel and those items become part of the agenda for the working groups in the interim between Solar Summits.

Organizers see this event fitting into a larger network of deliberations around energy by convening most of the relevant policy players and stakeholders in Arizona and some nationally to continue the dialogue on renewable energy policy that may be lacking at the state level. It fills a gap where the ACC and the AZ legislature should be making progress, by keeping dialogue going despite the political and economic climate of the state, region and the nation. The Solar Summit connects with energy networks on several levels: the Greater Phoenix Economic Council (GPEC), the Energy Consortium, ASU, UA, NAU, utilities, local policy makers and energy-industry insiders. There was a noted lack of state-level ACC commissioners and state legislators present at the Summit. Outcomes from the Solar Summit are seen primarily in the working groups' activity, a ten-minute solar documentary, the Solar Summit website, conference video and presentations of the panels. Legitimacy of these outcomes is questionable due to the lack of Republican legislators or commissioners and too many like-minded people (no dissenting opinions to solar energy were heard, Tea Partiers for example). There was, however, great representation across the solar industry and its stakeholders. A representative of the Republican governor was there as well as the head of the Governor's Office on Energy Policy. Due to the convergence of opinion around solar energy at the event there was not a large variation of views presented and only details were debated. Consensus was the goal around ways to move solar forward in Arizona and to identify steps to make that happen. It was felt that the discussions were somewhat superficial and that truly fleshing out what different solar futures might look like and the steps needed to achieve them were left unexplored.

4. FRAMEWORK ANALYSIS OF EVENTS

The public space was only represented at one event, the Town Hall, and this was still very limited in number so inclusiveness of this part of the deliberative system is certainly lacking across all three events. Democratic legitimacy requires that a deliberation be inclusive and consequential, so if the event is lacking both, legitimacy cannot be achieved; unfortunately the energy system as a whole and all three events failed to measure up to this standard across all categories. Authenticity was acceptable at both the Solar Summit and Emerge primarily because there were few public participants so discussions were tailored to the energy experts in the attendance. At the Town Hall, authenticity is brought into doubt due to the technical detail and complexity of the background material and the wanting participation of the public participants who were in attendance. Although public participation was low, all three events' deliberations produced outcomes that are consequential to the public via documentaries, public dissemination of reports and art exhibits.

All three events succeeded in authenticity within the empowered space as discussions were suited for energy insiders; this is an improvement over the whole deliberative system that sees deliberations within empowered stakeholder groups (usually divided along political lines) that is sometimes very coercive. Inclusion of the empowered stakeholders was good but due to the absence of state policy-makers at all three events, they cannot be categorized as inclusive.

Although the events were consequential in the public space, they have little impact on the outcomes that the empowered stakeholders are responsible for (primarily setting policy). Transmission from the public to the empowered space does not readily happen in the deliberative system as a whole but in the Town Hall, due to some public participation, the mini-Town Halls that happen around the state and the recommendations report being sent to stakeholders in the empowered space, it has both authentic and consequential transmission. The Solar Summit and Emerge did not achieve consequential transmission due to lack of public space stakeholders.

Accountability in the deliberative system is through the election of some empowered space stakeholders and public meetings if and when stakeholders from the public attend. Neither of the events in this study attempted to influence the authenticity or impact of that accountability of the empowered space to the public space. Deliberations at all the events were of a detailed and solution-oriented nature; consequently any type of meta-deliberation about how the deliberative system should be structured was absent. Decisiveness is the degree that the five categories on the left of the table influence collective outcomes of the deliberative system. While the three events looked at here do improve some of those categorical contributions to the system as a whole, those efforts are still insufficient to make the deliberative system decisive; inclusiveness, legitimacy, public spaces, transmission, accountability and meta-deliberation all need to be addressed to improve overall decisiveness.

Table 2: Deliberative Events: Effects Within the System

	Authentic	Inclusive	Consequential	Legitimate
Public Space	Yes	No	Yes	No
Town Hall	No	No	Yes	No
Emerge	Yes	No	Yes	No
Solar Summit	Yes	No	Yes	No
Empowered Space	No	No	Yes	No
Town Hall	*Yes	No	No	No
Emerge	*Yes	No	No	No
Solar Summit	*Yes	No	No	No
Transmission	No		No	No
Town Hall	*Yes		*Yes	No
Emerge	No		No	No
Solar Summit	No		No	No
Accountability	Voting only		Voting only	Voting only
Town Hall	No		No	No
Emerge	No		No	No
Solar Summit	No		No	No
Meta-Deliberation	No	No	No	No
Town Hall	No	No	No	No
Emerge	No	No	No	No
Solar Summit	No	No	No	No

Grayed boxes indicate where a deliberative capacity trait is not required within a certain part of the deliberative system (Dryzek, 2010).

Asterisk (*) indicates a positive change a specific event had to the deliberative energy system as a whole.

This leaves one question. Have the events looked at here fundamentally altered the deliberative system around energy in Arizona? In this context, a fundamental change to the deliberative system should make it decisive as a whole. Having just said that the three events looked at here do not improve decisiveness of the system; it follows that they have also not brought about a fundamental change to the deliberative system around energy in Arizona.

5. CONCLUSION

Concluding that there has been no fundamental change to the overall system is only one step. The question is how can we succeed where these events have failed? There is no silver bullet and it will take a concerted effort from stakeholders across all the categories of the system looking at the different capacity areas and striving to achieve them. Not every area of the system needs to be successful in all areas of deliberative capacity, nor is that likely possible, but certainly large improvements can be made from the current deliberative system. Inclusion is the most lacking capacity trait within the system. The Town Hall, Emerge and Solar Summit all need to reach out to people across all facets of the stakeholder groups in the system so they at least have the opportunity to participate if they so choose. In some cases, such as the Solar Summit and Town Hall, scholarships or discounted registration fees could be given to eliminate money as a barrier to participation. Lower socio-economic groups are traditionally under-represented in energy decision-making processes so this is a very critical part of inclusion. Consequently, this may require additional funding for the events or more economically priced events themselves. Policy makers need to be encouraged to participate; they have the final acting authority on energy policy in the state making them a critical component in the deliberations.

Once inclusion is improved and you have sufficient stakeholder representation, transmission and accountability become tangible. Events can be structured to provide veto power (much like the Town Hall) to everyone and voting can be easily implemented through cell phones. Public spaces can be developed through non-profits that encourage participation, progress and interaction building on momentum following each event, carrying it throughout the year and not just during the events. Greater media attention leading up to the events and in disseminating the tangible reports and recommendations that stem from the events will improve inclusion and transmission, respectively. Improving online resources and having interactive websites will drive participation and help transmit information. Events should be structured to foster debate and discussion, not just elaborate on an already existing consensus among stakeholders of like opinions. Assuming policy-maker participation, structuring events in this way can help fill the gap between deliberation and policy implementation and can help develop new policy innovations to address the myriad of issues related to energy in Arizona. Meta-deliberation on the structure of the deliberative system as a whole would be much easier to address with these improvements. This would also enhance the democratic legitimacy and decisiveness of the energy system and would make Arizona a leader in civic process innovation nationwide. To thrive, the deliberative system needs engagement and dissent, which is the beginning of the process of change, not the end.

Fundamentally, energy decision making is a political problem that rests with the officials we elect to set energy policy. Involving policy makers in the deliberative process will not only demonstrate consensus among stakeholders, but it will give them credit for the policies that work and political cover for ones that do not. This type of collective decision making removes individual culpability from the political process which is one of the largest barriers to policy change in Arizona.

In closing, we see that the Town Hall, Emerge and the Solar Summit did not fundamentally alter the deliberative system around energy in Arizona. However, they are a step in the direction of civic process and policy innovation that the system desperately needs to be legitimate and effective. Improvements to inclusion, public space participation, transmission, accountability and meta-deliberation will be difficult but can be achieved with cooperation across all stakeholder groups in the deliberative system currently. The consequences are too dire and the issues too important to continue with the status quo. Change is needed in the energy system in Arizona and the discussion presented here is a start down that path.

ACKNOWLEDGEMENTS: I would like to thank the Consortium for Science, Policy & Outcomes for providing an environment that fosters learning and collaboration. A special thanks also to Dr. Clark Miller for his help in creating this project and being on my committee, Dr. Jamey Wetmore for chairing my committee and giving me direction & Dr. Benjamin Broome for giving me a human perspective on dialogue.

REFERENCES

Abelson, J., Forest, P-G., Eyles, J., Smith, P., Martin, E., & Gauvin, F-P. (2003). Deliberations about deliberative methods: Issues in the design and evaluation of public participation processes. Social Science & Medicine, 57(2), 239–251.

Andersen, I-E., & Jaeger, B. (1999). Danish participatory models: Scenario workshops and consensus conferences: Towards more democratic decision-making. Science and Public Policy, 26(5), 331–340.

Arizona Solar Summit. (2012, March). Arizona Solar Summit. Retrieved from azsolarsummit.org.

Arizona Town Hall. (2012). Arizona town hall: Creating solutions for 50 years. Retrieved from aztownhall.org.

Button, M., & Mattson, K. (1999). Deliberative democracy in practice: Challenges and prospects for civic deliberation. Polity, 31(4), 609–637.

Cartwright, N. (1999). The dappled world: A study of the boundaries of science. New York, NY: Cambridge University Press.

Dryzek, J. S., & Niemeyer, S. (2010). Foundations and frontiers of deliberative governance. Oxford, UK: Oxford University Press.

Dupre, J. (1993). Methodological unity and science as a process. In The disorder of things (pp. 229–243) Cambridge, MA: Harvard University Press.

Emerge. (2012). Emerge: Artists & scientists redesign the future. Retrieved from emerge.asu.edu.

Hagendijk, R., & Irwin, A. (2006). Public deliberation and governance: Engaging with science and technology in contemporary Europe. Minerva, 44, 167–184.

Hendriks, C. M., Dryzek, J. S., & Hunold, C. (2007). Turning up the heat: Partisanship in deliberative innovation. Political Studies, 55, 362–383.

Kleinman, D. L., Delborne, J. A. & Anderson, A. A. (2009). Engaging citizens: The high cost of citizen participation in high technology. Public Understanding of Science, 20(2), 221–240.

Kleinman, D. L., Powell, M., Grice, J., Adrian, J., & Lobes, C. (2007). A toolkit for democratizing science and technology policy: The practical mechanics of organizing a consensus conference. Bulletin of Science Technology Society, 27(2), 154–169.

Pelletier, D., Kraak, V., McCullum, C., Uusitalo, U. & Rich, R. (1999). The shaping of collective values through deliberative democracy: An empirical study from New York's North Country. Policy Sciences, 32(2), 103–131.

Popper, K. (1968). A Survey of Some Fundamental Problems. In *The Logic of Scientific Discovery* (pp. 27–48). New York, NY: Routledge.

Powell, M. C., & Colin, M. (2008). Meaningful citizen engagement in science and technology: What would it really take? Science Communication, 30(1), 126–136.

Powell, M. C., & Colin, M. (2009). Participatory paradoxes: Facilitating citizen engagement in science and technology from the top-down? Bulletin of Science Technology Society, 29(4), 325–342.

Weeks, E. C. (2000). The practice of deliberative democracy: Results form four large-scale trials. Public Administration Review, 60(4), 360–372.

Wilson, E.O. (1998). The Natural Sciences. In *Consilience: The Unity of Knowledge* (pp. 45–6). New York, NY: Knopf.

A Pragmatic Paradox Inherent in Expert Reports Addressed to Lay Citizens

FRED J. KAUFFELD

Communication Studies
Edgewood College
1000 Edgewood College Dr.
Madison, WI 53711
USA
chris_fred@att.net

ABSTRACT : This paper addresses a problem inherent in *reporting* as a mode of communication between experts and lay citizens. The potential utility of such reports is obvious, but we commonly encounter critically debilitating frustration as experts, trained to address and to be accountable to other experts, attempt to report to citizens engaged in public decision making with proper regard for their own autonomy. We may move toward some resolution to these frustrations if we better understand the obligations inherent in the ordinary communicative act of *reporting*, which by its nature involves a delegation of responsibility.

KEYWORDS: expert reports, normative structure of reporting, autonomy, lay critical evaluation.

> No government by experts in which the masses do not have the chance to inform the experts as to their needs can be anything but an oligarch managed in the interest of the few. And the enlightenment must proceed in ways which force the administrative specialists to take account of the needs. The world has suffered more from leaders and authorities than from the masses.
>
> The essential need, in other words, is the improvement of the methods and conditions of debate, discussion and persuasion. That is *the* problem of the public. We have asserted that this improvement depends essentially upon freeing and perfecting the processes of inquiry and of dissemination of their conclusions. Inquiry, indeed, is a work which devolves upon experts. But their expertness is not shown in framing and executing policies, but in discovering and making known the facts upon which the former depend. They are technical experts in the sense that scientific investigator and artists manifest *expertise*. It is not necessary that the many should have the knowledge and skill to carry on the needed investigation; what is required is that they have the ability to judge of the bearing of the knowledge supplied by others upon common concerns.* (John Dewey, *The Public and Its Problems*, 1927).
>
> *This statement can be read as a call for training the many to critically assess expert statements. But it is, I submit, better understood as calling for a collaborative practice within which discourse from experts is profoundly calculated to enable the many to "judge of the bearing of the knowledge supplied" and in which the many are prepared to critically appreciate and utilized the "knowledge supplied."

1. INTRODUCTION

This paper addresses a problem inherent in *reporting* as a mode of communication between experts and lay citizens. The potential utility of such reports is obvious. Citizens, as lay persons attempting to participate responsibly in public decision-making activities, typically need to rely on reports from experts, and by reason of their epistemic superiority, experts often acquire knowledge relevant to public concerns, which they rightly take themselves to be responsible to report to the larger lay public. It seems, then, that delegating responsibility to experts for preparing reports for lay citizens would satisfy complementary needs and obligations.

However, the situation is complicated by the demands autonomy imposes on the lay citizen and by the nature of the expert's knowledge base. A morally competent citizen exercises self-reliance. In matters for which she bears responsibility, her decisions are to be based primarily on her own thought and experience. If she is to make responsible use of expert reports, she must in some sense be able to appropriate what they tell her, so that it becomes part of a matter which she has thought through. But the expert's report, presented in terms which would enable an expert colleague to assess its epistemic quality, typically is beyond the competence and practical capacity of ordinary citizens to fully evaluate. So, we commonly encounter critically debilitating frustration as experts, trained to address and to be accountable to other experts, attempt to report to self-reliant citizens engaged in public decision making and rightfully jealous their own autonomy.

Tensions inherent in communication between experts and lay persons have been productively studied from various perspectives. The predominant approach views the problem from the perspective of the addressees, i.e., lay citizens, and attempts to equip them with critical questions which they can utilize in evaluating statements from experts. This approach has produced important and useful protocols for critical evaluation of expert statements. Its focus on the lay addressee realistically reflects the preoccupation which reliable experts may have with the reception of their statement by peer experts. Moreover, it speaks to an inescapable aspect of the topic: how can the enlightened lay citizen acquire something *she* can rely upon as knowledge from the statements of experts? Consequently, it makes sense to focus on *training* the autonomous citizens to critically evaluate expert statements.

This essay takes a different, though complementary, approach. Working from a normative pragmatic perspective on discourse design, it focuses on a specific communicative activity and product utilized in addressing expert statements to lay decision makers, viz., reporting and reports. Keeping in mind what we have gained by working to educate a critically aware addressee, this study directs attention to how reports from experts can be designed to enable critical and responsible appropriation by autonomous, plain citizens.

2. BACKGROUND

A normative pragmatic focus on the strategic design of expert reports complements some recent work on their critical reception by lay addressees. Jean Goodwin has called attention to a close fit between work in Argumentation Studies and in Studies of Expertise and Experience which focus on the capacities of lay audience to evaluate experts on the basis of the audience's social knowledge (2011a, p. 289). Social knowledge, in these views, is seen to be "ubiquitous," composed of "abilities that people acquire as they learn to navigate their way through life (Collins & Evans, 2007, p. 16, as cited by Goodwin, p. 289). Thus ordinary citizens can document the purported expert's experience, check the internal consistency of his statements, their external consistency with evidence, their "scientificness," and the trustworthiness of the expert's demeanor (Goodwin, 2011a, p. 289). Similarly, as Douglas Walton observes, argumentation textbooks advise citizens to examine a purported expert's experience, education, access to the topic, credentials, peer recognition, track record, lack of bias or interest, and internal consistency of judgments, as well as recency, consistency with other experts and consistency with other evidence (1997, pp. 199–229).

However, Goodwin doubts whether this convergence between argumentation studies and studies of expertise has provided a full account of how relevant social knowledge is to be applied in practice (2011a). Focus on the discursive design of expert reports may enrich understanding of how common social knowledge can be critically applied in assessment and appropriation of expert reports. Presumably experts and lay persons share much the same social knowledge. That being the case, it should be possible for this shared social knowledge to be applied by experts in designing reports which facilitate critical evaluation by lay persons. (Indeed, as we will see, some expert reports do realize this possibility, at least in part.)

Support for this approach can be drawn from Axel Gelfert's adaptation of Robert Brandom's conception of "conversational score-keeping" to facilitate application of lay social knowledge to the appraisal of expert statements (2011). In Brandom's view, participants in a "discursive practice"—including experts—implicitly and explicitly undertake and attribute various commitments and/or entitlements. By tracking the commitments undertaken by experts in their statements and the corresponding entitlements, Gelfert suggests, lay persons can gain footing for appraisal of those statements (2011, p. 307). Rather than defending conversational scorekeeping as the basis of a universal theory, Gelfert suggests:

> [A] conscious effort at keeping track of one's interlocutor's (acknowledged and undertaken) commitments and entitlements, is a commendable strategy in dealing with appeals to expert opinion Irrespective of whether conversational scorekeeping generalizes to all kinds of speech acts and conversation moves, it may have its most natural application in the context of appeals to an external epistemic authority. (p. 308)

In support of this suggestion Gelfert points to the ways in which commitments undertaken in connection with testimony afford basis for its appraisal (p. 308).

Gelfert's suggestion implies a promising approach to the problem(s) of lay appropriation of expert testimony. But where Gelfert focuses on orienting the lay audience to the commitments undertaken by experts in making statements, I propose a complementary focus on the production of those statements, specifically on the production of expert reporting and reports designed to facilitate lay evaluation of their epistemic merit on the basis of ordinary social knowledge. It may be that expert reports can be designed to facilitate what Gelfert calls "scorekeeping" by careful attention to how epistemically relevant commitments are undertaken and discharged in the production of expert reports.

3. THE NORMATIVE PRAGMATIC PLATFORM

My approach requires a relatively rich understanding of how commitments are undertaken in communicative acts and how those commitments both enable addressee evaluation of the speaker's discourse and potentially provide the addressee with reason for placing confidence in that discourse. For this understanding I turn to a normative pragmatic view of discourse, which seems complementary to Brandom's but proceeds from an "assurance account" of the pragmatics underlying H. P. Grice's analysis of utterance-meaning, i.e., seriously saying and meaning something.

A basic principle of assurance accounts is that in the primary communicative act of seriously saying and meaning something and in speech acts performed by means of this primary act, speakers deliberately and openly manifest the intentions which constitute their speech acts and thereby conspicuously make themselves accountable for their primary communicative effort (Kauffeld, 2003, 2011; Moran, 2005; Stampe, 1967). Such manifestations of accountability are designed to provide addressees with assurance that the speaker is acting responsibly. The addressee is entitled to and intended to reason that the speaker would not make herself accountable for her communicative effort and, hence, risk resentment, retribution, etc., should she fail to live up to her openly incurred responsibilities. Given this presumption the addressee may find reason to respond as the speaker primarily intends.

In the primary case where speaking seriously and meaning what she says, a speaker says, e.g., that Uncle Bill has died, she openly and strategically takes responsibility for the veracity of her utterance. Accordingly, she makes herself inescapably vulnerable to criticism and resentment for mendacity should it turn out that she is speaking falsely. The speaker thereby generates a presumption of veracity on behalf of her utterance, which serves to provide her addressee with assurance that she is speaking truthfully. Given the speaker's openly incurred commitments, her addressee can reason (ceteris paribus) and is intended to reason that the speaker would not be manifestly willing to risk criticism for speaking falsely, were she not in fact speaking truthfully (Kauffeld, 2001; Stampe, 1967, 1975). This interpretation of the practical design underlying the constituents identified by Grice's analysis is a model of normative pragmatics. It exhibits the genesis of a normative obligation in a familiar communicative practice: in saying that p, the speaker openly incurs an obligation to speak truthfully. And it identifies the potential efficacy of that normative obligation, viz., by openly incurring an obligation to speak truthfully, the speaker generates reason to, e. g., believe what she says.

Variants of this basic assurance strategy for generating presumptions can be seen to be at work in the genesis of probative obligations in such speech acts as *accusing, proposing, praising,* etc. (Kauffeld, 1998, 2002). Similar strategies can also be seen to be at work in some communicative activities. Testifying and its product, testimony, provide examples similar to reporting and reports in some respects. Testimony is characteristically taken in the context of some larger inquiry involving serious decisions and/or judgments. Typically the situation is such that the recipient of the testimony (*TR*) needs to know something about a matter (*m*) to which he does not have direct access but which is (or may be) within the scope of the speaker providing the testimony (*TS*). Moreover, *TR*'s responsibilities are such that he cannot simply rely on *TS*'s presumed veracity; he needs to be able to test, as best he can, the grounds for her purported knowledge. Satisfying that need involves a breach in the trust which in normal circumstances would attend *TS*'s statements. To repair that breach, *TS* deliberately and openly speaks with the intention of answering *TR*'s questions regarding what *TS* knows about *m*, thereby manifestly consenting to being examined as to the grounds for her purported knowledge. At the same time, *TS* manifestly maintains her commitment to speak truthfully. By openly enlarging the commitments *TS* undertakes in saying what she knows regarding *m*, *TS* adjusts her discourse to accommodate *TR*'s need to examine her and also maintain an assurance of trust in what she says (Kauffeld & Fields, 2005). Our attention now turns to the discursive design of a somewhat similar communicative activity.

4. OPPORTUNITIES FOR CONSTRUCTIVE DISCOURSE DESIGN INHERENT IN THE COMMUNICATIVE STRUCTURE OF REPORTING

Like testimony, reports are produced in the course of a communicative activity, *reporting*. As with testimony, reports typically do their work in some larger activity involving decision, judgment, explanation and/or understanding. To apprehend the potentials for the discursive design of reports, we must first reflect on the dynamics of that activity. Within that dynamic we may then identify resources which can be used to enhance possibilities for lay critical reception of expert reports.

Reporting characteristically involves an authorization relating two (or more) parties. In its simplest paradigmatic structure, reporting involves some party (or parties) which authorize and ultimately receive the report. Let us designate the authorizing agent(s) and the party(s) to whom the report is addressed (*AA*)—as we will see the authorizing agent and the addressee may in many cases be distinct parties. Reporting further involves some party or parties who are authorized to investigate some matter (*m*) and provide *AA* with a statement of the results (findings) of that investigation, i.e., the report. Let us designate this party the investigator/reporter (*IR*)—noting again that the roles of investigator and reporter may fall to distinct parties. The situation in which reporting is undertaken typically has this structure:

(1) *AA* needs to know (learn) something about *m*, and *AA* is for some reason unable or unwilling to investigate the matter (perhaps *AA* lacks the time, the resources, ability, and/or inclination to conduct the inquiry herself).
(2) *AA* authorizes *IR* to investigate *m* and tell *AA* the results of that investigation.
(3) *IR* subsequently conducts the investigations and reports her findings to *AA*.

This structure is apparent in the ways we talk about reports. In a great many cases it would make sense (where the matter not entirely clear) to ask who or what authorized this report. To characterize a report as 'unauthorized' would be to imply a criticism of that "report" and in many cases that criticism would bring into question the discourse's claim to the status of a *report*.

From a normative pragmatic perspective, the fact that reports initially require a delegation of authority and responsibility is of considerable importance to the possibility of designing expert reports which can be assessed and interpreted by lay parties. Assurance accounts of the pragmatics of discourse direct our attention to those moments in a discourse in which commitments are manifestly undertaken in connection with the open identification of the intention/purpose constituting the central communicative act/activity (Jackson, 1992; Kauffeld, 2007). Commonly in connection with the activity of reporting those purposes and corresponding commitments are initially and openly identified in the course of delegating authority and responsibility for the investigation and preparation of the report. The construction of those purposes and corresponding commitments affords opportunities to design the activity of reporting in ways which can guide and give assurance of the production of well-founded reports which are intelligible to and can be evaluated by lay parties.

5. SITUATED DESIGNS FOR LAY-FRIENDLY EXPERT REPORTS

Strategic possibilities inherent in the delegation of authority and responsibility for designing lay-friendly expert reports can be identified in three of the many scenarios which give rise to expert- to-lay reports. In the first scenario the lay parties (lay *AA*s) directly authorize expert parties (expert *IR*s) to investigate *m* and report directly to the lay *AA*s. In the second scenario the report is prepared under the aegis of some authorizing institution (*A*) which delegates responsibility to some experts or body of experts to investigate and prepare a report (expert *IR*s) on *m* which is addressed to other interested experts and to relevant lay parties. In the third scenario, a primary report has been prepared by expert *IR*s, typically under the direction of some institution, and a second mediating IR_2 undertakes to interpret the report to lay addressees. These scenarios afford parallel but distinct opportunities for designing reports which facilitate critical appropriation by lay parties.

Consider first a not entirely hypothetical case in which lay *AA*s authorize selected experts *IR*s to investigate a matter and to prepare a report to the lay *AA*s. A small college is experiencing sudden, welcome growth in enrollment and corresponding program expansion. It needs to more carefully utilize the limited available physical space for facilities construction. The college constitutes a campus master plan committee, including representatives from various campus constituents and interests, which is to prepare an overall plan for future facilities development. Lacking the expertise, time, and inclination, the committee delegates to architectural specialists the task of investigating and preparing a report recommending a master plan for the campus. Here the task of investigating a matter and preparing a report has been delegated by a knowledgeable lay body to a small group of experts. The delegating body will in turn be responsible for evaluating their report, advocating for some or all of its findings, and ultimately utilizing the report as a basis for campus development.

Within this elementary reporting structure, the delegation of authority can in principle be directly negotiated between the lay authorizing agent, *AA*, and the expert investigative/reporter, *IR*. From a message design perspective, this initial delegation can (and should) be seen as an opportunity to allocate obligations such that the expert *IR* is openly committed to producing a report which autonomous lay decision makers can responsibly utilize. Ideally, then, *IR* would deliberately and openly give it to be believed that her investigation and report will be directed by the intention to enable *AA* to make a sound determination regarding *m*. In our example, the campus planning experts would openly commit themselves to investigate the college's needs and opportunities and to providing the planning committee with a report which would enable the committee to formulate a well-conceived campus master plan. This commitment entails not only the duty of providing a well-founded report, but also a larger responsibility to put the committee in a position to evaluate and appropriate the report. Failure to fulfill the first duty would subject the *IR* to criticism for negligence; one would say that the report was "shoddy." Failure to fulfill the second would subject the *IR*'s report to criticism as regards its intelligibility; one would say that the report did not make sense. The supposition that *IR* would not invite the resentment attending either of these complaints provides *AA* with an initial presumption regarding the expert's trustworthiness and would provide partial reason to delegate investigation and preparation of the report to these expert *IR*s (Also see: Goodwin, 2011b).

This initial delegation affords the expert *IR*s preliminary opportunity to frame their task in terms which facilitate providing an intelligible report. The interaction between *AA* and *IR* has a potential to establish a mutual understanding of what *AA* wants to know regarding *m* and why and, also, of the scope of the inquiry expected of *IR* and of the methods *IR* is to employ. Thus, in our hypothetical case, *IR* would learn from *AA* the situation which gives rise to the latter's need for a campus master plan and some of the restraints under which *AA* is working (the constituents, civic regulations, planning approval processes, etc.), and both can agree on the available best methods for conducting the inquiry. The extent to which this initial mutual determination of the case is possible will vary and may involve an extended conversation which corrects and adjusts previous understandings. The important point to notice here is that mutual understandings regarding rationale, scope and methods of the inquiry provide *AA* with subsequent basis for assessing and appropriating the expert's report and, importantly, provides *IR* with a framework for constructing a report which is both intelligible and assessable by the addressee. Such a report would conspicuously recall and address the initial mutual understandings regarding the rational, scope, and methods of investigation delegated to *IR*. A report constructed along these lines does not relieve *AA* of responsibility to critically evaluate the expert's statements, but it would facilitate that evaluation. In many cases, reports constructed along these lines would open to an ongoing conversation between *AA* and *IR* which facilitates the former's critical appropriation of the latter's report.

Consider now the second scenario in which the report is prepared under the aegis of some authorizing institution (*A*) which delegates to some experts or body of experts (expert *IRs*) responsibility for investigating and preparing a report regarding *m*, which is to be addressed to other interested experts and to relevant lay parties. Here the allocation of responsibilities is more complex than in the case where lay parties directly authorize experts to investigate and report on some matter; nevertheless, negotiation of that allocation affords opportunities for designing discourse calculated to assure lay-friendly reports. *The Miller Lite Report on American Attitudes toward Sports* affords an opportunity to reflect on some of that complexity and the available design opportunities.

The Miller Lite Report on American Attitudes toward Sports (hereafter, the Miller Report) is the product of a comprehensive study authorized by the Miller Brewing Company presented to "the general public" in 1983 (Miller Brewing & Research and Forecasts). In the introductory pages, Miller Brewing represents the presentation of the report as a benevolent act, a contribution, designed to "help the American public know more about itself." William K. Howell, then President and Chief Operating Officer of Miller Brewing, explains Miller's motivation in the following terms.

> It is logical for Lite beer from Miller to have commissioned this report because beer and sports go together in many ways. Many of American's sports fans are our consumers, and the Miller Brewing Company is becoming known as *the* sports company through sponsorship of sports broadcasts and thousands of events each year at all levels of competition. Few companies can claim such a close association with the world of sports and its millions of fans. (Miller Brewing & Research and Forecasts, 1983)

The report provided a platform for Miller's subsequent use of sports venues in very successful advertising campaigns (Sperber, 2001, pp. 39–40) and, as such, contains much in the way of precise data and interpretations of technical interest to Miller's marketing department. The report itself, however, is not addressed specifically to Miller's marketing and advertising experts; rather, it is openly addressed to a lay public and committed to assisting the public's self-understanding.

Given that commitment, the public is initially entitled to presume that Miller has authorized a report which may be well-founded, comprehensive and intelligible. The report's introductory discussion of its structure, methods, and aims reinforce that presumption. The preparation of the report, one is told, was under the supervision of an advisory panel consisting of distinguished scholars, physicians, and leaders qualified by their sports expertise. The study itself was conducted and the report prepared by a staff of experts using the latest systematic survey methods. The study purports to have been designed to provide a "modern full-scale national survey of sports attitudes and practices in America today" which can therefore be useful in several ways. It can:

- probe the opinions Americans hold toward activities that permeate all aspects of our national life;
- open up new issues for popular discussion and provide fresh information for familiar concerns;
- establish a benchmark reference point for sports professionals, managers, journalists, and enthusiasts; and
- compare public attitudes toward sports issues with those held by sports journalists, coaches, and sports doctors. (Miller Brewing & Research and Forecasts, 1983, p. 2)

The report is not only explicitly committed to providing publicly useful knowledge; the scope of the inquiry is specified by an inventory of questions the Miller Report is to address, e.g., what activities do American define as "sport"? How much are American's involved in sports, either as spectators or participants, on a weekly or daily basis? How closely are participation in and attendance at sports activities linked to the vitality of family life? Recalling public reports of serious misbehaviour in the sports world, how do the public and relevant experts view athletes as role models for children? etc. The introductory discussion of the structure, methods, aims and scope of the report strengthens the initial presumption that the report will provide reliable, intelligible, and useful information about American attitudes to sports.

The commitments openly undertaken in the Miller Report's discussion of the delegation of responsibilities related to its production not only provide lay addressees with a reasonable basis for initially presuming that the report will provide information that is reliable, intelligible and useful from a lay perspective, it also puts the authors of the report in a position to conspicuously discharge the authorizing commitments. And, by the same token, it positions the lay reader to make a reasonable assessment of the adequacy of the report.

The Miller Report conspicuously attempts to discharge its initially incurred commitment to provide a report which can be appropriated by lay readers as reliable, intelligible, and useful. The overall organization corresponds one-to-one on a chapter-by-chapter basis to the initial set of questions the report is (purportedly) designed to answer. Throughout, the report provides an ongoing lay-friendly discussion of it methods. Technical data is routinely interpreted in ordinary terms. Material accessible only with an expert's competence is relegated to appendices. In short, the report is structured such that lay persons can readily find their way through what is in fact a highly technical marketing study and can integrate into their thinking about sports in American life a presumptively reliable set of data and understandings.

In the third scenario a primary report has been prepared by expert IRs, typically under the direction of some institution, and a second mediating IR_2 undertakes to interpret the report to lay addressees. In this scenario the delegation of responsibilities and concomitant commitments is at once more complicated and simpler. Here we deal with two reports, the one originally prepared by experts and most likely primarily addressed to other experts; the other a report prepared by an intermediary agency which attempts to interpret the report (and its significance) for lay citizens. Nevertheless, here the possibilities for lay-friendly discourse design is fundamentally simple, and the potential for enabling critical appropriation of expert reports by lay citizens is reflected in the ubiquitous presence of such secondary reports in our public discourse.

A wide range of journalistic strategies designed to negotiate this third scenario have been developed; detailed examination of their strengths and liabilities lies outside the scope of the present study. Nevertheless, a particularly nice specimen from National Public Radio's programming illustrates the potentials here for lay critical appropriation of expert reports. The NPR report "Should we kill the Dollar Bill?" which aired on April 19, 2012 and is now archived on the internet at Planet Money Blog, provides a secondary report addressed to the lay public as a report on responses to a Government Accountability Office Report which recommends switching from paper one dollar bills to dollar coins. NPR's report covers the GAO report, the lobbying controversy it spawned, and responses from relevant economic experts. The program is explicitly addressed to a lay public and is openly designed to enable their decision making (Benincasa & Kestenbaum, 2012).

National Public Radio presents reports of this kind as a publicly funded service. NPR, that is, represents itself as an agent supported by the public to provide citizens with, among other things, reports that enable citizens to responsibly execute their duties. Accordingly, when NPR offers a secondary report of this kind it has a standing public commitment under that broad obligation, and its listeners not only have opportunity to air views critical of NPR's discharge of that duty, they effectively avail themselves of that opportunity on a daily basis. It is reasonable to presume that NPR tries to fulfill its duties rather than suffer adverse criticism from its listeners—a presumption which is strengthened by NPR's practice of airing critical, and sometimes answering, listener criticism. How then does NPR frame "Should We Kill the Dollar Bill?" (hereafter, Dollar Bill Report) so as to enable responsible decision making on the part of autonomous citizens?

NPR focuses its report on the controversy generated by the GAO report recommending a switch from dollar bills to dollar coins. Speaking in its capacity as a publicly supported agent of its addressees, NPR openly presents its report as the product of an investigation calculated to get to the bottom of the matter. The inquiry and its product are manifestly designed to enable lay persons to make sense of the controversy and to arrive at a well-founded judgment. The report unfolds in three stages corresponding to stages in NPR's investigation: the first provides reasons for and against the change as presented by the GAO and contending parties; the second provides an analysis of the case for change from the perspective of economic experts; and the third stage briefly provides evidence which tends to confirm the analysis offered by a leading economic expert. This progression enables lay persons to appreciate the scope of the inquiry and to understand each stage as moving closer to a bottom-line decision. Each stage is framed in terms of the methods of inquiry NPR used at that stage, explicitly addressing questions regarding their appropriateness. Throughout NPR conspicuously represents its inquiry as conducted fairly, recognizing the special qualifications of the sources interviewed and the special interests of the contending parties.

In order to get to the bottom of the matter NPR critically and with manifest fairness attempts to balance and comparatively evaluate the arguments presented by the contending parties. The GAO's report recommends the switch to dollar coins, we are told by NPR, primarily on the grounds that the change would make the federal government some $4.4 billion over the next thirty years. The change is supported by legislators who claim that coins are more efficient than paper dollars as evidenced by the experience of other countries which have made this kind of change. But NPR notes that these legislators represent constituents which have substantial interests in the production of metal coins. On the other side, the legislators vigorously opposing the change maintain that the American dollar is far more durable than the paper currency of other countries and is, hence, an economical basis for currency transactions. These legislators, NPR notes, represent constituents which have substantial interests in the production of paper currency. To adjudicate the controversy NPR then turns to academic economists, including a key voice recommended by the advocates for a switch to coins. These economists point to a part of the GAO's report which acknowledges that the $4.4 billion gain would come about as profit the government acquires because it makes money on the production of currency and would gain a profit from the need to produce more coins than bills because citizens leave coins lying about, out of circulation. Economists uniformly regard, we are told, this as a tax on the citizens (seigniorage), and, as one economist (scholar recommended to NPR by the pro-coin interests) tells the investigator (and us), "The government can make profits in all sorts of bad ways. People are going to be putting them (dollar coins) on top of their bureaus instead of spending them for transactions and that seems like a big waste of resources to me. This does not seem like a good way to raise money." To conclude its effort to reach the bottom of the matter, NPR presents independent evidence confirming the: economist's judgment: in the past when Congress has tried to get the public to use dollar coins, the coins remain languishing in vaults because the public had rejected them. By implication, languishing in vaults is a good predictor of coins languishing on the tops of bureaus, confirming an expectation already conceded in the GAO's report.

The reports generated across the three scenarios discussed above share certain features designed to enable lay addressees to appropriate and evaluate statements produced by experts. These features are realized differently in each of the three reports discussed; nevertheless, they reflect a common underlying design. In all three the expert *IR* is openly committed to providing a report which enables lay *A*s to understand and make decisions regarding *m*. In all three this commitment entails an effort to provide a report which is well-founded and intelligible to the lay *A*, which renders the expert *IR* vulnerable to criticism and resentment should negligent investigation and/or unintelligible findings emerge. Each report explicitly takes steps to enable lay addressees to interpret and evaluate the report and its findings; each provides a discussion of and justification for the methods used in the inquiry into *m*; each provides a discussion of the proper scope and importance of the inquiry; and each is organized in ways which explicitly connect discussion of the report's findings to its discussion of method, scope, and importance. Thus, each report enables the lay addressee to reasonably presume that the investigating expert is making a responsible effort to provide the lay addressee with an intelligible report and positions the addressee such that he can evaluate the report on the basis of its conformity to appropriate methods and proper scope.

5. CONCLUSION

This study directs attention to some ways in which reports produced by experts can be designed to facilitate their critical appropriation by self-reliant lay addressees. It has long been recognized that sound judgment and decision making by the latter often depends upon knowledge gained from the former, but the epistemic inferiority of lay agents can frustrate full comprehension of expert statements, impairing the capacity of lay agents to exercise self-reliant judgment. Standard approaches to this conundrum attempt to sharpen the critical skills which lay agents can bring to bear in evaluating expert statements. This essay points to complementary possibilities inherent in the nature of reporting for structuring reports which invite critical evaluation and enable the exercise of self-reliant judgment on the part of lay addressees. The design possibilities identified here emerge in light of a general account of how commitments which are openly incurred in seriously saying things provide addresses with reason to presume that speakers are acting responsibly. The communicative act of reporting typically requires a delegation of authority to one party, *IR,* to investigate some matter and to report to another party, the authorizing addressee, *AA.* Executing such delegations requires identification of the purposes for the report and a corresponding set of obligations incumbent on the *IR.* Expert reports then can be authorized in terms which charge the expert *IR* with an obligation to enable sound judgment and understanding on the part of the lay *AAs.* Discharging that obligation requires the production of a report which is well-grounded and intelligible to lay *AA.* In connection with the delegation of authority to *IR,* the scope of the investigation, the questions to be answered, and the methods to be used can be specified. The report itself can be constructed along lines which conspicuously discharge the duties initially identified, enabling understanding and critical evaluation on the part of lay addressees. Three examples involving different scenarios illustrate these potentials for designing lay-friendly expert reports.

The present essay is designed to open the door on a matter which merits further and fuller investigation. No attempt has been made here to exhaust the range and kinds of commitments which can be explicitly built into reporting activities to assure competence and intelligibility of expert reports addressed to lay citizens. A good deal of productive work has been done on the critical questions lay citizens can use in evaluating expert statements. Argumentation and critical-thinking scholars should now ask how expert reports can be designed to respond to those critical questions. And the entire matter needs study in terms of some good understanding of what understanding lay decision makers must require in order to exercise autonomy with respect to their use of expert reports.

REFERENCES

Benincasa, R., & Kestenbaum, D. (2012, April 19). This American life: Should we kill the dollar bill? Retrieved from www.npr/blogs/money

Collins, H., & Evans, R. (2007). *Rethinking expertise.* Chicago, IL: University of Chicago Press.

Dewey, J. (1927). *The public and its problems: An essay in political inquiry.* New York, NY: Henry Holt and Company.

Gelfert, A. (2011). Expertise, argumentation, and the end of inquiry. *Argumentation, 25*(3), 297–312.

Goodwin, J. (2011a). Accounting for the appeal to the authority of experts. *Argumentation, 25*(3), 285–296.

Goodwin, J. (2011b). Accounting for the force of the appeal to authority. In F. Zenker (Ed.), *Argumentation: Cognition and community: Proceedings of The 9th International Conference of the Ontario Society for the Study of Argumentation* (pp. 1–9). Windsor, ONT: OSSA CD-ROMM.

Jackson, S. (1992). Virual standpoints and the pragmatics of conversational argument. In F. H. v. Eemeren, R. Grootendorst, A. Blair & C. A. Willard (Eds.), *Argumentation illuminated* (pp. 260–269). Amsterdam: SicSat.

Kauffeld, F. J. (1998). Presumption and the distribution of argumentative burdens in acts of proposing and accusing. *Argumentation, 12*(2), 245–266.

Kauffeld, F. J. (2001). Argumentation, discourse, and the rationality underlying Grice's analysis of utterance-meaning. In E.T. Nemeth (Ed.), *Cognition in language use: Selected papers from the 7th International Pragmatics Conference* (Vol. 1, pp. 149–163). Antwerp: International Pragmatics Association.

Kauffeld, F. J. (2002). Pivotal issues and norms in rhetorical theories of argumentation. In F. H. v. Eemeren & P. Houtlosser (Eds.), *Dialectic and rhetoric: The warp and woof of argumentation analysis* (pp. 97–119). Dordrecht: Kluwer Academic Publishers.

Kauffeld, F. J. (2003). The ordinary practice of presuming and presumption with special attention to veracity and the burden of proof. In F. H. v. Eemeren, J. A. Blair, C. A. Willard, & A. F. S. Henkemans (Eds.), *Anyone who has a view: Theoretical contributions to the study of argumentation* (pp. 133–147). Dordrecht: Kluwer Academic.

Kauffeld, F. J. (2007). Keynote address: What are we learning about he pragmatics of the arguers' obligations? In S. Jacobs (Ed.), *Concerning argument: Selected papers from the 15th Biennial Conference on Argumentation* (pp. 1–33). Washington, DC: National Communication Association.

Kauffeld, F. J. (2011). *Strategies for strengthening presumptions and generating ethos by manifestly ensuring accountability.* Paper presented at the *Argumentation: Cognition & Community, Proceedings of the 9th Biannual Conference of the Ontario Society for the Study of Argumentation*, University of Windsor, Ontario, Canada.

Kauffeld, F. J., & Fields, J. (2005). The commitment speakers undertake in giving testimony. In D. Hitchcock & D. Farr (Eds.), *The uses of argument: Proceeding of a conference at McMaster University* (pp. 232–243). Hamilton, ONT: OSSA.

Miller Brewing & Research and Forecasts, I. (1983). *The Miller Lite report on American attitudes toward sports.* Milwaukee, WI: Miller Brewing Company.

Moran, R. (2005). Getting told and being believed. *Philosophers' Imprint, 5*(5). Retrieved from www.philosophersimprint.org/005005/

Sperber, M. (2001). *Beer and circus: How big-time college sports is crippling undergraduate education.* New York, NY: Henry Holt & Co.

Stampe, D. (1967). *On the acoustic behavior of rational animals.* Madison, WI: University of Wisconsin.

Stampe, D. (1975). Meaning and truth in the theory of speech acts. In P. C. a. J. Morgan (Ed.), *Speech Acts* (Vol. 3, pp. 25–38). New York, NY: Academic Press.

Walton, D. (1997). *Appeal to expert opinion: Arguments from authority.* University Park, PA: Pennsylvania University Press.

The Ambiguous Relationship between Expertise and Authority

MOIRA KLOSTER

Philosophy and Politics
University of the Fraser Valley
33844 King Road, Abbotsford, B.C.
Canada
Moira.Kloster@ufv.ca

ABSTRACT: What authority should experts have? The "authority" of "speaking with authority" is not the "authority of "acting with authority." In decisions shared with experts, we need clearer responsibilities and lines of authority for the non-experts as well as for the experts. A better balance between the experts' authority and our own, drawing on a wider variety of experts, promises much better-justified decisions.

KEYWORDS: authority, decision making, deep disagreement, experts.

1. INTRODUCTION

Our paradigm of the expert is the expert who is also an authority: the person whose knowledge can and should be the decisive factor in deliberations. What the expert advises is what should be done. The classic examples are the doctor whose medical advice should determine the correct treatment for what ails you, the lawyer whose legal advice will determine whether you have a case worth taking to trial, and the scientist whose knowledge of ecosystems will determine whether your community is drawing unsustainable amounts of water from the aquifer.

However, all of these classic examples also exhibit the problematic lack of authority that handicaps the expert, preventing his or her knowledge from being decisive. The decision is often not in the expert's hands. The patient can refuse treatment. The client, given deep enough pockets, can go ahead and sue anyway. The water-hogs might be stopped, but more likely by legislation than by scientific testimony alone. The expert has authority in the sense of "speaking with authority," but not in the sense of "signing authority."

When we worry about whether we can assess an expert's credibility accurately enough to make it safe to rely on the experts' advice or testimony, the expert has an equal and opposite worry: when can he or she rightly insist on having at least a voice in the final decision, or perhaps even cast the deciding vote?

2. THE GAP BETWEEN EXPERTISE AND AUTHORITY

Missing in most discussions of the wise use of experts is any explicit consideration of power. Power, in its simplest form, is the ability to get things done. All of us have at least some power, at least at the level of being able to make some decisions and carry out our intended actions.

"Positional power" (Fisher 1983, November/December)[1] is the added power of formal authority: the right to vote, the right to command, the right to make a binding decision: "signing authority." We typically do, and should, defer to legitimate authorities: those whose positions give them formal authority to make decisions which bind us. But in both our personal decisions, and in public policy decisions, it is not always clear who is the "legitimate authority."

For a personal medical decision such as whether to go ahead with surgery, am I the "legitimate authority" since only my signature on the consent form will count? For a public policy issue such as whether to build oil pipelines or permit tanker traffic down the British Columbia coast, how is authority shared? Partial or complete authority—in the sense of entitlement to make binding decisions—could be in the hands of voters, or the politicians they have already voted into public office. It could also be neither of those groups, since none of us have the same level of understanding of the consequences of our decisions as environmentalists, and, in British Columbia, the voters at large do not have a right to speak for the many aboriginal First Nations, who never signed any binding treaties with the Canadian Crown, and whose ancestral lands such as the World Heritage site, Haida Gwaii, are in the path of tanker traffic.

This range of potential authorities reveals the wide spectrum from the "authority" of experts to the "authority" of positional power and signing authority. The environmentalists bring scientific expertise: objective, predictive power about the likely risks of tanker traffic.[2] The aboriginal people bring lived experience, many generations of knowledge of how the land and sea behave. They and non-natives bring value systems, expertise in the form of setting priorities such as improving the economy, or guaranteeing sustainability. The politicians bring expertise in decision making, the back-room negotiations between interest groups that allow legislation to be crafted with the necessary compromises to get a majority vote. Theirs is the most clearly formal positional power, the power to set public policy, but even they are subject to the power of the courts to declare the legislation unconstitutional and to the power of the electorate to vote out any politician making too many unpopular decisions.

How do we place "expertise" in general, and "scientific expertise" in particular, into this power continuum? However much we worry in logic or rhetoric about the difficulties of weighing the relative merits of expert claims, we don't worry as often as we ought to about when we should defer to the experts and when they should defer to us. We drift along in a half-way state, confident that whoever isn't feeling sufficiently heard will speak up at some point, and at that point we can resolve whose knowledge is most accurate and should be given the greatest weight. By blurring together the "authority" of "speaking with authority" and the

---

[1] Original versions of the classic Harvard "win-win" model of negotiating assumed participants had equal power. In response to criticisms that this is unrealistic, Fisher identified six different types of power and suggested how each type of power plays a role in balancing the power of each negotiator.

[2] This is not a single type of expertise. As noted in Chantelle Marlor's (2009) Ph. D thesis, there are distinct differences between the ways research biologists and Department of Fisheries and Oceans biologists construct their knowledge of the abundance of clams. Both contrast sharply with the aboriginal clam diggers' own ways of knowing the abundance of clams, which is based on direct experience of their traditional patches of beach. One clam digger was incredulous that a scientist in Ottawa, thousands of miles away, could ever know anything at all about clams when the scientists only processed data collected by others and never saw or felt the beach and the clams himself.

"authority" of "acting with authority," we obscure the crucial variations along that spectrum from knowledge to power.

There is an ideal world in which knowledge and power are co-extensive. It's an Enlightenment world in which all people are formally equal, and only knowledge and reason can be used to tip the balance to one decision rather than another. In such a world, the expert, who is presumed not only to have knowledge but to reason correctly with that extra knowledge, would indeed be the authority whose word should direct the decisions of others. To the extent that this ideal is our paradigm, we assume that the expert not only can but probably should explain his or her judgment and its rationale well enough for the non-expert to have a crash course in the subject and feel able to understand.

In our less-than-ideal world, the expert's access to power is not as straightforward, either because he or she doesn't reason well enough, or because other people do not understand the information or the reasoning well enough to trust the expert's judgment, or because equally qualified experts disagree. If the defence calls 8 medical witnesses in a personal-injury case and the prosecution calls 9, none of whom can be readily understood by the jury, who should prevail? The standard questions then emerge: for example, how can we assess credibility in the absence of understanding? How can we protect ourselves from unfair liability in the event we rely on an expert who is later found to be at fault?[3]

We do have criteria for assessing the credibility of experts who speak at a level we cannot personally understand (Goodwin, 2011). The research in this area gives us a much better chance of being able to exercise our own authority with sufficient responsibility. We can at least follow the guidelines to be sure that we have chosen the most credible expert in the most relevant field, and even if we do not fully understand the expert's reasoning, we can weigh the expert's recommendation more heavily than our own preferences in reaching a decision.

However, the credibility of expert testimony or advice is a secondary question. The primary question remains the question of power. When should "authority" be located in the expert? Not all experts function in the same capacity, and so not all experts will have equal authority in relation to one another, or more authority than non-experts.

There are a considerable range of expert functions. The advisor/advisee relationship is only one. Others include judge, arbitrator, counsellor, coach, team leader, specialist employee, and provider of services for pay. What varies across the functions is not only the knowledge and the decision-making authority, but also the expectations of non-experts' entitlement to hear and understand the expert's judgment.

In some cases, the expert must make his or her knowledge and reasoning transparent to the non-expert. These would be the cases in which we would also be able to expect the non-expert to understand and be able to make wise use of the expert's judgment. In other cases, it is the expert's competence in performance that is really at issue, and only the final result or the person's track record need be evaluated before choosing to pay for expert services. The

---

[3] The studies revealing problems with expert decision-making are numerous. Experts are prone to all of the reasoning errors everyone else makes, even in their area of expertise. A particularly nice example is the study which showed that psychiatrists were less accurate than their clerical staff in making diagnoses. Philip Tetlock (2011), *Expert political judgment: How good is it? How can we know?* (Princeton, NJ: Princeton UP) reviews many of the experiments which showed that expert judgment is notoriously imprecise and inaccurate, especially when used to predict future events and trends. An equally fascinating and discouraging book is Carol Tavris and Elliot Arneson's (2008) *Mistakes were made (but not by me): Why We Justify Foolish Beliefs, Bad Decisions, and Hurtful Acts* (USA: Mariner Books).

contrast is shown, for example, by the difference between the expertise of a judge and the expertise of a counsellor. A judge or arbitrator has to make his or her reasons for decision clear to the parties. A counsellor does not: as long as the client begins to function in a more successful way, it may not matter what theoretical background the counsellor has brought to the therapy sessions. Similarly, a coach has to be able to share his or her performance knowledge explicitly enough that the athlete in training may surpass the expert, while an elite athlete is under no such obligation to share the secrets of top performance with competitors. A computer security expert employed to keep company data safe and a locksmith who is called in to change the locks are both free to keep their specialist knowledge to themselves—that is precisely what is earning them the fee for their services.

The teacher, professor, or coach whose expertise must be shared with the student/client often operates in a structure in which authority is gradually transferred to the student or client. The whole point of these people's expertise is precisely that they will be able to pass on the expertise, to ensure that the people they teach will understand all the key issues and reasoning, and will be able to perform up to the expected standard. As their performance hits the target levels, they reach the point where their own judgment becomes good enough for them to assume the responsibility for making their own decisions. They may in time even become experts in their own right, and their relationship with their former mentors may become equal and collegial rather than deferential.

In contrast, the chef or locksmith or electrician, whose livelihood depends precisely on not sharing expertise except as a paid-for finished product or service, operates in a structure in which the authority is always divided. The expert always retains authority over exactly how to provide the service, because in that respect his or her expertise is the relevant factor in ensuring not only quality of service but continued demand for the service. You won't come back to the restaurant if you can duplicate the recipe at home. The locksmith who shows how it's done will be out of business. The electrician dare not teach you unless there's some assurance that you will have to do your wiring safely to code. However, you as the customer or client always retain the authority to refuse the service if it does not meet your needs: the service cannot be imposed on you. It is this freedom to refuse that is often the authority we cling to when faced with decisions from authorities which are not to our liking.

Consequently, while it can be crucial to weigh the credibility of an expert, we are often not weighing only their knowledge. I can choose the "best chef" or "best locksmith" the same way I can choose the "best coach," by reputation and performance. What I don't know from these checks is when I should simply put myself in their hands, to get a good meal, get back into my house, or get a better athletic performance, or whether I should insist that I continue to have a say in the decisions they make. The "freedom to refuse" is a very limited amount of control over the decision making. It leaves me in or out: often a false dilemma. Often we want more authority that that, but less than a fully shared decision. We want the type of authority used in negotiation: the freedom to question the terms of the deal. Can I insist the chef leave out the salt, or the locksmith open only the side door instead of the one in full view of the street, or make the coach negotiate with me on what the training regimen will be? I cannot completely control the outcome of the negotiation. They have a responsibility to their own role and expertise to exercise at least some authority over how their expertise may be used. At the same time, I can expect at least the authority to expect an answer as to why my request cannot be granted. Will there be further discussion? Must we reach a consensus on whether a deal is or is not possible?

It will not always be obvious how much authority the expert and non-expert should each have. Our assessment of expertise should include not just when to value expert advice, but when our decision-making power as non-experts should be reduced accordingly. When can the expert make it "her way or the highway," and when can the non-expert cheerfully and justifiably ignore the expert? We have a number of practice-based guidelines, but no solid principles for power sharing. This is due in part to what seems to be an ongoing shift away from seeing experts as "authorities" who should be exercising decision-making power.

The shift might be due at least in part to the fact that it is scientific expertise—whether medical, environmental, or technological—which for most of the twentieth century was the respected form of expertise. Scientific expertise offers a specific and somewhat unusual place in the spectrum of expertise. Scientific expertise is supposed to be more objective than many other kinds of expertise. An astronomer assessing the risk of an asteroid hitting the earth is not expected to take a stance on whether the asteroid strike is or is not desirable. The expert chef assessing the taste of a new dish is expected to have a personal goal at stake: will this dish draw in more customers? The nutritionist assessing the nutritional value of this dish, however, might also be an advocate for healthier eating, and therefore be neither as purely objective as the scientist nor as clearly profit-driven as the chef.

Can the nutritionist have enough power to enforce a good dietary regime on you? In a hospital or care home, a nutritionist certainly can: if you have any menu choice at all, it will be between items the hospital dietician has decided are good for you. But are those choices optimal? The dietician may not have enough authority to compel the hospital to offer the best possible choices, if those choices are not as economical as some less nutritious but still healthy choices. The dietician has even less authority to help set public policy on vending machines in public schools or on menu choices in local restaurants. Are we eating ourselves into diabetes and heart attacks? Nutritional science can only make its message clear enough to be understood by the public; it cannot enforce compliance with a better diet. Our question about the authority of the expert is not just about the quality of the nutritional advice, but about the proper level of authority of the expert: the amount of power a dietician should have to enforce our dietary choices.

When we use scientists as the paradigm of experts, it seems strange that we would ignore their advice. Yes, as citizens we have the right to vote instead of letting them decide for us—but surely we would expect that at least a majority of us would be reasonable enough to listen to and vote based on their advice. But the paradox is already built into why we valued their expertise in the first place: they are, after all, objective in their advice, not pushing us in any direction the facts should not take us. It is not clear that "the facts" and the scientists' understanding of their significance, even assuming they reason correctly from their knowledge, should be enough to guide our decision: we have always had at least that authority of "freedom to choose," freedom to reject any advice.

There are two issues in balancing power. The first is that there is no straightforward way to balance the power between experts in different fields—to determine the relative "signing authority" of experts with equal "speaking authority" in their respective fields. The other is that the guidelines in place for making decisions may give signing authority to parties who hire the experts, and can contractually constrain the experts' authority.

Consider, for example, a patient-care team in a hospital where a team decision is required before surgery can be scheduled.[4] The team is chosen precisely in order to bring to bear a range of relevant expertise covering different factors in the case: for example, a decision on a child's surgery might include a paediatric surgeon, a child-welfare specialist, a nurse, and a physiotherapist. In a difficult case, where the surgery involves the brain or vital organs, the choice might be between surgery and palliative care (if the surgery has insufficient chance of success). In such circumstances, the surgeon may believe that understanding whether surgical intervention could help is the most crucial expertise in deciding whether to go ahead with the surgery. The nurse might argue that an understanding of how palliative care maximizes quality of life is the most crucial. The child welfare specialist might argue that it is crucial to understand whether the family's circumstances and values are more compatible with taking the risk of surgery or facing the finality of palliative care. Each could actively try to sway other members of the team and each could be correspondingly frustrated if they cannot persuade the others.

Understandable as their frustration might be, the patient-care team illustrates why decisions involving more than a single type of expertise often have no straightforward lines of authority. Nurses have expertise in patient care and education that is a different type of expertise from a surgeon's knowledge of the patient's medical problem. Child-welfare advocates understand the pressures on the family inside and outside the hospital in ways neither the nurse nor surgeon does. Ideally, they will share their expertise to reach consensus,[5] but if they don't, then what? A majority vote is a possibility, but the only reason to let "signing authority" lie with the majority is the assumption that the majority vote represents a decision that has more factors in its favour than any of the minority alternatives.

Typically, when an expert team cannot reach a decision, there is an authority structure that must step in: for a patient-care team, for example, it would likely be the hospital ethics board, which has principles it will apply as a matter of policy where the experts disagree. The ethics board is the *de facto* authority, but it stands in much the same relation to the experts and to the patients' families as the politicians do to their scientific advisors and to voters at large. It has the right to design a policy, and the considerations it is guided by might be completely unrelated to the medical expertise of the experts or the preferences of the patients' families. For example, it usually considers the legal liability for failing at "too risky" a surgery, and it probably also has an ethics policy like those of many large organizations which is explicitly utilitarian, to maximize results on a limited budget, and this ethical code may not at all accord with the actual beliefs and values of patients' families.

The patient-care team illustrates, therefore, not only the difficulty of spreading authority across multiple experts, but also the problem that their authority in total is still limited by the scope of their employment. They have the authority to use their expertise to its fullest in making a case for what is in their opinion the best decision. For some decisions, they will have signing authority; for others, they have at most one vote in a group decision; for still others, they have no vote at all. They are not employed to set the hospital's policies, and can at

---

[4]   The example is presented generically but the procedure is taken from a real example, examined in detail in a student research interview (personal communication, December 2011). All identifying details, including the interviewer's name, have been omitted to preserve confidentiality.

[5]   For example, the city of Abbotsford formed decision-making teams with police, social workers and a variety of other community services, to consult with each other on plans of action for the community, especially in its downtown core which had an exceptionally high crime rate.

most use their expertise to lobby the authorities above them to set policies they recommend. Is this the right scope of authority just because their employment contract is structured to limit their responsibilities to the cases they handle? I would consider this an open question. Medical decisions require such a complex blend of ethical and physiological considerations, it is by no means obvious that the top-level decision making on hospital policy is done by people with a better claim to expertise or wiser use of power.

Limited authority outside the scope of employment (and even within it) is not true of all occupations, and therefore not of all experts. For some, the authority that comes with their expertise and employment is "signing authority." For example, a fire chief or police officer will not take time to consult your wishes in ordering you out of a dangerous situation. As a matter of public policy, your safety has been determined to outrank your right to understand. Therefore these experts have been given the formal authority to make you obey first and question later, regardless of whether you personally would rank your safety as paramount. And they do have the knowledge base to make a better decision than you can on what is safe. You may have the right to refuse an evacuation order—Harry Truman on Mt. St. Helens being one of the best-known examples—but this is an example of rejecting expert advice, and resisting authority, not an example of the authorities having only expert knowledge with which to sway the non-expert's decision.

As with the patient's family in a paediatric surgery case, evacuees are not consulted about the policies with which they will be expected to comply. The homeowners might resist evacuation but must do so knowing no aid of any kind will be in place if they stay. The parents might refuse surgery if it is offered—but if surgery is not offered, typically they cannot compel it to happen. The most they can do is seek another surgeon, another team, or another hospital. In many cases, public policy has been decided far in advance of members of the public having any definite say in that decision. Neither they nor the most "objective" experts have any guarantee of having enough authority to sway a decision.

This shifts the nature of our quest. Finding the right signing authority may not be a matter of choosing between types of relevant expertise or between pre-existing rights to choose or to refuse. It may, instead, be a matter of attending to the process of choosing. If decision-making authority is to be shared, we presumably want to avoid closed-mindedness to one another's expertise just as much as we want to avoid being gullible and falling prey to an "investment guru" whose expertise is less than our own.

We like the notion of being well-informed but independent in our judgment, to the extent that when it comes to policy decisions, we are sceptical of policy set at the hospital administration level, and downright distrustful of policy set at the government level. We particularly want to avoid being controlled by politicians whose capacity to persuade us relies on the very deep pockets of the special interest groups whose motives we suspect.

This puts us in the position where the central question of authority arises in policy debates. Where should the authority be located? Unlike private enterprise, public organizations and government cannot claim that their senior executives are by definition the right authorities to make the decisions. In theory, policy debate is democratic and we all have influence as voters, but in practice the decisions are made after the politicians are elected and often in spite of what they heard in any public forum or from any scientific briefing paper. Unlike scientists, politicians do not have a reputation as objective fact-finders and decision-makers. However, politicians are no less "expert" in their own way. They have a form of expertise that is typically crucial to the development of policies: they know how to get things done. They have an expertise in bringing together people with disparate interests and skilfully negotiating sufficient common ground to develop legislation that will receive a majority vote.

This expertise is significant in the decision-making process, not just in the content of the decision. One of the frustrations of an objective expert—be she a scientist or a philosopher—is that people often "just won't listen" to the incontrovertible facts and the objective argument that ought, the expert believes, to be absolutely crucial in the formation of policy. Rejecting advice is common in corporate and political circles, and resisting authority is common among both the oppressed and the free. The shanty-town dweller, ordered to move for new development, can stay on until the bulldozer comes and the journalists with it. The telecommunications expert can recommend the best current technology to update the office phone system, and the boss can simply say, "But that's not Bell Canada," and reject the proposal.

The frustration with seeing good advice ignored often comes because the "objective" expert may lack, or even scorn, the politician's skills in figuring out how to make people listen. It's not uncommon, even, for such people to scorn the necessary negotiating skills—"that's all politics"; "I don't do politics"; "politics is a dirty business."

This is a matter not only of rhetoric but also of philosophy, because we are often dealing with "deep disagreements" (Fogelin, 1985/2005),[6] issues on which not only "office politics" but public policy must face fundamental differences in belief: about the importance of independence versus conformity, about the value of quality of life versus life itself, about the economy now versus the environment seven generations from now.

At this level of decision making we are in a realm where even expertise in ethics or philosophy cannot put forward a single best decision. There is no obvious "authority" on these issues—at best we have "signing authorities" whose decision making we may not agree with but whose entitlement to decide we can respect.

And here again is one of the difficulties of designing a better paradigm for decision-makers. We may not want to defer automatically to the expert, but nor do we necessarily know what it takes to assume responsibility ourselves. One component of decision making that is often under-examined is how much responsibility is expected of the non-expert.

For example, consider the current state of patient-doctor decision making. My mother had undergone several series of cancer treatments before she asked me—not the doctors!—"What is cancer, anyway?" There was a time when her behaviour would have been exactly what was expected of a patient. "Informed consent" in fact didn't mean much more than ensuring the patient was sufficiently oriented to time and place to count as rational, explaining the procedure, and getting the patient to sign on the dotted line where required. Even today, when I was given the consent form for "phacoemulsification cataract extraction with intraocular lens implant," I got impatient responses to my queries about details on the form. "It's just the standard form" meant "you're not supposed to think about it." But if I don't read the form, and the cancer patient doesn't ask, "What is cancer?" it's understandable behaviour, but it also seems like unjustifiable reluctance to take proper responsibility for one's own care.

The objective experts do help. That cataract surgery was explained in detail on reputable medical sites on the internet, and although there was nothing there related to the atypical features in my case, my ophthalmologist and opticians were both able to explain exactly why my experience of the surgery and recovery would be different from the norm. By

---

[6] This term originates with Robert Fogelin and has become a useful category to cover all disputes which have no apparent common ground on which to begin arguing towards agreement.

the time I got back to the corneal specialist to confirm the details of the operation, I was well enough informed to be confident I could make a sound decision.

However, there is one important element that is not enforced on me in exercising my authority in this situation. The surgeon is required to exercise due diligence and follow his professional code of ethics. I have no such constraint on me. As the medical examples illustrate, we are in an evolving paradigm of what "expert authority" is. Where once the expert would be making the decision and simply ensuring I consented, I am now more often in the position to make the decision myself based on expert advice. Similarly, in the public area, there is less authority granted automatically to science, and, in Canada in particular, greater authority granted to the aboriginal peoples' distinctive decision-making methods not only for themselves but also in increased consensus-based decision making by all stakeholders, not just by expert representatives.

As the paradigm shifts to greater consultation and a larger share in the decision making, the problem is that the down-grading of the experts' authority has not been matched by any corresponding pressure on me to exercise my authority more wisely when I must make a decision or cast a vote. Certainly, I am not prevented from researching diligently and consulting multiple sources for a variety of expertise. But I have to tell an embarrassing story against myself as an expert in decision-making theory to illustrate the point. It was not until I used the very situation of my impending eye surgery as the foundation for a class role-play on decision making that I even thought I ought to be asking myself what was important to me in this situation: what I wanted, expected, or needed, or how my options appeared from the broader perspective of my life and body changing as I age and not just the perspective of what would make it possible to grade papers by the end of term. In other words, I had still been construing my responsibility in the situation only in terms reading the consent form carefully and understanding the reasons to do the surgery now or wait until later—I had been considering only my "freedom to refuse."

Until a chance encounter with a "business coach" (anonymous source, personal communications, October 2011, February 2012), it would not even have occurred to me to add into my decision making anyone except people with some knowledge—expert or simply personal—of the surgery itself.

"Business coaching" illustrates nicely one of the forms of expertise that is unrelated to content-area knowledge.[7] It is one of the many forms of decision support. It provides expertise in the process of working through the stages of a decision, including where necessary the objective assessment of one's own reasons. What "business coaching" does is to hold people who have signing authority, such as CEOs, CFOs, and other senior executives, accountable for using wiser decision-making processes. The "coaches" do not offer business expertise of their own—they are not expert consultants. They are expert listeners, trained to ask questions which will make an executive think through his or her own reasons and expectations in far more detail than would normally happen in the rush of a business day. They ensure accountability by making the executive set deadlines to move events forward until a decision or plan for action is in place. In exercising his or her existing "signing authority," the executive can now be more confident that the authority was exercised responsibly. (The popularity of this form of coaching is that its results can be measured in efficiency and productivity, not merely personal satisfaction.)

---

[7] Sample texts in the area include *Coaching in Organizations*, Madeleine Homan and Linda Miller (John Wiley & Sons, 2008).

The presence of a coach imposes an obligation to recognize and carry out one's own responsibility with respect to how one makes a decision. If I am to play an equal or a decisive role in a decision, then I am supposed to exercise my authority responsibly. I had had plenty of opportunity in the real eye surgery situation to consult myself. In the role play, I first "consulted" my experts and then sat before a "life coach," who had been given instructions to ask me explicitly, "What is important to you?" Even knowing it was coming, the question startled me when I had to answer it. Then and only then did I begin to deal with this aspect of the decision—fortunately, before binding decisions had been made. When I did think about these things, the decision began for the first time to fall into a more manageable perspective. What could and should be done depended in part on how I saw the options in relation to my values.

Recognizing the need to consult my values explicitly instead of assuming they will simply come into play permits me to reconsider the nature of my own authority in this medical decision. What support do I need, from which experts, to exercise my signing authority wisely? It isn't just the effort of my own thinking; it likely needs the prompts from experts in reasoning and decision making. What should the expert surgeon bring, and what should the surgeon do in consulting with me, before lifting the scalpel? It isn't just to know what acuity of vision is possible and is normally recommended; it is also to find out whether I actually want maximum vision correction or not.

I need a much clearer vision of what authority I have and what responsibilities go with that. We in general need not only a clearer vision of what responsibilities come with authority but a more explicit discussion of how we want authority to be re-assigned in shared decision making.

3. CONCLUSION

How do we, in public policy decisions, include and properly weigh the advice of the right experts in decision making? We are often in the realm where no amount of objective scientific expertise—and for that matter, not even objective philosophical expertise in ethics—could suffice to guide our actions or policies. Where "truth" is not heard, or is heard but is rejected as untruth, we need additional sets of expert skills. These are the skills in decision-making processes that are still seldom called on, even though our decision-making paradigms are shifting away from reliance on the knowledge expert.

One such set of skills is direct expertise in decision-making processes, which for me must include both negotiation and dispute resolution as the components frequently required to reach a group decision. When we have authority, we need expertise in how to exercise it. And this is a form of expertise we cannot outsource completely to content-area experts.

We all, expert and non-expert alike, need to seek out and respect the advice of experts in decision making, so that our standard for accountability in decisions is not merely the common standard given in "Roberts' Rules" (parliamentary procedure): namely, being "sufficiently informed." We need the higher standard of "sufficiently skilled" decision making. This will necessarily include skills in weighing expertise, but also will include skills in consultation and co-operation, skills of negotiation, mediation, and peace-making.

However, even these skills are at most a partial answer to the question of how we allocate authority to experts and non-experts in decisions. We do not have a ready answer to when values must trump economic considerations, or when individual rights trump expert judgment even if the resulting decision is terrible.

I notice that the current term "Research Ethics" is gradually being replaced by "Responsibility in Research" ("Responsible," 2012). This suggests that we are indeed moving towards an "accountability" model of performance—not the compliance with some particular code of ethics, but the responsibility of carrying out tasks which assure the reliability and integrity of our conduct.

ACKNOWLEDGEMENTS: The final version of this paper benefited enormously from the feedback of Will Brooke, Jack Brown, and Rory Stevens. Their constructive suggestions shaped the direction of the argument. Larissa Oakey and Jaipreet Mattu helped to provide information that expanded my perspective on the use of different types of expertise.

REFERENCES

Collins, H., & Evans, R. (2009). *Rethinking expertise.* Chicago, IL: University of Chicago Press.

Darwall, S. (2006). The second-person standpoint: Morality, respect, and accountability. Cambridge, MA: Harvard University Press.

Fisher, R. (1983, November/December). Negotiating power. *American Behavioural Scientist, 27*(2), 149–166.

Fogelin, R. (2005). The logic of deep disagreements. *Informal Logic, 25,* 3–11. (Original work published 1985, *7,* 1–8).

Goodwin, J. (2011). Accounting for the force of the appeal to authority. In F. Zenker (Ed.), *Argument Cultures: Proceedings of the 8th International Conference of the Ontario Society for the Study of Argumentation* [CD-ROM]. Windsor, ONT: OSSA.

Haskell, T. L. (1984). *The authority of experts.* Bloomington, IN: Indiana University Press.

Kahneman, D. (2011). *Thinking, fast and slow.* New York, NY: Farrar, Strauss & Giroux.

Marlor, C. (2009). *Ways of knowing: Epistemology, ontology, and community among ecologists, biologists and First Nations clam diggers.* (Unpublished doctoral dissertation). Rutgers University, NJ.

Mirowski, P. (2004). The scientific dimensions of social knowledge and their distant echoes in twentieth-century American philosophy of science. *Studies in History and Philosophy of Science, 35,* 283–326.

Responsible professional practices in a changing research environment. (2012, February 16). Pre-conference workshop conducted at the annual meeting of the American Association for the Advancement of Science, Vancouver, B.C.

Woods, J. (2004). Standoffs in public policy. In *The death of argument: Fallacies in agent-based reasoning.* Applied Logic Series (Vol. 32). Dordrecht: Kluwer Academic Publishers.

Wright, G., & Bolger, F. (Eds.). (1992). *Expertise and decision support.* New York, NY: Plenum Press.

Professors and Scholars as Experts: Problem Setting and Methodological Considerations for the Examination of Newspaper Articles

ALAIN LETOURNEAU

Philosophy and Applied Ethics
Université de Sherbrooke, campus de Longueuil
150, Place Saint-Charles, Québec
Canada
Alain.Letourneau@USherbrooke.ca

ABSTRACT: Even if new media like wikis obtain some attention, classical media (radio, television, newspapers) still are important sources of information for the informed public, e.g. the public that is interested in what is happening in his/her environment and the world, and wants to understand events and situations. Authority still comes with specific media that have kept on establishing their credibility. But journalists themselves also need reliable sources of information, and they require regularly University professors and similar Scholars in renowned institutions. This allows them to include precisions and richer background in their papers to inform and help understand complex phenomena. After presenting the general context and the problem setting of this research both as a knowledge question and as a social issue, a few short articles will be examined in detail to see how this works in specific examples.

KEYWORDS: professors; experts; press; media; public opinion; newspaper articles; authority.

1. INTRODUCTION

It might be relevant to recall here in what research context this particular study is situated. It fits within the goal of a larger scope project that I aim of doing in the longer run. I collected systematically an important quantity of articles published during the year 2009 in three important Quebec newspapers, respectively *La Presse* (Montreal), *Le Devoir* (Montreal), *Le Soleil* (Québec city). With a competent research assistant, I found no less than 2400 occurrences, in the whole year of 2009, or precise references to professors giving opinions or information in the material content of the articles on different issues; these newspapers are published 6 days a week, for a possible sum of 302 issues per year, which means 906 issues total, without taking into accounts holiday breaks; this gives us roughly speaking a mean of 2.7 occurrences of professors by issue! These pieces all involve university professors, or similar research people at the same level, that are called to and intervene for discussing complex issues that sometimes are about policy controversies. We did not do a longitudinal study, only a brief verification outside of 2009; we saw a small decline of the total number of occurrences compared to the year 2004, of the order of 10%, but this would have to be looked at much more extensively to be decisive. At this point, the most we can say is that there is still a significantly important presence of professors in newspaper articles in recent years, even though there might be a slight decline in their use.

Letourneau, A. (2012). Professors and scholars as experts: Problem setting and methodological considerations for the examination of newspaper articles. In J. Goodwin (Ed.), *Between scientists & citizens: Proceedings of a conference at Iowa State University, June 1-2, 2012* (pp. 253-262). Ames, IA: Great Plains Society for the Study of Argumentation. Copyright © 2012 the author(s).

2. NEWSPAPERS AND PUBLIC OPINION

The newspapers and other media are obviously on the front scene of the relationship between scientists and citizens; and for any given problem or discussion, there is a university professor to be searched for and interviewed, that represents something akin to "scientific information" on a given issue. In the Quebec data at least, the thematic scope of their interventions is extremely large. Given the scope and complexity of this data set, it seemed to me probably preferable to furnish here a preliminary discussion establishing in particular the terms of the discussion, if this research is to be meaningful at all. After this discussion, it will be possible to look at a few newspaper articles to get a better grasp of what can actually be done.

In the past, I have been interested in the theory of the formation of public opinions, which goes far beyond opinion polls, since these processes of formation include a variety of factors, among which are basic education and continued processes of education, social circles, and also media material use, that all play an important part. When we look at specific pieces for expressing opinions, for instance opinion articles, we find specific elements that can be identified as contributing potentially to a process of formation of the opinion of persons. This process can be efficient or not, or to a degree; I have not invested myself into the process of trying to find how efficient or how convincing a message actually is by looking at the recipients of messages, for instance by trying to see in which way they were influenced or not by the media content. Even if I expressed in the past a proximity to cultivation analysis in general as a perspective for treating media effects, before considering this there is a basic general element that has been acquired by years of research in philosophy and social sciences since the linguistic turn. We understand more and more that thought comes in and through communication. It follows that a message with cognitive, performative or normative content can come to us with relative force, can be echoed also with force and importance by and through others; it can be seen by us as convincing or not, complete or not, sufficient or not to build an opinion. If it is relevant for our purpose and needs, the information available can be selected and become part of the process of reflection and decision making; of course its absence will deter any effect it might have had if it would have been available (Miller, 2005; Gerbner, 1998). Therefore, without denying the importance of looking at the receiver's end of the process, it is relevant and controllable to look at the actual messages that are produced, to see how they are structured, what is their relative force, how do they function as argumentative devices. And what is specifically interesting with newspaper articles is precisely their diversity, the way they mobilize a plurality of opinions and find a certain closure even if in many cases, this closure is uncertainty and the absence of decision. The simple fact that there is this significant number of articles and other media interventions of professors certainly "cultivates" the fact they have something to say on different matters; they are part of the debate.

Even if new media like wikis obtain lots of attention and get what we can call a good proportion of the whole of media uses, classical media (radio, television, newspapers) still are important sources of information. In any case, they are still so for the *informed* public, e.g., the public that is interested in what is happening in his/her environment and the world, and wants to understand events and situations. There is, in many measurements, reference to a regression of the number of readers of newspapers since 1990; the numbers differ from country to country, and in some cases the regression is quite important. But in some cases the free newspapers distributed in the metro station contribute to a bigger number of people reading the press everyday, which is the situation in Montreal between 2001 and 2011 (Université Laval, 2011, p. 4). The most important paying newspapers see their numbers declining while the free ones climb. And it seems that users of the internet edition of the newspaper generally speaking are still readers of the paper edition. On the business side, generally speaking in Quebec at least, the beneficiary margin for the published newspaper has diminished in recent years, losing between 2 and 3%, but they are currently around 10% of benefit margins. Even if the decline of the paying press is well- documented, it does not have to be equated with an upcoming disappearance of the written media or irreversible decline.

2.1 The continuation of the authority phenomena in the press

There are a few important elements here, onto which stress has to be placed: first, in the process of recognizing the role of the so-called new media, we have this tendency to underestimate the role and importance of what we can call in contrast the classical or already old media. We also miss the fact that in the process of the world wide web development, the classical media can and do seize the opportunity of asserting themselves as serious organizations, worthy of the attention and following of people. They develop while becoming integrative platforms with multimedia resources at the disposal of a mouse click for the interested reader or viewer. Second, we have a tendency to consider "the public" as a whole, as if there were no differences between publics, or as if there was a unity between the whole of the people considered as "the" public. Fragmentation of publics does not has to be equated with lack of strength of the publics; this impression sometimes goes with the nostalgia of the "one public" of which we could dream when there were very few media available. After the criticism of Habermas' first theory (1962/1990) about "the public" of the enlightenment period, the new situation of pluralism that has developed has yet to be fully accepted.

One possible way of constructing collective groupings of people is to group them in function of questions of perceived common interest. Some people do have a permanent interest into public affairs. We can wonder, in the enlarged discussion opened by the Lippman-Dewey controversy, to what point a critical discussion involving "the public" in a democratic life implying participation is realistically possible, considering the specialization required and the different types of expertise that are necessary to be able to validly discuss complex issues. But if we turn our backs on "the public" in general to consider instead the case of a very large diversity of publics sometimes regrouped around specific interests, we can also have specialization at the level of the different publics, and constitution of valid but limited publics.

As James R. Taylor explained in recent communications, there is still very little published work on authority as a communicative phenomenon (Taylor & Van Every, 2000; Taylor, 2012). Authority still comes with specific media that have kept on establishing their credibility. Even after the rise of the internet and Google, we still go back to classical media that are recognized as valid sources to get information, whether it be the *New York Times*, the *Economist*, and other written media that more and more include in their web versions videos and films that contribute to closing the gap that had separated them for some publics, from reputed TV networks. The text thus integrates images and filmed narratives, an element that for now is rarer on TV programming (even though with the WWW television, texts might also be available for reading through that medium). In the actual context, we can say that the written media that has successfully crossed over to the internet by introducing and using a rich array of media, is certainly in way of reviving its credibility and importance in the public life.

For all the classical media, including of course *El País, Die Welt, Le Monde* and many others including *Le Devoir, La Presse* or *The Globe and Mail*, authority has been established and is the result of years of journalistic work, and this does not mean that these journalists and media people can satisfy themselves with their reputation and not re-establish it continuously by keeping on doing a good job. But journalists need reliable sources of information, and they require regularly University professors and similar scholars in renowned institutions for getting precisions and richer background to understand complex phenomena. This contribution becomes then part of the larger phenomena that is the newspaper article. I propose to understand the newspaper article as a chorus of voices, among which we can count the journalist, important sources, actors, bystanders, but also knowledge experts that, most of the time, come from the scholarly world one way or another. Regarding the professors, for me what is interesting is to identify the importance of their contribution in the article, especially asking to what point what they say actually is the main message or part of the main message that is forwarded by the article in question. Being able to verify this would mean that they contribute strongly to the article's meaning and potential effect on eventual readers.

As we will see, this can happen and does happen in newspapers articles or other news media; let us think of debates on television with experts from renowned institutions, some of the times universities, some of the times similar venues; in the USA, PBS is the obvious example, but it is not to say that the other, private big media do not emulate them. Of course, some of the times neutralization of experts will also potentially take place (with people from both sides face to face, a phenomenon that is present). The situation is obviously always specific to the different countries, but this phenomenon (the professors used as expert sources by the media) is largely present at the international level and not limited to one country or province. In some countries, like France, the "intellectual" can be quite detached from the University, at least in some cases, but not always (Torck, 2012). Media actors, especially journalists, need reliable sources; the university professors in principle, but not in every instance, represent in principle the interest of being independent. At least they furnish the appearance of independence, even though in some cases this could be an illusion.

3. SPECIFIC QUESTIONS WITH ECONOMICAL DIMENSIONS

Before entering in the discussion of actual content, I would like to stress the fact that, in the Quebec province at least, professors and/or their employer, the universities which have a strong tendency to run deficits each year, do not receive compensation on a regular basis for their participation to media outlets, which has now been documented for the first time that I know of. Quite frankly here, I do not see the need for the professor to receive monetary compensation, even though in some cases of persons without tenure-track employment, the question could be seriously discussed. But with our universities struggling to get more money, the question arises of the price of this expertise that is freely offered by professors. One obvious trade-off that takes place is expertise for visibility, both for the institutions that see their credibility reaffirmed and for the professor. And a free participation to the public debate might be justified on the basis of the fact that, for the most part at least in Canada, it is public money that funds the universities, so it can be understood that this free service is part of what is required of them to justify publicly their very existence. Then again, some colleagues insist that this is not a necessary part of their workload as professors; it has to be a free contribution, not something mandatory. We can also express that we could compare ourselves to public and free access venues of knowledge like Wikipedia for instance, even though some would argue that universities are much more serious and important sources of knowledge! Others would say that there is a price for what we do and that price should be assumed, especially if it is for private profit (as in the private media). The discussion might stay at that level if there were no rights of access to pay for treating publicly and in published form newspaper articles, for instance. For the ends of this article, I submitted to the *New York Times* a request for the rights to use a very long quote from an article, just for having knowledge of the situation that would apply if I were to actually include such an extensive quote, e.g., how many rights would I have to pay; the answer referred me back to number of copies and extension of the distribution of the published work-to-come. The situation is obviously different for every newspaper in every country; in Canada, one of the authors in a book I am currently editing on this very same topic had to pay (I contributed) around $350 Canadian for a partial reproduction of an article from the newspaper *la Presse*, that publishes around 400,000 copies a day. The cost of using an entire article has a cost, and even to fully access the content sometimes is not free. There is in many cases the notion of fair use (among other elements, if no copying occurs and no revenue will be obtained). If the Universities were to apply a commercial kind of logic for dispensing public knowledge in newspapers, they would have to identify the number of uses of one of their personnel's expertise each year and charge the media for it...A possibility that should be considered more carefully. It might backfire and become an argument for further private uses of information!

It probably could be interesting to give a sampling of the work on the Québec newspaper articles, but with difficulties. Of course for the English-speaking hearer or reader to enter into the actual content of the analysis, this would require a translation of the articles. In fact, the actual treatment of the data accumulated is not completed as of yet, so it is not possible to just publish the result of the whole study. It is then preferable here to see what questions arise by looking at articles in English, published in well-known newspapers, and to discuss and show how they can be treated in a significant and enlightening way. Our point of view is argument-centered; we are mostly interested in how newspapers are using professors as sources of authoritative information, and how this might be in the long run, especially for decision makers, an important element to arrive to policy making. With a whole corpus thoroughly analyzed, if that were possible, it would be interesting to see what is the mean proportion of articles of which we can say that they give the opinion presented by the professor(s) called as expert(s) as the main opinion(s) on the topic presented by the article. What we can do here is to verify that this question can find a positive answer in one or two case studies. Some more specific studies can also be pursued later on, for instance to situate what can be said about the interventions by professors on environmental governance issues in particular, in communication technology issues, etc. One relevant question is to what point professors are only sources of relevant information that does have weight, or if they are really opinion makers by expressing positions on difficult issues; both situations exist (Schuetze, 2012; Lyall, 2012).

4. SOME INTERESTING ARTICLES

A first article can be looked at more closely; since we are freely giving publicity to this famous newspaper, as a subscriber I share only the analysis of this document with colleagues and graduate students doing research, for discussion. The selected newspaper article, published in the *New York Times* and then republished in the *International Herald Tribune* for people living abroad, is treating costs for the upcoming summer Olympics in London, more precisely the

British discussion that is going on about the price tag involved by this big event, scheduled between July 27 and August 12 in 2012. This happens of course in the context of the British government's recent budgetary rigour and deep cuts. The newspaper article can be seen as a series of quotes by different sources, which are also different opinions or data on the subject (Lyall, 2012, p. B10). The article is particularly rich, referring and using quite diversified sources. As for the University or independent researchers that are called into play, there are two of them; they come towards the end of the article and their opinion can easily be seen here as closing the discussion that was opened by the article in the beginning and intermediary part of the text.

The paper starts with two paragraphs recalling the costs of the last London Olympics, in 1948, and the poverty at that period is stressed, contrasting the £760,000 budget of the time (which was paid by sponsors) with the "£9.3 billion (and counting)" (Lyall, 2012, p. 1) that is estimated at the present moment, of course before the Olympics. The first source mentioned is then-prime minister Clement Attlee; the numbers given, for instance the fact that there is a 8.4 percent unemployment rate in Great Britain in 2012, "the highest in 17 years," are given without referring to sources. Sources quoted in the government are actual Prime Minister David Cameron, with the desire expressed in November 2011 to "showcase the best of Britain to a massive audience" (Lyall, 2012, p. 1), then Jeremy Hunt, Culture Secretary, who talks of the event as an opportunity to harness in a time of crisis. To support the article's claim that this expected high cost "has not been universally popular" (Lyall, 2012, p. 1) in a place where the government wants a solution to the financial crisis in terms of deep cuts, the paper refers to an article in *The Guardian* by Richard Williams; in the web edition, it provides a direct link to the article. As quoted in the *New York Times* article, Williams comments on the addition of budget that was given to the person responsible for the opening event, Danny Boyle, famous for realising *Trainspotting* and more recently the celebrated *Slumdog Millionnaire*. Here the article potentially touches the film amateur and cultural critic's interest for high-level media culture. The three next paragraphs of the article refer to other countries' difficulties and criticism towards their own Olympics and what were the costs: Barcelona, Athens, Montreal, then two paragraphs are given to the present Italian government's attitude; considering to apply for the 2020 games, they decided not to; here the quote is given to Mario Monti, Prime Minister of Italy who said so much. In a movement in the opposing direction, two paragraphs are then again given to the British Government's justifications in this dire economic context, and here it is Hugh Robertson, the Sports Minister, that is quoted: the decision to hold the Olympics was taken before the recent cuts, and they are presenting London not as a superpower but as a place where it is good to come to spend money (Lyall, 2012, p. 2). It is after all this that the professor is called in, as we might say. Let us take then a closer look introducing and giving the contribution of the University professor, called as an expert to discuss the issue for the public of the *NY Times* (a thing he probably did for free, as most professors do).

> Tony Travers, a professor in the government department at the London School of Economics, said that it would take decades before the long-term financial implications of the Olympics became clear. But, he said, the great Olympic-related achievement has been the regeneration of a huge tract of derelict urban land in East London, site of the Olympic village, into a viable community with enough housing and infrastructure to carry it forward.
>
> Planning and building have been carried out with ruthless efficiency, he said, so that an effort that would normally have taken decades has been achieved in little more than five years. After a string of construction embarrassments like Wembley Stadium and the Millenium Dome,

both of which were marred by trouble and went wildly over budget, the project will help transform Britain's reputation abroad, he said.

"The advertisement for British planning, architecture, design, project management and building is extraordinarily good," Travers said. "This will undoubtedly send a message that Britain is good at delivering big projects on time." (Lyall, 2012, p. 2)

Clearly, here the professor gives the limelight to two sets of very positive side effects of the Olympic process. One is very recent, and it gets us out of the discussion focusing only on costs too high in the context of cost-cutting everywhere else from the government. He is asserting that some very important social goods derive or will derive from all this investment, which is clearly a positive element in the evaluation, even though he expresses some reservations as to what will be the actual cost of all that. One positive side effect is expected in the future: this will be excellent for the businesses of Britain and for their reputation, another enviable side effect of the upcoming games. These games will then have both public investment and private investment positive side effects; what could be better!

Let us continue to review the article, towards its conclusion now. This positive assertion from the professor is rapidly put in question again, with another series of interventions from a diversity of sources. A discussion of pre-games grumbling as something that was there from the start is asserted, this time with the support of an article published in the Bagehot column of the *Economist*. Here again, an important hyperlink leads us directly to this other publication. The article is recalled in the *New York Times* as showing both the project as "a juggernaut controlled by an unaccountable sporting elite," and as opening the possibility that will give Britons to "feel only pride at hosting a spending games, fueling new confidence in Britain's future," while asserting that for now, the discussion is more about money than about glory (Lyall, 2012, p. 2). Many other critics are listed then: some have compared the logo to badly drawn male genitalia; the American comedian Jackie Mason, summarising the national mood, said that is all a question of if you don't mind paying higher taxes for ten years; and Leon McCluskey, an important union leader, uttered a menace of a strike during the Olympics, on the basis that the idea to see people celebrating these wonderful Olympic games in those dire economic moments "is unthinkable"; we learn then that these menaces have been then criticized as unpatriotic (Lyall, 2012, p.2).

It is interesting to note that here, at the very end of the article, the University professor has an ally (or a competitor, depending on how you see it) as an authoritative source that comes to close the debate in a way. It is worth another longest quote:

Meanwhile, Joe Twyman, director of political and social research at YouGov, a polling organization, said that griping is almost second nature in Britain. Britons were not particularly positive before the royal wedding last year, he said, but changed their mind once they saw how well it went. He predicted that the same thing would happen with the Olympics.

"It's one of those things that you would describe as 'being British,'" Twyman said in an interview. "People don't get massively enthusiastic until the time comes, and even if the whole country's not on board, enough people are enthusiastic enough to make it a memorable event." (Lyall, 2012, p. 2)

This other source is asserted by the article itself as being of the "research" kind; this "director of political and social research," comes to give the *coup de grâce* to the critics; it is only a national mood, the spirits of the Britons are taken by regular griping phases, but they get over it, as they did with the royal wedding. We then have good reason to think, after having reviewed a series of criticisms and complaints, that it will probably be a positive experience, so it is asserted under and by the authority of the professor and of the private researcher! The article as such does not take position; it can let the experts do the job, but the questionings that surface in the first half of the article, and that sustains probably the interest of the reader, do find some positive resolution and answers by the professor and expert's contribution.

We have relatively simple cases where (again) the university professor or equivalent sets the tone (in a recent paper on oil price increases, where, as in the case just reviewed, one private research institute and a university professor are both placed in the forefront), among cases where their contribution is more anecdotal or complementary to the main content of the article (Davidson, 2012).[1] But sometimes things can also become complicated, when an important number of expert sources of different kinds come into play. Even then, a thorough analysis might surprise us by its results.

In some settings, professors do not get a very important exposure in the context of the articles; their institution, name and sometimes title is mentioned. But with the internet links, in some cases now we can find explicit links to pages on the experts quoted in the articles, which permits to the reader to help him assess the value of the expert in question, at least minimally. An example of this is in an extended recent article about the BRICS, concerning this alliance of countries that still have some growth in their economy, namely Brazil, Russia, India, China and the more recently added South Africa (Yardley, 2012). A number of important scholars are quoted here in an article that is essentially a presentation and a political analysis of what BRICS is and what it is not, e.g., the diversity and lack of unity between its members, qualified of being a "photo op" by one of the experts. Successively we see intervening Brahma Chellaney, of the Centre for Policy Research in New Delhi, Yahend Huan, professor of global economics and management at MIT. There is also C. Raja Mohan, presented as "a leading strategic affairs analyst in New Delhi" (Yardley, 2012, p. A4), but when we follow the internet link furnished in the article, we learn he previously was named for 2009 the Henry Alfred Kissinger Scholar in the John W. Kluge Center at the Library of Congress; we also learn that a Goldman Sachs economist, Jim O'Neill, is the person that first identified with the acronym (BRIC at the time) this group of fast growing countries. There is also Sreeram Chaulia, who "teaches at the Jindal School of International Affairs in Sonipat, India" (Yardley, 2012, p. 2). In that piece, as can be verified, the most relevant and strategic information is given at the end, by the MIT professor Huand and by Chellaney, from the Centre from New Delhi, a research hub (it certainly could be qualified as a think thank) that has both private and public sources of financing, and has ex-government officials or private researchers on its board. The first argues that BRICS is first and foremost China's way of getting closer to natural resources in Africa and Brazil, whereas Chellaney argues that each of the countries of the BRICS group has its most important business relationship as a country with the United States. This article is interesting obviously by the number of resources mobilized, their diversity of origins and for the portrait they are helping the journalist to give of an uncommon alliance between very diverse countries. But it still showcases a professor and a scholarly "independent" expert as being the most relevant sources for the reader to make him/herself an opinion on a given issue.

---

[1] Retrieved from http://www.nytimes.com/2012/04/01/magazine/rising-gas-prices-dont-actually-affect-americans-behavior.html. The online editions gives the following detail: "A version of this article appeared in print on April 1, 2012, on page MM12 of the Sunday Magazine with the headline: The Real Oil Shock"; the experts called here are "the economists Lutz Kilian at the University of Michigan and Paul Edelstein of the consulting firm IHS Global Insight."

5. CONCLUSION

If we get back to the previous discussion about costs, here the simple fact, in the last example, that a link is given inside the newspaper-internet article to the web page of the professor or expert, gives some better and more detailed publicity and information in exchange for the participation by the professor to the debate and article information, considered as an opinion-building process. This audience giving is much richer than the simple reference to the name and University of the colleague. In a context where budget cuts are everywhere, universities in many cases come to depend more and more on private funds to operate; here we have uses of university resources that might be better recognized; in some cases at least, they get a good advertisement in exchange for their services, but is it enough considering the value of the service rendered? In many places we see young people fighting to keep an access to university services for less cost for themselves and their children (for instance, the recent protests (since February 2012 and continuing to this date but seeming to be slowing, June 2012) in Quebec province's southern cities. That was the case of students on strike against the government's decision to augment the fees; these kinds of protests are certainly not limited to that place, let us recall recently the UK). And in the perspective of keeping the quality of what can be offered without sacrificing the freedom of research, it might be appropriate to rethink the relationship of universities as public bodies that need to stay that way, with private firms who sell information. Obviously, we would need a very large sample of articles to verify the relatively important exposure of professors in the media of different countries, and ask ourselves more broadly what is the authority game that they are actually playing; here we just saw a few examples that at least in some cases in a prestigious United States newspaper (and the documentation about the Quebec case abounds in the same direction), they do play an important part in building the meaning of the articles themselves at least in specific cases and probably much more often, when their intervention is not giving its whole direction to the opinion piece.

If a professor with little media exposure like myself managed to be interviewed 14 times in something like 17 years of career, by generalizing the example we can arrive at striking and important numbers; of course, there again, broader empirical studies would be required. I do not see why as members of a well-known and respected profession we should ignore those kinds of facts; it is certainly required that a collective reflection come into play at one point on those issues.

REFERENCES

Calhoun, C. (Ed.). (1993). *Habermas and the public sphere*. Cambridge, MA: MIT Press.
Davidson, A. (2012, March 27). The real oil shock: Rising gas prices don't actually affect Americans' behavior. *The New York Times*, pp. MM12.
Dewey, J. (1927). *The public and its problems: An essay in political inquiry*. New York, NY: Holt.
Gerbner, G. (1998). Cultivation analysis: An overview. *Mass Communication and Society, I*(3/4), 175–194.
Habermas, J. (1962; 1990). *Strukturwandel der Öffentlichkeit*. Untersuchungen zu einer kategorie der bürgerlichen Gesellschaft (mit einem Vorwort zur Neuauflage 1990). Frankfurt: Suhrkamp Verlag. (Original work published 1962).
Lippman, W. (1925). *The phantom public*. New York, NY: Harcourt & Brace.

Lyall, S. (2012, March 28 [online March 27]). Costs complaints, an Olympic rite, voiced in Britain. *The New York Times*, p. B10. Retrieved from http://www.nytimes.com/2012/03/28/sports/before-the-london-games-the-grumbling-about-money.html?pagewanted=all

Miller, K. (2005). Communications theories: Perspectives, processes, and contexts. New York, NY: McGraw-Hill.

Schuetze, C. F. (2012, March 27). Smart shoppers in global market. *The New York Times*. Retrieved from http://www.nytimes.com/2012/03/28/world/europe/iht-smart-shoppers-in-global-market.html?pagewanted=all

Splichal, S. (1999). *Public opinion: Developments and controversies in the twentieth century*. Lanham, MD: Rowan & Littlefield.

Taylor, J. R., & Van Every, E. (2000). *The emergent organization: Communication as its site and surface*. Mahwah, NJ: Lawrence Erlbaum Associates.

Taylor, J. R. (2012), *The question of authority in communication processes*. Paper presented at the National Communication Association Convention, New Orleans.

Torck, D. (2012). The *arroseur arrosé* or the misfortunes of pathos in a media dialogue. *Language and Dialogue* 2 (1), 80-104.

Toulmin, S. E. (1958). *The uses of argument*. Cambridge: Cambridge University Press.

Université Laval. (2011). La presse quotidienne: Titres et tirages. Centre d'études sur les medias, p. 4. Retrieved from http://www.cem.ulaval.ca/pdf/pressequotidienne.pdf

Walton, D. (2007). *Media argumentation: Dialectic, rhetoric and persuasion*. Cambridge: Cambridge University Press.

Yardley, J. (2012, March 29 [online March 28]). For group of 5 nations, acronym is easy, but common ground is hard. *The New York Times*, p. A4. Retrieved from http://www.nytimes.com/2012/03/29/world/asia/plan-of-action-proves-elusive-for-emerging-economies-in-brics.html?pagewanted=all

Should Scientists Communicate Uncertainty to the Public in Health Controversies? The Case of Endocrine Disrupters' Effects on Male Fertility

LAURA MAXIM

Institut des Sciences de la Communication
CNRS UPS 3088
20 Rue Berbier du Mets, 75013 Paris
France
laura.maxim@iscc.cnrs.fr

MARTINE CADOT

Laboratoire Lorrain de Recherche en Informatique et ses Applications
CNRS UMR 7503, Université de Nancy 1
Campus scientifique, BP 239, F-54506, Vandœuvre lès Nancy
France
martine.cadot@loria.fr

PASCALE MANSIER

Laboratoire Communication et Politique
CNRS UPR 3255
20 Rue Berbier du Mets, 75013 Paris
France
pascale.mansier@yahoo.fr

ABSTRACT: Uncertain knowledge must be communicated to the public, as environmental problems can potentially reach many people. Uncertainty communication is assumed by some to increase public trust in science and policy makers, by others to produce public panic. We have used focus groups for getting insights about this assumption and more generally about peoples' attitudes following uncertainty communication, for the controversy on the effects of endocrine disrupters (EDs) on human male fertility.

KEYWORDS: uncertainty, communication, endocrine disrupter, controversy, perception, chemical risk.

1. INTRODUCTION

Scientific controversies related to public health issues are often characterized by a significant level of uncertainty. Despite this uncertainty, available knowledge must be communicated to the public, who is potentially in danger. Nevertheless, all science communicators do not share the conviction that uncertainty should be communicated to the public. Empirical work in the realm of science communication reveals skepticism among scientists, who assume that communicating uncertainty will result in a negative public reaction. Many of the scientists surveyed by Frewer et al. (2003) thought that informing the public of uncertainty would lead to increased distrust in science and scientific institutions, as well as cause panic and confusion regarding the extent and impact of a particular hazard. The general public might perceive

Maxim, L., Cadot, M., & Mansier, P. (2012). Should scientists communicate uncertainty to the public in health controversies? The case of endocrine disrupters' effects on male fertility. In J. Goodwin (Ed.), *Between scientists & citizens: Proceedings of a conference at Iowa State University, June 1-2, 2012* (pp. 263-274). Ames, IA: Great Plains Society for the Study of Argumentation. Copyright © 2012 the author(s).

reports of uncertainty within risk communications as evasiveness or as an admission of ignorance (Fessenden-Raden, Fitchen, & Heath, 1987). Lay people are supposed to expect certainty and to be disappointed by uncertainty communication, which could leave them an impression of arbitrariness (Renn, 2011).

Empirical appraisal of peoples' attitudes and feelings, as they receive messages about scientific uncertainty, is currently needed, as previous research showed contradictory results. Many studies cite beneficial effects, such as reducing public perception of risks and increasing the credibility of scientific and/or risk-assessment agencies (Funtowicz & Ravetz, 1990; Habicht, 1992; Van der Sluijs, 2002; Patt & Schrag, 2003). Citizens who are aware of uncertainty are thought to make more informed decisions (Carnegie Commission, 2011) and to be more willing to reduce catastrophic risks (Slovic, Lichtenstein, & Fischhoff, 1984).

Other results show that uncertainty can be disturbing, which can lead to denial (Slovic Slovic, Fischhoff, & Lichtenstein, 1982; Weinstein, 1987) or even outrage (Slovic, 1993). Paradoxically, desire for certainty is nevertheless not universal—only a third of the sample in one study expressed a desire for certainty (Johnson & Slovic, 1996). Johnson and Slovic (1995) showed that communicating uncertainty has ambiguous consequences, signalling honesty to some and dishonesty to others (Johnson & Slovic, 1995, 1996).

Much of the literature in the field of experimental economics has focused on how people react to uncertainty and on whether/how the way this uncertainty is communicated influences decision making. A central result is that, in most situations, individuals tend to adopt a much more risk-averse decision stance when faced with ambiguity and poorly-defined risk (Chow & Sarin, 2001), potentially due to feeling a lack of control (Heath & Tversky, 1991). Kahneman, Slovic, & Tversky, (1982) analyzed how people assess the probability that uncertain events will occur and showed that a limited number of heuristic processes are involved; for example, representativeness, availability, and anchoring. The way in which uncertainty communication elicits these heuristics may influence people's judgments. The perception of uncertainty was influenced by whether it was communicated within a positive or negative framework (Kuhn, 1997).

The case of endocrine disrupters (EDs) and their effects on male fertility provide a good example of uncertain and controversial science; this is therefore a good case study for our theoretical investigation. EDs are "exogenous substances that alter function(s) of the endocrine system and consequently cause adverse health effects in an intact organism, or its progeny, or (sub)populations" (Commission of the European Communities, 1999). EDs are thought to contribute to the incidence of diseases such as cancer, diabetes, obesity, and reproductive disorders (decline in number and quality of sperm, testicular cancer, earlier puberty, etc.). Most, if not all the population, is exposed to them, but the ED issue is relatively new and currently controversial.

2. METHODS

In line with previous literature (Brashers, 2001, Powell, Dunwoody, Griffin, & Neuwirth, 2007), we distinguish *expressed* from *received* (or *perceived*) uncertainty, according to the assumption that the scientific message transforms during the communication process between the emitter (i.e., the scientist) and the receptor (i.e., lay public). Expressed and perceived uncertainty must be analyzed as two interrelated but different entities.

Eleven focus groups consisting of five to twelve laypeople, were organized between October 2010 and May 2011. Each group was homogenous for the following criteria, and the wide variety between the groups was intended: 1. High-revenue[1] mothers of children younger than three; 2. Low-revenue mothers of children younger than three; 3. Men and women with advanced scientific education; 4. Low-revenue men under 30 (young), without children; 5 and 6. High-revenue young men, without children (2 groups) 7. High-revenue young women, without children 8. Religious women (practicing); 9. High-revenue men and women over 40; 10. Low-revenue man and women over 40; 11. Farmers.

Each group participated in a three-hour meeting and each meeting followed the same protocol. The meetings started with a ten-minute introduction to the topic. Participants were then invited to read a one-page text, watch a short video[2] and then discuss both. This text + video + discussion sequence was repeated four times. The first four sequences were organized in two pairs of two; each pair included one text and one video *without (expressed) uncertainty* and *with (expressed) uncertainty*. The fifth contained uncertainty expressed by industry scientists, instead of academic scientists (as the first four).

If additional time remained, the groups would watch two or three more videos. The four texts were selected from a single popular science book and the videos were all extracted from a documentary or from scientific videos freely available on the Internet. A 20- to 30-minute group discussion followed each text + video sequence.

The meetings were video-recorded and discussions were transcribed. The transcripts were coded by two (9) researchers or three (2) researchers; each individual coding was then discussed. This process led to the identification of the following six categories of discussions:

- the reference science model (frame);
- sociopolitical and economic framing of science;
- non-scientific references for assessing the relevance of the message;
- characteristics of the communication format (text or video);
- perceived uncertainty sources;
- feelings.

Coding revealed several classes for each category.

Two categories (perceived uncertainty sources and one feeling, i.e., fear) are analyzed in the present communication.

Deviant-case analysis (Silverman, 2011) allowed us to systematically analyze the transcripts and ensured that we interpreted them as objectively as possible. The purpose of the analysis was to derive propositions that applied to all the data to arise from the focus groups. The analyst progressively modified the expression of his/her results to incorporate deviant cases.

After the presentation of each set of text and video, the participants were invited to express their judgment on the statement: *Some substances present in our environment produce a decline in male fertility, in humans.* This judgment had to be expressed using a figure on the scale proposed by Weiss (2003) (table I), as a communication tool in controversies, when

---

[1]   We define revenue above 2000 € per household as *high* and revenue below this amount as *low*.

[2]   The Eurobarometer report "Scientific research in the media" (2007) showed that most people in the EU get their scientific information from television.

generalists untrained in sciences must understand the merits of opposing arguments in disputes among scientific experts.

Table I. Scale of perceived degrees of uncertainty (modified).

Level (Score)	Convincingness of the evidence (Standard of proof)
10	Beyond any doubt
9	Beyond a reasonable doubt
8	Clear and convincing evidence
7	Clear showing
6	Substantial and credible evidence
5	Preponderance of the evidence
4	Clear indication
3	Probable cause: reasonable grounds for belief
2	Reasonable, articulable grounds for suspicion
1	No reasonable grounds for suspicion
0	Impossible
PP	Cannot express

At the end of each sequence (video and text), each subject chose a value from 0 to 10 corresponding to her/his PDU (Perceived Degree of Uncertainty). The statistical analysis aimed at answering two questions:

- Is there a global increase or decrease of uncertainty judgments when a researcher communicates or not uncertainty?
- Are there different patterns of change in uncertainty judgments, depending on people's socio-economic characteristics?

Two tests are used to establish the significance or not of the results:

- the Sign test, which indicates the significance of a general decrease or increase;
- the Chi2 test of independence, which indicates the significance or not of differences between subsets of subjects, e.g., the subsets of age groups.

3. UNCERTAINTY RECEPTION: A QUALITATIVE ANALYSIS

Our definition of uncertainty is inspired by post-normal science, which distinguishes several dimensions: technical (inexactness), methodological (unreliability), epistemological (ignorance) and societal ((un)robustness) (Funtowicz & Ravetz, 1990).

According to this choice, the messages used in our empirical setting contained different types of uncertainty in the different sequences. The test messages have been structured by pair. The first sequence (text + video) did not contain uncertainty but presented epidemiological data about sperm decline. The second sequence did contain uncertainty related to the information presented in the first video, from the epistemological (referring to the strength of the causal relationship and to the available scientific knowledge about the human body), the methodological and the technical classes.

The second pair included a sequence presenting toxicological results on animal studies together with data about human exposures to EDs, and a sequence containing epistemological uncertainty associated to extrapolation from animals to humans, and epistemological uncertainty related to the causal relationship and the form of the dose-response relationship.

We found that, in reaction to scientific messages, laypeople raise more and different uncertainties than those contained in the original message communicated by researchers. In reaction to the sequences (text + video) in which uncertainty was not included in the communication, group discussions nevertheless highlighted a significant number of perceived uncertainties. Thus, even if the first sequence did not address uncertainty, during the discussions participants extensively questioned this causal relationship—they were particularly concerned about the link being weak. Many participants formulated their own multi-causal hypotheses to explain the epidemiological data presented.

Though participants received information about repeated studies that link reproductive disorders to EDs, they systematically questioned whether the data was complete. This technical uncertainty in the data was the most frequent type highlighted by participants for all the sequences of text and video, with and without uncertainty. Participants either explicitly felt that the information they received lacked precise details, or that it was simply not enough, though they did not indicate what they felt was missing. The data chosen by science communicators from all was a debated question. Participants concluded that selectively choosing data to communicate can either lead to concerns about the unavoidable simplification needed to popularize science or to suspicions about the intentions behind this selection.

Methodological uncertainty was also brought up by participants during discussions that followed sequences both with and without uncertainty. For example, the choice of parameters used to measure male fertility was questioned, in particular the choice of the sperm count as a measure of male fertility or of the pesticide content in urine as a parameter for determining the causal origin of the observed decline in sperm counts.

Several participants insisted on the importance of knowing the details of the methods together with the results themselves. The willingness of researchers to communicate details of their protocols seemed to be more relevant than the technical content of the methods itself.

Among all the types of uncertainty communicated during the focus groups, extrapolation uncertainty raised the strongest reaction. The message communicated by scientists in the second sequence about uncertainty related to extrapolation from animals to humans generated significant confusion. Arguments brought by participants for such a reaction focused the fact that animal studies have been validated—by extensive previous experience—as a viable replacement for ethically impossible human experimentation. Therefore, challenging animal studies raises radical questions both about the possibility or not to test toxicological properties of contaminants in laboratory, and about the real relevance of extrapolated results that led to marketed chemicals currently present in consumer products.

4. PANIC ABOUT CERTAINTY AND UNCERTAINTY

As shown in the introduction, some science communicators assume that communicating uncertainty would cause panic among laypeople. Our results invalidate this assumption.

After receiving scientific messages both with our without uncertainty, participants expressed panic when they perceived lack of control over the negative effects:

- messages without uncertainty that indicated a causal relationship between EDs and male reproductive disorders induced fear because of the ubiquitous nature of ED exposure, which makes it impossible to control their (known) dangers
- uncertainty communication elicited relief rather than fear, except when participants associated uncertainty with an inability to precisely identify the cause of the decline in male fertility and hence the ability to control it

All but one participant[3] who expressed anxiety and fear following sequences with uncertainty also expressed the same feelings after videos without uncertainty. They associated these feelings with the inability to control potential negative health effects when the causative agent is not precisely known. These participants completely dismissed the causal relationship between EDs and male reproductive disorders when there was associated uncertainty. For this category of participants, uncertainty was enough to dismiss the scientific messages that were not associated with uncertainty.

Citizen: "At the same time, this is not necessarily more reassuring because, ultimately, the first [video] provided a cause, so one knows where to act; but here, see that this range [of factors] is open."

Participants who felt alarmed,[4] anxious or frightened after sequences without uncertainty associated these feelings with lack of control over the (known) effects of EDs on health because of their ubiquitous presence in daily life, their invisibility, past experience of risks, and lack of trust that policy makers can adequately control them. These feelings were also elicited by epidemiologic information on male reproductive disorders. Even though the messages transmitted by scientists were questioned on several points, they were sufficiently trusted to produce strong feelings about the negative effects of EDs.

Citizen: "Yes, this scared me a lot."

Citizen: "Perfume fixatives are worrying. This does not make us feel like using [perfumes] anymore."

Comparatively, women expressed fear more often than men.

5. RELATIONSHIP BETWEEN PERCEIVED UNCERTAINTY AND DEMOGRAPHIC PARAMETERS

The analysis used a total of 455 PDU (Perceived Degree of Uncertainty) score values, i.e., 5 successive scores for the question corresponding to each sequence with/without uncertainty (Q1, Q2, …, Q5), for each of the 91 subjects, except 4 subjects who chose no value for 1 or 2 sequences.

We have examined the relation between the increase or decrease in perceived uncertainty and several demographic variables, i.e., gender (F/M), monthly household income (around 1000 Euros, 1500, 2500, 4000), post-graduate (Yes/No), scientific studies (Yes/No, with 7 non-response), practice of a religion (Yes/No, with 15 non-response), occupation (student, employee, liberal profession/senior executive, farmer or other), age group (21–29, 30–40, 41–69), and having children (Yes/No).

---

[3] Five participants explicitly expressed this feeling after sequences 2 and 4, of which 3 were women.

[4] 30 participants explicitly expressed this feeling after sequences 1 and 3, of which 22 were women.

Among the PDU-score differences, Q3–Q4 (message with/without extrapolation uncertainty) is the one most related to socio-economic variables.

The increase in uncertainty level following uncertainty communication is more often found for:

- older people (compared to younger ones)
- employees and liberal professions/senior executives (compared to other professions)
- low education level (compared to higher levels)

In other words, people increase easier their degree of perceived uncertainty following uncertainty communication (i.e., are more easily responsive to messages containing uncertainty), if they are in one of these categories.

The Q2–Q1 difference (i.e., perceived uncertainty following uncertainty communication in the sequence Q2) is significantly related to a few socio-economic variables, with a more important decrease from Q1 to Q2, for:

- employees (compared to other professions),
- non-scientific studies (compared to scientific studies)

The practice of religion is significantly associated to repetitive uncertainty communication, i.e., Q24–Q13.

6. A MODEL FOR THE RELATIONSHIP BETWEEN EMITTED AND PERCEIVED UNCERTAINTY

Based on existing literature and our own results, we propose a model (Fig. 1) for the communication of uncertainty associated with environmental and health risks.

This model highlights the role of each of the three components of any communication process:

- the transmitter (which is, in our case, the scientist, but can be a regulatory agency, a policy maker, a representative of an NGO or professional organization, etc.)
- the receiver (in our case, the general public, but may also be policy makers, stakeholders, etc.)
- the communication process itself, having its own characteristics. Indeed, this component is sometimes forgotten from risk or uncertainty studies, leading to incomplete focus either on the transmitter (assuming that the receiver will receive the message as intended by the transmitter, e.g., the deficit model) or on the receiver (leading to narrow formulations of the source message, in discrepancy with the diversity of uncertainty types and framings in the real world communication)

There is a process of transformation of the message during the communication process, which depends in our case not only on the substantive features embodied in the message by the transmitter (e.g., with or without uncertainty), but also on:

- the characteristics of the process (e.g., the nature of the communication support, video, text or other),

- the characteristics of the context in which the message is positioned (e.g., socio-economic, political and cultural connotations of science and/or of the particular scientific topics addressed)
- the features of the transmitter him/herself (e.g., academia or industry researcher),
- the features of the audience (e.g., socio-economic status, level of education, etc.) (Fig. 1)

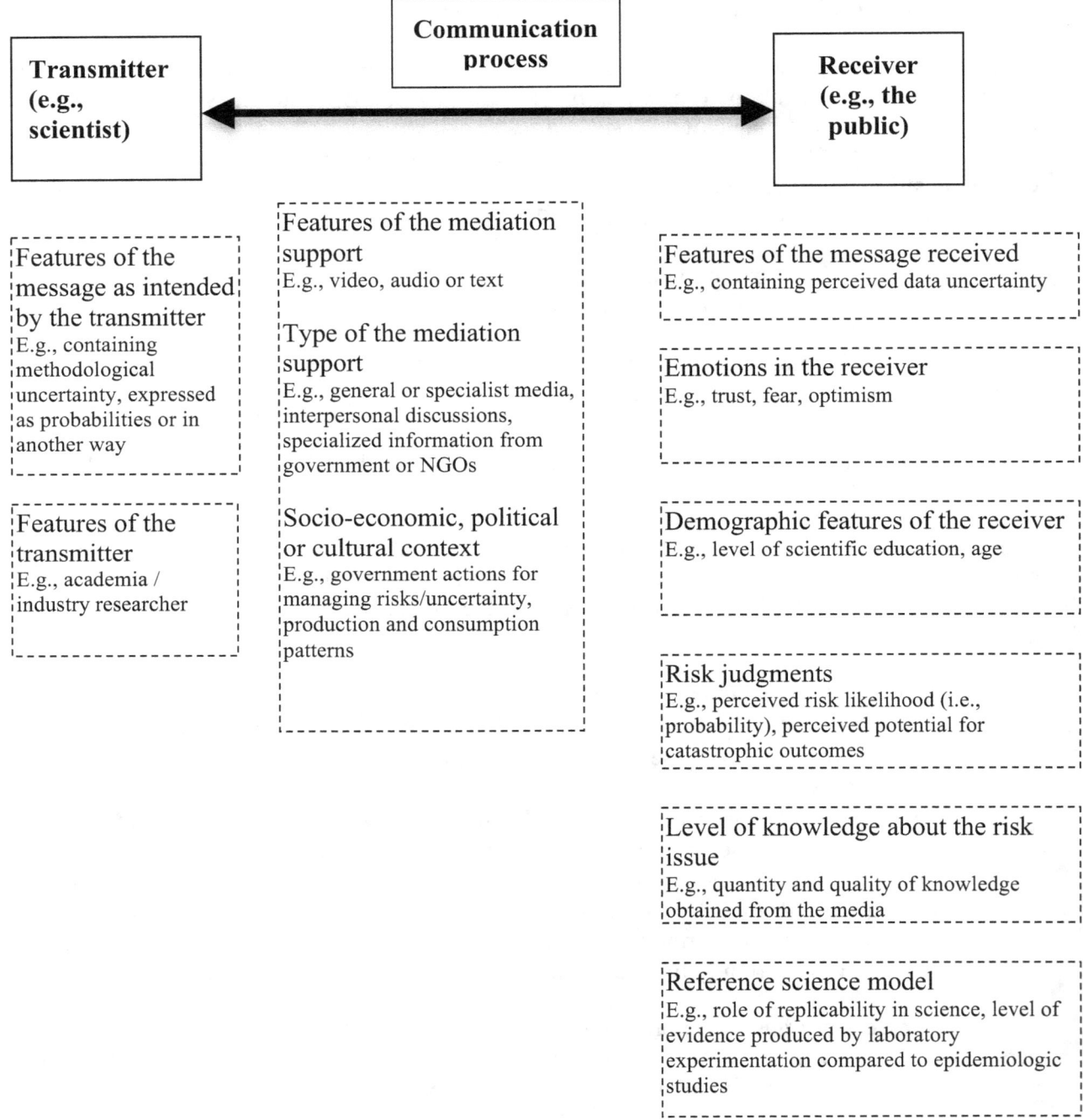

Fig.1. A model for the communication of uncertainty about health and environmental risks

The transmitter will draft his/her message according to his/her knowledge or assumptions about each of the components of the model. Anyway, some of the components remain unknown to the emitter, due to their complex influence on the communication process (e.g., the socio-economic and political context). The transmitter cannot control all the aspects influencing the reception of his/her message. These components might equally remain unknown to the receiver, who might not be conscious of (or able to explain) all the reasons of his perception of the message. Indeed, the receiver will perceive the message according to the more or less conscious appraisal of the different parts of the model. The role of each component in the reception of the uncertain message has already been or can be highlighted by research in psychology, communication, etc.

As regards to uncertainty communication, this representation helps at better separating emitted from perceived uncertainty, understanding that each of them should be defined and characterized in its own way, and at distinguishing the contextual characteristics of the communication process itself.

7. CONCLUSION

There is currently relatively little research about perceived uncertainty (Powell et al., 2007) and almost any literature about how it relates to different patterns of emitted uncertainty. Nevertheless, there is an increasing need to understand how the public perceives the uncertainty communicated by scientists or by other science communicators. Indeed, many current risks must be communicated to the public potentially exposed to important hazards, when uncertainty about them is still present and sometimes important (e.g., risks from nanotechnologies, EDs, etc.). Furthermore, there is an increasing tendency from expert bodies like regulatory agencies (e.g., European Food Safety Authority, European Chemicals Agency) to communicate uncertainty associated with their risk assessments.

Research is needed for bridging the gap between increasingly abundant practices and research on expressing uncertainty (e.g., post-normal school) and relatively scarce research about how the audience (i.e., the public, the risk managers, the regulated industry, etc.) perceive these messages.

Using a quasi-realistic experimental setting and the case study of ED effects on male fertility, we have investigated the assumptions currently made by some scientists about the potential negative effects of uncertainty communication on lay publics. In particular, we have tested the assumption found in scientists by Frewer et al. (2003) that communicating uncertainty causes panic.

Our results contradict this assumption. Participants expressed relief rather than fear in reaction to uncertainty communication, except when they associated it with an inability to precisely identify the cause of, and therefore control, the decline in male fertility. Anyway, people also highlighted that the "dangers of relief" include reduced attention to and protective measures against a risk that may ultimately prove to be real.

Messages that indicated a causal relationship between EDs and male reproductive disorders without addressing uncertainty induced fear related to lack of control, because ED exposure is ubiquitous, making it impossible for people to control the (known) dangers. This confirms previous results on the central role of lack of control in lay people risk and uncertainty perceptions (Heath & Tversky, 1991; Siegrist, 2008). Fear is consistently associated with feeling unable to act on his/her own life, and also with demands that relevant information about risks should be easily available to everyone. Though experts disagree and policy-makers may fail to protect public health, participants demand the right and the capability to make their own choices based on the available knowledge; they do not want to be artificially protected though lack of transparency.

We found that, in reaction to scientific messages, laypeople raised more and different uncertainties than the researchers originally raised. Causal, data, and methodological uncertainty were those most often highlighted by participants.

Perceived uncertainty was systematically associated with its sources, instead of being treated as *global* uncertainty. This confirms previous work (Rogers, 1999) indicating the differential perception of various types of uncertainty. From a regulatory point of view, this is important because it indicates that communication might be more effective if each of the relevant sources of uncertainty is addressed separately. Nevertheless, current policy initiatives such as the REACH regulation, dealing with risks from industrial chemicals (among which are EDs), propose approaches expressing cumulative uncertainty, in particular probabilistic risk assessments. Such tools might create a potential of miscommunication of uncertainty and their effects on the intended audience (i.e., risk managers) should be first tested before giving them regulatory weight.

About the relationship between demographic variables and perceived uncertainty, statistical analysis showed that:

- uncertainty communication by academic researchers significantly increases the uncertainty perceived by people, on average;
- the effect of repetitive uncertainty communication (e.g., Q24–Q13) is significantly more important for people who do not practice religion;
- the higher the age and the household income, the easier uncertainty communication produces perceived uncertainty.

Also, the lower the education level and in absence of scientific studies, the easier uncertainty communication produces a change of judgment (i.e., perceived uncertainty). This facility to change opinion (i.e., perceive uncertainty) also seem to be more important for women, employees and liberal professions/senior executives, but this has to be confirmed in further experiments.

Based on these findings, our model for the communication of scientific uncertainty highlights all the complexity of the science and uncertainty communication process, which cannot be conceived as a linear transmission of information reaching the audience as the transmitter intends.

Previous literature showed an important influence of the source of uncertainty communication (i.e., public or industry scientists). Results previously reported for uncertainty communication by regulatory agencies (Johnson & Slovic, 1995, 1996) are therefore not necessarily comparable with our results obtained for uncertainty communication by researchers. We have the intuition that the reception of uncertainty will depend on the disciplines of the science communicator, and that our results obtained on communication by (eco-)toxicologists cannot necessarily be extrapolated without critical analysis to other domains of research (e.g., nuclear physicists).

Our results give a positive experimental feed-back for policies related to the public right to know, reinforcing the access to health and environmental information. They show that lay publics are able to intelligently deal with scientific uncertainty, all by acknowledging and managing their own feelings related to it.

ACKNOWLEDGEMENTS: This work has been funded by the Interdisciplinary Programme *Communication Sciences* of the National Center for Scientific Research (CNRS), and within the framework of the PRO-MALE project.

REFERENCES

Brashers, D.E. (2001). Communication and uncertainty management. *Journal of Communication, 51(3)*, 477–497.

Carnegie Commission (2011, December 4). *Risk and the environment, improving regulatory decision making.* Retrieved from www.ccstg.org/pdfs/RiskEnvironment0693.pdf

Chow, C.C., & Sarin, R.K. (2001). Comparative ignorance and the Ellsberg paradox. *Journal of Risk and Uncertainty, 22*(2), 129–139.

Commission of the European Communities (1999). *Community Strategy for Endocrine Disrupters, COM 706 final,* Brussels, Belgium.

Fessenden-Raden, F., Fitchen, J.M., & Heath, J.S. (1987). Providing risk information in communities: Factors influencing what is heard and accepted. *Science, Technology and Human Values, 12*, 94–101.

Frewer, L.J., Hunt, S., Brennan, M., Kuznesof, S., Ness, M., & Ritson, C. (2003). The view of scientific experts on how the public conceptualize uncertainty. *Journal of risk research, 6*(1), 75–85.

Funtowicz, S.O., & Ravetz, J.R. (1990). *Uncertainty and quality in science for policy.* Dordrech: Kluwer.

Habicht, F.H. (1992). *Guidance to risk characterization for risk managers and risk assessors.* Washington , DC: U.S. Environmental Protection Agency, Office of the Administrator.

Heath, C., & Tversky, A. (1991). Preference and belief: Ambiguity and competence in choice under uncertainty. *Journal of Risk and Uncertainty, 4*, 5–28.

Johnson, B.B., & Slovic, P. (1995). Presenting uncertainty in health risk assessment: Initial studies on its effects on risk perception and trust. *Risk analysis, 15*(4), 485–494.

Johnson, B.B., & Slovic, P. (1996). *Lay views on uncertainty in health risk assessment: A report on phase II research.* U.S. Environmental Agency for Environmental Protection.

Kahneman, D., Slovic, P., & Tversky, A. (Eds.). (1982). *Judgment under uncertainty: Heuristics and biases.* New York, NY: Cambridge University Press.

Kuhn, K.M. (1997). Communicating uncertainty: Framing effects on responses to vague probabilities. *Organizational Behavior and Human Decision Processes, 71*(1), 55–83.

Patt, A.G., & Schrag, D.P. (2003). Using specific language to describe risk and probability. *Climatic Change, 61*, 17–30.

Powell, M., Dunwoody, S., Griffin, R. & Neuwirth, K. (2007). Exploring lay uncertainty about an environmental health risk. *Public Understanding of Science, 16*, 323–343.

Renn, O. (2011). *Science and society: Public engagement and confidence in science.* Paper presented at *The EFSA Consultative Workshop on Independence and Scientific Decision-making,* Brussels, Belgium.

Rogers, C.L. (1999). The importance of understanding audiences. In S.M. Friedman, S. Dunwoody and C.L. Rogers (Eds.), *Communicating uncertainty: Media coverage of new and controversial science* (pp. 179–200). Mahwah, NJ: Lawrence Erlbaum.

Siegrist, M., Stampfli, N., Kastenholz, H., & Keller, C. 2008. Perceived risk and perceived benefits of different nanotechnology food packaging. *Appetite, 51*, 283–290.

Slovic, P. (1993). Perceived risk, trust and democracy: A systems perspective. *Risk Analysis, 13*, 675–682.

Slovic, P., Fischhoff, B., & Lichtenstein, S. (1982). Response mode, framing and information-processing effects in risk assessment. In R. Hogarth (Ed.), *New directions for methodology of social and behavioral science: Question framing and response consistency* (pp. 21–36). San Francisco, CA: Jossey-Bass.

Slovic, P., Lichtenstein, S., & Fischhoff, B. (1984). Modeling the societal impact of fatal accidents. *Management Science, 30*, 464–474.

Van der Sluijs, J. (2002). A way out of the credibility crisis of models used in integrated environmental assessment. *Futures, 34*, 133–146.

Weinstein, N.D. (1987). *Public perceptions of environmental hazards: Statewide poll of environmental perceptions*. Final report to the New Jersey Department of Environmental Protection. New Brunswick, NJ: Rutgers University.

Weiss, C. (2003). Expressing scientific uncertainty. *Law, Probability and Risk, 2*, 25–46.

The Ethos of Expertise: How Social Conservatives Use Scientific Rhetoric

JAMIE MCAFEE

Rhetoric and Professional Communication
Iowa State University
Ames, Iowa
United States
Jamiemc@iastate.edu

ABSTRACT: *Dare to Discipline* (1970), a polemic that links "permissive" parenting to social disorder, is an important text in the history of the conservative movement. This paper uses rhetorical analysis of *Dare to Discipline* to argue that an ethos of expertise is a professionally informed response to "ubiquitous" commonsense knowledge.

KEYWORDS: conservatism, evangelicalism, expertise, family, hegemony, parenting, political rhetoric.

1. INTRODUCTION

For more than three decades, social conservative activist Dr. James Dobson has campaigned on behalf of a concept of a healthy family drawn from both conservative evangelical Christianity and from contemporary psychology. Dobson, a pediatric psychologist (he has a PhD from University of Southern California), is one of the leading voices of the religious right, and the organizations he founded, Focus on the Family and the Family Research Council, are two of the most frequently cited representatives of conservative evangelicalism in public policy debates that concern sexuality, abortion, or marriage. Dobson's most important contribution to the social conservative movement, however, was his pioneering use of therapeutic rhetoric in political discourse.

Dobson's career as a public figure began with the publication of *Dare to Discipline* (1970), a polemic that articulates a behaviorist response to contemporaneous debates within developmental psychology to a conservative vision of moral order. This paper will use a rhetorical analysis of this important work to argue that an ethos of expertise can be understood as a professionally informed response to what Harry Collins and Robert Evans (2007) call "ubiquitous" commonsense knowledge. *Dare to Discipline* is an important rhetorical artifact because it was the text through which James Dobson transformed his professional status into political capital, and it offers a useful site for examining how disciplinary knowledge can be used to create an appropriate ethos for public debate.

2. WHO IS JAMES DOBSON, AND WHY DOES HE MATTER?

James Dobson's early career was that of a therapist, not that of an activist, minister, or polemicist, and even during his career at Focus on the Family, his writing and speaking was primarily about parenting and marriage, not politics. Dobson, the son of a traveling Nazarene Evangelist, earned a PhD from USC in 1967. That choice was somewhat unusual at the time,

McAfee, J. (2012). The ethos of expertise: How social conservatives use scientific rhetoric. In J. Goodwin (Ed.), *Between scientists & citizens: Proceedings of a conference at Iowa State University, June 1-2, 2012* (pp. 275-283). Ames, IA: Great Plains Society for the Study of Argumentation. Copyright © 2012 the author(s).

as during the fifties and sixties, evangelical Christians disapproved of psychology. After receiving his degree, he taught at USC's Keck School of Medicine for over a decade. During his career as a psychologist he worked as an assistant to Paul Popenoe (an important figure in the history of therapy who helped to invent marriage counseling), published research about children with developmental difficulties, and worked as a pediatric counselor.

With the 1970 publication *Dare to Discipline*, Dobson began a career as an advocate of discipline-oriented parenting. After the book's initial success (it sold three million copies during its first printing), he began booking speaking engagements, often at PTAs and churches, and in 1977, he founded Focus on the Family as a brand name for a series of videotapes of his lectures. In 1978 left his position at USC and began a syndicated radio show, also called Focus on the Family. Dobson was an important figure during the Reagan administration, serving on Reagan's National Advisory Commission to the Office of Juvenile Justice and Delinquency Prevention from 1982–1984, on the Citizens Advisory Panel for Tax Reform as a Co-Chair, on the United States Army's Family Initiative from 1986–1988, and on the Meese Commission on Pornography from 1985–1986. By 1987, Focus on the Family filled eight buildings in Pomona, California (Buss, 2006, p. 114). In the early 90s, the organization moved to a campus in Colorado Springs. By that point Focus on the Family was an evangelical media empire with a budget of approximately 80 million dollars (Buss, 2006, pp. 118–119). During the past two decades Dobson has participated in more overtly political activity, including campaigning for George Bush and penning a widely discussed letter about the dangers posed by the election of Barack Obama.

Dobson's ascendance as a public figure paralleled the rise of the influence of conservative evangelical culture in American politics. Ernesto Laclau's and Chantal Mouffe's *Hegemony and Socialist Strategy* (1985) discusses the conservatism of the 1980s as a hegemonic realignment that broadly rearticulated the norms of political discourse in the United States, and we can understand Dobson, who was pioneer for the religious right, to be an important part of this realignment. Laclau and Mouffe argue that during the late seventies and early eighties, relationships between various strands of conservative thought were reimagined and articulated into a robust conservative hegemony, creating a new conservative common sense that redefined the ground upon which American policy debates happened. The sort of seemingly incongruous connections that we see in Dobson's career (his professional expertise is about developmental psychology, but he speaks and writes about a about a wide variety of social issues, and he has been called upon to serve as an expert about media culture and tax policy) are what Laclau's and Mouffe's version of hegemonic theory predicts: "we will call *articulation* any practice establishing a relation among elements such that their identity is modified as a result of the articulatory practice. The structured totality resulting from the articulatory practice, we will call a *discourse*" (1985, p. 105). We might label the discursive project that Dobson participated in during the late seventies and eighties "popular American conservatism."

Laclau's and Mouffe's argument that conservatism succeeded by articulating together various cultural strands and political perspectives matches perfectly the strategies explicitly endorsed by some conservative intellectuals as early as the 1960s, particularly *National Review* editors William Buckley and Frank Meyers. Meyers coined the term "fusionism" to describe a coalition of libertarians, pro-business conservatives, and evangelical Christians (along with conservative Catholics like Buckley). The similarity of "fusionism" to "articulation" suggests something of a match between the deliberate strategies of the conservative thinkers who helped pave the way for the political realignment of the 1980s and the version of hegemonic theory that Laclau and Mouffe designed to explain these political shifts.

3. CONCEPTUALIZING AN ETHOS OF EXPERTISE

Dare to Discipline was Dobson's first contribution toward designing the conservative project that bloomed in the early eighties. It is a particularly valuable book for thinking about political rhetoric because it is the work of a rhetor using specific disciplinary expertise to participate in a broad ideological project.

In order to participate in this project, Dobson had to construct an appropriate ethos for public debate from his professional knowledge. In Aristotelian rhetorical theory, ethos is the term used to describe the goodwill, character, and common sense that makes a speaker persuasive. In his survey of rhetorical scholarship about ethos, James Jasinski notes that contemporary rhetoricians have added to those Aristotelian personal virtues qualities like competence and knowledge (20001, p. 230). We might label these kinds of qualities that have become a part of credibility "expertise." It is important to note that it is not expertise *itself* that has become important, but the ability to create an *ethos* of expertise. When we talk about ethos we are not talking about the speaker's "real" self; instead, we are talking about a speaker's ability to effectively perform character and expertise through rhetorical strategy (Jasinski, 2001, pp. 229–230). Rhetoric is an art, not a knack, and so ethos should be understood as an aspect of performance, not as an essential quality.

How then, do rhetors effectively "perform" expertise? Responses to this question might involve looking at strategies rhetors use to demonstrate their professional authority or experience. However, while such demonstrations *are* necessary for a rhetor to demonstrate his or her competence to understand a problem, identifying the ways that speakers demonstrate authority does not explain why an audience might acknowledge and accept that authority. George Yoos points out that "the speaker cannot adopt a superior role toward his audience unless the audience concedes that role" (1970, p. 52). Harry Collins's and Robert Evans's *Rethinking Expertise* (2007), an effort to reimagine expertise after the destabilizing work done by what they call "second wave" science studies, offers a complex meditation on the nature of expertise and a useful starting point for talking about how disciplinary authority might be used in public debate.

Before I can discuss the ways that Collins and Evans are helpful, I need to acknowledge that I am, to some degree, reading against the grain of the text. The expressed purpose of *Rethinking Expertise* is to find new ground for understanding scientific knowledge. Collins and Evans write: "in this book we move from evaluating science as a provider of truth to analyzing the meaning of the expertise upon which the practice of science and technology rests" (2007, p. 2). Their solution to this problem is to locate expertise in experience and tacit knowledge. They write: "Acquiring expertise is . . . a matter of socialization into the practices of an expert group . . . and expertise can be lost if time is spent away from the group" (2007, p. 3). As useful as this social vision of expertise is, particularly as an alternative to debating the nature of *knowledge*, it fails to provide an immediately useful description of how expertise can inform public debate. Collins's and Evans's critics have gone so far as to argue that the arhetorical nature of their work is a fundamental flaw. In a response to the 2002 article where Collins and Evans first proposed their expertise-based vision of science studies, Bryan Wynne complains that Collins and Evans fail to consider the discursive spaces in which debate happens:

> They thus define the public domain to be only about whether or not something is true. They entirely ignore that public policy processes, and public reactions to scientific discourses of intervention in, and attempted management of nature and society, are processes of (often implicit and oblique) negotiation of public meanings. (2002, p. 404)

Wynn describes their project as a "decisionist" program in which "policy and political processes are conceptualized exclusively as a series of completely unrelated specific decisions, each one of which has no interaction with any other (2002, p. 410).

However flawed the technical bias that underpins Collins's and Evans's response to second wave science studies might be, their work, particularly as it is elaborated upon in *Rethinking Expertise*, allows for rich discussions of how expertise can inform public debate, *if* we pay particular attention to relationships between expert knowledge and everyday knowledge. *Rethinking Expertise* is concerned not just with validating the expertise of disciplined professionals and experienced laypeople, but also with describing different kinds of knowledge through a "periodic table of expertise" (2007, p. 14). For example, Collins and Evans distinguish between "internationalist expertise," which describes those who are able to participate in a discourse (although it is not the focus of this presentation, this category might of particular interest for rhetoric) with "contributory expertise," which describes the ability to make decisions or to create new knowledge. The most important parts of their taxonomy for thinking about ethos in public debate are the category of expertise labeled "ubiquitous tacit knowledge" and the category of "meta expertise" labeled "ubiquitous discrimination." "Ubiquitous" is the terms that Collins and Evans use to describe the accumulation of every day, "common sense" experience:

> What we will call "ubiquitous expertise" includes all the endlessly indescribable skills it takes to live in a human society. . . . For any specific society, its "form of life" or "culture" provides, and is enabled by, the content of the ubiquitous expertises of its members. (2007, p. 45)

Ubiquitous meta-expertise is, similarly, the socially prescribed ways that we recognize markers of expert authority. It is not unlike Pierre Bourdieu's notion of "cultural capital" in that it describes our implicit awareness of authority: "Ubiquitous discrimination, like other ubiquitous expertise, is acquired as part-and-parcel of living in our society" (Collins & Evans, 2007, p. 45).

My contention is that we can discuss an ethos of expertise, by which I mean the ability for an expert to participate in public conversation, as a response to this kind of "ubiquitous" expertise and discrimination. Because Collins and Evans are particularly concerned about peer-to-peer discrimination or "downward" (expert-to-non-expert) discrimination, they do not devote much attention to what, for me, is perhaps *the* crucial issue in thinking about expertise: how do experts persuasively advertise their knowledge to non-experts? The concept of "meta-expertise" begins to addresses the problem of describing the ability of non-experts to recognize experts (although non-experts can be fooled), but it does not help us understand why non-experts might be willing to accept expert authority. I believe that answers to this problem can be found by looking for ways that experts can *articulate* their expert opinions to commonsense arguments and popular opinion. *Dare to Discipline* is a text where an expert used professional knowledge to successfully participate in a political project, and it is a therefore a text we can examine what a strategy for accomplishing this task might look like.

4. DARING TO DISCIPLINE

Dare to Discipline is billed as a book about parenting, and its popular reputation is as a controversial pro-corporal punishment manifesto, but it is actually a somewhat unwieldy collection of essays that touch on many topics, including Dobson's anger at Benjamin Spock's contemporaneous status as *the* parenting expert, his dissatisfaction with the muddle of psychiatry at the time, his concern about education, and his hatred of the counterculture. Biographer Dale Buss writes:

> There really was no mistaking Dobson's clarion call to parents, educators, policy makers, and the culture as a whole. He wanted parents to reverse much of the child-rearing advice they'd been given since World War II—all of which, he believed, had come home to roost in the turbulence of the sixties. His was still a rather lonely voice at that point. . . (2006, p. 45)

Buss writes,

> [E]ven Dobson agrees that *Dare to Discipline* was far from a literary tour de force. . . . The book is actually a somewhat loosely packaged collection of essays and snippets of advice whose subjects are mainly united by being Dobson's passions in those days. (2006, p. 45).

While I agree with Buss's (and Dobson's) assessment that the book is a strangely organized outpouring with a wildly shifting authorial voice, I disagree with the idea that the book is "united" only as a snapshot of Dobson's concerns. Or, rather, I think that Dobson successfully defines these seemingly disparate issues as a part of the same overarching narrative about a disintegrating social order. Richard Vatz argues that rhetorical situations are defined not through the intrinsic properties of the facts of a case, but through rhetorical activity: "When political commentators talk about issues they are talking about situations made salient, not something that became important because of its intrinsic predominance" (1975, p. 160).

Dare to Discipline is an important book not only because it was Dobson's debut as a popular author, but also because it was the text in which he first attempted to understand "the family" in a broad political context. Throughout *Dare to Discipline*, Dobson's authority as a pediatric expert is used to construct a powerful vision of social conservatism as common sense:

> Children thrive best in an atmosphere of genuine love, undergirded by reasonable, consistent discipline. In a day of widespread drug usage, immorality, civil disobedience, vandalism, and violence, we must not depend on hope and luck to fashion the critical attitudes we value in our children. That unstructured technique was applied during the childhood of the generation which is now in college, and the outcome has been quite discouraging. Permissiveness has not just been a failure; it's been a disaster! (1970, p. 13–14)

Dobson's war against "permissiveness" is both an argument by an expert against a specific school of "progressive" parenting and an effort to contrast the values of the new left with commonsense conservative prudence. Dobson's claim is that experiments in social values that took place during the 1960s have created a moral vacuum in which a generation of young people has grown up. In the catalogue of social ills Dobson presents ("widespread drug usage, immorality, civil disobedience, vandalism, and violence") progressive parenting is linked to social protest, which is linked to petty criminality, addiction, and meaningless violence. This passage's primary purpose is to discuss different schools of parenting, but Dobson makes his argument by contrasting parenting as common sense to parenting as a received "technique."

One of the key paradoxes of *Dare to Discipline* is that Dobson creates identification with his audience by positioning himself against the elite opinions of the mental health establishment. Dobson begins the book with a horror story of an unruly, intolerable child and an ineffectual mother shackled to the "unworkable" and "permissive" philosophy of child care that foregoes the use of discipline:

> Mrs. Nichols and her little daughter are among the many casualties of an unworkable, illogical philosophy of child management which has dominated the literature on this subject during the past twenty years. This mother had read that a child will eventually respond to patience and tolerance, ruling out the need for discipline (1970, p. 10).

The persona Dobson adopts here is that of an expert—he is qualified to comment that a school of parenting that has "dominated the literature . . . during the last twenty years" is "illogical." As the passage continues, Dobson's complaint implicitly transforms from being a strictly professional opinion into being a commonsense observation and moral judgment:

> She has been taught that conflicts between parent and child were to be perceived as inevitable misunderstandings or differences in viewpoint. Unfortunately, Mrs. Nichols and her advisors were wrong! She and her child were involved in no simple difference of opinion; she was being challenged, mocked, and defied by her daughter. (1970, p. 11)

Dobson characterizes progressive "permissive" parenting as being an acceptance of a peer relationship between a child and an adult, and he argues that the unruly child is not just being disruptive, but actively challenging the social order. Dobson concludes the anecdote with a reiteration of this grim take on the situation:

> [T]he real issue was totally unrelated to the water or the nap or other aspects of the particular circumstances. The actual meaning behind this conflict and a hundred others was simply this: Sandy was brazenly rejecting the authority of her mother. (1970, p. 12).

Dobson's argument is that the faddish consensus of mainstream pediatric psychology (or at least, he makes an "expert" claim about what this consensus is) should be done away with and replaced by a school of parenting that conforms to conservative prudence. He is not only an expert; he is a wise expert, capable of critiquing mainstream medicine through the common sense of the superior moral grounding of the everyday social order. Parents and children cannot be peers, or society will fall apart. The reader is invited not only to consider Dobson's practical advice, but to take comfort in the acceptable common sense and moral certainty to which it is articulated. Dobson writes, "I reject [laissez-faire parenting] and I have considerable evidence to refute it" (1970, p. 13). This rhetoric divides the practical, ethical, morally grounded "we" from an aloof elite who would risk families and societies for their ideological commitment to permissiveness.

Dobson begins the first chapter of the book with a biological analogy describing the dangers of drug use and sexual freedom. He begins by setting up his story as a description of scientific realism: "Nature has generously equipped most animals with a fear of things that could be harmful to them. Their survival depends on recognition of a particular danger in time to avoid it" (1970, p. 15). He quickly shifts into a more colloquial voice and describes the phenomena he is using as his example:

> But good old Mother Nature did not protect the frog quite so well; she overlooked a serious flaw in his early warning system that sometimes proves fatal. If a frog is placed in a pan of warm water under which the heat is being increased very gradually, he will typically show no inclination to escape. . . . He will just sit there, contentedly peering over the edge of the pan while the steam curls ominously around his nostrils. Eventually, the boiling frog will pass on to his reward, having succumbed to an unnecessary misfortune that he could easily have avoided. (1970, p. 15)

Dobson admits that he has moved from a naturalist description of biology (explaining that fear is a survival mechanism) to using frogs as a metaphor, but he does so without abandoning his "naturalist" grounding:

> Now obviously, this is a book about parents and children, not frogs. But human beings have some of the same perceptual inadequacies as their little green friends. We have passively accepted a slowly deteriorating "youth scene" without uttering a croak of protest. (1970, p. 16).

This move is particularly rich, as it admits to the extravagance of Dobson's allegory (the cultural changes of the sixties and seventies were like a *frog being slowly boiled alive!?*) while holding onto the authority of expertise. He argues that we cannot assess our degraded state because we suffer from "perceptual inadequacies" that are cognitively similar to an animal in an unnatural situation for which its warning mechanisms are ill-equipped.

The passage is at once a simple fable that espouses common sense through a tall tale-like colloquial story, an expert's explanation of a particular cognitive phenomenon, and a blistering assessment of contemporary America. The frog serves as a funny, accessible story, but also a rather grim example. It isn't that Dobson has a difference of opinion—it is that we are being boiled alive and do not know it. Contemporaneous attitudes about drug and sex are articulated to the naturalist argument that people become acclimated to their environments to the warning that our cultural norms have become dangerous. Again, we see ordinary common sense and expert opinion intertwined.

The various contexts that Dobson has put his argument in—professional judgment, common sense, timeless moral law, social critique—invite the audience to share not just Dobson's narrow opinions about techniques of raising a child, but an overarching worldview. Dobson's argument is both "my professional advice will help you navigate conflicts in your home" and "our moral responsibility is to establish a social order and quell dissent." The strangely hostile language Dobson uses to describe children makes sense only when we see his arguments in the light of the political and moral contexts in which he places them. Dobson writes, in one of the most famous and controversial passages from the book:

> When a youngster tries this kind of stiff-necked rebellion, you had better take it out of him, and pain is a marvelous purifier . . . . You have drawn a line in the dirt, and the child has deliberately flopped his big hairy toe across it. Who is going to win? Who has the most courage? Who is in charge here? (1970, p. 27)

Dobson's style here might be read as a challenge to the "permissive" school of parenting or as an effort to steel the noncommittal parent for her (Dobson uses feminine pronouns and mothers as his representative weak parents) responsibilities, but the evocation of "hairy" rebellion is an almost overt evocation of the sixties counterculture. This strange language choice (hairy toes on a toddler?) is also suggestive of threatening unsubdued wildness. The child threatens to become a dissident hippie or a demonic beast. Dobson argues that if rebellion is not quelled at the earliest possible stage in a child's development schism and disorder are inevitable:

> A controlling but patient hand will eventually succeed in settling the little tyrant, but probably not until he is about four years of age. Unfortunately, however, the child's attitude toward authority can be severely damaged during his toddler years. The parent who loves her cute little butterball so much that she cannot risk antagonizing him, may lose and never regain his control . . .
> The proper time to begin disarming the teen-age time-bomb is twelve years before it arrives. Perhaps the most difficult problems referred to me occur with the rebellious, hostile teen-ager for whom the parents have done everything wrong since he was born. . . . For a psychologist, this problem must be approached as a physician views terminal cancer: "I can't cure it now; it's too late. Perhaps I can make its consequences less painful." (1970, p. 33–34)

Dobson ends this narrative that connects disciplining small children to juvenile delinquency to moral chaos with a medical analogy. Even though he has gone to the trouble of transforming his professional judgments into common sense and then into large-scale social commentary, he resolves the incongruities of the political and social conflicts he is using to frame his warnings about disciplining toddlers by steering the audience back into a therapeutic context. Dobson frames his argument with appeals to conservative prudence, with appeals to timeless wisdom, with invocations of moral panic, and with appeals to a "commonsense" distrust of elitist, faddish science, but ultimately, an ethos of expertise ties these various frames together.

V. CONCLUSION

While religious figures like Pat Robertson and brash media personalities like Ann Coulter have attracted more attention for their occasionally outrageous public statements, James Dobson has less conspicuously, but more consistently, articulated the concerns of conservative evangelicalism to various contexts, and Focus on the Family has established a kind of institutional permanence and popularity that no other social conservative organization has. The therapeutic rhetoric project that Dobson initiated with *Dare to Discipline* has been tremendously important to the politicized evangelical movement. When mainstream journalists want a spokesperson quote representing the opposition to same-sex marriage or to abortion rights, they often contact a representative of Focus on the Family or its spin-off organization, the Family Research Council.

The fusion between the therapeutic industry and conservative politics that Dobson pioneered has become a resilient and important part of the infrastructure of the American right. In addition to its advocacy and media work, Focus on the Family has continued to participate in mental health care by providing access to advice from professional counselors through a helpline and through a website, by organizing conferences and seminars about marriage and parenting, and by maintaining a database of Christian counselors. A visit to a Christian book store or an afternoon listening to a Christian radio station will reveal both the pervasiveness of therapeutic rhetoric in contemporary evangelical culture and the carefully elaborated political perspective to which this rhetoric is connected. Since Dobson pioneered the use of therapeutic rhetoric in conservative evangelical discourse, many professional organizations for Christian therapists have formed to organize evangelical therapists, including both apolitical organizations that exist to help ministers or counselors keep up with contemporary therapeutic practices and highly politicized organizations that understand conservative Christianity as a political movement. Participants in the conservative evangelical project can be found among the more psychodynamically inclined therapists (like marriage counselors), but also among the most biomedical practitioners (like psychiatrists who work in hospitals).

The question we must ask of social conservative rhetoric is not how it manages to unite such disparate discourses, but how this fusion operates, both in terms of understanding the political consequences of this fusion and in terms of understanding the rhetorical strategies that are used to create it. The ethos of expertise that Dobson constructed through his somewhat paradoxical "commonsense-oriented-ethos-of-expertise" is an example of this kind of strategy. I do not want to argue that the kind of populism Dobson uses is the *only way* that a rhetor might articulate his or her arguments to ubiquitous knowledge, but the paradox of the strange figure, the anti-elitist expert, that he became in his writing suggests how serious the challenges inherent in using expert knowledge to find political agency are.

ACKNOWLEDGEMENTS: This paper is my first effort to participate in the important ongoing interdisciplinary conversation about expertise. Thank you to Jean Goodwin for introducing me to that conversation.

REFERENCES

Bourdieu, P. (1991). *Language and symbolic power*. (G. Raymond & M. Adamson, Trans.) Cambridge, MA: Harvard University Press.
Buss, D. (2006). *Family man: The life of James Dobson*. Wheaton, IL: Tyndale House.
Chodoff, P. (2002).The medicalization of the human condition. *Psychiatric Services, 53*, 627–628.
Colllins, H., & Evans, R. (2002). The third wave of science studies: Studies in expertise and experience. *Social Studies of Science, 33*, 235–296.
Colllins, H., & Evans, R. (2007). *Rethinking expertise*. Chicago, IL: University of Chicago Press.
Dobson, J. (1970). *Dare to discipline*. Wheaton, IL: Tyndale House.
Jasinski, J. (2001). *Sourcebook on rhetoric: Key concepts in contemporary rhetorical studies*. Thousand Oaks, CA: Sage.
Laclau, E., & Mouffe, C. (1985). Hegemony *and socialist strategy: Towards a radical democratic politics*. Norfolk: Verso.
Vatz, R. E. (1975). The myth of the rhetorical situation. *Philosophy and Rhetoric, 6*, 154–161.
Wynne, B. (2003). Seasick on the third wave? Subverting the hegemony of propositionalism: A response to Collins and Evans. *Social Studies of Science, 33*, 401–417.
Yoos, G.E. (1970). A revision of the concept of ethical appeal. *Philosophy and Rhetoric, 12*, 41–58.

Examining News Coverage and Framing: The Case Study of Sea Lion Management at the Bonneville Dam

TESS MCBRIDE & CYNTHIA-LOU COLEMAN

Department of Communications
Portland State University
1600 Southwest 4th Avenue Portland, OR 97201
USA
TessMcBride19@gmail.com
CColeman@pdx.edu

ABSTRACT: This study examines how the construction of news stories reveals relationships among groups of stakeholders and how their views unfold within environmental conflict coverage. We look at framing of news stories to asses which voices are heard, how blame is leveraged, which solutions are proposed and how failures are framed.

KEYWORDS: American Indian, authority sources, blame, Bonneville Dam, environmental conflicts, framing, news bias, salmon, sea lions, solutions.

1. INTRODUCTION

Hard news coverage, particularly issues that fall under the rubric of environmental journalism, often stems from conflicts, accidents and disasters—what some critics refer to as event-driven or episodic coverage. The current study is concerned with a case that engendered such coverage: the encroachment of sea lions on the Pacific Northwest landscape and their impact on salmon populations and stakeholder groups.

The conflict regarding how to manage the sea lions reveals, on the surface, the divide between groups of stakeholders with competing, disparate agendas and the vagaries of laws and policies, while revealing fissures in the construction of objective and balanced information from news writers. But the conflict also reveals deeper divides that emerge from the values and ideological perspectives of those who are granted the authority to speak in the public arenas of debate and those responsible for deciding whose voices are heard and whose are silenced.

At its core, the current study examines how coverage of an environmental, political, legal, and moral conflict illuminates qualities of news construction by examining how arguments are framed, whose voices are heard, how blame, successes and failures are attributed, and which solutions gain traction. We demonstrate that, despite journalistic goals to provide information that is balanced and objective, some perspectives are given a wider berth than others and some solutions gain greater legitimacy in news coverage.

McBride, T., & Coleman, C.L. (2012). Examining news coverage and framing: The case study of sea lion management at the Bonneville Dam. In J. Goodwin (Ed.), *Between scientists & citizens: Proceedings of a conference at Iowa State University, June 1-2, 2012* (pp. 285-296). Ames, IA: Great Plains Society for the Study of Argumentation. Copyright © 2012 the authors.

2. LITERATURE REVIEW

2.1 Environmental News Reporting

We situate our study in the arena of environmental coverage, arguing that such coverage frequently follows accidents and disasters. Reporting in a conflict mode often results in less thoughtful and less rich coverage (Bendix & Liebler, 1999; Friedman, 2004; McPherson & Shaw, 1994; Nisbet, 2009; Schoenfeld, Meier, & Griffin, 1979). Conflicting moral views feed into the staple of news reporting where two sides are pitted against each other. Such dramatic coverage has been described as "episodic" by Shanto Iyengar (1991), who argued news narratives report on issues as concrete, single events, resulting in simplistic coverage. Deeper, more thoughtful coverage, which he labeled "thematic," is much less likely to occur, particularly in television news. Iyengar argued the result is that readers and viewers become less informed, and simplified coverage (episodic framing) results in more superficial thinking by publics.

Environmental reporting often suffers from episodic coverage, as noted by Boykoff and Boykoff (2007), who wrote, "dramatized news tends to eschew significant and more comprehensive analysis of the enduring problems, in favor of covering the spectacular machinations that sit at the surface of events" (Boykoff & Boykoff, 2007, p. 1192). Critics insist that important issues often receive superficial treatment, including climate change (Boykoff & Boykoff, 2007; Gordon, Deines, & Havice, 2010; Jones, 2006; Takahashi, 2010) and endangered species issues (Bendix & Liebler, 1999; Carolan, 2008; McPherson & Shaw, 1994; Peterson, Peterson, Peterson, Lopez, & Silvy, 2002).

2.1.1 Reliance on authority

The literature has clearly established dependence of reporters on government sources, particularly during environmental crisis (Bendix & Liebler, 1999; Bennett, 1997; Lacy & Coulson, 2000; McPherson & Shaw, 1994; Molotch & Lester, 1975). Grassroots groups and protest organizations receive fewer opportunities to voice their opinions and are often de-legitimized for their perspectives (Coleman, 1996; Lacy & Coulson, 2000). For example, McPherson and Shaw's 1994 study of coverage of the Yellowstone fires and Endangered Species Act showed that elected officials and local merchants were favored sources, rather than ecologists and scientists. The researchers argued that coverage was overly simplified and lop-sided because "reporters misled the public through reliance upon people and organizations with vested interests rather than upon scientific researchers investigating long-term policies" (p. 337).

2.2 Framing

Many framing studies examine how media organizations and individuals select and then report on events, while trying to unpack how information is presented and where we derive our knowledge. Frames live a double-life in that they also refer to the ways in which messages are constructed—cognitively and in texts (Coleman, Ritchie & Hartley, 2008; Dardis, 2007; Druckman, 2001; Entman, 1993; Haider-Markel & Joslyn, 2001; Iyengar, 1991; Semetke & Valkenburg, 2000). Gamson and Modigliani considered a frame "a central organizing idea or story that provides meaning to an unfolding strip of events, weaving a connection among them" (1987, p. 143). The assumption is that the act of framing serves to bundle certain definitions and perspectives, and that frames themselves are infused with meanings that are bounded and thus limited.

When Robert Entman noted that frames present a story "in such a way to promote a particular problem definition, causal interpretation, moral evaluation and/or treatment recommendation," his implication was that frames and framing actions reduce the scope of meanings and interpretations (1993, p. 52). As a result, news discourse has the potential to limit critical interpretation because framing casts meaning within the more narrow bounds of journalistic practices.

2.2.1 Blame Frames

Scholars who have studied frame construction note that, in addition to presenting information as episodic or thematic, frames often assign blame. Blame frames construct an issue by assigning responsibility to an individual, or group, or rarely, to a social structure or macro-social cause. In its simplest form, blame is considered a behavioral reflex when something goes wrong—a negative event (Anderson, 1991). Shaver's 1985 model divides blame attributions into stages of casual attribution, responsibility and blameworthiness. Blame shapes our views on the world, including issues surrounding public policy, politics and international relations (Haider-Markel & Joslyn, 2001; Iyengar & Kinder, 1987).

Iyengar and Kinder (1987) examined how individuals' perceptions of political responsibility and accountability were affected by television news framing. They argued attributions of blame affect formation of political opinions and evaluations of public policy and thus provide a link to understanding the formation of social knowledge. News media also attribute responsibility, which can play a powerful role in influencing opinions and policies (Haider-Markel & Joslyn, 2001). Blame has been attributed to shifts in public thought (Knoblotch-Westerwick & Taylor, 2008), political decision-making (Iyengar & Kinder, 1987; Gomez & Wilson, 2001), and policy changes (Haider-Markel & Joslyn, 2001).

Environmental disasters, unlike natural disasters, rarely occur on their own, and blame is laid on someone or something (Luke, 1987). When an oil tanker spills, companies, manufacturers, governments and sometimes individuals are blamed. In his examination of Eastern and Western media coverage of the 1986 Chernobyl nuclear disaster, Luke (1987) noted that U.S. news packaged the event by claiming the Soviets had "no one to blame but themselves," while the Soviets shifted blame for the accident to delays, inefficiencies and to "the Brezhnev appointees in the local and regional party apparatus" (p. 359). Blame frames may obviate the need for deeper discussions about the role of social structures in environmental disasters by securing a villain, thus ending speculation about causal effects.

2.2.2 Solution Frames

News stories often weave together blame with solutions (Coleman & Corbitt, 2003). Benford and Snow (2000) argued that such frames have the potential to serve as a call to arms for social change within communities and governments. Gamson (1992) suggested frames identify, evaluate and seek solutions relating to a particular issue, while Entman (1993) noted that frames define problems, diagnose causes, make moral judgments and suggest remedies. This suggests solution frames are an essential aspect of media framing, and may arrive on the heels of blame frames, which define a problem, identify the cause, and make moral judgments, according to Shaver's (1985) model of responsibility and causality evaluations.

Because of the preponderance of episodic news coverage, solution frames provide the opportunity for balance to be achieved within topical, episodic coverage. In his analysis of communication campaigns, Dardis (2007) found that messages are most effective when they present a public concern and then offer credible solutions, thus providing audiences with the total frame package of problem, cause and solution. Dardis invoked Gleicher and Petty's (1992) claim that offering a solution provides individuals an opportunity to accept solutions, whether they agree or not.

Yet solutions presented in coverage may be vague, as researchers have discovered in analyses of climate change reporting (Gordon et al., 2010; Jones, 2006; Pellow, 1999; Takahashi, 2010). The lack of specific and clear methods to solve problems may instill uncertainty on the part of publics, resulting in confusion and mistrust, often aimed at government officials. News coverage has the ability to fuse problems with solutions, thus reassuring audiences, while simultaneously presenting puzzling problems with indeterminate solutions.

2.3 Summary

In the last 40 years the growth of the environmental news has proven its place as a bona fide news beat, but critiques of coverage remain. Journalistic routines, hegemonic structures and profit motives continue to unfold in the newsroom, influencing how information is framed and passed on to readers. By deconstructing how information is framed, we can gain insight into shifts in public thought (Knoblotch-Westerwick & Taylor, 2008) and influences on political decisions (Iyengar & Kinder, 1987; Gomez & Wilson, 2001), policy making (Medler & Medler, 1993) and public opinion about the environment (Molotch & Lester, 1975).

3. HISTORICAL OVERVIEW: THE SEA LION AND SALMON CONFLICT

The Columbia River has long been central to the livelihood of indigenous peoples of the Pacific Northwest. The construction of dams along the Columbia River indelibly changed local ecosystems and economies. The Bonneville Dam, completed in 1938, diminished salmon populations dramatically. Adding insult to injury, groups of adult male sea lions are annually swimming from California to the Columbia River to hunt threatened or endangered salmon.

The current controversy regarding sea lions at the Dam has embroiled stakeholders in a struggle regarding how to manage these non-native pinniped populations. After the failure of non-lethal hazing efforts (e.g., physical barriers, underwater firecrackers, rubber bullets and buckshot), state agencies in Oregon, Washington and Idaho, with the blessing from local Native American tribes, asked for permission to use lethal means to remove sea lions. Permission for the request was granted and then rescinded multiple times over the last several years.

To complicate the story, some sea lion populations receive federal protection. California sea lions are protected under the Marine Mammal Protection Act, which makes killing or lethally harassing them illegal. While California sea lion populations dwindled to about 1,000 in the 1930s, they have since rebounded with a record number of 238,000

(NMFW, 2008). Most scientists consider this number to be stable, and the Marine Mammal Protection Act does not have a process for de-listing animals.

Several agencies have been involved in the effort to assess the sea lion presence at Bonneville Dam, including the Oregon and Washington Departments of Fish and Wildlife, the U.S. Army Corps of Engineers and the National Marine Fisheries Service (NMFS). Between 2002 and 2007, a total of 267 identified sea lions were present at the Dam (NMFW, 2008). The amount of California sea lions spotted at the Dam rose to 89 in 2010.

Officials have also tried to quantify how much of an impact the pinnipeds have on fish populations. Looking at the total catch in 2010, scientists estimated that between two and three percent of the total salmonid run was consumed annually by pinnipeds (Stansell, Gibbons, & Naggy, 2010, p. iii).

Removal measures can only be sanctioned if sea lions are having a significant negative impact on the protected salmonid populations. The U.S. Humane Society seized on the language of the act and protested plans to lethally remove sea lions in 2007. The judge ruled that the states could begin killing as many as 85 California sea lions annually for five years—as a last resort—if animals could not be relocated to zoos, aquariums or wildlife parks. This permitted the National Marine Fisheries Service and the states of Oregon, Washington and Idaho to trap and relocate "repeat offenders," and, if necessary, kill them.

4. HYPOTHESES AND RESEARCH QUESTIONS

Bearing in mind the literature concerning source use and framing in environmental news coverage, we expected sea lion coverage in the Pacific Northwest would demonstrate a heavy reliance on authority sources, particularly governmental officials. With this in mind, we predicted that:

> *Hypothesis 1*: Looking at the mainstream news stories concerning the sea lion and salmon conflict at the Bonneville Dam, the coverage, on average, will reflect a statistically greater number of governmental sources than advocacy and tribal sources

Our quest was to ascertain whether frames expressed in stakeholder platforms were reflected in news coverage. Moreover, we were interested in the frequency of blame frames in comparison to successful solution frames. We proposed:

> *Hypothesis 2*: When blame frames and successful solution frames appear in the news coverage of the sea lion/salmon controversy, the coverage, on average, will reflect a statistically significant greater number of blame frames than successful solution frames

Due to the lack of previous research examining failed solutions, we wanted to know how often failed solutions occurred in comparison to blame frames. We asked:

> *Research Question 1*: When looking at the blame frames and failed solution frames that appear in the news coverage of the sea lion/salmon controversy, will news stories, on average, include more instances of blame frames or failed solution frames?

To dig deeper into the characteristics of frames, we asked:

> *Research Question 2*: Among the blame frames, who or what is the focus of blame?
> *Research Question 3*: Which stakeholder groups are associated with discussing blame?

In addition, the literature demonstrated that offering solutions is typical in news coverage of environmental conflicts, and we therefore asked two descriptive questions:

> *Research Question 4*: In news coverage of the sea lion and salmon conflict at the Bonneville Dam, which successful and failed solution frames occur?
> *Research Question 5*: Which stakeholder groups are associated with solutions?

5. METHODOLOGY

Timeline for the study was January 1, 2003, to June 21, 2010. The beginning date marked the second season of sea lion monitoring at Bonneville Dam, which allowed officials to compare sea lion presence from the prior year. The study's end date marked the conclusion of the sea lion season at Bonneville Dam in 2010. For the pool of news articles we searched the Lexis-Nexis database, limiting the search to the six states in the Pacific Northwest with government, tribal or commercial ties to the salmon and sea lion issue (Alaska, California, Idaho, Montana, Oregon and Washington). We searched using key words "sea lion" and "Bonneville" during the six and one-half year timeframe. The search yielded a total of 451 articles from 15 newspapers (333 stories) and wire services (118 stories). In total, 67 articles were duplicates, so we randomly selected only one of any duplicates for the study. This reduced the sample to 384 articles.

Because the theoretical underpinnings for the study emerge from news studies, we omitted editorials, opinions and letters to the editor, leaving a sample size of 161 articles published in nine newspapers from five states. Oregon had the most articles with 88, followed by Washington with 56, Idaho with ten, California with six, and Montana with one.

5.1 Coding

In order to address the hypotheses and research questions, we developed both quantitative and qualitative content analysis techniques to assess newspaper coverage (Coleman, Hartley, & Kennamer, 2006; Krippendorff, 2004; Weber, 1990). A codebook was created to identify obvious (manifest) information, borrowing heavily from prior research on content analysis and framing (Benoit, Brazeal, & Airne, 2007; Gamson & Modigliani, 1989; Semetke & Valkenburg, 2000; Tankard, 2001).

For sources, coders were asked to identify all sources mentioned in the publication and their organizational affiliation. In addition to recording the source name, title and organization, we determined whether the organization was government (local, state, federal, international or tribal), a non-governmental agency, a for-profit organization, an advocacy organization, or unknown.

To assess frames, we attended to stakeholder platforms and the environmental framing literature to establish a priori the types of frames. We defined frames from a social constructionist perspective, which conceptualizes frames as "central organizing ideas" that promote "a particular problem definition, causal interpretation, moral evaluation and/or treatment recommendation" (Entman, 1993, p. 52). Coders were instructed that frames must be independent from one another. In cases where frames overlapped or could not be determined, coders met to decide how to determine a frame's prominence.

A team of two graduate students, one undergraduate student and an instructor created the codebook over a six-month period in 2010. Following Weber (1990), we created and tested a coding scheme, where researchers must define the frames they are attempting to code and then test the codes with group members presenting, evaluating and discussing frames or topics that should be included in the codebook.

5.2 Operationalizations

The majority of the coding required a manifest reading of the content. Determining frames, however, required both manifest and latent readings. Coders were instructed to code conservatively, relying on manifest words, rather than latent meanings. We established the following frames in advance: blame, war, solution, intrinsic values (ethics), extrinsic values (economy), balance, harmony, laws and policies, and politics, based on environmental reporting and framing literatures. Coders also noted frames that were emergent.

5.3 Intercoder reliability

In order to determine intercoder reliability, 10 percent of the sample was extracted randomly (Benoit et al., 2007; Lacy & Riffe, 1996; Neuendorf, 2002; Wimmer & Dominick, 1991). We used Cohen's Kappa to calculate intercoder reliability for the manifest content. We found 99 percent agreement on variables such as article date, newspaper name, etc. Turning to frames, the unweighted Kappa was determined for frame type and the weighted Kappa was calculated for number of frames because it accounts for differences between the responses in disagreement. High agreement was found with type of blame ($k = .96$) and the lowest level of agreement was found when coders were asked which organization was linked to the blame frame ($k=.69$). Based on Landis and Koch's (1977) interpretation of Kappa agreement, half of the calculated Kappas were considered to be in *almost perfect agreement* (.81-1.00), and the others were considered to have *substantial agreement* (.61-.80).

6. RESULTS

6.1 Sources

Across all stories 244 sources were mentioned, with the majority (75%) from government organizations, followed by advocacy/non-profit and then tribal organizations. Governmental sources (both state and federal) appeared far more frequently than others, with a total of 182 (M=1.13, SD=1.19). Advocacy/non-profit sources appeared less frequently with 43 source mentions (M=.27, SD=.48). Tribal sources appeared a total of 19 times (M=.19, SD=.34). Governmental sources were significantly more likely to be quoted than other sources, as hypothesized. Paired sample t-tests revealed a statistically significant difference between federal sources and advocacy sources [(t(384)=6.097, p< .000)], and between local government sources and advocacy sources [(t(384)=4.816, p< .000)], thus supporting Hypothesis 1.

6.2 Frames

Blame frames were presented the most frequently in the data, with a total of 279 mentions (*M*=1.73, *SD*=1.36) in the 161 articles. Successful solution frames occurred 211 times

(M=1.31, SD=.69) while failed solution frames were presented 96 times (M=.60, SD=.71).Turning to Hypothesis 2, which predicted that the coverage would reflect a greater number of blame frames than successful solution frames, we found a statistically significant difference between blame frames and successful solution frames [(t(160)=3.92, p< .000)].

Because we found little in the literature regarding failed solutions, we asked whether the coverage would reflect more instances of blame frames or failed solution frames. Failed solution frames appeared a total of 96 times (M=.60, SD=.71). When compared with blame, we discovered a statistically significant difference between blame frames and failed solution frames [(t(160)= 10.96, p < .000)], in response to Research Question 1.

The next task was to look more deeply at the types of blame frames that appeared in the news coverage and address Research Question 2. Most of the 279 blame frames centered on the decline in salmon: 261 of the total (94%). The most common blame was directed towards sea lions, which is not surprising as "sea lions" was among the two search terms used to select news stories. Sea lions were blamed for killing salmon in nearly half of all blame mentions (N=133; 47%). Dams were blamed for killing salmon with much less frequency: 39 mentions out of the set of 279 blame frames (14%). Fishing was blamed for salmon decline in 9 percent of the blame frames (N=25). Environmental factors, such as habitat loss, were noted in 8 percent of blame frames while birds were blamed in 5 percent of stories (N=14). Miscellaneous attributions of blame occurred in the remaining 4 percent of stories (N=12).

We also asked coders to note whether the attributions of blame were linked to a particular source and, if mentioned, his or her organization, to address Research Question 3. Recall the organizational categories included local or state government; federal or international government; tribal government; non-governmental agency; for-profit organization; advocacy or non-profit group; and other. Overall, the data showed that not every instance of blame had an organization linked to the discussion. Of the 279 instances of blame, organizations were linked to blame 147 times (53% of all blame frames); 132 discussions of blame were not presented by an organization (47%). Advocacy and non-profit groups were linked to the discussion of blame with the greatest frequency, for a total of 109 mentions (74%). Federal and international governmental organizations presented 18 (12%) discussions of blame, followed by state and local governments with nine (6%), and tribal governments with seven (5%). The "other" category yielded three instances of blame and the non-governmental/for-profit organizational category was linked to two instances of blame.

We also asked what type of solution frames were constructed in the news coverage (Research Question 4). Recall that we created two categories for solution frames: a successful solution and a failed solution, which occurred 211 times. We found six categories best describe successful solutions: Trapping, relocating, or lethal removal of sea lions was the most common category, accounting for more than three-quarters of the solution frames (N=163). Other solutions included hazing and other non-lethal methods (N=26, 12%), physical and electronic/sonar barrier solutions (N=6, 3%) and tracking, monitoring, and branding (N=6, 3%). The remaining solution frames (N=10, 5%) varied from using electrical shocks in the water to drive away sea lions to removing dams altogether.

We also tracked failed solutions. Most of 96 mentions of failures described either the failure of hazing and non-lethal methods (N=51) or trapping, relocating or lethal methods (N=41), representing 96% of the failed solutions. Remaining frames included unsuccessful attempts at electronic or sonar barriers, and branding, monitoring and tracking of sea lions.

In addition to looking at the types of successful solutions that appeared in the coverage, we examined which organizations were linked to solutions (Research Question 5). Note that not all solutions were tied to an organization if the coder could not determine the organization through a manifest reading. Of the 211 successful solutions presented, we could identify only 78 organizations linked to the solutions. When they were mentioned, government stakeholders were mentioned in most of the cases (N=61, or 78%). For-profit organizations were associated with seven solutions (9%) while advocacy/non-profit agencies were linked to five solutions (6%). Tribal governments presented four solutions (5%) and non-governmental organizations presented one solution (1%).

In a similar vein we looked at the sources associated with failed solutions. We were able to identify 49 organizations associated with the 96 failed solution frames. In contrast to the solution frames—where government sources dominated—advocacy and non-profit groups were linked to about half (N=24, 49%) of the frames. Governmental organizations represented 39% (N=19) of the organizations that were mentioned with failed solutions. The remaining six frames linked for-profit organizations with failures (N=3, 6%), tribal governments (N=2, 4%) and non-governmental agencies (N=1). We also noted that the difference between advocacy and non-profit groups as sources differed significantly from the linkage of government sources $[(t(160) = 2.45, p< .015)]$.

7. DISCUSSION

Our results align with prior empirical work on environmental communication and framing while offering some insights into what constitutes "balanced" coverage, source use and stakeholder message framing—specifically blame frames and their attendant solutions.

The overreliance on government sources had the effect of framing the sea lion issue from the vantage point that favored lethal removal of the invading pinnipeds as a solution to sustaining salmon populations. Criticism of lethal removal—voiced predominately by the Humane Society—meant re-framing the focus of blame away from the sea lions and onto other culprits, such as the dams, over-fishing and poor science. We argue that coverage shows that government officials stuck to a common message: lethal removal was the best method for solving the current crisis of salmon depletion. As a result, blame was more likely to focus on sea lions, and sources most readily available to speak to the issue—government workers—favored sea lion removal.

Yet critics of the government solution (lethal removal) countered solutions with a "failed solution" frame that also had currency in coverage. Advocacy groups such as the Humane Society employed two types of frames: they blamed other causes and described current solutions as failures. Such techniques provided something of a counter-weight to lethal removal of sea lions. Rather than blaming sea lions for salmon depletion, such groups argued that dams, overfishing, habitat loss, birds and human actions were to blame. Moreover, they noted that solutions advocated by government sources—trapping, relocating and lethal removal—failed to solve the problem. One spokesperson noted: "My frustration is there is no point in killing them if it isn't going to make a difference, and it isn't going to make a difference" (McCall, 2007).

The coverage suffered from a heavy dose of blame, regardless of the source of the disparagement. As a result, environmental coverage of this issue continues to be plagued with stakeholders pointing fingers at others more often than offering solutions. Our findings also reinforce the literature on source reliance in environmental reporting, and show the need for news reporters to branch out from dependence on authority sources in order to seek the opinions of non-mainstream and alternative viewpoints that provide readers with a fuller range of perspectives. With governmental sources appearing more than four times as often as advocacy or nonprofit sources, and nearly ten times more often than tribal sources, there is clearly a reliance on official sources. Perhaps reporters felt governmental representatives were a neutral authority on the matter. We contend, however, that for the current conflict, governmental sources were heavily invested in solutions to the sea lion encroachment. Government organizations were responsible for monitoring, hazing, trapping, relocating and lethally removing sea lions, in addition to being involved in salmon production and protection. While their involvement in sea lion management was often front and center in the news articles, their connection and responsibilities relating to sustaining salmon populations were not as prominently recorded.

7.1 Limitations and Future Directions

While our findings provide insight into framing of the sea lion management issue, we recognize that our choice of mainstream print media limits the generalizability of our findings to one dimension of news production. Because our study was designed to examine message frames rather than public opinion, we are hopeful that future studies will embark on gauging the effects of such frames on judgment. Do lay publics endorse the decision to kill sea lions at Bonneville Dam? Are publics likely to blame sea lions rather than dams for the depletion of salmon runs? How does public judgment align with the Humane Society's contention that humans, rather than mammals, are to blame for salmon loss? Attending to such questions will help us better understand the effects of message framing on audience framing, thus adding to the body of literature on framing.

8. CONCLUSION

Reporting on environmental conflicts was called one of the greatest challenges of the 21st century (Peterson et. al, 2002). Yet, the roles and functions of the news media have undergone dramatic changes in the last decade, cutting a deep swath through print news organizations and thus limiting their reach and impact. Can mainstream reporters provide an arena of rich, balanced and nuanced coverage for reading publics? Such an approach demands pluralistic actions, which eschew privileging of some sources over others, and thus some perspectives over others (Christians, Fackler, Rotzoll, & McKee, 2001). Pluralism, McQuail noted, offers a "complex of groups and interests, none of them predominant all the time" (1987, p. 85). With shrinking news budgets, we can't help but wonder how contemporary news organizations can continue covering information in a pluralistic vein.

ACKNOWLEDGEMENTS: Meghan Kearney, Ayshlee Koontz, Celeste Moser, Sean Rains and Dila Altin Sonmez. The authors received partial funding for the study from the Portland State University Office of Research and Sponsored Projects.

REFERENCES

Anderson, N. H. (1991). Psychodynamics of everyday life: Blaming and avoiding blame. In N. H. Anderson (Ed.), *Contributions to information integration theory* (Vol. 2, pp. 243–275). Hillside, NJ: Lawrence Erlbaum Associates.

Bendix, J., & Liebler, C. M. (1999). Place, distance and environmental news: Geographic variation in newspaper coverage of the spotted owl conflict. *Annuals of the Association of American Geographers, 89*(4), 658–676.

Benford, R. D., & Snow, D. A. (2000). Framing processes and social movements: An overview and assessment. *Annual Review of Sociology, 26*(1), 611–639.

Benoit, W. L., Brazeal, L. M., & Airne, D. (2007). A functional analysis of televised U.S. Senate and gubernatorial campaign debates. *Argumentation and Advocacy, 44*(Fall), 75–89.

Boykoff, M., & Boykoff, J. (2007). Climate change and journalistic norms: A case-study of US mass-media coverage. *Geoforum, 38*(1), 1190–1204.

Carolan, M. S. (2008). The politics in environmental science: the Endangered Species Act and the pebble's mouse controversy. *Environmental Politics, 17*(3), 449–465.

Christians, C., Fackler, M., Rotzoll, K. B., & McKee, K. B. (2001). *Media ethics: Cases and moral reasoning.* New York, NY: Longman.

Coleman, C. L. (1996). A war of words: How news frames define legitimacy in a native conflict. In S. E. Bird (Ed.), Dressing in feathers: The construction of the Indian in American popular culture (pp. 181–194). Boulder, CO: Westview Press.

Coleman, C.L., & Corbitt, J.A. (2003). The social construction of mental health anddepression. *Ecquid Novi, 24*(1), 99–114.

Coleman, C. L., Hartley, H., & Kennamer, D. J. (2006). Examining claimsmakers' frame in news coverage of direct-to-consumer advertising. *Journalism & Mass Communication Quarterly, 83*(3), 547–562.

Coleman, C. L., Ritchie, L. D., & Hartley, H. (2008). Assessing frames and metaphors in newscoverage of prescription drug advertising. *Journal of Health and Mass Communication, 1*(1/2), 108–127.

Dardis, F. E. (2007). The role of issue-framing functions in affecting beliefs and opinions about a sociopolitical issue. *Communication Quarterly, 55*(2), 247–265.

Druckman, J. D. (2001). The implications of framing effects for citizen competence. *Political Behavior, 23*(3), 225–256.

Entman, R. (1993). Framing: Toward clarification of a fractured paradigm. *Journal of Communication, 43*(4), 51–58.

Frazier, J. B. (2007, August 25). Task force in Ore. to recommend fate of predatory sea lions. The Associated Press. Retrieved from Lexisnexis.com

Friedman, S. M. (2004). And the beat goes on: The third decade of environmental journalism. In S.L Senecah, (Ed.), *The environmental communication yearbook* (Chap. 9). Malwah, NJ: Lawrence Erlbaum Associates.

Gamson, W. A. (1992). *Talking politics.* New York, NY: Cambridge University Press.

Gamson, W. A., & Modigliani, A. (1987). The changing culture of affirmative action. *Research in Political Sociology, 3*(1), 137–177.

Gamson, W. A., & Modigliani, A. (1989). Media discourse and public opinion on nuclear power: A constructionist approach. *The American Journal of Sociology, 95*(1), 1–37.

Gomez, B. T., & Wilson, M. J. (2001). Political sophistication and economic voting in the American electorate: A theory of heterogeneous attribution. *American Journal of Political Science, 45*(4), 899–914.

Gordon, J. C., Deines, T., & Havice, J. (2010). Global warming coverage in the media: Trends in a Mexico City newspaper. *Science Communication, 32*(2), 143–170.

Haider-Markel, D. P., & Joslyn, M. R. (2001). Gun policy, opinion, tragedy, and blame attribution: The conditional influence of issue frames. *The Journal of Politics, 63*(2), 20–543.

Iyengar, S. (1991). *Is anyone responsible? How television frames political issues.* Chicago, IL: The University of Chicago Press.

Iyenger, S., & Kinder, D. R. (1987). *News that matters: Television and American opinion.* Chicago, IL: University of Chicago Press.

Jones, A. J. (2006). How the media frame global warming: A harbinger of human extinction or endless summer fun? (Unpublished doctoral dissertation). University of Oregon, Eugene, OR.

Knoblotch-Westerwick, S., & Taylor, L. D. (2008). The blame game: Elements of causal attribution and its impact on siding with agents in the news. *Communication Research, 35*(6), 723–744.

295

Krippendorff, K. (2004). *Content analysis: An introduction to its methodology*. (2nd) Thousand Oaks, CA: Sage Publications.

Lacy, S., & Coulson, D. C. (2000). Comparative case study: Newspaper source use on the environmental beat. *Newspaper Research Journal, 21*(1), 13–25.

Lacy, S., & Riffe, D. (1996). Sampling error and selecting intercoder reliability samples for nominal content categories: Sins of omission and commission in mass communication quantitative research. *Journalism & Mass Communication Quarterly, 73*(1), 969–973.

Landis, J. R., & Koch, G. G. (1977). The measurement of observer agreement for categorical data. *Biometrics, 33*(1), 159–174.

Luke, T. W. (1987). Chernobyl: The packaging of transnational ecological disaster. *Critical Studies in Mass Communication, 4*, 351–375.

McCall, W. (2007, November 3). Task force leans toward killing nuisance sea lions in Oregon. The Associated Press. Retrieved from Lexisnexis.com

McQuail, D. (1987). *Mass communication theory*. Beverly Hills, CA: Sage.

McPherson, J. E., & Shaw, J. H. (1994). More accurate media coverage of wild-life related issues will require our active involvement. *Wildlife Bulletin, 22*(1), 336–338.

Medler, F., & Medler, J. (1993). Media images and environmental policy. In Spitzer, R.J. (Ed.), *Media and Public Policy* (Chap. 8). Westport, CN: Praeger Publishers.

Spitzer, R. J. (Ed.). *Media and public policy*. Westport, CT: Praeger.

Molotch, H., & Lester, M. (1975). Accidental news: The great oil spill as local occurrence and national event. *The American Journal of Sociology, 81*(2), 235¬260.

National Marine Fisheries Service (Northwest Region). (2008). *Final environmental assessment: Reducing the impact of at-risk salmon and steelhead by California sea lions in the area downstream of Bonneville Dam, on the Columbia River, Oregon and Washington.* National Marine Fisheries Service.

Neuendorf, K. A. (2002). *The content analysis guidebook*. Thousand Oaks, CA: Sage.

Nisbet, M. C. (2009). Communicating climate change: Why frames matter for public engagement; Report. *Environment, 2*, 1–20.

Pellow, D. N. (1999). Framing emerging environmental movement tactics: Mobilizing consensus, demobilizing conflict. *Sociological Forum, 14*(4), 659¬683. Retrieved from http://www.jstor.org/stable/685078

Peterson, M. N., Peterson, T. R., Peterson, M. J., Lopez, R. R., Silvy, N. J. (2002). Cultural conflict and the endangered Florida Key deer. *The Journal of Wildlife Management, 66*(4), 947–968.

Reese, S. D. (1990). The news paradigm and the ideology of objectivity: A socialist at the *Wall Street Journal*. *Critical Studies in Mass Communication, 7*, 390–409.

Robinson, E. (2006, March 8). New Bonneville fish ladder barrier no match for determined sea lion. *The Columbian*, p. 1A.

Schoenfeld, A. C., Meier, R. F., & Griffin, R. J. (1979). Constructing a social problem: The press and the environment. *Social Problems, 27*(1), 38–61.

Semetke, H. A., & Valkenburg, P. M. (2000). Framing European politics: A content analysis of press and television news. *Journal of Communication, 50*(2), 93¬109.

Shaver, K. G. (1985). The Attribution of Blame: Causality, Responsibility and Blameworthiness. New York, NY: Springer-Verlag.

Stansell, R., Gibbons, K. M, & Naggy, W. T. (2010). Evaluation of predation on adult salmonids and other fish at Bonneville Dam Tailrace, 2008–2010. U.S. Army Corps of Engineers. Retrieved from http://www.dfw.state.or.us/ fish/sealion/index.asp

Takahashi, B. (2010). Framing and sources: A study of mass media coverage of climate change in Peru during the V ALCUE. *Public Understanding of Science, 1*(10), 115.

Tankard Jr., J. W. (2001). Empirical approach to the study of media framing. In S. D. Reese, Jr., O. H. Gandy, & A. E. Grant (Eds.), *Framing public life: Perspectives on media and our understanding of the world* (pp. 95¬106). Malwah, NJ: Lawrence Erlbaum Associates.

Weber, R. P. (1990). *Basic content analysis*. Sage University Paper Series on Quantitative Applications in the Social Sciences. (2nd) . Newbury Park, CA: Sage Publications.

Wimmer, R. D, & Dominick, J. R. (1991). Mass media research: An introduction (3rd ed.). Belmont, CA: Wadsworth.

Scientists, Other Citizens, and the Art of Practical Reasoning

GITTE MEYER

Department of Learning and Philosophy
Aalborg University
Denmark
gm@learning.aau.dk
gitte@gittemeyer.eu

ABSTRACT: Inspired by the Arendtian distinction between the social and the political, this essay offers a critique of the tendency to frame the relationship between (scientific) expert knowledge and (political) democracy as a social issue or conflict between 'ordinary citizens' and 'scientific experts' as social groups. A tentative analysis explores the role of scientific expertise in democracies viewed as a practical issue in the classical, Aristotelian sense. It is suggested that the notions of *praxis* and practical reasoning as *phronesis* offer a framework that allows citizenship to scientists and might facilitate the integration of scientific knowledge into public deliberation on public affairs, but also would direct attention to the limitations of science.

KEYWORDS: dichotomies, *phronesis*, *praxis*, public deliberation, science communication.

1. INTRODUCTION: AN ANOMALY

The current discourse on the role of science in society is marked by a tendency to frame the relationship between (scientific) expert knowledge and (political) democracy as a social issue, or even as a social conflict between 'ordinary citizens' and 'scientific experts' as social groups. As a probably widely unrecognised and unintended consequence of that framing, scientists appear—in their capacity as scientists—to be excluded from the citizenry and the civic responsibility that citizenship implies. Inspired by the Arendtian distinction between the social and the political, this essay offers a critique of such framing.

Slaves and women were excluded from citizenship in the classical *polis*. In principle, modern democracies grant citizenship to all adults, and the tendency in the science-society discourse to exclude scientists from the citizenry, or to regard them as extraordinary citizens— whatever that might imply—appears as a thought-provoking anomaly. Responsibility, it has been argued, is a key ethical concept of technological civilisations (Jonas, 1984). Seen in that light, the anomaly is also rather worrying.

The following brief and tentative inquiry is primarily intended to direct attention to, and raise questions about, the anomaly. It is suggested that the science-society discourse may have become deadlocked in a framework of thought that combines two tendencies. One is a tendency to think in terms of dichotomies; the other is a tendency to constrain analysis of societal issues to the use of a social perspective, resulting in a focus on status and power relations.

It is furthermore suggested that the classical, Aristotelian notions of *praxis* and practical reasoning as *phronesis* offer a framework that—because it does not take science and politics to constitute a dichotomy, but to be substantially different—allows citizenship to scientists and might support the integration, case by case, of scientific knowledge into public deliberation on public affairs. The framework would, however, also likely bring the issue of

Meyer, G. (2012). Scientists, other citizens, and the art of practical reasoning. In J. Goodwin (Ed.), *Between scientists & citizens: Proceedings of a conference at Iowa State University, June 1-2, 2012* (pp. 297-306). Ames, IA: Great Plains Society for the Study of Argumentation. Copyright © 2012 the author(s).

the limitations of science to the forefront and should not be perceived as a solution to problems concerning the role of science in society.

2. SCIENTIST OR CITIZEN: A CRITIQUE

Why is the science-society relationship often seen as a *conflict* between seemingly irreconcilable spheres or activities? And why is that supposed conflict often seen as *social* conflict? These two huge, separate, but interrelated questions must be opened up to reflection.

2.1 The Assumption of a Science Versus Politics Dichotomy

Interpretations of the science-society relationship as a fundamental conflict can be seen as outcomes of dichotomic—polarised and polarising—frameworks of and for thought. More specifically, approaches that radically separate and oppose the roles of the scientist and the citizen, respectively, may be viewed as reflections of a widespread assumption of a dichotomy of science versus politics. That assumed dichotomy, in turn, is an instance of an entire range of assumed dichotomies that tend to inform, not least, academic discourse and enquiry: truth versus power; objectivity versus subjectivity; observation versus participation; and the spiritual versus the material are a few examples of such assumed dichotomies. Indeed, the dichotomy appears to be a dominant figure of thought in academic work, originating, it seems, in the notion of universal truth and the corresponding arch-dichotomy of truth versus falsity (Meyer & Lund, 2008b).

An antagonistic force has been ascribed to the monotheistic or secondary religions that brought the truth versus falsity dichotomy into the domain of human beliefs (Assmann, 2010). There are of course important differences between the religious and scientific idea(l)s of universal truth, but there are also some striking similarities and a shared history of interaction, as evidenced, for instance, in the significance of the English civil, confessional wars to the early development of science (Sprat, 1667/1734). Not only monotheistic religions, but also scientific monism may inspire dualism and be disinclined to acknowledge or even consider its own boundaries and/or limitations. Perhaps, one of the most striking similarities is the above antagonistic force, expressed as a capacity to generate dichotomies, and as a general inclination to make dichotomic distinctions.

Dichotomies, however, represent a particular variety of distinctions. They express opposite valuations of things, phenomena or qualities that seem to be taken, as the point of departure, to be substantially similar and to represent the two sides of the same coin. The two sides of an assumed dichotomy are mutually exclusive and interdependent: truth is defined by not being false, objectivity by not being subjectivity, and vice-versa.

The phenomenon of normative inversions (Assmann, 2010) may bring about re-valuations—or, if you prefer, 're-volutions.' A positive valuation of objectivity and a negative valuation of subjectivity may be supplanted by a positive valuation of a re-interpretation of subjectivity and a negative valuation of a re-interpretation of objectivity. A new school, or theory, or '-ism' is founded without affecting the basic assumption of a dichotomy. Thus, a normative inversion that shifts the balance from objectivity to subjectivity, or from subjectivity to objectivity, does not affect the very assumption of an objectivity versus subjectivity dichotomy, and does not raise questions concerning the proper reach and applicability of that dichotomy. Actually, the accumulated effect, over the centuries, of series of normative inversions could be that the dichotomy acquires the appearance of a general, or even natural, figure of and for thought, rather than a particular one.

If so, dichotomic forms of distinction may be applied indiscriminately to all kinds of differences. Even the capacity for critical judgement—the very ability, that is, to make distinctions—may be ascribed the quality of being negative and in opposition to something as opposed to being positive toward and supportive of it (Marcuse, 1968). What is more, attempts to escape dichotomic deadlocks may take the form of general assaults on the very practice of making distinctions at all (for possible examples of this see for instance Callon, 1986; Latour, 1993).

Our question, now, concerns the assumption of a science versus politics dichotomy. Taking into account a possible origin in a dichotomic framework, it can be linked to several other assumed dichotomies: truth versus power; facts versus emotions; and facts versus values spring to mind. These connect science to truth and facts, and politics to power, emotions and values. In order, however, for the assumption of a fundamental conflict between science and politics to make sense, they must be perceived to be substantially similar and, thus, to be concerned with similar questions. But *are* they substantially similar in that sense, or should they rather be considered to be concerned with different questions? Monistic—and thus potentially dualistic—frameworks of and for thought seem to hamper rather than facilitate reflections along those lines.

2.2 The Interpretation of the Science-society Relationship as a Social Conflict

Our second question concerns the transformation of the assumed fundamental conflict between science and politics into a social conflict between scientific experts and (ordinary) citizens as social categories. Although some of the major themes of the science-society discourse relate to political notions such as 'citizen' and 'democracy,' a social perspective is widely applied and seems almost to be taken for granted. Where does that perspective take us?

The social perspective represents a view on humans as one of those animal species that live in groups. In order to study humans from that perspective, one has to adopt the position of an outside observer. The position facilitates that social groups or categories may be identified by the criterion of homogeneity. Patterns of resemblances and differences become visible. Status and power relations, and the degree of distance or intimacy within or among groups, come into focus. Furthermore, the objects of study appear to the observer as possible targets of technical intervention aimed at affecting the social relationships or mechanisms of or among groups. The social perspective, thus, can be characterised as an offspring of the classical notion of *techne*, extended and applied to human beings and human affairs.

Techne belongs in the sphere of production. Correspondingly, interpretations of social relationships as those between producers and consumers are widespread. Scientists, for instance, may be seen as producers of scientific knowledge. Knowledge, then, comes to be seen as a good for possession and/or distribution and consumption, and other citizens appear as consumers of knowledge. Interpretations along such lines have belonged to the staples of the science communication discourse for decades (Friedman, Dunwoody & Rogers, 1986). Alternatively, knowledge may be seen as a tool for power holders, and connected to social conflicts between scientists, as holders of knowledge power, and other citizens, as knowledge-have-nots (Felt, 2003; Goede, 2002).

In general terms, the social perspective directs attention to hierarchies and social (in)equalities, to the fair distribution of goods, and to the fair representation of different social groups. It also diverts attention from the substance of issues. Precisely for that reason, it has been argued, references to social distinctions were disapproved of in the coffee houses of the early enlightenment (Sennett, 1986).

When seen from the social perspective, thus, the role of science in public deliberation on public affairs is transformed into an issue of status and power relations. Moreover, the classical understanding of the public or citizenry is marginalised. The *diverse* group of citizens who are bound together merely by co-responsibility for public affairs does not constitute a social group. Strictly speaking, it has no place in social reality.

Another ancient figure of thought is more compatible with the social perspective: the assumed dichotomy of the masses versus the elites.

The notions of the masses and the elites have been significant in modern, Western social thought (see for instance Bottomore, 1964/1971; Carey, 1992; Mills, 1956/2000; Ortega y Gasset, 1930/1993; Veblen, 1899) but are in fact neither particularly modern nor particularly Western. Thus, the idea that members of a society are divided into the masses and the elites has been influential also in pre-modern times (Hill, 1961/2010) and non-Western cultures (Hourani, 2002). The positive valuation of the notion of masses, on the other hand, is probably predominantly a modern idea or, if you prefer, represents a modern normative inversion, connected to what has aptly been termed the invention of the people (Morgan, 1989).

The two notions can be seen as pre-modern exemplars of social categories or groups. Each group is characterised by homogenous features. The elites occupy power positions in the economic, political and intellectual systems; the masses do not. Looking back, for example, to seventeenth-century English discourse, the latter group was freely talked about as "the rabble that cannot read" (Morgan, 1989) or "the unknowing multitude," but positive references to "the people" also gained some momentum (Hill, 1961/2010). In current usage, less immediately demeaning expressions, such as 'the average citizen' or 'ordinary people' are common.

A positive valuation of the masses became manifest during and in the wake of the American War of Independence and has been connected to a wave of fascination with quantitative science:

> People now [early nineteenth century] described society more and more as a 'mass' and for the first time began using this term in reference to 'almost innumerable wills' in a positive, nonpejorative sense. The individual was weak and blind, said George Bancroft in a common reckoning, but the mass of people was strong and wise. From all this followed, too, a new appreciation of statistics: in 1803 the word 'statisticks' first appeared in American dictionaries. (Wood, 1993, p. 360).

Along related lines, the rise of quantitative science has been connected to anti-elitism (Porter, 1995).

The view of society as an entity divided into the masses and the elites has remained a staple of social thought and continues to give rise to conflicting interpretations and valuations. The Dewey-Lippmann controversy of the 1920s can be seen as a model of such conflicts (Dewey, 1927/1991; Lippmann, 1922/1997). But the very notion of the masses—and, thereby, the assumed dichotomy that it forms part of—has also been subjected to critique. It is a contested concept (Collier, Hidalgo & Maciuceanu, 2008).

As a quantitative concept, the notion of the masses simply signifies the many, the majority. As a qualitative term it has been connected to a kind of person—that has come to be, or to be perceived to be common—who is motivated primarily by the immediate prospects of pain, pleasure and gain, who is caught up in concerns with his or her private affairs, and who is easily manipulated and disinclined to engage in any kind of abstract thinking (Arendt, 1958–59). The rise, on a grand scale, of such assumptions about the general public, perceived as a mass audience, seems expressed in journalistic criteria that stress the importance of dramatisation, emotional appeal and what's-in-it-for-me approaches (Meyer & Lund, 2008a).

A pertinent question to our issue concerns whether and how the discourse on the science-society relationship may have been affected by the assumed dichotomy between the social categories of the masses and the elites. The discourse draws heavily on the notion of the layperson, inherited from the medieval church (Meyer & Sandøe, in press). At the same time, the social categories of the scientific experts and the ordinary citizens can be seen as representatives of an (intellectual) elite and the (lay) masses.

Transformed into a social concept, thus, the notion of the *citizen* seems to have come to signify a *subject* who is excluded from positions of power, and who lacks (scientific) knowledge. This interpretation, in turn, takes us some way toward understanding the puzzling tendency to exclude scientists from (ordinary) citizenship. But there is more to it than that. The predominance in the science-society discourse of the social perspective may be doing away altogether with the classical idea of the citizen. Actually, neither scientists nor other citizens appear to be regarded as citizens in that sense.

Against that background, critical reflection among participants in the science-society discourse might be directed to questions such as: Do terms such as 'average citizens' and 'ordinary people' come with tacit and potentially self-fulfilling assumptions concerning, among other things, the absence of intellectual capacity in the public? If so, how might that affect the general ability of democratic knowledge societies, that are pervaded by scientific enquiries and knowledge claims, to deal with the outcomes of such enquiries and to assess such claims? How might individual scientists and the scientific community as a whole be affected by a discourse that excludes scientists from (ordinary) citizenship? Have we somehow become locked into a framework of thought that has—through its combination of inherent dualism and a tendency to constrain analysis of societal issues to the use of a social perspective—transformed the debate over the role of science in public deliberation on public affairs into a fundamental conflict between scientific experts and (ordinary) citizens perceived as social groups?

3. A WIDER FRAMEWORK: AN EXPLORATION

Using the classical, Aristotelian approach to the public or citizenry as our point of departure might take us in other directions that could lead to other tentative answers to the science-society question—and to other possible problems.

In its modernised version, the classical definition of the citizenry in principle grants citizenship to all adults. It comes with the advantage that it does not exclude scientists from citizenship. It seems, therefore, worthwhile to briefly explore the background of the understanding of the citizenry as a diverse group that is bound together by the co-responsibility for public affairs, and by the capacities for thought and speech that identify human beings as political animals. This idea of the public is much older than modern science. Does it make sense in and to the knowledge societies of today?

3.1 The Classical Notions of Praxis and Practical Reasoning

The Aristotelian understanding of the citizenry belongs in a three-dimensional, non-dichotomic framework of thought that operates with a particularly human dimension of reality —*praxis*— and regards politics as constituting the highest form of that dimension. Life, according to Aristotle, is action (*praxis*), not production (Aristotle, trans. 1992, I.iv). The notion of *praxis*, thus, seems to be the proper starting point for possible re-interpretations.

Praxis differs from the mechanics of nature and the unlimited universe in the same way that human beings differ from other animals and gods. Human life as *praxis* is marked by unpredictability, uncertainty and diversity. It is also characterised by the human capacity to act—and, thus, to deliberate on action—on those conditions (Arendt, 1969). Humankind is composed of a multitude of different humans, and all have different perspectives on human affairs, but as political animals they are able to deal with the uncertainty of those affairs in a specifically human way: exchange between different points of view is the practical-political mode (Crick, 1962/2005).

The political institution of public discussion or deliberation is concerned with proper and rightful action, not with questions relating to universal truth. It is preconditioned by the existence of citizens who are both willing and able to participate critically—and, thus, to pay thorough attention to the substance of issues (Aristotle, trans. 2002, A.III, IV; Hastrup 2002)— in public deliberation on public affairs, as distinct from the private affairs of households and matters of religion. There is, in other words, no compatibility with ideas about laypersons in political life, nor with assumptions of a dichotomic relationship between participation and deliberation (Mutz, 2006), nor with other assumed dichotomies relating to human affairs.

The view of human life as *praxis* is accompanied by a concept of practical reason or *phronesis* (Arendt, 1969; Gadamer, 2001; MacIntyre, 1984; Schnädelbach, 2007). It differs from technical rationality and from the contemplation of universal truth in much the same way that humans differ from other animals and from gods, and that life as *praxis* differs from the unlimited and the mechanical dimensions of reality. *Phronesis* is a worldly, temporal and personal kind of reason. Aimed at proper and rightful action, and suited to the conditions of limitations, diversity and uncertainty, it has purposes, but no objects or aims of control. Practical reasoners make assessments case by case—including, at the same time, factual and normative aspects of individual cases—while drawing on personal experience and taking other points of view into account.

The political institution of public discussion, with its inherent pluralism, makes sense only in regard to those assumptions that connect the human condition with limitations *and* with the human capacities of thought and speech. It is an institution for enquiry into practical, political problems that can neither be answered by religion nor be solved by technical means, but which may include elements that relate to specialised knowledge (Aristotle, trans. 2002, A. II). There is uncertainty and diversity. Reasonable argumentation and critical assessments are possible; proof is not. It is, therefore, necessary to include multiple points of view in discussions and to avoid allowing anyone a monopoly on reason with respect to practical issues. Such multiple viewpoints must be dealt with through discussion among citizens who are co-responsible for public affairs and represent different perspectives on issues.

The classical understanding of politics, thus, marks it out as substantially different from modern science, circling the concept of universal truth and operating along the lines of technical rationality. At the same time, however, the practice of *doing* science appears as an instance of *praxis*; it is a human endeavour, subject to the practical conditions of uncertainty, unpredictability and human diversity.

Of immediate relevance to the current science-society discourse is the assumption that uncertainty is fundamental to the human condition. During recent decades, uncertainty has been seen to be re-discovered. Scientific uncertainty has become a key term in the science-society discourse, and attempts have been made to understand this disturbing aspect of modern or post-modern science (among numerous possible examples, see for instance Beck, 1992; Friedman, Dunwoody, & Rogers, 1999). If viewed, however, from a classical, practical perspective, the specificity of the notion of *scientific* uncertainty is unhelpful, and attempts to solve the problem by technical-scientific means are counterproductive.

As the use of scientific methods and approaches has expanded into ever more walks of life, scientific enquiry has come to be increasingly concerned with human affairs and practical, political issues. No wonder, then, that the condition of uncertainty increasingly makes itself felt. It is a general feature of *praxis* and an expression of those limitations that form part of the human condition. It is not a technical problem that can be solved, but an indication of basic conditions that should be recognised.

From this perspective, the expansion of science also increases the need to consider its limitations, not least when scientific experts participate in public exchanges as citizens with specialised knowledge. There is a place for scientific rationality within the wider framework of practical knowledge pluralism and critical, practical reasoning. Scientific knowledge may be integrated (Gadamer, 2001) into public deliberation on public affairs. But the place comes, as places do, with boundaries. They do not follow the lines of a dichotomy of facts versus values and cannot be defined once and for all, but need continuous attention. That challenge, however, connected as it is to a distinction between technical and practical issues which is no longer in vogue, might be perceived by many as alien and perhaps even as hostile to science.

3.2 Practical Reason and Scientific Rationality: Conflict and Complementarity

With their revised translations of classical texts and with their often keen interest in societal debate, Renaissance humanists gave the Aristotelian approach to ethics, politics and rhetoric a new lease of life (Kristeller, 1961). Against that background, the rise of modern science has been described as a Counter-Renaissance (Toulmin, 1990), and it has been noted that the distinction between *praxis* and production already was rejected by Hobbes in the early seventeenth century (Höffe, 2010). Indeed, both to the early and the later development of modern science, many are likely to have regarded the notion of a particularly human sphere of action as *praxis*, with its pluralism and its emphasis on the limitations of human endeavours, as an inappropriate frame that stood in the way of the (unlimited) progress of mankind.

Evolving in a climate of confessional warfare—related at the same time to politics and religion (Schorn-Schütte, 2010; Worden, 2009) —some of the founding features of modern science can be seen as aiming to escape the dangerous sphere of conflicting confessions (Sprat, 1667/1734). Somewhat paradoxically, however, the development of ideas of science may still have been informed by the very mental climate they were actually intended to counteract. To some extent, confessional features, connected to the notion of universal truth, may have been mimed and carried on, including such tendencies as to think in stark terms of pro- versus anti-science attitudes and to generate science wars.

More subtle approaches may be inspired by the history of the various enlightenment movements that contributed to the continuous development of ideas about science and modern democracy during the seventeenth and eighteenth centuries. Recent decades have witnessed considerable numbers of accounts of those movements. Together these accounts portray the enlightenment as a tradition of multiple strains, and of tensions (see for instance Bahr, 2002; Jacob, 2006; Porter, 2001). Some of those tensions are encapsulated in the enlightenment motto: *sapere aude* which can be translated into *dare to know* (Kramnick, 1995) and into *have courage to use your own reason* (Kant, 1784/1995).

Equally valid, the two translations can be seen as representing a conflict between phronetic reasoning and the authority of scientific rationality, and, at the same time, as indicating a complementary relationship between those two varieties of reason. Along that line, the indiscriminate use of, and appeal to, scientific expertise would be considered a fallacy, as would the indiscriminate rejection of such uses and appeals. One task of practical reasoning would be to facilitate critical reflection and discussion case by case of whether or not, or to what extent, scientific approaches should be deemed appropriate to the issue in question. One should dare to know, and to use one's own reason. The very relationship between scientific rationality and practical reasoning can be seen as constituted by conflict and complementarity and, thus, by a combined capacity to keep each other in check.

4. CONCLUSION: COPING WITH THE EXPANSION OF SCIENCE

It is the tentative conclusion of this brief exploration that a re-introduction of the classical notions of *praxis* and *phronesis* into the science-society discourse not only offers a framework that allows ordinary citizenship to scientists and includes them in the co-responsibility for public affairs that citizenship implies. The framework also comes with the demand that the limitations of science be considered case by case as part of a continuous public discussion in the shape of practical reasoning among citizens that represent multiple points of view. Therefore, the framework possibly could be perceived as an affront to science, but it can also be seen as offering a possibility for scientists to combine the role of scientist with that of the co-responsible citizen. And it can be seen as a possibility for modern knowledge democracies to cope with the expansion of science in a reasonable way, steering clear of the pitfalls of populism and technocracy.

ACKNOWLEDGEMENTS: This work has been made possible by a grant from The Danish Council for Independent Research | Humanities.

REFERENCES

Arendt, H. (1958–59). Kultur und Politik. Merkur: Deutsche Zeitschrift für Europäisches Denkens, 12, 1122–1145.

Arendt, H. (1969). *The human condition.* Chicago, IL & London: The University of Chicago Press.

Aristotle. (2002). *Retorik.* (Hastrup, T., Trans.) Copenhagen: Museum Tusculanums Forlag.

Aristotle. (1992). . *The Politics.* (Sinclair, T.A., Trans., Saunders, T.J., Ed.). London: Penguin Books.

Assmann, J. (2010). *The price of monotheism.* Stanford, CA: Stanford University Press.

Bahr, E. (Ed.). (2002). *Was ist aufklärung? Thesen und definitionen.* Stuttgart: Reclam.

Beck, U. (1992). *The risk society: Towards a new modernity.* London: Sage.

Bottomore, T. B. (1971). *Elites and society*. Harmondsworth, Baltimore & Ringwood: Pelican Books. (Original work published 1964).

Callon, M. (1986). Some elements of a sociology of translation. In J. Law (Ed.), *Power, action and belief: A new sociology of knowledge?* (pp. 196–223). London: Routledge.

Carey, J. (1992). *The intellectuals and the masses: Pride and prejudice among the literary intelligentsia 1880–1939*. London: Faber and Faber.

Collier, D., Hidalgo, F. D., & Maciuceanu, A. O. (2008). Essentially contested concepts: Debates and applications. *Journal of Political Ideologies, 11*, 211–246.

Crick, B. (2005). *In defence of politics*. London & New York: Continuum. (Original work published 1962).

Dewey, J. (1991). *The public and its problems*. Athens, OH: Swallow Press & Ohio University Press. Original work published 1927).

Felt, U. (Ed.). (2003). *O.P.U.S. Optimising public understanding of science and technology: Final report*. Retrieved from http://www.univie.ac.at/virusss/opus/mpapers.html

Friedman, S. M., Dunwoody, S., & Rogers, C. L. (Eds.). (1986). *Scientists and journalists: Reporting science as news*. New York & London: The Free Press.

Friedman, S. M., Dunwoody, S., & Rogers, C. L. (Eds.). (1999). *Communicating uncertainty: Media coverage of new and controversial science*. Mahwah, NJ & London: Lawrence Erlbaum Associates.

Gadamer, H-G. (2001). *Truth and method*. London: Sheed & Ward.

Goede, W. (2002, November). *Civil journalism & scientific citizenship*. The keynote address to the *Third World Conference of Science Journalists*. Retrieved from http://www.forum-community-organizing.de/goede_journalism.htm

Hastrup, Thure (2002). Introduktion. In *Retorik*. Translated by Thure Hastrup. Copenhagen: Museum Tusculanums Forlag.

Hill, C. (2010). *The century of revolution: 1603–1714*. London & New York: Routledge. (Original work published 1961).

Hourani, A. (2002). *A history of the Arab peoples*. London: Faber and Faber.

Höffe, O. (2010). *Thomas Hobbes*. München: Verlag C.H. Beck.

Jacob, M. C. (2006). *The radical Enlightenment: Pantheists, Freemasons and republicans*. Lafayette, LA: Cornerstone.

Jonas, H. (1984). *Das prinzip verantwortung: Versuch einer ethik für die technologische zivilisation*. Frankfurt am Main: Suhrkamp.

Kant, I. (1995). What is Enlightenment? In Kramnick (Ed.), *The Portable Enlightenment Reader* (pp. 1–7). New York, NY: Penguin Books. (Original work published 1784).

Kramnick, I. (Ed.). (1995). *The Portable Enlightenment Reader*. New York, NY: Penguin Books.

Kristeller, P. O. (1961). *Renaissance thought: The classic, scholastic and humanist strains*. New York, NY, Evanston, & London: Harper Torchbooks.

Latour, B. (1993). *We have never been modern*. Cambridge, MA: Harvard University Press.

Lippmann, W. (1997). *Public opinion*. New York, NY: Simon & Schuster. (Original work published 1922).

MacIntyre, A. (1984). *After virtue: A study in moral theory*. Notre Dame, IN: University of Notre Dame Press.

Marcuse, H. (1968). *One-dimensional man*. Boston, MA: Beacon Press.

Meyer, G., & Lund, A. B. (2008a). International language monism and homogenisation of journalism. *Javnost: The Public, 15*(4), 73–86.

Meyer, G., & Lund, A. B. (2008b). Spiral of cynicism: Are media researchers mere observers? *Ethical Space: The International Journal of Communication Ethics, 5*(3), 33–42.

Meyer, G., & Sandøe, P. (in press). Going public: Good scientific conduct. *Science and Engineering Ethics*. doi:10.1007/s11948-010-9247-x

Mills, C. W. (2000). *The power elite*. New York, NY: Oxford University Press. (Original work published 1956).

Morgan, E. S. (1989). *Inventing the people: The rise of popular sovereignty in England and America*. New York, NY, & London: W.W. Norton & Company.

Mutz, D. C. (2006). *Hearing the other side: Deliberative versus participatory democracy*. New York, NY: Cambridge University Press.

Ortega y Gasset, J. (1993). *The revolt of the masses*. New York, NY: Norton. (Original work published 1930).

Porter, T. M. (1995). *Trust in numbers: The pursuit of objectivity in science and public life*. Princeton, NJ: Princeton University Press.

Porter, R. (2001). *Enlightenment: Britain and the creation of the modern world*. London: Penguin Books.

Schnädelbach, Herbert (2007). *Vernunft*. Stuttgart: Reclam.

Schorn-Schütte, L. (2010). *Konfessionskrige und europäische expansion: Europa 1500–1648*. München: Verlag C.H. Beck.

Sennett, R. (1986). *The fall of public man*. London: Faber and Faber.

Sprat, T. (1734). *The history of the Royal Society of London, for the improving of natural knowledge* (4th ed.). London. (Original work published 1667).

Toulmin, S. (1990). *Cosmopolis: The hidden agenda of modernity*. New York, NY: The Free Press.

Veblen, T. (1899). *The theory of the leisure class*. Retrieved from http://xroads.virginia.edu/~HYPER/VEBLEN/veblenhp.html

Wood, G. S. (1993). *The radicalism of the American Revolution*. New York, NY: Vintage Books.

Worden, B. (2009). *The English civil wars: 1640–1660*. London: Weidenfeld & Nicolson.

Mork and Mindy, Canola Oil and Mustard Gas: The Dilemma of Scientific Illiteracy in Decisions about Food and Health

MARY L. NUCCI

Department of Human Ecology
Rutgers, The State University of New Jersey
Cook Office Building, 55 Dudley Road
New Brunswick, NJ 08902
USA
mnucci@rutgers.edu

WILLIAM K. HALLMAN

Department of Human Ecology
Rutgers, The State University of New Jersey
Cook Office Building, 55 Dudley Road
New Brunswick, NJ 08902
USA
hallman@aesop.rutgers.edu

ABSTRACT: People often use food to represent and communicate their role in society, or political or ideological beliefs. Food consumption is thus laden with meaning beyond health or nutrition. Multiple audience studies examining perceptions about food/technologies/health have shown that scientific illiteracy and confusion are key to decision making about food.

KEYWORDS: food, food technologies, knowledge, science constructs, science literacy.

1. INTRODUCTION

People use food choices to represent and communicate who they are as individuals (Sadalla & Burroughs, 1981), their roles in society, or to express their political or ideological beliefs. Feasting, fasting, ritual preparation, and taboos or restrictions regarding the touching or eating of certain foods play crucial roles in religious and cultural practices and identities (Bynum, 1985; Douglas, 1966; Douglas, 1972; Fiddes, 1994; Levi-Strauss, 1966; Levi-Strauss, 1970). Giving or sharing food with others is considered crucial to creating and maintaining bonds between people (Miller, Rozin, & Fiske, 1998).

In studies examining consumer perceptions of new food technologies, perceived costs and benefits (Frewer, Scholderer, & Lambert, 2003; Ronteltap, van Trijp, Renes, & Frewer, 2007), the technology used (Lahteenmaki, Lyly, & Urala, 2007), the manufacturing process involved (Caporale & Monteleone, 2004), and issues of morality, democracy, and uncertainty (Brown & Ping, 2003; Hallman, Adelaja, Schilling, & Lang, 2002; Siegrist, 2000) have all been shown to be major factors in determining consumer acceptance. Critically, however, the public perception of new food technologies is often exacerbated by the mental models consumers have about the new technology. For example, the lack of consumer acceptance of food irradiation technology is often blamed, in part, on the public's inability to separate the concept of irradiation from that of radiation. The negative affective responses many have

Nucci, M.L., & Hallman, W.K. (2012). *Mork and Mindy,* canola oil and mustard gas: The dilemma of scientific illiteracy in decisions about food and health. In J. Goodwin (Ed.), *Between scientists & citizens: Proceedings of a conference at Iowa State University, June 1-2, 2012* (pp. 307-314). Ames, IA: Great Plains Society for the Study of Argumentation. Copyright © 2012 the author(s).

towards radiation, and specifically towards the thought of food potentially contaminated by radiation, led many consumers to reject the idea of irradiated foods (Resurreccion, Galvez, Fletcher, & Misra, 1995). Similarly, although the use of carbon monoxide in modified atmosphere meat and seafood packaging is recognized as safe by the FDA, many consumers perceived it as an unacceptable practice not only as it might mask food spoilage (Boyle, 2006; Weiss, 2006) but because of the knowledge that carbon monoxide is a poisonous and deadly gas (Health Sciences Institute, 2012).

2. THE PUBLIC AND FOOD: KNOWLEDGE AND PERCEPTIONS

In a series of audience studies over the last fifteen years, we have examined public perceptions of a variety of food-related issues, including genetic modification (GM), animal cloning, food safety, health-related food claims and nanotechnology. In 2004, when presented with stories about GM food taken from the media, the majority of those questioned found every story to be somewhat believable. This included two stories relating false information that had been circulated by the media and the Internet: that people had allergic reactions to GM foods, and that a large fast-food chain was selling chicken products "so altered by genetic modification that they can't be called 'chicken' anymore" (Hallman, Hebden, Cuite, Aquino, & Lang, 2004, p. 6).

When asked what ideas or concepts came to mind when they heard the terms genetic modification, genetic engineering or biotechnology, the term genetic modification yielded images of Frankenstein, test-tube babies, mutants or monsters. Genetic engineering evoked references to sheep, lambs or names that rhymed with Dolly, the first cloned sheep (e.g., Polly, Molly, Golly). Biotechnology was associated with new medicines, new foods, the future or progress. Of all the terms, biotechnology was linked most to science terms, such as test-tubes, laboratories, DNA or chemicals (Hallman, Adelaja, Schilling, & Lang, 2002).

When opinion leaders were asked about their understanding of animal cloning, many were unable to distinguish between cloning and genetic modification. Initial responses focused more on whether they (or others) thought it was a good or bad idea rather than describing what cloning is and how it is accomplished. In spite of the lack of knowledge about the science, they were more interested in questions such as who is doing cloning, what are the goals, what is the current status of the research, are the cloned animals normal, who is regulating the technology and is the technology safe, and not on the science behind animal cloning (Hallman & Condry, 2006).

In 2010 a series of interviews assessed consumers' perceptions and acceptance/rejection of nanotechnology, endeavoring to mimic the path towards decision making about an unknown field: exposure to the subject, information gain about the subject, and access to the subject (through descriptions of hypothetical nanotechnology-enabled food products). Consistent with previous studies of consumer's knowledge of nanotechnology (Cobb & Macoubrie, 2004; Gaskell, Ten Eyck, Jackson, & Veltri, 2005; Lee, Scheufele, & Lewenstein, 2005; Macoubrie, 2005; Vandermoere, Blanchemanche, Bieberstein, Marette, & Roosen, 2009) we found that very few of the participants reported knowing anything about nanotechnology prior to reading the NNI brochure, noting comments such as they "honestly can't think of anything," "don't know technology," "it doesn't ring a bell," "I've heard of it but I really don't know anything. I'm sure someone will go 'nano nano,'" or "this would be something I couldn't figure out and understand. I'm not technologically savvy."

When asked what came to mind when they heard the word 'nanotechnology,' participants mentioned technologies such as computers, IPods, chips, junk drives, or lasers; medical therapies such as miniature cameras for in vitro examination and monitoring or repairing human cells, and popular culture, including *Mork and Mindy* ("nanu nanu": the character Mork used the phrase *nanu nanu* to say hello), James Bond, and *Stargate*. Only three participants linked nanotechnology to food and food preparation. One participant felt that nanotechnology was used for "Instant fractions of a second, computer-generated devices to help make decisions about food quality or food analysis," while another commented that it had something to do with food composition. Referencing their mention of *Mork and Mindy*, the third participant commented that "I think of a planet like ours with fields of vegetables really, and—something flat like a satellite flying over and just, like, getting images of these elements that, whatever, vegetables or what have you." Notably, later in the interview, two of the three participants acknowledged that they mentioned food in reference to their description of nanotechnology because they were influenced by the flyer used to recruit participants that specifically mentioned that the study concerned food and new food technologies.

Many knew that the term nanotechnology was somehow linked to concepts of size ("Because I just think of something little when you say 'nano.' I think my kids have something that's 'nano' and its small"; "Well, I don't know much about it, but 'nano' usually means something small"; "The smaller it is, the better"; "Very small but intelligent"), often associating it with the benefits of miniaturization. Participants commented that "I guess it's compressing more and more and smaller and smaller and smaller"; "it can do all the same stuff, but it comes in a much smaller package, like almost like a fifth of the size or less"; "Nanotechnology is just making things smaller"; and "Assuming I'm correct about the miniaturization, then I would say that it's—you know, an evolution of an existing technology that has, you know, somehow become miniaturized."

However, no participant was able to describe the key attributes of nanotechnology identified in most authoritative definitions; that is, the ability to understand and manipulate matter at the nanoscale, and the capability to generate and make use of the novel properties of materials at the molecular level.[1] When asked specifically about familiar products that currently used nanotechnology, very few were able to provide any examples; the majority could not name *any* products. Categories of products thought to incorporate nanotechnology included clothing (shirts), electronic (microwaves, mp3 players, computer chips, cell phones, video games, hard drives, computers, iPods), medical (miniature cameras, pharmaceuticals) or industrial applications (robotics, carbon fiber tubes) were mentioned by several participants.

Participants had more difficulty when asked about the connection between nanotechnology and food, often struggling to come up with a linkage. Several mentioned processed, engineered or efficient food as an outcome of using nanotechnology, while others felt that the use of nanotechnology was associated with healthy foods, making foods better, downsizing meals, nutraceuticals, or low calorie foods. Focusing on the technology aspect, a

---

[1] Nanotechnology has multiple definitions, including: "Nanotechnology is the engineering of functional systems at the molecular scale" (Center for Responsible Nanotechnology, 2008); "Nanotechnology is the understanding and control of matter at dimensions between approximately 1 and 100 nanometers, where unique phenomena enable novel applications" (National Nanotechnology Initiative, 2011a); "Nanotechnology allows scientists to create, explore, and manipulate materials measured in nanometers (billionths of a meter). Such materials can have chemical, physical, and biological properties that differ from those of their larger counterparts" (Food and Drug Administration, 2010).

number of participants felt that nanotechnology would be connected to food through new or novel cooking technologies, microwaves or cooking equipment; or improvements in agricultural production through increased yield, hybridization of plants or animals, food safety/food quality monitoring, prevention of illness in animals or replacement of pesticides. Quick meals, futuristic foods, unhealthy/processed foods, food production, or food safety were each mentioned by a single participant. One participant commented:

> I really doubt there would be too much nanotechnology in the food we actually eat. That I think would be pushing the boundaries a little. . . . If people don't know what's inside their food—you know, that's technically, I think, breaking the law.

After reviewing the NNI brochure, slightly less than half of the participants used terms about size (small, minimize, shrink or miniaturization), instead mentioning green concepts related to cleaning up the environment or nature, or concepts such as cost-effectiveness, improving products, greater efficiency, better solutions, better materials, higher standards, stronger, lighter, or safer. One participant noted that there was "not much about food" in the brochure. In fact, there were only three references to food/water applications of nanotechnology in the brochure: "Cosmetics and food producers are 'nano-sizing' some ingredients, claiming that improves their effectiveness" (National Nanotechnology Initiative, 2011b, p. 4); "which has been shown to neutralize bacteria, including *E.coli*, in water" (National Nanotechnology Initiative, 2011b, p. 6); and "low cost technology for cleaning arsenic from drinking water" (National Nanotechnology Initiative, 2011b, p. 8).

It was not surprising therefore, that there was no significant increase in the number of participants who mentioned associations between food and nanotechnology after reading. Following on the concepts of efficiency and cost-effectiveness concepts prevalent in the brochure, participants noted that nanotechnology was linked to food through time and size concepts such as growing food faster or easier, growing bigger fruits and vegetables, or increasing yield, concepts of purification, or detection or prevention of contamination. Other participants linked nanotechnology to food through food engineering (taste/color/quality and creating new foods), production of healthier food, new and improved cooking technologies and packaging.

Significantly, before reading the brochure, nearly all of the associations that participants mentioned with regard to nanotechnology were either positive or neutral. Only one individual commented in the negative, saying, "Because I don't want to think about eating technology. I mean, I think it should be a choice if you're going to put something in you." After reading the brochure, only about a third raised unprompted concerns about the potential health impacts of nanotechnology; primarily concerns about side effects, long-term impacts, and overall safety. Moreover, the percentage of unprompted statements expressing concern was essentially unchanged after tasting the food products. However, when prompted by specific questions as to whether the participant had any health, religious, ethical/moral, environmental or other concerns about nanotechnology and food, 84% of participants noted issues with the application of the technology to food products.

In current studies, consumers are being asked about their use and knowledge of health claims on food. Single-gender focus groups showed that regardless of gender,[2] consumers'

---

[2] This is in spite of the fact that food knowledge and food purchase are usually the responsibility of the female partner in a relationship.

determination that a series of food products (generic and brand samples of tomato sauce, nuts, and green tea), were healthy was as much a function of nutritional factors, as issues including packaging (canned foods not healthy), manufacturer (store brands not as good for you), label aesthetics (Asian theme connotes longevity and better health; tomatoes didn't look healthy; colors clash), cost (the more expensive product was healthier), cultural linkages ("Arthur Godfrey drank tea," "If Joe Torre drinks it, it has to be good."), as was incorrect knowledge (green tea is a potent anti-cancer food, and for those who drank it regularly, they drank it in spite of disliking the taste; sea salt is better than table salt) about the food.

When specifically asked about qualified health claims[3] consumers were often confused by the content of the claim. In some cases, they said would go against FDA recommendations if they had other evidence. Incorrect knowledge about health claims was not uncommon, such as assuming that canola oil was toxic as it was related to the chemical weapon, mustard gas[4]; or a lack of information about a health claim which is interpreted as the claim possibly being dangerous ("If I don't know what it is, it can't be effective. It could be poison."). Concerns about the amount of information ("Personally, if there is a lot of information, I skip it.") or the language used (layman's language versus scientific terms) would lead consumers to avoid foods with health benefits.

3. CONCLUSIONS

Over the past decade our research has consistently demonstrated that a lack of familiarity—knowledge and mis-knowledge—plays an important role in public perceptions and attitudes about food decision, and that the consumer may often be laboring under false knowledge or false constructs in decision making. For most Americans, television is the primary source of information for science and technology (National Science Board, 2012) as well as food and nutrition (American Dietetic Association, 2002; Hoban & Kendall, 1993; IFIC, 2005; Verbeke, 2005). However, our team and others have demonstrated a lack of coverage and content in television information about science (Nucci, Cuite, & Hallman, 2009; Nucci & Kubey, 2007; Project for Excellence in Journalism, 2007[5]).

Our research has pointed out that relevancy, rhetoric, linkage to false constructs ("irrational thinking," Valdecasas & Correas, 2010), the need to tailor messages to specific audiences and a range of educational levels are all key to learning about food. Differences in age, education, gender (Hallman, Hebden, Aquino, Cuite, & Lang, 2003), ethnicity, religion, and trust in scientists, corporations and government (Hallman, Adelaja, Schilling, & Lang, 2002; Siegrist, 1999; Siegrist, 2000) are all important factors in decision making about science. Whether it is the lack of knowledge or irrational knowledge, such as conflating mustard gas with canola oil, or "nanu nanu" with nanotechnology, heuristics related to other technologies

---

[3] "Health claims characterize a relationship between a substance (specific food or food component) and a disease or health-related condition (see 21 CFR 101.14). Both elements of 1) a substance and 2) a disease are present in a health claim" (Food and Drug Administration, 2011).

[4] Canola oil comes from either the rapeseed or mustard plant. Mustard gas, so-called because of its yellow color and an aroma like mustard, is sulfur mustard (Canola Council of Canada, 2007). See also http://www.snopes.com/medical/toxins/canola.asp

[5] In 2007 science stories accounted for only 1% of the total news time on the morning news television network shows and only 2% of news time on the evening television network news shows (Project for Excellence in Journalism, 2007).

or other knowledges must be carefully considered. As in the case with food irradiation, public opinion can as easily be likely to be based on objective science as on negative constructs, but is less easily displaced when science is considered difficult or confusing.

ACKNOWLEDGEMENTS: Research described here was funded by Grant 2008-01415 to the Food Policy Institute, Rutgers, the State University of New Jersey, from the Cooperative State Research, Education, and Extension Service of the United States Department of Agriculture (USDA): *Food Nanotechnology: Understanding the Parameters of Consumer Acceptance*, Dr. William K. Hallman, principal investigator; Grant 2002-52100-11203 from the U.S. Department of Agriculture (USDA), under the Initiative for the Future of Agricultural Food Systems (IFAFS): *Evaluating Consumer Acceptance of Food Biotechnology in the United States*, Dr. William K. Hallman, principal investigator; and Grant 66487 from the Robert Wood Johnson Foundation: *The Diet-Health Nexus: Communicating Emerging Evidence*, Dr. William K. Hallman, principal investigator. The opinions expressed are those of the authors and do not necessarily reflect official positions or policies of the USDA, the New Jersey Agricultural Experiment Station, the Robert Wood Johnson Foundation, or the Food Policy Institute, Rutgers, the State University of NJ.

REFERENCES

Boyle, T. (2006, February 21). Groups protest use of carbon monoxide in meat packaging. *USA Today*. Retrieved from http://www.usatoday.com/news/health/2006-02-21-carbon-monoxide-meat_x.htm

Bynum, C. W. (1985). Fast, feast, & flesh: The religious significance of food to medieval women. *Representations, 11*, 1–25.

Brown, J. L., & Ping, Y. (2003). Consumer perception of risk associated with eating genetically engineered (GE) soybeans is less in the presence of a perceived consumer benefit. *Journal of the American Dietetic Association, 103*(2), 208–214.

Canola Council of Canada. (2007). What is canola oil? Retrieved from http://canolainfo.org/canola/index.php

Caporale, G., & Monteleone, E. (2004). Influence of information about manufacturing process on beer acceptability. *Food Quality and Preference, 15*, 271–278.

Center for Responsible Nanotechnology. (2008). What is nanotechnology? Retrieved from http://www.crnano.org/whatis.htm

Douglas, M. (1966). *Purity and danger*. London: Routledge.

Douglas, M. (1972). Deciphering a meal. *Daedalus, 10*, 61–81.

Fenichel, M., & Schweingruber, H. A. (2010). *Surrounded by science: Learning science in informal environments*. Board on Science Education, Center for Education, Division of Behavioral and Social Sciences and Education. Washington, DC: The National Academies Press.

Fiddes, N. (1994). Social aspects of meat eating. *Proceedings of the Nutrition Society, 53*(2), 271–279.

Food and Drug Administration. (2010). Nanotechnology. Retrieved from http://www.fda.gov/ScienceResearch/SpecialTopics/Nanotechnology/default.htm

Food and Drug Administration. (2011). Guidance for industry: FDA's implementation of "qualified health claims": Questions and answers; final guidance. Retrieved from http://www.fda.gov/Food/GuidanceComplianceRegulatoryInformation/GuidanceDocuments/FoodLabelingNutrition/ucm053843.htm

Frewer, L., Scholderer, J., & Lambert, N. (2003). Consumer acceptance of functional foods: Issues for the future. *British Food Journal, 105*, 714–731.

Gaskell, G., Allum, N. & Stares, S. (2003). Europeans and biotechnology in 2002: Eurobarometer 58.0. Brussels: European Commission.

Gaskell, G., Ten Eyck, T., Jackson, J., & Veltri, G. (2005). Imagining nanotechnology: Cultural support for technological innovation in Europe and the United States. *Public Understanding of Science, 14*(1), 81–90.

Hallman, W. K. (2008). Communicating about microbial risks in foods. In D. W. Schaffner (Ed.), *Microbial Risk Analysis of Foods* (pp. 205–206). Washington, DC: American Society for Microbiology (ASM) Press.

Hallman, W. K., Adelaja, A. O., Schilling, B. J., & Lang, J. (2002). *Public perceptions of genetically modified foods: Americans know not what they eat* (Food Policy Institute Research Report No. RR-0302-001). New Brunswick, New Jersey: Rutgers, the State University of New Jersey, Food Policy Institute.

Hallman, W. K., & Condry, S. C. (2006). *Public opinion and media coverage of animal cloning and the food supply* (Food Policy Institute Research Report No. RR-1106-011). New Brunswick, New Jersey: Rutgers, the State University of New Jersey, Food Policy Institute.

Hallman, W. K., Cuite, C. L., & Hooker, N. H. (2009). *Consumer responses to food recalls: 2008 National Survey Report* (Food Policy Institute Research Report No. RR-0109-018). New Brunswick, New Jersey: Rutgers, the State University of New Jersey, Food Policy Institute.

Hallman, W. K., Hebden, W., Aquino, H., Cuite, C., & Lang, J. (2003). *Public perceptions of genetically modified foods: A national study of American knowledge and opinion* (Food Policy Institute Research Report No. RR-1003-004). New Brunswick, New Jersey: Rutgers, the State University of New Jersey, Food Policy Institute.

Hallman, W. K., Hebden, W. C., Cuite, C. L., Aquino, H. L., & Lang, J. T. (2004). *Americans and GM food: Knowledge, opinion & interest in 2004* (Food Policy Institute Research Report No. RR-1104-007). New Brunswick, New Jersey: Rutgers, the State University of New Jersey, Food Policy Institute.

Health Sciences Institute (2012). *Carbon monoxide as a meat preservative.* Retrieved from http://hsionline.com/2006/09/05/carbon-monoxide-as-a-meat-preservative/

Hoban, T. J., & Kendall, P. A. (1993). *Consumer attitudes about food biotechnology.* Raleigh, NC: North Carolina Cooperative Extension Service.

Lahteenmaki, L., Lyly, M., & Urala, N. (2007). Consumer attitudes towards functional foods. In L. Frewer & H. van Trijp (Eds.), *Understanding consumers of food products* (pp. 412-427). Cambridge, UK: Woodhead.

Levi-Strauss, C. (1966). The culinary triangle. *New Society, 166,* 937–940.

Levi-Strauss, C. (1970). *The raw and the cooked.* London: Cape.

Miller, L., Rozin, P., & Fiske, A. P. (1998). Food sharing and feeding another person suggest intimacy; two studies of American college students. *European Journal of Social Psychology, 28,* 423–436.

National Nanotechnology Initiative. (2011a). Nanotechnology: Big things from a tiny world. Retrieved from http://www.nano.gov/node/240

National Nanotechnology Initiative. (2011b). What is nanotechnology? Retrieved from http://www.nano.gov/nanotech-101/what

National Research Council. (2009). *Learning science in informal environments: People, places, and pursuits.* P. Bell, B. Lewenstein, A. W. Shouse, & M. A. Feder (Eds.). Committee on Learning Science in Informal Environments. Board on Science Education, Center for Education. Division of Behavioral and Social Sciences and Education. Washington, DC: The National Academies Press.

National Science Board. (2012). *Science and engineering indicators 2012.* Retrieved from http://www.nsf.gov/statistics/seind12/start.htm

Nucci, M. L., Cuite, C. L., & Hallman, W. K. (2009). When good food goes bad: Television network news and the spinach outbreak of 2006. *Science Communication, 31*(2), 238–265.

Nucci, M. L., & Kubey, R. (2007). "We begin tonight with fruits and vegetables": Genetically modified (GM) food on the evening news 1980–2003. *Science Communication, 29,* 147–176.

Onyango, B., Miljkovic, D., Hallman, W., Nganje, W., Condry, S., & Cuite, C. (2007, August). Food recalls and food safety perceptions: The September 2006 spinach recall case. Paper presented at the annual joint meetings of the American Agricultural Economics Association (AAEA), Western Agricultural Economics Association (WAEA), and Canadian Agricultural Economics Society (CAES), Portland, Oregon.

Project for Excellence in Journalism. (2007). *The state of the news media 2007: An annual report on American journalism.* Accessed from http://www.stateofthenewsmedia.org/2007/

Resurreccion, A. V. A., Galvez, F. C. F., Fletcher, S. M., & Misra, S. K. (1995). Consumer attitudes toward irradiated food: Results of a new study. *Journal of Food Protection, 58*(2), 193–196.

Sadalla, E. K., & Burroughs, W. J. (1981). Profiles in eating. *Psychology Today, 15*(10), 51–57.

Siegrist, M. (1999). A causal model explaining the perception and acceptance of gene technology. *Journal of Applied Social Psychology, 29*(10), 2093–2106.

Siegrist, M. (2000). The influence of trust and perceptions of risks and benefits on the acceptance of gene technology. *Risk Analysis, 20*(2), 195–204.

Valdecasas, A. G., & Correas, A. M. (2010). Science literacy and natural history museums. *Journal of Bioscience. 35,* 507–514.

Vandermoere, F., Blanchemanche, S., Bieberstein, A., Marette, S., & Roosen, J. (2009). The public understanding of nanotechnology in the food domain: The hidden role of views on science, technology and nature. *Public Understanding of Science, 1,* 1-12.

Verbeke, W. (2005). Agriculture and the food industry in the information age. *European Review of Agricultural Economics, 32,* 347–368.

Weiss, R. (2006, October 19). Religion a prominent cloned-food issue. *The Washington Post,* p. A9.

Stephen Jay Gould and *McLean v. Arkansas*: Scientific Expertise and the Nature of Science in American Culture 1980–1985

MYRNA PEREZ

History of Science
Harvard University
Science Center 371, Cambridge MA 02138
United States
mlperez@fas.harvard.edu

ABSTRACT: Stephen Jay Gould was a Harvard evolutionary biologist who became a prominent media figure and commentator on evolution through the 1970s–1990s. In 1981 Gould was called as an expert witness on behalf of the plaintiffs in the court case *McLean v. Arkansas*. The lawsuit filed against the state claimed that a state law which mandated the teaching of creation science in Arkansas public schools was unconstitutional because it violated the Establishment Clause of the US Constitution. Therefore it was crucial for the plaintiffs to establish that creation science was not really 'science' but in fact a religious belief. Gould's testimony along with other scientists, theologians and philosophers helped the plaintiffs win the case, which set a precedent for banning the teaching of creation science in public schools across the country.

KEYWORDS: scientific expertise, evolution and creationism, science and the law, New Left, New Right, Stephen Jay Gould.

1. INTRODUCTION

There is a long history of popular interest generated by evolutionary biology. From the speculations over the authorship of *The Vestiges of the Natural History of Creation* in 1844, to Darwin's "bulldog," T.H. Huxley, the public debates between Louis Agassiz and Asa Gray, and the Scopes "Monkey Trial" of 1925, the history of evolutionary biology is full of examples of evolution in the public frame (Shapin, 1990). Biologists have taken on the role of public intellectuals to comment on the relationship between scientific knowledge, religious belief and evolution, and human progress (Wigam, 1922). In the twentieth century, those involved in evolutionary discourse had various visions of this public role. In his 1936 novel *The Shape of Things to Come* H.G. Wells created a futuristic society in which biologists formed a more intelligent minority that used their knowledge of genetics to further the progress of society. Wells's good friend Julian Huxley capitalized on this idea and in newspapers, popular books and radio shows promoted an evolutionary ethic for social progress to a general audience from the 1920s to the 1940s (Huxley, 1964). By 1932 at the American Museum of Natural History, Henry Osborn designed displays intended to educate a general audience on the evolution of vertebrates, through visualizations of evolutionary trees and the famous African Mammal Hall (Rainger, 1991). The eugenics movement, the publicity surrounding the elucidation of the structure of DNA, the development of evolutionary psychology in the 1980s and 1990s, and the sequencing of the human genome in the 1990s and 2000s, further highlighted that 'evolution' and genetics were fundamental to questions of human nature and origins. Throughout these many episodes biologists, academics, politicians, and journalists

Perez, M. (2012). Stephen Jay Gould and *McLean v. Arkansas*: Scientific expertise and the nature of science in American culture 1980–1985. In J. Goodwin (Ed.), *Between scientists & citizens: Proceedings of a conference at Iowa State University, June 1-2, 2012* (pp. 315-323). Ames, IA: Great Plains Society for the Study of Argumentation. Copyright © 2012 the author(s).

alternatively proclaimed and questioned the efficacy of evolutionary processes to reveal human nature or characterize social progress.

This paper takes up the themes of this longer history while examining the interplay between evolutionary biology, popular science, and various publics in the context of late twentieth-century America. In my larger dissertation project I investigate Stephen Jay Gould's role as a public intellectual in the United States during the period 1974 to 2002. Gould was a Harvard evolutionary biologist and paleontologist with a notable scientific career, who became a prominent media commentator on evolution. His column "This View of Life" ran from 1974 to 2001 in *Natural History*, the magazine for the American Natural History Museum. Gould was known to a large readership through this column, and subsequent bound editions (including *Ever Since Darwin, The Panda's Thumb* and *The Flamingo's Smile*) (Gould, 1977, 1980, & 1987). Gould contributed enormously to the formation of the discipline of evolutionary paleobiology—his 1972 paper, "Punctuated equilibria: the tempo and mode of evolution reconsidered" co-authored with Niles Eldredge has been cited in the journal *Paleobiology* more than 3,000 times (Gould & Eldredge, 1972; Ruse & Sepkoski, 2010). In 1977 Gould published an influential paper with Richard Lewontin titled "The Spandrels of San Marcos and the Panglossian Paradigm," which argued that evolutionary biologists were mistaken to understand all biological features as atomized entities produced by natural selection (Gould & Lewontin, 1979). Gould himself argued that his 'popular' and 'professional' work existed in the same sphere (Allmon, 2009). In his later career this popular work would include involvement in public television programs such as *NOVA* and the *Children's Television Workshop,* and appearances on television news shows including *Charlie Rose, 60 Minutes* and *CNN Talkback Live.* Gould even appeared as a guest on *The Simpsons.* But Gould was also an accomplished and lauded academic scientist—over his career he received forty-four honorary degrees and sixty-six major fellowships, medals and awards. And in 2001 the Library of Congress honored Gould as one of their "Living Legends," as a figure who had "made significant contributions to America's diverse cultural, scientific and social heritage" (Living Legends, 2000). During his lifetime Gould was widely regarded as one of the most distinguished and well-known American scientists.

The rising fame and influence of public scientific intellectuals such as Gould reveals much about the role of scientific expertise in late twentieth-century American culture. Did American publics turn to science as a source of objective information that effectively explained the natural world? Did different publics utilize evolutionary knowledge as a normative narrative to construct personal ethics and order human society? In 1981 Gould was called as an expert witness on behalf of the plaintiffs in the court case *McLean v. Arkansas.* The paper explores these questions through an examination of the court documents, media coverage and Gould's correspondence with other biologists and fans of his popular column in *Natural History* magazine during and around the *McLean v. Arkansas* creation science trial. Ultimately I suggest that Gould came to have the legitimacy to authoritatively comment on the nature of science and the proper teaching of evolution by deploying both his status as a technical practitioner of biology as well as his identity as a popular science writer.

2. THE CASE

In December of 1981, Gould testified before the Arkansas state Supreme Court as an expert witness on behalf of the plaintiffs. The plaintiffs, a number of clergy members and educators, together with the ACLU and the pro bono assistance of an Arkansas law firm (Skadden, Arps, Slate Meagher & Flom), brought a suit against the state of Arkansas for its passage of Act 590, "The Balanced Treatment for Creation-Science and Evolution-Science Act" (LaFollette, 1983). Act 590 was a legislative measure introduced by creationist organizations which mandated equal time for creation-science along with evolutionary accounts for what the act termed 'origins' in public science classrooms. The plaintiffs argued that creation-science was in fact primarily motivated to encourage religion, and therefore the bill violated the establishment clause of the first amendment (Larson, 2003).

The trial lasted for two weeks, but Gould was so confident in victory for his side, that he wrote an op-editorial for the *New York Times* about the significance of the trial before the verdict was delivered. A letter to Tamar Jacobsen, the editor of the op-ed section, captured Gould's sentiments at that moment:

> But I feel so confident of [Judge Overton's] favorable ruling…, that I am submitting this Op-Ed comment now to minimize delays after the actual decision comes down. Of course, the piece will have to be scrapped if he rules against us (nearly inconceivable), or if he rules favorably but narrowly (also unlikely). So do read the ruling before running the piece, though I am confident that I have foreseen the actual course of events. (Gould, 1981e)

Overton did in fact rule in favor of the plaintiffs, and his decision, though it was only binding in one district of Arkansas, was influential in the 1987 US Supreme Court case which ruled the teaching of creation-science in public schools unconstitutional for the whole United States.[1] This case is part of a long list of trials that encapsulated the legal issues of the American evolution-creation controversy during the twentieth century. It has featured in the most important histories of the American creationist movement, including work by Ronald Numbers (2006), Edward Larson (2003), Dorothy Nelkin (1982) and Michael Ruse (2005). (Nelkin and Ruse both served as expert witnesses in the case itself). Additionally, the trial appeared in both national and regional newspapers of the day, as well as well as popular books documenting the creation-evolution debate in the intervening decades. With so much attention given to this trial in particular and to the controversy with creationism generally, my interest is to add a fresh perspective by focusing on Gould's identity and subsequent activity as an expert witness in the trial. I suggest there are two reasons that this trial deserves renewed attention.

First, because more concentrated attention needs to be paid as to how the creation controversy connects to the general discourse about the ethical responsibility of biologists to participate in the project of American society. The stakes in the McLean trial stand in direct contrast to another public controversy over biology that Gould had been involved in just a few years earlier. This was the sociobiology debate, which erupted out of Cambridge, Massachusetts in the mid 1970s and captured with it all the most potent issues of the day: racial tension, gender relations and the trustworthiness of science (Jumonville, 2002). Gould and other critics of sociobiology believed strongly that the existing conditions in American society were unfair, and more that they could be *changed.* They did not want a biology that described the universal reasons for human behavior, for human action and human social relations. They were mistrustful of biologists who claimed to speak for other groups and other categories. And they felt that that type of thinking only hurt social progress. There the stakes of

---

[1] The case, *Edward v. Aguillard* (1987) centered on a Louisiana 'equal time' law very similar to the Arkansas Act 590. The case went to the United States Supreme court, and ruled that 'equal time' laws were generally unconstitutional because of the violation of the establishment clause of the First Amendment.

the debate were in what way evolutionary scientists could ethically contribute to a progressive liberal society. Famously, Gould and Lewontin had very different views from E.O. Wilson (whose text *Sociobiology* sparked the public debate) as to how this was properly to be undertaken. However, they all share the presumption that evolutionary biology could positively shape American society. But by the time of the McLean trial, the stakes were quite different. With the emergence of creationism with the broader political context of the New Right and the rise of the moral majority, it was no longer simply a question of how evolutionary science was to participate in American society, but whether it could at all (Larson, 2003).

Secondly, this episode highlights and crystallizes the identities that make up Gould's persona as a "scientific expert." During the trial he was called upon as a professional scientist, an evolutionary popularizer and as an evolutionary historian. Each of these aspects of his identity were brought out explicitly in both his trial testimony and in his subsequent writing about his experience testifying. As a professional biologist he was asked to comment on what he saw as the misuse by creationists of his theory of punctuated equilibrium and his discussion of uniformitarianism. For instance, his article on "Punctuated Equilibrium" was entered into evidence: "Q. Your article from *Paleobiology* which is entitled "Punctuated Equilibria, the tempo and Mode of Evolution Reconsidered," do you plan to rely on this in your testimony? A. Yes. Well, I plan to use it" (Gould, 1981b, p. 176). Two of Gould's popular books, *Ever Since Darwin* and *The Panda's Thumb* were entered into the official court record along with his curriculum vitae as an acknowledgement of his expertise. And finally he was also called upon as a historian of science: "Q: You've been offered as an expert also, Doctor Gould, on the history of evolutionary theory or evolutionary thought. A: Yes" (Gould, 1981f, p. 621).

These three aspects of Gould's identity are clearly interrelated, but it is this third aspect that I focus on in this paper, because it was as an expert in the history of evolutionary theory that Gould was most directly able to comment on the state of evolutionary knowledge within the trial. It was also as an *historian* that he was made sense of the events of the trial to his professional colleagues and to his popular audience. He did this by setting the McLean trial into a larger historical narrative of the evolution-creation controversy in the twentieth century. Particularly he compared McLean to the Scopes Monkey Trial of 1925 both in his testimony and in his writing on the trial. Edward Larson, in his book *Trial and Error*, has pointed out that McLean had nothing like the "profound social experiences" of the *Scopes* "monkey trial" of 1925 (2003, p. 159). Indeed the McLean case relied on comparisons to *Scopes* in much of the press coverage to help generate popular interest in the 1980s press coverage of the trial. Thus, by comparing McLean to the Scopes trial, Gould was able to use history in order to create a rallying cry against the creationist movement both for his professional colleagues and for the readers of his popular writing.

3. SCOPES IN TESTIMONY AND WRITING

During his testimony Gould returned many times to the Scopes trial as a way to contextualize his experience as an expert witness. For instance, he called up on the 1925 trial as a way of arguing that the creationists' arguments were tired, worn-out and dogmatic in their unchanging nature.

> Q: I think you earlier stated that as far as you know, there is no new evidence and no new idea for creation science in the past one hundred years; is that true? Gould: I think I said since William

Jennings Bryan and the Scopes trial I have seen no new arguments from the creationists. (Gould, 1981f,, p. 629)

Similarly when asked to explain his views on the central issue of the trial—namely the appropriate use of education—he again used Scope as his reference.

"Gould: "Education, you know, means broadening, advancing, and if you limit a teacher to only one side of anything, the whole country will eventually have one thought, be one individual. I believe in teaching every aspect of every problem or theory." Q: Who is the source of that quote? A: John Thomas Scopes." (Gould, 1981f, p. 606)

Interestingly, Gould's use of a quote from John Scopes himself highlighted the *changes* since the 1925 Monkey Trial. The litigious situation had completely reversed—in 1925 it was the legality of evolution in the classroom that was under question. Almost 60 years later, the place of science in public schools had been secured, now it was religion that had to work on science's terms to enter into the public education space.

The Scopes trial was such a predominant theme for Gould in his preparation leading to the trial that he took a family trip to Dayton, Tennessee in order to visit the site of the Monkey Trial. The visit was featured in a piece, simply titled "A Visit to Dayton" for his monthly column in *Natural History* magazine. This photo, taken by his first wife, shows Gould walking away from the Dayton courthouse. The central message of Gould's essay was that American evolutionists and liberals had been sadly misinterpreting the events of the Scopes trial for the last five decades. Due to the popularity of fictional depictions of the trial (most notably the play and later film *Inherit the Wind*), many Americans had believed that the events of Scopes spelled the end of fundamentalism in American culture. (Larson, 2003 & 2008; Gould, 1981a) In "A visit to Dayton" Gould attempted to disabuse his readers of this comfortable notion, citing recent historical work on the subject; he argued that Scopes actually strengthened the fundamentalist movement—but that the Depression and the Second World War had merely diverted the movement's efforts away from the issue of evolution (Gould, 1981g). A resurgence of interest in the topic in the late 1970s, however, belied "any hope that the issues of Scopes' trial had been banished to the realm of nostalgic Americana have been swept aside by our current creationist resurgence—the climate that inspired my own detour across the Tennessee River" (Gould, 1981g, p. 9; see also Numbers, 2006).

Not only had Scopes not ended fundamentalism, according to Gould it halted the incorporation of evolution into high school biology textbooks. Gould made this point in another piece written for *Natural History*, by vividly weaving in his own personal narrative into the historical events:

Now, more than half a life later (I studied high school biology in 1956), I finally understand why Mrs. Blenderman had neglected the subject that so passionately interested me. I had been a victim of Scopes' ghost . . . Most people view the Scopes trial as a victory for evolution, if only because Paul Muni and Spencer Tracy served Clarence Darrow so well in theatrical and film versions of *Inherit the Wind*, and because the trial triggered an outpouring of popular literature by aggrieved and outraged evolutionists. Scopes' conviction had been a mere formality; the battle for evolution had been won in the court of public opinion. Would it were so. As several historians have shown, the Scopes trial was a rousing defeat. It abetted a growing fundamentalist movement and led directly to the dilution or elimination of evolution from all popular high school texts in the United States . . . The situation did not change until 1957, a year too late for me, when the Russian Sputnik provoked a searching inquiry into the shameful state of science education in America's high schools. (Gould, 1982, p. 5)

Gould did not confine his comments about McLean and its relevance to Scopes to his column for *Natural History* magazine. His most famous piece from the period, "Evolution as Fact and Theory," appeared in *Discover* Magazine in May of 1981. As in other instances, Gould chose to emphasize that the creationist movement had changed little, but had lay sleeping, gathering power against the progressive, liberal establishment.

> But nothing has changed; the creationists have presented not a single new fact or argument. Darrow and Bryan were at least more entertaining than we lesser antagonists today. The rise of creationism is politics, pure and simple; it represents one issue (and by no means the major concern of) the resurgent evangelical right. Arguments that seemed kooky just a decade ago have re-entered the mainstream. (Gould, 1981a, p. 34)

The piece was written before the trial, but appeared in a number of collected volumes after the McLean case. One was an edited volume by Ashley Montague, the British anthropologist and scientific activist. His 1984 volume, *Science and Creationism*, was a collection of essays by scientists in direct rebuttal to the claims of scientific creationism. Gould's essay was also featured in a volume *Speak Out Against the New Right* that explicitly connected the evolution-creation controversy to other political issues between the Left and the New Right (including abortion, the economy and gender roles). It was the language of Scopes in the *Discover* piece that once again allowed Gould to make the events of the McLean trial sensible. In all of Gould's writing in response to the trial, creationism figured as a kind of slumbering evil that had recently gained new blood. Gould gave his readers a sense that evolutionists have been asleep at the helm—that they had not been paying attention to what mattered. By relating McLean to the Scopes-era history, Gould cast the trial into a larger narrative about the contests for science and progressive liberalism in American society.

4. GOULD'S CALL TO ARMS

This perspective might be dismissed as a quirk of Gould's personality, as he was liable to relate current events to historical examples in many of his popular publications. However, in his correspondence, both in the incoming and outgoing letters there is a sense of a call to arms to a new professional duty—the campaign against the creationists. And it was the historical record that provided the framework for articulating this professional duty. One letter, to Edward Linenthal, a biology professor at Indiana University, captures this perspective:

> I am surprised that your biological colleagues have been so unhelpful. I had thought that the general awakening among biologists had occurred and much as we still don't know whether to laugh or cry, were at least committed to some activity against creationism. . . . With best wishes in a common fight." (Gould, 1981c, p. 1)

Additionally, many of Gould's exchanges in this period coincided with another important evolutionary event in 1982—the centennial of Darwin's death. Historians such as Betty Smocovitis (1999) have illustrated the importance of the 1959 Darwin centennial in the intellectual and institutional work that went into unifying the biological disciplines after the modern synthesis. In her article Smocovitis explores the role that the centennial took in shoring up biological identity, but also notes that there were hardly any historians involved in the proceedings (Smocovitis, 1999). The centennial celebrations of 1959 had been an opportunity for biologists to articulate the intellectual framework of the modern synthesis and secure their disciplinary identity. However, in 1982 many more historians were involved in the celebrations of Darwinian evolution. Due to this, the 1982 commemorative activities were also an opportunity to address concerns over creationists (Wassersug & Rose, 1984).

And for Gould this meant that much of his professional correspondence directly after the McLean trial was directly connected to 1982 centennial celebrations. One letter to Paul Kurtz December 21, 1981, a prominent American skeptic, explicitly tied these two themes together:

> I am, indeed, well aware—especially after a fascinating day spent testifying in Arkansas last week. Although the judge will not render his decision until later this week, I can state with confidence that we routed the creationists. This, of course, will not end the issue, and I applaud you for the topic of your Darwin centennial meeting. (Gould, 1981d)

Indeed, I suggest that the McLean trial, even more than the legal issues involved, helped to solidify for Gould and other evolutionary biologists an ethical responsibility to narrow the definition of science and publicly defeat the American creationist movement.

This new public responsibility was seen most clearly in the formation of a new committee within the professional academic organization, the Society for the Study of Evolution (SSE). An internal memo, sent to Gould and other prominent evolutionary biologists called for the formation of an 'education committee'—an internal group whose explicit purpose was to deal with creationists.

> The Society for the Study of Evolution has recently formed Education Committee. As you might suspect, this euphemism hides the fact that it is to deal with the current contention of the creationists. . . . There is ample evidence that our problems with the creationists, while presently acute are chronic. (SSE internal memo, 1982, p. 1)

This committee eventually became the National Center for Science Education (NCSE). It was, and remains, the most prominent anti-creationist organization in the United States. It began during the vents of the McLean trial as a way for professional biologists, including the very public Gould, to organize against creationism during the 1980s and 1990s.

5. CONCLUSION

There was an air of mobilization in the correspondence, publications and trial preparations by Gould and other evolutionists in the early 1980s. History had snuck up on evolutionary biologists, and they were going to band together to do something about it. The questions I have attempted to answer in this paper are: Why did it become a perceived public duty for Gould and for other evolutionary thinkers and biologists to publicly combat creationists? And how did controversy with creationists change the perceived ethical responsibilities for biologists to engage politically? Eventually I aim to suggest that public debate with creationists made political involvement much less professionally or personally problematic for evolutionists than it had been over topics such as sociobiology just a few years prior. And particularly, this trial meant that part of being an evolutionary biologist was to be public anti-creationist in a way that was not conceivable a decade earlier. I would like to end with a quote from one of the trial lawyers from the firm that led the McLean case. It nicely summarizes the terms on which scientific expertise was deployed in this context:

The issues presented in McLean, however, occasioned an interesting variation on the usual role of the expert. Particularly in the case of the science and education experts, the opinions and conclusions offered on behalf of the plaintiffs were not the opinions of experts applying their experience to a particular set of facts significant to the legal issues at trial, nor were they those of experts testifying about the absolute truth or falsity of any particular scientific fact or conclusion regarding the origin or development of the universe. Rather, those experts presented to the court the results of twenty years of the scientific and educational communities' evaluation of the scientific and educational claims of the creationists, set against one hundred years of experience since the Darwinian revolution. (LaFollette, 1983, p. 99)

REFERENCES

Allmon, W. D. (2009). The structure of Gould. In Allmon, W.D., Kelley, P.H., & Ross, R.M. (Eds.). *Stephen Jay Gould: Reflections on his view of life* (pp. 3-68). Oxford: Oxford University Press.

Gould, S. J., & Eldredge, N. (1972). Punctuated equilibria: an alternative to phyletic gradualism. *Models in Paleobiology,* 82–115.

Gould, S. J., & Lewontin, R. (1979). The spandrels of San Marco and the Panglossian paradigm: A critique of the adaptationist programme. *Proceedings of the Royal Society of London B, Biological Sciences*, 205(1161), 5.

Gould, S. J. (1977). *Ever since Darwin: Reflections in natural history.* New York, NY: W.W. Norton & Company.

Gould, S. J. (1980). *The panda's thumb: More reflections in natural history.* New York, NY: W.W. Norton & Company.

Gould, S. J. (1981a). Evolution as fact and theory. *Discover, 2*(5), 34–37.

Gould, S. J. (1981b.) Gould Deposition in McLean v. Arkansas Trial, 27 November 1981.

Gould, S. J. (1981c). Letter to Edward Linenthal March 1, 1982 in Box 109, Folder 1 in Stephen Jay Gould Papers, M143. Dept. of Special Collections, Stanford University Libraries. Stanford, Calif. University.

Gould, S. J. (1981d). Letter to Paul Kurtz December 21, 1981 in Box 109, Folder 1 in Stephen Jay Gould Papers, M1437. Dept. of Special Collections, Stanford University Libraries. Stanford, Calif. University.

Gould, S. J. (1981e). Letter to Tamar Jacobsen in Box 109, Folder 1 in Stephen Jay Gould Papers, M1437. Dept. of Special Collections, Stanford University Libraries. Stanford, Calif. University.

Gould, S. J. (1981f). Testimony in McLean v. Arkansas Trial.

Gould, S. J. (1981g). A visit to Dayton. *Natural History, 90*(10), *8–22.*

Gould, S. J. (1982). Moon, Mann and Otto. *Natural History, 91*(3), 4–10.

Gould, S. J. (1987). *The flamingo's smile: Reflections in natural history.* New York, NY: W.W. Norton & Company.

Huxley, T. H. (1964). *Essays of a humanist.* New York, NY: Harper and Row.

Jumonville, N. (2002). The cultural politics of the sociobiology debate. *Journal of the History of Biology, 35*(3), 569–593.

LaFollette, M. C. (1983). *Creationism, science, and the law: The Arkansas case.* Cambridge, MA: MIT Press.

Larson, E. J. (2003). *Trial and error: The American controversy over creation and evolution.* Oxford: Oxford University Press.

Larson, E. J. (2006). *Summer for the gods: The Scopes trial and America's continuing debate over science and religion.* Basic Books.

Lambert, F. (2010). *Religion in American politics: A short history.* New Jersey: Princeton University Press.

Living Legends. (2000). *Library of Congress Website.* Retrieved from www.loc.gov/about/awardshonors/livinglegends/bio/goulds.html

Montagu, A. (Ed.). (1984). *Science and creationism.* Oxford: Oxford University Press.

Nelkin, D. (1982). *The creation controversy: Science or scripture in schools.* New York, NY: W.W. Norton.

Numbers, R. L. (2006). *The creationists: From scientific creationism to intelligent design* (2nd ed.). Cambridge, MA: Harvard University Press.

Rainger, R. (1991). *An agenda for antiquity: Henry Fairfield Osborn and vertebrate paleontology at the American Museum of Natural History, 1890–1935.* Tuscaloosa, AL: University of Alabama Press.

Ruse, M. (2005). *The evolution-creation struggle.* Cambridge, MA: Harvard University Press.

Ruse, M., & Seposki, D. (2010). *The paleontological revolution: Essays on the growth of modern paleontology.* Chicago, IL: Chicago University Press.

Shapin, S. (1990). Science and the public. In R.C. Olby, G.N. Cantor, J.R.R. Christie, & MJS Hodge (Eds.), *Companion to the history of modern science* (pp. 990–1007). New York, NY: Routledge.

Smocovitis, V. B. (1999). The 1959 Darwin centennial celebration in America. *Osiris, 14*, 274–323.

SSE internal memo. (1982). In Stephen Jay Gould Papers, M1437. Dept. of Special Collections, Stanford University Libraries. Stanford, Calif., Calif. Library Archive.

Wassersug, R. J., & Rose, M.R. (1984). A reader's guide and retrospective to the 1982 Darwin centennial. *The Quarterly Review of Biology, 59*(4), 417–437.

Wiggam, A. E. (1922). The new decalogue of science: An open letter from the biologist to the statesman. *The Century Magazine. 103*(5), 643-50.

Assessing Bias Charges against Collaborative Expertise, with an Application to the IPCC

WILLIAM REHG

Department of Philosophy
Saint Louis University
3800 Lindell Blvd., St. Louis, MO 63108-3414
USA
rehgsp@slu.edu

ABSTRACT: In controversial science-intensive policy debates, charges of expert bias often arise. How does one sort out such charges—especially when expertise is interdisciplinary and collaborative? In this paper I address the problem of collaborative expert bias at the level of group process. Identification of bias is complicated not only by interdisciplinary complexity, but also by the ubiquity of bias, some of which can be fruitful for scientific discovery. Drawing on the Intergovernmental Panel on Climate Change (IPCC) for illustration, I distinguish different kinds of group-level bias in the sciences and propose ways of identifying bad bias.

KEYWORDS: bias, expertise, collaboration, climate change, IPCC.

1. INTRODUCTION

When scientific experts find themselves entangled in controversial policy debates, charges of bias often arise. How does one sort out such charges—especially when expertise is interdisciplinary and collaborative? In such cases, the alleged bias often operates at the group level. The idea of group bias is hardly new. A substantial literature on group bias now encompasses a range of disciplines and approaches, including cognitive and social psychology, feminism, and science studies. Nonetheless, treatments of bias by informal logicians and critical thinking scholars have focused mainly on bias as a problem for individual arguers. To be sure, scholars of argument have long recognized how group membership can bias an arguer in a particular direction. But the "bearer" of bias remains the individual. Policy-relevant expertise raises the question of bias at the level of the group itself.

This paper addresses the problem of collaborative expert bias in contexts that meet three conditions: (i) a group of experts deliver advice on policy-relevant natural-scientific or health questions that admit of objectively correct and incorrect answers; (ii) reliable answers to such questions require the input of multiple scientific disciplines; (iii) the alleged bias at issue lies at the level of group process. Assessing group-level bias charges becomes especially important in such contexts, given that complex collaborative expert arguments outstrip the competence of any single person to assess. Thus our confidence in the quality of expert advice rests largely on the quality of the process itself.

But how should we make that assessment, given the ubiquity of bias and complexity of technical content? For bias occurs in different forms, and not all bias is bad. After some initial orientation, I distinguish two kinds of group-level bias (secs. 2, 3). I then describe two apparently ubiquitous biases in science: confirmation bias and "preference" bias. Because these can be fruitful for research, it is important to develop criteria for sorting good from bad cases

Rehg, W. (2012). Assessing bias charges against collaborative expertise, with an application to the IPCC. In J. Goodwin (Ed.), *Between scientists & citizens: Proceedings of a conference at Iowa State University, June 1-2, 2012* (pp. 325-334). Ames, IA: Great Plains Society for the Study of Argumentation. Copyright © 2012 the author(s).

(sec. 4). To illustrate these issues, I refer throughout to the Intergovernmental Panel on Climate Change (IPCC), and I close with a particular bias charge against that panel (sec. 5).

2. ORIENTATION

The literature on group bias is both sizable and multi-disciplinary, and so some initial sorting is necessary to delimit the scope of my analysis. Notice first the difference between two uses of the word "bias," which we might label "technical" and "partisan" (cf. Walton, 1999, p. 224ff). In some fields the "bias" label refers to technical probabilistic and statistical errors of reasoning. Statistical biases involve errors in data sampling. Cognitive psychologists use the term "cognitive bias" for certain types of widespread errors in human reasoning, such as the "base-rate fallacy," and errors in conditional reasoning (Evans & Over, 1996). Statistical and cognitive biases reflect tendencies in human cognition that make certain kinds of technical mistakes likely. Understood merely as a tendency to mistaken reasoning, or as the mistake itself, such biases do not stem from a flawed character or a partisan commitment to defending one's position at all costs. In this paper I am not so concerned with basic technical mistakes, as with bias in the partisan sense. (To be sure, a partisan bias in disposition can lead to sloppy reasoning, thus to technical biases in content.)

Whereas technical bias appears primarily in the content of an argument, partisan bias can appear in both content and process. At the level of *content*, Douglas Walton defines partisan bias "simply as a one-sided argument—an argument that lacks the balance necessary for it to be two-sided" (Walton, 1999, p. 76). This definition is helpful for its breadth, capturing a wide range of cases of bias. Indeed, partisanship implies a kind of lack of balance, which might or might not be appropriate. In agonistic settings, such as law courts, we expect each side to produce an unbalanced argument for its position. Balance is achieved through the division of labor. Thus a biased argument is not necessarily a bad argument in context.

The level of *process* potentially includes everything in the social context that affects the quality of argument-making: the participants' background and dispositions, the formal and informal procedures they employ, how their personalities interact, the institutional setting, and so on. I am concerned here with group settings, more precisely with "transactional contexts," that is, the contexts in which members of a group are engaged in person-to-person exchange of arguments. For present purposes, two main aspects of transactional contexts are crucial for the analysis of bias: the participants' dispositions and the structure of their transaction.

The literature on bias tends to identify partisan bias in process with the arguer's disposition. Insofar as argument-making presupposes a personal disposition to favor one position over another, hence a kind of partisanship, argumentation seems to require bias. To that extent, we may regard bias as normal or good (Blair, 1988). Walton concurs: "Bias or dialectical slanting in argumentation . . . is not inherently bad"; rather, "if you equate it with advocacy, partisanship, or point of view in argumentation, it may be, in many instances, a good thing" (Walton, 1999, p. xviii). Bad bias occurs when the commitment necessary to engaged argument-making undermines the quality of the transaction or the argument content in context. Although one might associate bias with the arguer's close-mindedness or lack of impartiality, in some contexts these do not undermine the transactional goal. Thus, a public debate between two close-minded opponents might prove helpful for an audience trying to make up its mind.

3. GROUP-LEVEL PROCESS BIASES

At the group level we can identity distinctively group-level dispositions and procedures as potential sources of bias. To my knowledge, neither of these has received much attention from argumentation theorists.

3.1 Group Dispositions

The prime example of a group-level disposition in the psychological sense is "groupthink." This phenomenon, identified by Irving Janis (1972), has received considerable attention from social psychologists, though doubts remain about its empirical support, scope, and usefulness as an explanatory model (Esser, 1998; Turner & Pratkanis, 1998). According to Baron's "ubiquity model," however, groupthink symptoms—suppression of dissent in groups, tendency to conformity (polarization of attitudes in a group), self-censorship, and the illusion of consensus—typically arise in groups that meet three antecedent conditions: members identify with their group; group discussion gives rise to attitudes that have a normative character for members; and the "situational self-efficacy" of individual members—their confidence in dealing with group tasks effectively—is low (Baron, 2005, pp. 238–244).

Groupthink presents a potentially serious problem in science. Scientific specialties exhibit at least two of Baron's antecedent conditions. Both empirically informed social studies of science, as well as research on collective intentionality, suggest that (a) scientific cooperation depends on some level of group identification (understanding oneself as a competent member of a discipline), and (b) scientific discussion presupposes and generates empirical and theoretical results that count as normative standards for competent reasoning.[1] When the group faces problems that exhibit the daunting complexity of climate science, we might expect the third condition to appear as well, low situational self-efficacy.

As it turns out, recent IPCC procedural revisions reveal a concern for groupthink in author teams:

> Be aware of a tendency for a group to converge on an expressed view and become overconfident in it. Views and estimates can also become anchored on previous versions or values to a greater extent than is justified. One possible way to avoid this would be to ask each member of the author team to write down his or her individual assessments of the level of uncertainty before entering into a group discussion. If this is not done before group discussion, important views may be inadequately discussed and assessed ranges of uncertainty may be overly narrow. Recognize when individual views are adjusting as a result of group interactions and allow adequate time for such changes in viewpoint to be reviewed. (IPCC, 2010, Appendix 4, Treatment of Uncertainty, no. 3, draft)

This statement responds to recommendations from the InterAcademy Council (IAC, 2010) regarding the estimation of uncertainty—more on which below.

3.2 Procedures

In considering group-level sources of expert bias, we must also look to the design and execution of committee procedures. Agencies responsible for expert advice pay close attention

---

[1]   For relevant work in the social studies of science, see Kuhn (1996); Ziman (1968); for worries about groupthink based on collective intentionality, see Tollefsen (2006).

to committee procedures, whose design and psychology is now the subject of a growing empirical literature (e.g., Bijker, Bal, & Hendriks, 2009; Hilgartner, 2000; Jasanoff, 2005). What we want, presumably, are committees that

- do not dissolve into contention but reach robust conclusions,
- which are scientifically accurate or reliable,
- sufficiently clear to guide policymakers (rather than so hedged with second-guessing as to provide no support for decisions),
- democratically accountable,
- yet not unduly politicized.[2]

Behind the last condition lies the idea that reliable science for policy should not substantively depend on political partisanship, though it might have partisan political *implications*. After all, the natural and health sciences—my focus in this paper—reach conclusions about an objective reality, knowledge of which is independent of partisan allegiance, right?

Not quite. In the health sciences, conclusions can hardly avoid cultural assumptions about human well-being. Even when dealing with the non-human world, descriptive categories and risk estimates often involve cultural values; when such values are controversial, politicization is practically unavoidable (Pielke, 2007). In such contexts, we do better to understand "unduly politicized" conclusions as conclusions that do not suffer from bad bias— committee dispositions and procedures foster a balanced assessment of the available science, which takes into account different cultural perspectives and values as appropriate. For that to occur, process designs must succeed in combining rigorous technical analysis with deliberative forums that include the full range of stakeholders, expert and lay.[3]

In any case, bad procedural biases can arise in a number of ways, all of which involve a lack of appropriate balance in committee deliberation. A committee might lack representation from relevant disciplines for the policy-relevant topic. Or it might lack the wider sources of input for reflecting on social and moral values that condition scientific reasoning and judgment at different points (Douglas, 2009).

IPCC procedures incorporate a number of mechanisms that foster wider input and deliberation. The IPCC is designed to include not only all the relevant disciplines but also scientists from different regions of the world. Moreover, government representatives as well as other stakeholders (business interests, NGOs, etc.) are involved at certain stages of the multi-step report process. That process begins with the selection of authors and determination of report scope; moves through the various stages of drafting, reviewing, and revising of reports; and concludes with plenary sessions in which reports are finally accepted, which is to say: accepted both by scientists and governments (see IPCC, 2008; Bolin, 2007).

---

[2] This list expands on Douglas (2009), p. 134; for a history of science advising, see Douglas (2009), chaps. 2, 7–8.

[3] On the analytic-deliberative approach to science advising, see Douglas (2009), chap. 8; note that inclusion need not require sunshine rules for all expert discussion in committee; see Rehg (2009), chap. 8; Bijker et al., (2009).

4. BIASES INHERENT IN SCIENTIFIC ARGUMENTATION

We should not be surprised if scientific argumentation, like argumentation in general, is inherently biased in certain ways. Here I examine two such biases: confirmation bias and "preference bias." These appear ubiquitous, but they are not always bad. Consequently, critical assessment of scientific bias requires an eye for the bad cases. Both have their most obvious source in dispositions, but at the group level they can also arise through flawed procedures.

4.1 Confirmation Bias.

Psychological research shows that human beings are inherently prone to favor considerations and evidence that support their position, and to disregard or discount counter-evidence (Mercier & Sperber, 2011; for doubts, see Evans & Over, 1996, pp. 103–109). Up to a point, this kind of bias might be fruitful for scientific discovery. As critics of Popper's falsificationism have pointed out, great discoveries can require perseverance in the face of apparently contrary evidence. To put it in Kuhn's terms, good science requires an ability to live with a certain level of "anomalies" in one's approach, on the assumption that one will eventually be able to resolve inconsistencies between one's hypothesis and other considerations. Contrary empirical measurements, for example, may simply prove mistaken or insufficiently accurate. Confirmation bias is bad, however, when it leads a researcher to cling to a favored hypothesis beyond all reason, or even worse, when it becomes so channeled and rampant in the community that progress falters.

Mercier and Sperber (2011) suggest that confirmation bias has an evolutionary basis. If human reasoning evolved for its use in social processes of argumentation, then we should not be surprised by widespread confirmation bias in individual reasoning. Though often a weakness at the individual level, confirmation bias can strengthen social argument-making inasmuch as (a) each participant strives to convince others by putting forward the best case for his or her own argument, but (b) by doing this together, participants provide a natural check against each other's confirmation bias.

As with bias in general, then, context matters in distinguishing good from bad confirmation bias. This is especially true for policy-relevant expertise. Because expert knowledge confers confidence in one's judgments, expertise can actually accentuate confirmation bias (Mercier, 2011). Even if such confidence can foster discovery in research contexts, it poses a serious problem in policy contexts. Experts, after all, are charged with assessing the overall state of the literature, not pursuing their own research agenda. Thus committee venues are especially important for countering such bias. To do so, the committee must be sufficiently heterogeneous with respect to viewpoints on the issue at hand. Thus composition of the committee and the provision for outside review play important roles in countering confirmation bias. At least at first glance, the size of author teams responsible for IPCC chapters, along with review procedures, should provide some check on expert confirmation bias at the level of individual reasoning.

4.2 Preference (Inductive-risk) Bias

Torsten Wilholt (2009) has argued that "preference bias" is ubiquitous in scientific research. Preference bias "occurs when a research result unduly reflects the researchers' preference for it over other possible results" (Wilholt, 2009, p. 92). Thus defined, preference bias coincides with bad partisan bias in general. However, Wilholt is after something more specific: the kind of bias that involves "tampering with the balance of inductive risk" (2009, p. 94). Thus the more accurate term is "inductive-risk bias."

In testing hypotheses, scientists must make value-laden judgments about the relative costs of two kinds of error: not accepting true hypotheses, and accepting false hypotheses. The widely used 95% confidence level reflects a risk-averse valuation: one would prefer to delay acceptance of a true hypothesis rather than advance a false claim. Decisions that flow from such value-judgments have consequences, not only for the researcher's career, but in some cases for groups affected by the social consequences of the decision—for example, the delay of new treatments for the critically ill. But, Wilholt maintains, there is no principled yet feasible method for impartially balancing the costs and risks involved in the two kinds of error. The reason is that we cannot distinguish honest disagreement in judgments of value from genuinely bad bias. Inductive-risk bias, then, is both ubiquitous and ambiguous—it "will simply be part of the scientific condition" (Wilholt, 2009, p. 95; also Douglas, 2009).

Inductive-risk bias can arise at various stages of research: in experimental design, interpretation of data, and communication of results. Biased experimental design, for example, makes it less likely that one will falsify one's favored hypothesis. Bias can affect interpretation of data in various ways, such as through one's choice of statistical standards for rejecting data. Some feminist critiques of masculinist bias in science lie at this level (e.g., Longino, 1990). Finally, publication bias is an example of inductive-risk bias that affects communication of results. Publication bias arises from the fact that researchers and journals systematically prefer positive findings over research that fails to confirm a hypothesis. For example, if climate-journal editors believe the social-environmental risks of underestimating the effects of climate change are greater than the risks of overestimation, they might preferentially publish studies that confirm the threats of climate change. Climate change skeptics have charged climate science with such bias. But without a record of high-quality unpublished findings, the charge is speculative at best (for a reply to skeptics, see Nordhaus, 2012).

Drawing on social epistemology, Wilholt identifies disciplinary conventions as one solution to the problems posed by the ambiguity of inductive-risk bias. The key problem lies in the corrosion of trust necessary for fruitful scientific collaboration. Scientific practices heavily depend on the mutual trust scientists place in their peer-reviewed arguments. Conventional standards in effect remove some judgments about inductive risk from the individual's purview. The result is a better coordination in the normative standards scientists use in evaluating results. Insofar as their papers adhere to conventional standards of experimental design, confidence levels, and interpretation, differences in individuals' judgments about the value of this or that hypothesis and relative costs of error have less effect on results, and epistemic trust is preserved across the community. And insofar as unpublished negative research findings are made available for scrutiny (as in some medical fields), suspicions of publication bias can be confirmed or rejected (Wilholt, 2009, p. 97). This analysis leads Wilholt to a usable concept of inductive-risk bias: "the infringement of an explicit or implicit conventional standard of the respective research community in order to increase the likelihood of arriving at a preferred result" (2009, p. 99).

Nonetheless, context matters here too. Inductive-risk bias is not necessarily bad, inasmuch as different risk preferences in the community can feed competing research programs and therewith the exploration of more avenues of discovery. But expert committees pose distinctive challenges. Similar to scientists' use of conventional research standards, expert committees, the IPCC included, develop their reports using procedures that are supposed to foster the audience's trust in the report content (Hilgartner, 2000). Unlike research standards, however, committee procedures are not designed simply to coordinate practices across a discipline by removing value-judgments from individual discretion. Here the difference between long-term and short-term timeframes is decisive. As long as members of a discipline follow the same conventions, whether those conventions favor risk-taking or risk-averse strategies is less important than the fact that they do not squelch discovery and correction of error over time. In science advising, a *genuinely* impartial balance between risk and caution becomes crucial for the trustworthiness of policy-relevant advice, which is inherently value-laden and thus potentially politically contentious. Consequently, expert procedures must reflect such impartiality, both in the estimation of technical uncertainties and in their responsiveness to democratically accountable deliberation over the relevant values.

I thus suggest that bad inductive-risk bias in science advising occurs when an expert committee infringes publicly accepted, democratically accountable guidelines for assessing relevant literature, estimating uncertainties, and characterizing risk in an impartial manner. Note that this definition does not depend on any attribution of suspect motives to committee members; the bias might involve such dispositions, but it can also arise from unintentional procedural violations. Adequate procedures must ensure both technical reliability and, at some level, democratic accountability—demands that federal agencies and lawmakers have recognized for some time now (Douglas, 2009, chap. 7).

5. BIAS IN THE IPCC

I close by applying the above analysis to an actual bias charge against the IPCC. Critics have leveled a range of bias charges against the IPCC: ignoring relevant literature; groupthink; political bias; and publication bias in climate science as a whole (resp., Pielke & Staley, 2007; Lemonick, 2010; Horner, 2007; Michaels & Balling, 2009, chap. 7). Given the inclusive, multi-stage design of IPCC report-writing procedures, bias charges face a significant burden of proof. In general, a strong bias charge combines multiple indicators of possible bias (Rehg, 2011, pp. 396–397). In the case of the IPCC, the stronger charges link (1) technical problems in report content, (2) flaws in procedures or their execution, and (3) evidence of close-mindedness.

Judith Curry, an active climate researcher, has issued a string of criticisms of the IPCC that link these elements (see Lemonick, 2010; Curry, 2010; Curry, n.d.). What makes her critique interesting for my purposes is her suggested tie between the IPCC's problematic uncertainty estimates and groupthink bias, more precisely a defensive mentality in the IPCC that cuts off potentially fruitful dialogue with more responsible critics. This charge fits with the contents of the purloined emails of some IPCC scientists, whose remarks indeed display a certain defensiveness (Powell, 2011, chap. 14). Moreover, the IPCC apparently meets the antecedent conditions that make groupthink likely, and both IPCC scientists and critics have concerns about procedural flaws (see IAC, 2010; Hulme, Zorita, Stocker, Price, & Christy, 2010). So we should not simply dismiss Curry's critique.

But we should hesitate before accepting Curry's defensiveness charge. The IPCC procedural revisions reflect a concern not with defensiveness, but with factual errors and group dynamics that engender a kind of team-level confirmation bias. Nor is it clear that IPCC scientists are defensive about their process. Besides accepting certain IAC recommendations (IPCC, 2010), IPCC scientists have engaged criticisms online (at realclimate.org), and have publicly entertained a range of structural revisions to the IPCC (Hulme et al., 2010).

What is more, both Curry's technical objections about uncertainty and her defensiveness charge fall under the ambiguous category of inductive-risk bias. We should thus ask, first, to what extent the technical objections simply express different value-judgments of the costs of different kinds of error. This question lies at the heart of the political controversy over climate projections. Before we can connect uncertainty estimates with bad bias, we must clarify the values that underlie competing assessments of the possible impacts of climate change and response strategies on the economy, environment, and public health.

Second, we should ask to what extent the defensiveness charge boils down to an honest disagreement over the merits of two transactional strategies. Some of Curry's own remarks support this reading:

> There seems to be some sort of unwritten rule by the IPCC scientists and their defenders not to engage with critics/skeptics, since they think that such engagement legitimizes the skeptics. Personally, I think that the almost total lack of "mainstream" climate scientists engaging with skeptics has resulted in a loss of the moral high ground in the public's view, and has acted to increase the public credibility of the skeptics. (2010)

At this point we do not have a straightforward bias charge, but a genuine dispute over transactional merits. At issue are the proper conduct of science advisory panels and their mode of public engagement.

More precisely, we have two procedural alternatives—engaging critics/skeptics more widely or more narrowly—that incorporate different value-judgments about the costs of different kinds of error. As some analysts have pointed out, opening modeling data to critical scrutiny can assist in the correction of errors (Edwards, 2010, pp. 421–427). However, too wide an engagement with skeptics can retard argument-making—in effect, one raises the bar on acceptable conclusions, similar to raising the confidence level for acceptable hypotheses. If one judges the risks of inaction as the more serious (compared to the risk of unnecessarily aggressive measures), then a more controlled admission of criticism has some justification.

The narrower strategy has its roots in the history of climate-change skepticism, whose ties to business interests are well-documented (e.g., Oreskes & Conway, 2010; Lahsen, 2008). I think the record shows that the IPCC did not initially dismiss industry-sponsored and libertarian skeptics; rather, scientists were unconvinced by skeptical doubts, and as typically happens in science, unconvinced scientists eventually stop listening to holdouts. The field moves on, presuming that the truth will prove itself at the level of practice, in the comparative fruitfulness of competing research programs. The failure of skeptics to display the appropriate ethos of scientific argumentation only hastened their dismissal: many of them lacked credentials in climate science, and others argued in a way that points to bad faith (see Oreskes & Conway, 2010). Moreover, skeptics often tend to repeat old objections to climate science, but without any mention of the strong rebuttals on record.[4] These ethotic failures partly stem from differences between the ethoi of political and scientific argumentation. Because politics, more than science, is characterized by mistrust, stricter adherence to predetermined procedures becomes crucial to legitimate outcomes, something IPCC scientists did not seem fully to

---

[4] I have noticed this in my own research on the history of the controversy; I also draw here on conversations with a climate scientist, Benjamin de Foy. But see also Powell (2011).

appreciate (see Edwards & Schneider, 2001); In addition, the self-expressive character of political argument allows losing parties to repeat their position far longer than would be appropriate in science.

In support of the more open approach, Curry points out that climate-change skepticism has evolved over the last decade from industry-sponsored attacks to decentered, blog-based challenges that do not stem from special interests (Curry, n.d.). Thus climate scientists' wariness toward outsider critique no longer obviously holds. There may be advantages, both for the public credibility of climate science and for its technical merits, in developing some wider venues of public engagement that could positively interact with expert forums such as the IPCC.

However one settles this question of procedure, one should do so in the honest recognition that in the United States, the skeptics have won the rhetorical contest over inductive-risk bias, for they have cast sufficient doubt on the impartiality of IPCC process to undermine trust where it matters most: in the halls of Congress and in the portion of the population that supports skeptical legislators. In winning this victory, skeptics were assisted by well-publicized infringements of IPCC procedures designed to safeguard impartiality. It is time for public reflection on how to improve the communicative dimensions of IPCC process.

REFERENCES

Baron, R. S. (2005). So right it's wrong: Groupthink and the ubiquitous nature of polarized group decision making. In M. P. Zanna (Ed.), *Advances in Experimental Social Psychology* (Vol. 37, pp. 219–253). New York, NY: Elsevier.

Bijker, W. E., Bal, R., & Hendriks, R. (2009). *The paradox of scientific authority*. Cambridge, MA: MIT Press.

Blair, J. A. (1988). What is bias? In T. Govier (Ed.), *Selected Issues in Logic and Communication* (pp. 93–103). Belmont, CA: Wadsworth.

Bolin, B. (2007). *The history of the science and politics of climate change*. Cambridge, MA: Cambridge University Press.

Curry, J. (2010, April 23). An inconvenient provocateur. Interview with Keith Kloor, *Collide-a-Scape*. Retrieved from http://www.collide-a-scape.com/2010/04/23/an-inconvenient-provocateur/

Curry, J. (n.d.). On the credibility of climate research, part II: Toward rebuilding trust [Personal web log]. Retrieved from http://curry.eas.gatech.edu/climate/towards_rebuilding_trust.html

Douglas, H. E. (2009). *Science, policy, and the value-free ideal*. Pittsburgh, PA: University of Pittsburgh Press.

Edwards, P. N. (2010). *A vast machine*. Cambridge, MA: MIT Press.

Edwards, P. N., & Schneider, S. H. (2001). Self-governance and peer review in science-for-policy: The case of the IPCC second assessment report. In C. A. Miller & P. N. Edwards (Eds.), *Changing the Atmosphere* (chap. 7). Cambridge, MA: MIT Press.

Esser, J. K. (1998). Alive and well after 25 years: A review of groupthink research. *Organizational Behavior and Human Decision Processes, 78*(2–3), 116–141.

Evans, J. St. B. T., & Over, D. E. (1996). *Rationality and reasoning*. East Sussex, UK: Psychology Press.

Horner, C. C. (2007). *The politically incorrect guide to global warming and environmentalism*. Washington, DC: Regnery.

Hulme, M., Zorita, E., Stocker, T. F., Price, J., & Christy, J. R. (2010, February). IPCC: Cherish it, tweak it or scrap it? *Nature, 463*(11), 730–732.

IAC. (2010). *Climate change assessments: Review of the processes and procedures of the IPCC* [Copy, PDF document, Prepublication]. Retrieved from http://reviewipcc.interacademycouncil.net/report.html

IPCC. (2008). Procedures for the preparation, review, acceptance, adoption, approval, and publication of IPCC Reports, Appendix A to the principles governing IPCC Work. Retrieved from www.ipcc.ch/organization/organization_procedures.htm

IPCC. (2010). Decisions taken by the Panel at its 32nd session, with regard to recommendations resulting from the review of the IPCC processes and procedures by the InterAcademy Council (IAC). Draft. Retrieved from http://www.ipcc.ch/meetings/session32/ipcc_IACreview_decisions.pdf

Janis, I. L. (1972). *Groupthink*. Boston, MA: Houghton-Mifflin.

Jasanoff, S. (2005). *Designs on nature*. Princeton, NJ: Princeton University Press.

Kuhn, T. S. (1996). *The structure of scientific revolutions* (3rd ed.). Chicago, IL: University of Chicago Press.

Lahsen, M. (2008). Experiences of modernity in the greenhouse: A cultural analysis of the physicists' "trio" supporting the backlash against global warming. *Global Environmental Change, 18*, 204–219.

Lemonick, M. D. (2010). Climate heretic: Judith Curry turns on her colleagues. *Scientific American. 303*(5): 78-83. Retrieved from http://www.scientificamerican.com/article.cfm?id=climate-heretic

Longino, H. E. (1990). *Science as social knowledge*. Princeton, NJ: Princeton University Press.

Mercier, H. (2011). When experts argue: Explaining the best and the worst of reasoning. *Argumentation, 25*, 313–327.

Mercier, H., & Sperber, D. (2011). Why do humans reason? Arguments for an argumentative theory. *Behavioral and Brain Sciences, 34*, 57–111.

Michaels, P. J., & Balling, R. C., Jr. (2009). *Climate of extremes*. Washington, DC: Cato Institute.

Nordhaus, W. (2011, March 22). Why the global warming skeptics are wrong. *The New York Review of Books, 59*(5), 32–34.

Oreskes, N., & Conway, E. M. (2010). *Merchants of doubt*. New York, NY: Bloomsbury.

Pielke, R. A., Jr. (2007). *The honest broker*. Cambridge, MA: Cambridge University Press.

Pielke, R. A., Sr., & Staley, D. (2007). Documentation of IPCC WG1 bias (part I). *Climate science: Roger Pielke, Sr.* Retrieved from http://pielkeclimatesci.wordpress.com/2007/06/20/documentation-of-ipcc-wg1-bias-by-roger-a-pielke-sr-and-dallas-staley-part-i/

Powell, J. L. (2011). *The inquisition of climate science*. New York, NY: Columbia University Press.

Rehg, W. (2009). *Cogent science in context*. Cambridge, MA: MIT Press.

Rehg, W. (2011). Evaluating complex collaborative expertise: The case of climate change. *Argumentation, 25*, 385–400.

Tollefsen, D. P. (2006). Group deliberation, social cohesion, and scientific teamwork: Is there room for dissent? *Episteme, 3*(1–2), 37–51.

Turner, M. E., & Pratkanis, A. R. (1998). Twenty-five years of groupthink theory and research: Lessons from the evaluation of a theory. *Organizational Behavior and Human Decision Processes, 78*(2–3), 105–115.

Walton, D. (1997). *Appeal to expert opinion*. University Park, PA: Pennsylvania State University Press.

Walton, D. (1999). *One-sided arguments*. Albany, NY: SUNY Press.

Wilholt, T. (2009). Bias and values in scientific research. *Studies in History and Philosophy of Science, 40*, 92–101.

Ziman, J. (1968). *Public knowledge*. Cambridge, MA: Cambridge University Press.

Objectivity vs. Advocacy: Newspaper Rhetoric during the "Bemis Affair" and the "Oleomargarine Controversy"

DAVID SEIM

Department of Social Science
University of Wisconsin-Stout
Menomonie, WI 54751
USA
seimd@uwstout.edu

ABSTRACT: I introduce two important case studies of media roles in public debate over the nature of economics as a science. These cases, one during the 1890s and the other in the 1940s, reveal uncertainty among the American citizenry concerning what kind of science economics is. The question—a long-running one—was this: How pure and detached from policy advocacy must economists be?

1. INTRODUCTION

During 1894 and 1895, at a time of U.S. economic depression and industrial unrest, an event known as the "Bemis Affair" unfolded at the University of Chicago, a grand private institution. In 1892, in the school's first year in existence, the president of the university hired Edward W. Bemis, a well-regarded teacher and researcher, as the school's first tenure-track extension economist. Bemis came from Vanderbilt University, where he was known for holding anti-capitalist leanings. Bemis taught and published a kind of economics known as "historical" economics, which differed from a mainstream emphasis on abstract, ahistorical theory. He had engaged in dialogue with the university about whether to accept school's job offer, which he did, only after feeling reassured that it was understood that his approach to economics differed markedly from economic views held by the chair of the school's department of political economy, J. Laurence Laughlin, as well as the school's founder, John D. Rockefeller.

Bemis's dismissal from Chicago, three years later, received much news attention. The university, without firm supporting evidence, said that Bemis needed to be fired for incompetence—that he simply did not teach well. The media, without firm supporting evidence, argued that Bemis was fired for his economic views, as disapproved of by Laughlin and Rockefeller.

Understanding the Bemis affair requires multiple levels of analysis. One factor for analysis stems from pressures within a new university attempting to achieve rapid ascent to a highest echelon. Another relevant factor occurred at the level of an economics profession working its way through intellectual disagreements between "pure" methods and "reform" activities for social scientists. A third factor involved uncertainties about Rockefeller as the school's funding source, and how the school's president, William Rainey Harper, might perceive Bemis's impact on Rockefeller's willingness to donate funds.

Let's begin with Bemis's hiring. He came to Chicago recognized as a teacher and a scholar. His published works were solid, and those who knew them could recognize Bemis as a kind of economist who dealt with real-world facts so as to understand complex problems and potential policies for fixing runaway capitalism. Bemis's publications focused on monopolies, especially so-called "gas trusts." He was trained by economists in a reformist tradition, which

Seim, D. (2012). Objectivity vs. advocacy: Newspaper rhetoric during the "Bemis Affair" and the "Oleomargarine Controversy". In J. Goodwin (Ed.), *Between scientists & citizens: Proceedings of a conference at Iowa State University, June 1-2, 2012* (pp. 335-344). Ames, IA: Great Plains Society for the Study of Argumentation. Copyright © 2012 the author(s).

believed that meaningful social science must willingly recommend policies. But such a willingness to recommend policies could cause troubles for a social scientist at a time when many social scientists sought to achieve a more "scientific" status for their profession.

When considering the Bemis affair from a point of view of citizens' engagement with science, we can identify a variety of public attempts to understand how social science works and how it might relate to policy making. Yet when they paid attention to the Bemis affair, the public often simply reduced it to a single question: was academic freedom violated when Bemis was fired, or did a financial supporter of a university have a right for some specific economic viewpoint to be supported by the school?

Historians have studied the Bemis affair and agree that Bemis's first 'infraction' was (supposedly) clear by the end of his first year at the school: it was his expressed critical stance toward Chicago's "gas trust." At some point, gas industry complaints against Bemis began arriving at the school.

Bemis's job expectations were a bit different from most professors at Chicago, in that he was hired as an extension economist—what the school called an "outside" faculty. He was to teach real-world economic issues to the public (issues such as high gas prices in Chicago, or labor discontent in Pullman, IL!), and it was of the very nature of extension lecturing that his instruction was to minimally employ abstract, rigorous theory.

In our search for any sort of more specific infraction that might have been asserted against Bemis, one possibility is that local donor money potentially was becoming withheld from the school until the school did something about Bemis. Another possibility could be that Chicago's gas companies decided to withhold any favorable rates from the university until the school dismissed Bemis. Also possible is that Rockefeller personally withheld financial support until the occasion of Bemis's dismissal. There has been some discussion amongst historians of all these possibilities—however there is no known traceable record of any of them.

On January 15, 1894, fifteen months after Bemis started at Chicago, President William Rainey Harper informed Bemis that he would not be reappointed. Bemis did not see this coming. His quality teaching and research had attained for him a tenured status at Vanderbilt, and he had brought this status with him to Chicago. Bemis was even informed by Harper in the summer of 1892 that he was wanted *specifically* because his economic thought was so different from Laughlin's. Bemis is also on record as drawing respectable attendance numbers for his extension lectures—even though the university would, in time, claim otherwise. Bemis simply did *not* see the dismissal letter coming.

Yet Harper's letter could not actually announce a dismissal; he could only request a resignation. He offered that if Bemis did just this, then there should be no great difficulty in finding him another fine job somewhere, indeed with Harper's support. Bemis decided not to resign.

The burden thus fell upon the university to create pressures enough to obtain Bemis's resignation, or if not successful with that, to identify a basis for firing a tenured professor. The new situation began with months passing and nothing obvious happening. But this was only how things seemed. Both sides actually moved quietly forward in accordance with their opposed beliefs about what seemed possible with respect to the idea of firing Bemis. For Bemis, he expected that his tenured status coupled with successful meeting of job requirements ought to mean job security. As to the university, for some reason they wanted Bemis gone.

For awhile each side backed away from each other, with each side instead communicating with persons outside the school. Harper occasionally shared thoughts in correspondence with friends, including with a least two persons from the press. Bemis shared his concerns with economist friends, most notably Richard Ely at Wisconsin and H.C. Adams at Johns Hopkins.

There are various methods for piecing together a history of the Bemis affair, extending from the requested resignation (Jan. 1894) to his resignation and departure (Oct. 1895). One method focuses on archives, where various bits of useful information exist. A history of the Bemis affair can also be traced in the media—which is our particular approach herein.

2. THE BEMIS AFFAIR AS TOLD BY THE MEDIA

On June 30, 1894, about half a year after Harper's private request for Bemis's resignation, the *Chicago Tribune* announced that Bemis planned to leave the school upon completion of his contract for 1894–95. The writer for the *Tribune* tended to distrust Rockefeller and emphasized that Bemis's approach to economics was simply unwanted at Chicago.

> For two years Prof. Bemis has suffered the embarrassment and disadvantage of being at swords' points with Head Professor J. Laurence Laughlin of the economic[s] department. That a breach would eventually ensure there was little doubt. Indeed, the differences of opinion which separated Profs. Laughlin and Bemis were so decided that the former refused to recognize the latter on the streets."

The article saw the issue as a disagreement about which kind of economics was true science—Laughlin's general theorizing or Bemis's case studies, such as trade unionism, factory legislation, and the like. The writer added how it "has frequently been told that Prof. Laughlin openly advised students in economics not to take Bemis' courses if they desired to do scientific work."

This paper—the *Chicago Tribune*—continued with the story over time, as did the *Chicago Daily News*, doing so with at least five articles by August 7 of *the following year*. In the *Daily News* on Aug. 7, 1895, for example, it was written how "it is generally conceded that the professor's study of economics is not prolific of agreeable results to the 'authorities' at Mr. Rockefeller's fane of learning."

Many other news outlets covered the story by the summer of 1895—owing to the fact that Bemis was still not making it clear that he was willing to go. In New York, the new home city of Rockefeller (who moved there from Cleveland), *The World* covered it (August 14), as did *The Voice* (August 17). The *New York Evening Post* (August 12) strongly impugned Rockefeller:

> Any suggestion that capital in the form of a university is oppressing labor in the person of a professor is capable of indefinite expansion. It happens that Mr. John D. Rockefeller has given a large sum of money to this university, and the conclusion is obvious that the Standard Oil Company has built up an institution of learning to promote its own theories of political economy.

In Great Britain, the *Bath Times* (August 8) covered the story, with its writer arguing that maybe different kinds of truths are attainable in social science, when compared to truths attained by other sciences. Also offered was that

[t]he removal of Prof. Bemis, of the Chicago University, for teaching his students the truth in political economy is not surprising. Every institution of learning dependent for its existence on the bounty of millionaire monopolists is in constant liability to similar calamities. It is quite natural that Mr. Rockefeller should not relish having a teacher, paid with his money, expose the essentially unjust and iniquitous way in which that money was obtained.

Looking westward from Chicago, the *San Francisco Bulletin* (August 19) suggested, at length, that Rockefeller must not like Bemis:

> It is known that during his residence in Chicago Mr. Bemis has been active in his opposition to monopolies. One particular object of his hostility was the Chicago Gas Company, a corporation which makes gas at 6 cents a thousand feet and sells it at $1.10. The Chicago Gas Company's stock was known to be inflated, and it was also known that enormous dividends were paid on this inflated capital. Professor Bemis thought that some method should be devised by which the profits of making gas should be distributed among the people . . . The inference is that Mr. Rockefeller did not approve of the position Professor Bemis took and tenaciously held.

The University of Chicago, on August 20th, released a statement expressing that any matter between the university and Bemis had nothing to do with Rockefeller. And also it should be known, according to the statement, that the university does not violate academic freedom.

The *Boston Herald* (August 22) figured that Bemis was discharged at the "instigation" of Rockefeller.

> In all the letters sent out from this institution the heading reads that Mr. John D. Rockefeller is the founder, and it is impossible at the present knowledge to avoid the inference that Prof. Bemis has been discharged at his instigation.

The *Philadelphia Inquirer* (August 26) suggested that Bemis's "scalp now dangles at the breast of monopoly," but that this was fine somehow, in that Rockefeller's money had left their state—and so it was "just as well after all that the Standard Oil Money made in Pennsylvania went to no Pennsylvania College."

The *Ft. Worth Gazette* (August 31) —in the midst of the great 1895 Texas oil rush— saw Bemis as the scientific one. "The institution was chiefly endowed by John D. Rockafeller (sic), assisted by Charles T. Yerkes—the Standard Oil magnate of the continent, and the street railway potentate of Chicago," the writer observed, and then added that Bemis's "conclusions as to the danger from natural monopolies passing into private control are derived from scientific and historic study....He treats the matter as a college professor should treat it." The *Montreal Witness* (September 3), citing also an article in the *New York Recorder*, opined *that Rockefeller* should be able to have his views supported by social science:

> Mr. Rockefeller and his associates have a perfect right to employ professors and pay them roundly for teaching the Rockefeller views of political economy, and if they do not get the views they pay for they have also a clear right to stop the teaching and discharge the teacher.

In September 1895, University of Chicago sociologist Albion Small used the school's *American Journal of Sociology* to critically comment in a way evidently designed to clamp down on anyone who might again express support for Bemis. Small reprinted a letter, supposedly typical of many letters, which he described as flawed in its logic for supporting Bemis. Small described the reprinted letter as an attempt to describe

> an educational institution founded by the arch-robber of America and which already, by its treatment of Professor Bemis, exhibits a determination to throttle free investigation of sociological or economic subjects wherever there is any danger of running counter to plutocratic interests.

Small then proceeded to destroy such a view.[1]

But more media coverage came. In October, *The Voice* (NYC) aligned itself with Bemis, who was respected for his efforts which have "contributed more than the efforts of any other individual to the municipalization of gas and cognate commodities in this country." The *Chicago Daily Tribune* (October 3) doubted the university, primarily because of its continued silence. The *Chicago Daily News* (October 9) published Bemis's belief that bad word about him must have been leaked by the university. Bemis challenged Harper to explain what he meant in his original letter back on January 15, 1894, when he allowed that "peculiar circumstances" existed at Chicago which, evidently, would make it tough for Bemis to remain there. The *Chicago Daily Tribune* (October 10) wondered if there was solid reason for the university's firing of Bemis, while *The Kingdom* (Minneapolis) (October 11) supported Bemis yet wondered the same thing. *The Kingdom* asked what kind of social science the university might allow.

News outlets soon had new material to consider, when the university released its press statement explaining why Bemis was being let go: not because of any specific expressed viewpoint, but because of his supposedly poor teaching; he was dismissed for "incompetence." The university's press release, including a statement from President Harper, was published in the *Chicago Record* (October 18), and then in that evening's *Chicago Daily News*.

The *Chicago Daily Tribune* the next day introduced Bemis's words in his defense, including his belief that his economics is good science. Bemis cited an opinion previously shared by Small, to the effect that Bemis is

> the best man in the country to write books on many of the following: immigration, population, cooperation, profit-sharing, building & loan associations, life insurance, labor organization, arbitration, factory and other labor legislation but those subjects were too specialized for university instruction.

Even after he no longer worked for the school (as of October 1, 1895), Bemis strengthened his counterpoint, as was published in the likes of the *Boston Transcript* (October 29). Also critically exploring the matter was a piece in *City and State*, edited by Herbert Welsh, an important, Philadelphia-based political reformer. Welsh editorialized about the gist of the university's public statement "that"—in Welsh's summarizing words—"Professor Bemis was not much of a teacher any way."

---

[1] Small argued that a capitalist economy is substantially more complex than the writer imagines, as it is in fact "possible to serve the cause of justice and to promote the common weal without begging social questions, and without joining in vulgar denunciations of social factors which after all may prove to be social blessings." Small explained his concern at least to "oppose to the assumption that industrial combination is robbery, the counter assumption that industrial combination is progress." Neutral science is what is needed to determine which assumption is the more correct of the two, and doubtless "the final truth" will, upon pure and undetached scientific arbitration, be found to "lie somewhere between these two extremes." Persons inside the university needed to be trusted in these affairs, as not only are they "the only persons who know the facts," but more importantly, they "have repeatedly assured representatives of the press that nothing in the case is of any public interest to the public, because no principle in which the public is concerned is in any way involved." This, Small explained, should put the whole matter to rest.

It is known in private circles that a feud has long existed in the university between some of these professors and Professor Bemis. If incompetency on the part of the latter was the real difficulty, it is strange that the public has not heard of that very much, if at all, until just now.

An opinion piece came the next month in the journal *Social Economist*, which declared that "The University of Chicago is a private, not a socialistic, institution." The article added a question quoted from the *New York Times*: "Why should an institution pay a professor to teach social doctrines which are contrary to the consensus of opinions of the faculty, the supporters of the institution, and of the general community?" The writer of the piece believed there was no reason for any public explanation from the University of Chicago. In his seeing capitalism and socialism as competing systems of faith, so incompatible with each other as to preclude their coexistence, the writer—George Gunton—added his analogy:

There is no more reason why those who believe in the present industrial institutions should be surreptitiously made to support teachers of Socialism than there is that Catholics should employ Protestants as priests or Jews install Christians to preside in their synagogues.

As to Bemis, within about a year of his departure from Chicago he was hired by a public land-grant school, Kansas State Agricultural College.

3. SAME STORY, DIFFERENT SETTING

A half century later, during 1943 and 1944, economists, college administrators and citizens of Iowa waged a battle over the purposes of social science. Iowa's debate was a frontline event in a struggle to establish safeguards allowing policy research at public educational institutions. In a conflict over a proposed policy to temporarily produce less butter, one side declared that economists at land-grant Iowa State College must limit themselves to advocating policies directly supporting Iowa interests, while an opposing group advocated policy research to win the war. What eventually happened was that financial donors and college administrators insisted on social science devoid of policy arguments. However, Iowa State's social scientists wanted what they believed was a more realistic standard. The debate resulted in an unresolved conflict over the potentials and limitations for policy-oriented social science.

The controversy at Iowa State began with a policy proposal from the school's economics department that more consumption of margarine would help make more milk products available for soldiers. Iowa interest groups promptly objected to such a policy recommendation. Much bad press then ensued with respect to how Iowans felt about their economists at Iowa State. In the end, there were numerous resignations by these economists. We can look back and interpret the controversy as a test case for the validity of allowing social scientists at public institutions to advocate public policy.

Much that happened during Iowa State's butter-margarine war was because there was a strong leader in the school's economics department, namely, Theodore W. Schultz. By the 1940s, Schultz held views well-known within the profession, and also often introduced into public discussion—certainly, at least, in the state of Iowa.

With the coming of U.S. involvement in the Second World War, a request came to Iowa State, in October 1942, from the U.S. Department of Agriculture. The USDA asked Iowa State to create a more efficient food policy to help win the war. In agreement with the USDA, Iowa State expressed willingness to do so, and they requested a grant from the Rockefeller Foundation to support production of about a dozen pamphlets on national food policy. In the request for the grant, Schultz stated the pressing importance of "a study of governmental policies affecting production and distribution of food." He assured the foundation that the project was cleared with Iowa State President Charles E. Friley, and was encouraged by the USDA's "urgent" belief in "the need for critical appraisals made by persons outside of government, evaluations which will point out the merits and limitations of current policies and programs." Friley and others at Iowa State provided letters of support.

The Rockefeller Foundation itself stated an expectation to see "recommendations as to food production, distribution and consumption policies." Iowa State's economists produced outlines for fifteen pamphlets, and by the end of March, four pamphlets were in print. It was so-called "Pamphlet No. 5," published the first week of April, 1943, that brought trouble. The pamphlet was produced by economics graduate student Oswald H. Brownlee, and was titled "Putting Dairying on a War Footing." The chief goal of all policy recommendations in the pamphlet was to make more milk products available to soldiers. Brownlee recommended that American households use more margarine instead of butter.

Iowa's dairy industry responded severely to Brownlee's recommendation. Industry representatives said that as Iowa taxpayers, they had been betrayed. Any policies that were to come from an institution so heavily supported by Iowa's taxpayers should favor Iowans; citizens of Iowa saw the school as their personal advocate. Publication outlets for such views included the *Dairy Record* (St. Paul, MN), the *Creamery Journal* (Waterloo, IA), and especially the major newspaper in the state, the *Des Moines Register*. In at least one case an early news item opined about the quality of science coming from an Iowa social scientist who would attempt to help formulate public policy; this was the *Dairy Record* (April 28), which stated that the economist authoring the pamphlet was a "sadistic" person who

> has a false notion that he pursues a calling that is, of itself, a science. The very fact that the author…fails to take cognizance of the economic importance of the butter industry to the state he is supposed to serve seems to indicate that, in his search for the profound, he has forgotten the simplest definition of his vocation.

The author, like all economists, must be an "unstable" person troubled by an "inferiority complex," who during college days was "unwilling or unable to provide the concentration needed to master the exact sciences."

Media coverage of the controversy was severe, and by May 19, a decision was made to retract "Pamphlet No. 5." The media covered the meeting whereupon this decision was made, as well as covered the school's announced plans to critically study contents of the pamphlet to identify and remove all that was not scientifically certain in the pamphlet. Dairy interests communicated their view of the relationship between social science, public policy, and the public—with an example being Francis Johnson, president of the Iowa Farm Bureau Federation. Johnson made it into the news for drawing attention to farming interests "alarmed over the apparent tendency to make over Iowa State College into a tax-supported blueprint of Harvard University." Iowa State is different from Harvard by not being a "free-lance" institution, and the school has no right to risk making "impractical suggestions or recommendations" on policy matters. "The true test of the value of most research on matters of public policy," Johnson pronounced, "is determined by the eventual acceptance and use of the recommendations. The college cannot justify its existence on the basis on mere 'irrational value.'" A central question in this 1940s debate was clear: what kinds of policy research by social scientists would be allowed at a *taxpayer*-supported school?[2]

The press regularly took time to report on the story. Newspapers debated whether Iowa's social scientists should be permitted to make policy arguments. An editorial in the *Des Moines Register* framed some of the issues. In expressing "devotion to 'the scientific approach'" to social research, the *Register* opined that "as a democratic people we are trying to thrash the thing out, in the light of all the facts and interpretations that we can get, so as to arrive eventually at the right answer." Yet the editorial added that "the issue is not one of the right and duty of professors to try to serve the public interest." Many continuing letters and editorials revealed complex thinking on both sides of the issue.

Media coverage of Iowa's citizen-based debate about the purpose of social science at a taxpayer-supported institution turned rough. The *Register* published excerpts from Pamphlet No. 5, to which dairy interests responded with a full-page advertisement nationally sponsored by the American Dairy Association (ADA). The ad accused Iowa State's economists of proposing "that the housewives of America be denied butter and be forced to accept a product they have refused on its own merits." The ADA depicted Iowa State as subverting the war effort by "taking a stand against the Government's Wartime Food Production Program." The group claimed that no fewer than "five million dairy farmers are shocked at the rumpus created by the much-discussed Pamphlet No. 5," which "rocks the very foundation of diversified farming" and "challenges the dairy farmer's way of life." Iowa dairy representatives, following in the slipstream of the ADA's advertisement, passed a resolution declaring that the pamphlet "jeopardizes the national war food program," and "has done untold injury to a basic industry which means an annual income to the state of more than 100 million dollars per year."

Reports by two committees came in July. First to report was the Joint Committee, which got their views into the press. There was also a Special Committee that reported only to Friley, and any of their actual findings stayed out of the press. The Joint Committee's meeting on July 12 gained heavy press coverage reporting an agreement by all committee members that so many elements in the pamphlet were incorrect, or at least misinterpreted, that the pamphlet required retraction.

The Iowa press extensively covered the story. Iowans also wrote letters to Iowa State, mostly accusing the school of capitulating to special interests. Even the national press took an interest in the controversy, at a time when Americans had bigger worries on their minds. In the midst of war coverage, *Time Magazine* published a brief article titled "The Butter Atheist," while *Newsweek* facetiously reported that Iowa's dairy leaders had "found a traitor in their ranks"—the traitor being Iowa State College, "for years tax-supported by the farmers." Even the *Chicago Journal of Commerce* expressed disbelief that President Friley and the Iowa State administration were trying to "bamboozle" the public with "puerile actions" that have "cast suspicion on all future publications coming from faculty members" at Iowa State. As seen through the eyes of the nation's business leaders, the problem was clear:

---

[2] Another view was expressed by Schultz, on multiple occasions. For example, during a meeting with Friley, Schultz pondered whether a professor at Iowa State can, if under an arrangement to serve some industry's own interests, "stay wholly impartial, unbiased and objective?" Schultz (although referencing an example of a professor writing advertisement language to support a cattle breed association) asked Friley: "Will not other special interest groups, seeing arrangements of this type, quite properly come to expect similar personal services on their behalf?" Schultz believed that any such ties to special interests necessarily will lead to a loss of public confidence in research findings.

> If the pressure groups like the dairymen in Iowa get research conclusions revised merely by putting the squeeze on the college president and threatening to have the legislature cut the college's appropriations, why should anyone believe that any of the college's future research publications are impartial and not written with an eye to catering to the prejudices of the producers around the state?

Leadership at Iowa State moved to begin the revision process—a protracted effort that would witness the cancellation of other planned pamphlets, a reorganization of the editorial board of Iowa State's college press that oversaw publication of the pamphlets, the creation of a "Committee to Reorganize the Department of Economics and Sociology," creation of a "Committee on Sponsorship of Publications," multiple resignations from the President's Special Committee (and also a heart-exhaustion related death of a committee member), public statements by the governor of the state, and the beginnings of national-level investigations by the ACLU and other groups.

Media coverage especially attended to Schultz's resignation from the land-grant school, on September 15, 1943f—or a job at the University of Chicago. And, when no revised pamphlet was forthcoming by October, news outlets began questioning what kind of science social science is?

Wallaces' Farmer (Des Moines) recognized not only Schultz's resignation as a "great loss" for Iowa State as well as "in the larger field of public affairs," but tried to find the central issue in the controversy: it was what kind of social science would be allowed at Iowa State. To be useful, economists "must deal with pressing and controversial issues." Iowa State's economists are expected to be as impartial as possible, and to present facts as they see them. "But so long as it bases its conclusions on the best evidence it can find, nobody should object, altho some may squirm and altho others may—quite properly—ask for further investigation into the facts." The editorial added,

> You can't cure cancer by telling the doctor you don't believe in it, and that he is to find another diagnosis. So, in economic diseases, we need to let the economists do the best they can without any orders as to what the diagnosis should be.

Another news commentary suggested that the central issue that

> must be taken into account…is that the social sciences are not precise sciences. On many questions it is possible for another person in the field to take a different—even an opposite—position from Doctor Schultz and still be considered as competent an economist as he.

In the *Des Moines Register*, a particularly vocal supporter of Schultz, one Thomas Keenan, offered historical comparison:

> When Copernicus reported his conclusion that the sun did not revolve around the earth but that the earth revolved around the sun, there was plenty of 'studied judgment by qualified authorities' to the effect that he was a heretic. If he had been on the faculty at [Iowa State] I.S.C. would the college have refused to assist in publicizing that report? When Harvey reported that the blood in our bodies circulated through the veins and arteries there was plenty of 'studied judgment by qualified authorities' to the effect that he was crazy; so we would have turned thumbs down on him and he would have gone to the University of Chicago.

Time Magazine (October 11, 1943) reported that an Iowa State graduate student produced a "disinterested oleopus," but the Iowa Farm Bureau declared it foul. Such a pamphlet "might befit scholarly Harvard," the article reported the Farm Bureau crying, "but was disloyal in a cow college." According to *Reader's Digest*, Brownlee had published an informative and balanced collection of facts only to discover that "there was the very devil to pay." Dairy interests "demanded Brownlee's scalp"; Schultz then "chucked his job and escaped to Chicago," while President Friley "placated the dairy interests by disowning the heretical tract." *Harper's Magazine* reported that margarine, suddenly charged with "the power of dynamite," had ignited an explosion that has "blown up the works at Iowa State College of Agriculture— through the suppression of a pamphlet enumerating the virtues of margarine during the wartime butter shortage."

Temporary relief came with the arrival of the holiday season, which in a college environment can slow down the pace of meetings, at least just a bit. But by late January, with no revision of Pamphlet No. 5 forthcoming, some real possibility existed that investigators might visit Ames on behalf of the ACLU or other groups. The possibility of investigation by the ACLU, in particular, was announced to a national readership in an article in *The New Republic*. J.M. O'Neill, who chaired the ACLU's Committee on Academic Freedom, identified the Iowa case as concerning "the freedom to speak, to teach, to publish the truth as he sees it on the part of the teacher and research scholar." President Friley, O'Neill charged, had "given up without a fight the fortress for truth and the public interest."

Approval of revised Pamphlet No. 5 came on March 16, 1944. Iowa State released a publicity notice the following day, and the revised pamphlet was printed and mailed on May 2, 1944. During the time of the controversy or soon thereafter, sixteen of Iowa State's twenty-six economists resigned.

3. CONCLUSION

To conclude: I believe these two cases work together when we seek better understanding of public conflict over the place of economic science in society. Do economists only discover objective scientific truths that, in turn, political and governmental processes must evaluate for any policy implications? Or, do economists themselves have permission to explore—and even advocate—particular policy changes in response to their economic analysis? This debate did not begin in 1894 and it did not end in 1944. Yet with respect to a history of contention between objectivity and advocacy, the two episodes of conflict, while not wholly unique, are perhaps as central as they come.

REFERENCES

Beneke, R. R. (1998, Summer). T.W. Schultz and Pamphlet No. 5: The oleo margarine war and academic freedom. *Choices*, 2nd quarter, 4–8.
Bergquist, H. E., Jr. (1972, December). The Edward W. Bemis controversy at the University of Chicago. *AAUP Bulletin*, 384–393.
Furner, M. O. (1976). *Advocacy & objectivity: A crisis in the professionalization of American social science, 1865–1905*. Lexington, KY: The University Press of Kentucky.
Seim, D. L. (2008, Winter). The butter-margarine controversy and 'two cultures' at Iowa State College. *The Annals of Iowa*, 1–50.

The Reasonableness of Argumentation from Expert Opinion in Medical Discussions: Institutional Safeguards for the Quality of Shared Decision Making

A. F. SNOECK HENKEMANS

Department of Speech Communication, Argumentation Theory, and Rhetoric
University of Amsterdam
Spuistraat 134
1012 VB Amsterdam
The Netherlands
a.f.snoeckhenkemans@uva.nl

J. H. M. WAGEMANS

Department of Speech Communication, Argumentation Theory, and Rhetoric
University of Amsterdam
Spuistraat 134
1012 VB Amsterdam
The Netherlands
j.h.m.wagemans@uva.nl

ABSTRACT: The ideal of shared decision making starts from the assumption that physicians and patients are able to take a joint decision as to what is the best treatment. However, since medical consultations are to be viewed as discussions between an expert and a layman, in practice it will often be the case that the patient has to rely on the physician's expertise. In this article we examine the extent to which the Dutch laws, guidelines and professional conventions within the medical domain positively influence the quality of the process of shared decision making, even in cases where the physician makes use of an argument from expert opinion. To this end, we will chart some of the most important institutional safeguards for the quality of medical decisions and analyze how these safeguards relate to the critical questions associated with the argument scheme of argumentation from expert opinion.

KEYWORDS: argument scheme, argumentation from expert opinion, critical questions, shared decision making, medical discussion, institutional safeguards.

1. INTRODUCTION

For the last ten years it has been increasingly regarded desirable that the physician and the patient take a joint decision as to a safe and acceptable treatment for the patient by means of conducting a discussion. This process is also known as 'shared decision making.' In its ideal form, shared decision making is:

> [a] decision-making process jointly shared by patients and their health care provider, [which] relies on the best evidence about risks and benefits associated with all available options (including doing nothing) and on the values and preferences of patients, without excluding those of health professionals. (Légaré et al., 2008, p. 1)

Snoeck Henkemans, A.F., & Wagemans, J.H.M. (2012). The reasonableness of argumentation from expert opinion in medical discussions: Institutional safeguards for the quality of shared decision making. In J. Goodwin (Ed.), *Between scientists & citizens: Proceedings of a conference at Iowa State University, June 1-2, 2012* (pp. 345-354). Ames, IA: Great Plains Society for the Study of Argumentation. Copyright © 2012 the author(s).

The process of shared decision making is one possible way to meet the legal requirement of informed consent. It is legally required for any treatment that the physician obtains permission from the patient, after having given the patient enough information for taking a decision on the matter. The new consultation format of shared decision making does not prevent the patient from having to rely on the expert opinion of the physician. Since in general, a medical consultation is to be characterized as a discussion between an expert and a layman, patients will rarely be able to assess the quality of the information and opinions of the physician directly. In the communicative activity type of a medical consultation, the expertise of the physician still plays a decisive role.

It is therefore important to analyze to what extent the laws, guidelines and professional conventions that relate to these consultations, offer a safeguard for the quality of the expert opinion put forward by the physician. In this article we will provide such an analysis from the perspective of argumentation theory. We reconstruct the appeal to expert opinion as 'argumentation from expert opinion' and relate the abovementioned laws, guidelines and professional conventions to the different critical questions that are associated with this type of argumentation.

First, we will indicate how the asymmetrical relationship between physician and patient relates to the ideal of shared decision making (section 2). Then we will give a brief overview of the key critical questions that play a role in the assessment of so-called 'argumentation from expert opinion' (section 3). Taking these critical questions as a starting point, we will then list a number of institutional safeguards for the quality of medical decisions (section 4). Finally, we will summarize our findings (section 5).

2. SHARED DECISION MAKING

In the literature it is generally assumed that shared decision making has a positive influence on the quality of medical decisions. Rather than confining themselves to informing the patient about the various treatment options and their pros and cons, physicians actually discuss the options with the patient (Charles, Gafni, & Whelan, 1997). Since during the process physicians are also supposed to tell their patients which treatment they prefer and why, in the process of shared decision making the physician not only provides information, but also argumentation. As a result, the expertise of the physician is more fully utilized. In addition, shared decision making is recommended because it often leads to more satisfaction with the consultation and improved therapy compliance.[1] Patients feel more involved in the decision about their treatment because the physician allows them to participate in the discussion and to put forward their own preferences.

From an argumentation theoretical point of view, the requirement that the physician should discuss the available treatment options with the patient can be seen as an institutional obligation with respect to the burden of proof regarding medical decisions (Goodnight, 2006; Mohammed & Snoeck Henkemans, 2012). The main reason to impose this burden of proof upon the physician is that in medical consultations there usually is an 'asymmetric'

---

[1] This is especially the case in decisions about long-term treatments, like chronic diseases. See Joosten et al., 2008, p. 224.

relationship between the discussants.[2] This means that the discussants do not have the same knowledge of the topic of discussion: the physician is an expert, and the patient is a layman.[3]

In some cases, this asymmetry makes it impossible for the ideal of shared decision making to be realized completely, because the physician will not be able to comply with the institutional burden of proof in all respects. Depending on the degree of difference in expertise, at some point in the discussion the physician will have to refrain from providing a substantive or 'direct' defense of his position and will have to appeal to his expertise.[4]

In an indefinite context, contributions to the discussion can be analyzed as argumentation from expert opinion when there is an *explicit* appeal to expertise. In the more specific context of the medical consultation, we believe also other contributions to the discussion can under certain conditions be reconstructed as this type of argumentation. This is the case when the physician chooses not to further defend a (sub) standpoint—e.g., by merely repeating or confirming his standpoint—while the responses from the patient indicate that the standpoint does need further support. An example of such a situation would be the following fragment adapted from Ariss (2009, p. 914):

> D: And I don't want to see your blood pressure for six months, I don't wanna know about it.
> P: Ohf. Are yuh sure?
> D: Absolu- Yes absolutely fine.

In an indefinite context, the reply of the physician (D) in turn 3 to the question of the patient (P) in turn 2 would be evaluated as an evasion of the burden of proof. In the specific context of the medical consultation, however, it is more appropriate to reconstruct the physician's response as an *implicit* argument from expert opinion. For the patient, by having requested the consultation in the first place, has already indicated that he is prepared to rely on the physician's expertise.

The patient's lack of expertise may render it impossible for him at some point in the discussion to determine the acceptability of the physician's standpoint in a direct, substantive manner. This, however, does not mean that the patient is forced to accept the physician's standpoint regarding the diagnosis, prognosis or treatment without further consideration. Apart from in a direct, substantive way, an expert opinion can also be assessed in an indirect way. According to Goldman (2001, p. 93), a layman may check the extent to which the expert opinion is consistent with that of other experts, what results have been achieved by the expert so far, and whether there is a conflict of interests.

We believe that these indirect assessment possibilities can be transformed into criteria for assessing the reasonableness of argumentation from expert opinion. Within the field of argumentation theory, such criteria generally take the form of a series of critical questions (see for instance Walton, 1997; Walton, Reed, & Macagno, 2008; Wagemans, 2011). In the context of the medical consultation, the general rule is that the more opportunity there is for the patient

---

[2] According to Ariss (2009) it is not only the factual difference in knowledge that hinders an equal participation to the decision process, but also the view of both doctors and patients that the doctor has more epistemic authority.

[3] By 'expert' we mean someone who is a professional expert and by 'layman' we mean someone who is not a professional expert. Of course, a layman can be an expert by expertise.

[4] Goodwin & Honeycutt (2009, pp. 27–28) say that whenever scientists in the context of a public debate choose to not give arguments but appeal to their authority, the laity does not have enough incentive to draw a conclusion that is based on their own analysis of the material.

to determine whether the criteria for the reasonableness of argumentation from expert opinion have been met, the more fully the ideal of shared decision making can be realized—even in cases where the physician explicitly or implicitly appeals to his expertise.

3. CRITICAL QUESTIONS

Argumentation from expert opinion is a type of argumentation in which the protagonist supports the standpoint that a certain opinion is acceptable (A is true of O) with the argument that the opinion at issue has been put forward by an expert (P is true of O).[5] The standpoint (1), argument (1.1), and acceptability transfer principle (1.1′) involved in this type of argumentation can be represented in the following way:

> 1   A is true of O
> 1.1   P is true of O
> 1.1′   The fact that P is true of O renders acceptable that A is true of O

> O = opinion
> A = being acceptable
> P = being put forward by an expert in the relevant field

Viewed from a pragma-dialectical perspective, the antagonist in response to an argument from expert opinion may call the propositional content of the argument (1.1) as well as the justificatory force of the argument (1.1′) into question. The first type of criticism can be represented as a question in the following way:

> 1.1?   Has the opinion at issue indeed been put forward by an expert in the relevant field?

This critical question concerning the propositional content of the argument can be further differentiated. A first sub-question is whether the person who has expressed the opinion is indeed an expert in the relevant field. It may be the case that he is not in fact an expert, or in a different field than that to which the opinion belongs. The second issue is whether the person in question has indeed put forward the opinion mentioned in the standpoint.

The second type of criticism that the antagonist may put forward in response to argumentation from expert opinion relates to the justificatory force of the argument at issue (1.1′). This type of criticism can be formulated as follows:

> 1.1′?   Does the fact that the opinion has been put forward by an expert in the relevant field indeed render the opinion acceptable?

This critical question can be further differentiated as well. A first sub-question is whether it is indeed the case that the expert has voiced his opinion primarily from his own expertise and not from his personal interest. A second issue is whether the expert is able to defend his opinion in a way different from referring to his expertise. And a third sub-question is whether experts in the same field agree as to the acceptability of the opinion expressed in the standpoint.

---

[5] This section is based on Wagemans (2011), who takes 'argumentation from expert opinion' to be a type of 'argumentation from authority' and specifies the associated critical questions by incorporating Walton's (1997) critical questions into a pragma-dialectical framework.

Summarizing, the sub-questions regarding the propositional content of the argument raise doubt with respect to the expertise of the person and the accuracy of the representation of his opinion. The sub-questions regarding the justificatory force of the argument respectively raise doubt about the personal reliability of the expert, the presence of further evidence for the acceptability of the opinion, and the consistency of the opinion with that of other experts in the field. In practice, of course, the antagonist may also express doubt regarding a combination of these issues.

4. INSTITUTIONAL SAFEGUARDS

In the previous section we indicated that the acceptability of argumentation from expert opinion can be established by checking whether the relevant critical questions for this type of argumentation can be answered in the affirmative. If this is the case, accepting the standpoint defended by the physician may in principle be considered as reasonable, which is beneficial to the realization of the ideal of shared decision making.

In this section, we will give a number of examples of institutional safeguards for the reasonableness of argumentation from expert opinion. It is our aim to show that these safeguards may be interpreted as an institutionalized anticipation of the critical questions pertaining to argumentation from expert opinion.

In what follows, we propose to make a distinction between 'direct' and 'indirect' institutional safeguards. A safeguard is 'direct' if it provides the patient with some assurance that the answer to a particular critical question will be affirmative. A safeguard is 'indirect' if it offers patients the possibility to investigate themselves whether the answer to a certain critical question is affirmative or not.

4.1 The Physician's Expertise

The first critical question that can be raised concerning argumentation from expert opinion is: 'Has the opinion at issue indeed been put forward by an expert in the relevant field?' Since the type of discussions that are at issue in this paper generally speaking involve a reference to the speaker's own opinion, and not to that of another expert, the question whether or not the person concerned has really put forward the opinion is not of importance to our analysis.[6] In what follows we will therefore concentrate on those safeguards that can be related to the first sub-question: 'Is the person who put forward the opinion indeed an expert in the relevant field?'

According to Goldman (2001, p. 93) a novice can evaluate the expertise of an expert by relying on the judgments of 'meta-experts.' In this category of judgments Goldman includes formal forms of recognition, such as certificates and diplomas. Another way in which the novice can evaluate the expertise of an expert is by gathering information about the expert's track record.

Some institutional rules and guidelines within the Dutch healthcare system offer a number of direct safeguards that are comparable to the judgments of meta-experts. One

---

[6] See Pilgram (2012), who makes a distinction between argumentation from authority (when reference is made to someone else's authority), and argumentation by authority (when reference is made to the speaker's own authority). In the case of the latter type of argumentation by authority, Pilgram regards the question of whether the authority has been correctly represented only relevant in cases in which the doctor refers to statements made by him or herself at an earlier occasion.

example is the Dutch law for professions in the individual health care (BIG). This law aims to protect patients against incompetent and negligent treatment by health care providers. Healthcare providers are obliged to enroll in an official register and may only carry a protected medical title and practice, if they have been registered. It is checked by a special committee so that only medical specialists in the possession of a "recent competence" are registered. Professionals registered in the medical register are governed by the relevant disciplinary rules.[7]

In order to be able to enroll in the medical register, the physician must satisfy the relevant medical requirements for his or her own specialty. Roughly speaking these requirements encompass that the specialist must have followed an education for a couple of years in an authorized hospital. Medical specialists—as of 2012, other health care professionals as well—are obliged to renew their registration every five years. In order to be able to do so, the specialist must meet the minimum criteria of having regularly cared for patients, of having participated in the relevant inspection programs, and of having followed a minimum number of hours of accredited post-graduate courses and refresher trainings.

A second direct safeguard for the expertise of the medical expert is the Dutch law concerning the medical treatment contract (WGBO). This law became valid in 1995 and aims to strengthen the patient's position. Article 453 of this law runs as follows (our translation):

> Health care providers should provide good medical care and should act in accordance with their responsibilities that follow from the professional standard of health care professionals.

The code of conduct for physicians of the Royal Dutch Society of Medicine (KNMG) provide a further specification of what it means to provide good medical care. For the expertise of the physician rule I.3 and I.5 in particular are relevant:

> I.3 The care that is provided should be of good quality. Relevant aspects in this connection are: expertise, efficacy and efficiency, patient centeredness, accuracy and safety. The physician should keep the medical knowledge and skills of his own specialization up to standard. Postgraduate education and refresher courses are a necessity in this respect.
> I.5 The physician should take care not to cross the boundaries of the execution of his professional duty. He should refrain from performing actions and making statements that fall outside the scope of his own expertise.

In combination with the law itself, the code of conduct for physicians ensures that the expertise of the physician is a legal imperative.

Apart from these direct safeguards, there are also a number of indirect institutional guarantees that can be seen as relating to the sub-question about expertise. One example is that patients may check in the medical register whether their health care provider is indeed registered. They also have access to a so called 'black list' which contains the names of physicians and other health care professionals that have been suspended or have been expelled from their profession by the disciplinary judge.

In recent years more and more initiatives have been taken to give patients instruments with which they can check the quality of health care. Examples are internet sites on which comparisons between various health care providers are published, and sites in which patients' experiences are made public.

---

[7] This section is based on information drawn from various websites concerning the Dutch health care system, a.o. http://orde.artsennet.nl/Opleiding-4/Registratie_en_herregistratie.htm; http://knmg.artsennet.nl/Nieuws/ Nieuwsarchief/Nieuwsbericht-1/Relaties-transparant.htm.

4.2 The Physician's Reliability

The second critical question that can be raised in the case of argumentation from expert opinion is: "Does the fact that the opinion has been put forward by an expert in the relevant field indeed render the opinion acceptable?" The first relevant sub-question is whether or not the judgment of the expert is unbiased.

There are several factors that can endanger the independence and integrity of the physician, such as financial and other types of reward, research interests, pressure from the organization for which the physician works and personal contacts. These factors can influence the treatment given to the patient, but also the type of research that is carried out and the presentation of the results of this research.

An example of a *direct* institutional safeguard related to this sub-question is the oath that most physicians and health care professionals have to take when they receive their medical qualification. This oath can be seen as a standard for the moral self-regulation of the professional group. In the Netherlands, physicians are no longer legally required to take the oath, nor does not taking the oath have consequences for the inscription in the medical register. Nonetheless, the oath is still seen as decisive for physicians' decisions and for the patients' trust in their physicians (CHA, 2009).

Originally, the Physicians' oath was based on the Hippocratic Oath, but since 2003, in the Netherlands, the Hippocratic Oath has been replaced by a more modern version. In a commentary on this new oath, van Everdingen and Horstmanshoff make the following observations:

> Apart from a personal declaration about the physician's devotion to the patient, the text of the new physicians' oath also refers to aspects of the relationship with society and contains a number of new elements that are related to present-day discussions about professional ethics. Examples of such elements are the testable attitude of the physician (openness about data concerning the performance and about complaints and errors) and the recognition of one's own limitations (referring to other specialists on time). On the other hand, there are a number of actual problems that are not raised, such as the pressure of free market processes on professional ethics and the execution of scientific research in relation with the pharmaceutical industry (2005, p. 1066, our translation).

There are also a number of legal rules that aim to prevent different forms of conflicts of interest. The Dutch law concerning the medical treatment contract (WGBO), for instance, does pay attention to the actual problems that were just mentioned. This becomes clear if one looks at two articles of this law which aim to prevent conflicts of interest in carrying out scientific research and in maintaining contacts with the business world:

> IV.4 When doing scientific research, the physician always puts the patient's interest before his research interest so as to avoid any conflict of interest that may harm the patient. The physician only accepts recompense for the research in so far as this is proportional to the efforts that have been put in.
> V.I The physician maintains an open and honest relationship with the business world and prevents conflicts of interest that may harm the patient. Accepting favors is only acceptable to a limited extent, in accordance with the standards in the Code of conduct of the Foundation Code drugs advertisements.

Apart from such direct safeguards, there are also *indirect* institutional guarantees for the reliability of the physician. One example is the initiative taken by a group of organizations in health care in October 2011 to develop a uniform code in order to prevent both improper influencing in cases of medical advice and development of protocols that may result from conflicts of interest between physicians and the pharmaceutical industry. Since it is unavoidable that there will be interests at stake, the organizations concerned believe that optimal transparency is the most appropriate means to combat inappropriate influencing: in this way it becomes possible to make the interests visible and checkable.

In 2005 a study group of the Royal Netherlands Academy of Arts and Sciences (KNAW) published a report with recommendations for doing commissioned research. One of the most important recommendations of the study group was to sign a "Declaration of independence":

> With this written declaration, client and researchers promise to stick to a number of rules that will guarantee the independence of the scientific research. A person acting contrary to this declaration breaks his public and explicit promise, which must lead to sanctions after this has been reported to a national body (2005, p. 2, our translation).

This declaration is a direct safeguard for the independence of scientific research. The same report also mentions a measure that could be seen as an example of an indirect guarantee. In a combined editorial, a number of leading international medical journals have laid down that they will require all authors to sign a declaration in which they promise to mention their potentially conflicting interests in their scientific publications.

4.3 Additional Evidence and Consistency

The second and third sub-question with respect to the justificatory force of argumentation from expert opinion are the question whether the expert has further evidence for the opinion and the question of whether the expert's judgment is consistent with that of other experts. In this section we will give a number of examples of direct and indirect institutional safeguards that can be related to these sub-questions.

An important *direct* guarantee that is related to *both* sub-questions is the rules concerning Evidence Based Medicine (EBM). These rules are aimed at ensuring that the patient has some guarantee that the opinion of the consulted expert is in accordance with the current knowledge of experts in the same field. EBM is "the conscientious, explicit and judicious use of current best evidence in making decisions about the care of individual patients" (Sackett, Rosenberg, Muir Gray, Haynes, & Scott Richardson, 1996, p. 71). In the Dutch law concerning the medical treatment contract (WGBO), the following stipulation concerning EBM can be found:

> I.6 The physician is prepared to account for his opinions and to adopt a testable attitude. Guiding principle for this test is the criterion "customary practice among professional colleagues." The implementation of this criterion should be by an accredited scientific association.

This quote makes clear that the rules governing EBM do not just constitute a direct safeguard for an affirmative answer to the question about further evidence for the physician's opinion, but also for the question about whether the expert's opinion is consistent with that of other experts.

As regards the indirect safeguards, the asymmetrical relationship between physician and patient will in many cases make it impossible for the patient himself to check whether the physician's opinion is based on further evidence. The recent publication of summaries of guidelines for medical specialists in non-technical language by a cooperative of the Dutch scientific associations for medical specialists may be seen as an attempt at giving the patient the opportunity to check whether his physician's opinions are in accordance with the medical evidence, and consistent with the opinions of other experts in the relevant field.

By far the most important indirect safeguard for the consistency of the expert's opinion with that of other experts are the regulations with respect to the so-called 'second opinion.' In the aforementioned Dutch law concerning the medical treatment contract (WGBO, II.19), it is specified that a physician should comply with the patient's request to be referred to another health professional for a second opinion, unless there are weighty considerations against doing so, which should then be made explicit and motivated. Since patients thus have the right to ask for a second opinion, they have the possibility of checking whether the first expert's opinion is consistent with the second expert's opinion.

5. CONCLUSION

In this paper we have shown that the institutional guidelines and procedures within the medical field can be related in a meaningful way to argumentation theoretical standards for the reasonableness of argumentation from expert opinion. On the basis of this research it may be concluded that the asymmetrical relationship between the physician (the expert) and the patient (the layman) does not necessarily put the ideal of shared decision making at risk. Even in cases where a direct assessment of the physician's opinion is not possible and the patient has to rely solely on the physician's expertise, the reasonableness of the judgment is to a large extent guaranteed. This is done both by direct safeguards, which can be viewed as assessments of argumentation from expert opinion that have been delegated to the institution, and by indirect safeguards, which enable patients to evaluate the reasonableness of this type of argumentation themselves. Of course, physicians will always have to meet the minimum requirements for informed consent, which means that they should allow their patients to give their consent for the treatment on the basis of an understanding of the facts, implications and consequences of the treatment proposed. Whenever the explicit or implicit appeal to expertise obstructs this understanding, without there being an adequate justification for it, the appeal is not only contrary to the ideal of shared decision making, but also contrary to the legal requirement of informed consent.

ACKNOWLEDGEMENTS: We would like to thank Bart Garssen, Nynke Kalkers and an anonymous referee for their critical comments on an earlier version of this contribution.

REFERENCES

Ariss, S. M. (2009). Asymmetrical knowledge claims in general practice consultations with frequently attending patients: Limitations and opportunities for patient participation. *Social Science & Medicine, 69*, 908–919.
CHA—Commissie Herziening Artseneed (2009). *Nederlandse artseneed.* [The Dutch physicians' oath]. Badoux: Houten.
Charles, C., Gafni, A., & Whelan, T. (1997). Shared decision-making in the medical encounter: What does it mean? (or it takes at least two to tango). *Social Science & Medicine, 44*(5), 681–692.
Goldman, A. I. (2001). Experts: Which ones should you trust? *Philosophy and Phenomenological Research, 63*(1), 85–110.

Goodnight, T. G. (2006). When reasons matter most: Pragma-dialectics and the problem of informed consent. In P. Houtlosser & M.A. van Rees (Eds.), *Considering pragma-dialectics: A festschrift for Frans H. van Eemeren on the occasion of his 60th birthday* (pp. 75–85). Mahwah, NJ: Lawrence Erlbaum Associates.

Goodwin, J., & Honeycutt, L. (2009). When science goes public: From technical arguments to appeals to authority. *Studies in Communication Sciences, 9*(2), 19–30.

Joosten, E. A. G., DeFuentes-Merillas, L., de Weert, G. H., Sensky, T., van der Staak, C. P. F., & de Jong, C. A. J. (2008). Systematic review of the effects of shared decision-making on patient satisfaction, treatment adherence and health status. *Psychotherapy and Psychosomatics, 77*, 219–226.

KNAW Werkgroep Opdrachtonderzoek (2005). *Wetenschap op bestelling: Over de omgang tussen wetenschappelijk onderzoekers en hun opdrachtgevers* [Science on command: About the contact between researchers and their clients]. Amsterdam: Koninklijke Nederlandse Akademie van Wetenschappen.

KNMG (2002). *Gedragsregels voor artsen. Richtlijn II.01.* [Code of conduct for physicians]. Utrecht: KNMG.

Légaré, F., Elwyn, G., Fishbein, M., Frémont, P., Frosch, D., Gagnon, . . . van der Weijden, T. (2008). Translating shared decision-making into health care clinical practices: Proof of concepts. *Implementation Science, 3*(2), 1–6.

Pilgram, R. (2012). Reasonableness of a physician's argument by authority: A pragma-dialectical analysis of the specific soundness conditions. *The Journal of Argumentation in Context, 1*(1), 33–50.

Sackett, D. L, Rosenberg, W. M. C., Muir Gray, J. A., Haynes, R. B., & Scott Richardson, W. (1996). Evidence based medicine: What it is and what it isn't. *British Medical Journal, 312*, 71–72.

Snoeck Henkemans, A. F. & Mohammed, D. (2012). Institutional constraints on strategic maneuvering in shared medical decision-making. *The Journal of Argumentation in Context, 1*(1), 19–32.

van Everdingen, J. J. E., & Horstmanshoff, H. F. J. (2005). De nieuwe Nederlandse artseneed. [The new Dutch physicians' oath). *Nederlands Tijdschrift voor Geneeskunde, 149*, 1062–1067.

Wagemans, J. H. M. (2011). The assessment of argumentation from expert opinion. *Argumentation, 25*, 329–339.

Walton, D. N. (1997). *Appeal to expert opinion: Arguments from authority*. University Park, PA: Penn State University Press.

Walton, D. N., Reed, C., & Macagno, F. (2008). *Argumentation schemes.* Cambridge: Cambridge University Press.

Do Experts Help or Hinder? An Empirical Examination of Experts and Expertise during Public Deliberation

LEAH SPRAIN
ANDY M. MEROLLA
MARTÍN CARCASSON

Department of Communication Studies
Colorado State University
Fort Collins, CO
USA
Leah.Sprain@colostate.edu
Andy.Merolla@colostate.edu
Martin.Carcasson@colostate.edu

ABSTRACT: We consider expertise in interaction during small group public deliberations. Taking communication as design, we analyze the intentional design of deliberative format using invited experts to support public discussions. Through discourse analysis of one expert's interventions into the group discussion, we suggest how expertise might best contribute to public deliberation.

KEYWORDS: experts, expertise, public deliberation, discourse analysis, communication as design.

1. INTRODUCTION

In the Mountain West, water is a wicked problem (Rittel & Webber, 1973), a problem with dimensions that are "ill-formulated, involve uncertainty and confusing information, have many decision-makers and affected parties with different and conflicting values, and promise ramifications for the whole system" (Ferkany & Whyte, 2011, p. 3). Lacking technical solutions, wicked problems require new relationships between citizens and experts as they require broad citizen participation since science cannot resolve value dilemmas (Fischer, 1993). Wicked problems require moving beyond the "cult of the expert" (Boyte, 2009) and the "culture of technical control" (Yankelovich, 1991) to reimagine the nature and role of expertise within public deliberation. As Fischer (2000) argues, "experts . . . possess no analytical wizardry capable of resolving our pressing societal problems. Expert judgment, we come to recognize, provides few uncontested solutions or answers . . . while we still need experts, expertise cannot stand alone" (p. 41).

In this paper, we consider expertise in interaction during small group public deliberations about water in Northern Colorado. Taking communication as design (Aakhus, 2007; Aakhus & Jackson, 2005), we describe a deliberative format designed to use outside experts to support public discussions and analyze when one expert did not follow the design. Through discourse analysis of this expert's interventions into the group discussion, we suggest how expertise might best contribute to public deliberation.

Sprain, L., Merolla, A.M., & Carcasson, M. (2012). Do experts help or hinder? An empirical examination of experts and expertise during public deliberation. In J. Goodwin (Ed.), *Between scientists & citizens: Proceedings of a conference at Iowa State University, June 1-2, 2012* (pp. 355-364). Ames, IA: Great Plains Society for the Study of Argumentation. Copyright © 2012 the author(s).

2. PUBLIC DELIBERATION DESIGN CHALLENGES

Several disciplines have started reconsidering the role of scientific experts and values in public problem-solving under several names: civic science (Bäckstrand, 2003), post-normal science (Functowicz & Ravetz, 1993), and post empirical policy analysis (Fischer, 2003). These efforts emphasize the importance of a more informed citizenry who have weighed multiple options with opportunities for interaction and exchange with decision makers (Abelson et al., 2003). But the harder questions center on how interactions between experts and nonexperts should take place (Goodwin, 2011).

Discussing experts raises the question of how to conceptualize experts and expertise. Experts can be knowledgeable contributors to the topic at hand; the nature of their contribution can lead to different categories of experts, such as Collins and Weinel's (2011) schema of beer-mat knowledge, popular understanding, primary source knowledge, interactional expertise, and contributory expertise. Expert status can also be a social identity connected to institutional positions, professional roles, or even location (e.g. the local expert). From a discourse perspective, the social position of being an expert is constituted by utterances in interaction rather than a quality of an individual (Hartelius, 2011; Jacoby & Gonzales, 1991). Our orientation to communication as design focuses our attention to the ways that expertise was operationalized in a particular set of deliberative meetings. Nonetheless, understanding multiple ways of conceptualizing expertise can open up new design possibilities for the future.

In the move to rethink how to best solve public problems through interaction between experts and citizens, three important challenges related to expertise must be considered. First, many scientists still support the informational deficit model of communication (Petts & Brooks, 2006), which suggests that it is the lack of information that leads to current public opinion. This reinforces the belief that the solutions to various social problems lie in information dissemination from experts and, by extension, knowledge acquisition by citizens, voters and decision-makers. This challenge suggests the need to cultivate new understandings of the role of information and data in decision-making, not just expertise.

Second, the public often wants experts to tell them what to do. In complex public issues, the public will often look for easy solutions rather than work through complex value dilemmas. Experts can represent a shortcut, an alternative to the difficult work of deliberation. Yet despite looking to experts for answers the presence of experts during a public discussion can also create a chilling effect on the public who become scared to say the wrong thing.

Third, some deliberative practitioners want to exclude experts and facts, often in response to the second challenge. This move is not just a way to get rid of an easy excuse for citizens. As Mahdik and Keith (2011) argue, "expertise is a kind of authority, and so stands in contrast to liberal democratic values; at its core, a democratic polity depends on its ability to keep a check on authority" (p. 371). Yet a particularly important problem is created by this move: without good information deliberation can be easily dismissed. In particular, experts who listen to deliberation and only hear factual errors become convinced (or more convinced) that deliberation promotes misinformation, and the public is unable to understand complex technical issues.

In response to these challenges, deliberation practitioners must design processes that equip the public to work through a range of options and fully consider their implications. In this paper, we follow Aakus and Jackson's (2005) call to open up intentional design as an object of inquiry. In our case, deliberation is a designed context that uses techniques, devices, and procedures to shape new possibilities for communication during public meetings. Using discourse analysis to attend to the micro-matters of language and interaction, empirical analysis enables reflection of a particular design, detecting surprises, flaws, and opportunities for redesign (Aakus, 2007).

3. DELIBERATIVE DESIGN AND METHODS

The transcripts for this paper come from a public series on water in Northern Colorado. A coalition of local organizations joined with Colorado State University to help improve the community's conversation about water, a polarized topic in the arid Mountain West. The public series included a kick-off event focused on values, three public education sessions featuring water, and public deliberations capped the series.

The deliberation process was designed by the Center for Public Deliberation (CPD) working with the Colorado Water Institute (CWI). It used a modified National Issues Forum format focused on the central question: how should we meet our future water supply needs? Groups worked through four approaches to answering this question, focusing on addressing growth, urban conservation, storage projects, and agricultural conservation and transfers. The background materials were initially developed by a graduate seminar on water conflict, piloted, and revised by the CPD and CWI. Local groups for and against specific water projects vetted drafts of the background materials, and their feedback was integrated into the final materials.

The public deliberation consisted of small group discussions of six to nine people (14 total groups) facilitated by a CPD student-facilitator. CPD facilitators were trained on the process and background information on water issues, but they were experts in the process, not water issues. Given the technical nature of this issue, the CWI raised concerns that groups may need outside experts who know more about water law, engineering, agriculture, and the like in order to have productive conversations. The CPD suggested having experts "on tap but not on top." In practice, this meant that individuals with relevant expertise (e.g., Director of the Colorado Water Institute, an Environmental Protection Agency staff member, a Sociology professor) were invited to serve as outside experts. They were given a separate nametag that marked their status, and they were asked by the CWI if they would be available to answer questions if needed. The design was to have them walking around so that a facilitator could bring one of them over to a group if needed, but otherwise the group would focus on talking with each other rather than continuing to ask questions of the expert.

Overall this design made an official designation between *invited experts* and expertise within the small groups. To be clear, several participants in the groups also had relevant expertise. For example, one group included the chair of an activist organization to save the local river, a citizen who kayaked the entire river from the headwaters to the Gulf of Mexico, a water engineer for a nearby city, and an employee of the Colorado Water Institute. In this paper, we focus on the design of invited experts, but we recognize concerns about expertise within these public deliberations extend beyond invited experts.

4. EXPERTS IN INTERACTION

At the beginning of the small group deliberation, the facilitator initiated a round to have each person in the group introduce themselves and explain what brought them to the meeting (sometimes called a personal stake). An invited expert explained to one group that she was invited as an expert and may go around to other groups, but she is there to "basically observe, listen, and clarify anything about the EIS process." The Director of the CWI introduced himself by saying "I'm not really at this table." Instead, he would be going around to "listen" and "answer questions." Both of these invited experts followed the design to have experts available to support deliberation if needed.

One of the invited experts did not follow this design. Instead, he walked around to listen to several groups, eventually intervening in one conversation multiple times. Our analysis focuses on three of his extended interventions to understand the interaction created through his interventions.

4.1 Intervention 1

> *Man 2*: … so, again I bring up the timing issue that it's not so much that we need these reservoirs because people are consuming all that but it's just different times of the year we have the water and then it needs to be redistributed that's one of the fundamental things which you get to also
>
> *Expert*: if I may offer just a quick comment you are quite correct it seemed to me to mention that it is a question of timing now the problem is cities use water very differently than agriculture
>
> *Man 1*: that's right
>
> *Expert*: agriculture if you build a dam "x" and you release that water to agriculture, you can pretty well figure the farmers are going to use their water they're going to spend their water every year they're going to drop that reservoir and they're going to grow crops with it and if they're sufficiently inefficient and I hope they are then a good fraction of that water will return to the river by return flows now when the city builds a dam for urban purposes the city has got to keep industry going the Kodak plant has to run 365 days a year you can't shut things down and for that matter other enterprises and our households
>
> *Man 1*: OK
>
> *Expert*: so cities hold water and spread it out across the area and they don't send down [indecipherable] but cities are a key watershed for generating return flows you want to look at the pattern of return flows and are you just creating a canal which would be not ecologically very healthy or are you creating a variable river where you're filling out the banks and you're creating a variable flow and I think that's where your conversation has to go is how can a city be a better watershed through its return flows than it otherwise might be if you connect your dam to the city to the way the city organizes its return flows then you're getting somewhere I don't know if I'm making sense to you or not
>
> *Man 3*: that sounds like a very complex thing to actually implement in you know both politically and just hydrologically
>
> *Expert*: agriculture shifts water to the cities the river can be hurt very badly unless the cities undertake specific programs of making sure the river doesn't turn into a canal=
>
> *Man 1*: yeah
>
> *Expert*: =which is just constant flows
>
> *Man 3*: are there good examples of that having been done
>
> *Expert*: actually, yes but I don't think you want me to go on [chuckle; laughter from facilitator] on to that little
>
> *Man 1*: schedule it
>
> *Expert*: I think that's where the thinking has to be headed when you raise your point perfectly good point just don't stop there
>
> *Facilitator*: well thank you for that I'm sorry we don't have more time to discuss this but if you could fill out your yellow sheets once again

The expert intervenes in the discussion using a politeness token "if I may" and labels his move a "quick comment." He ends up talking for almost two minutes before being challenged by a participant, which is an extended turn in this conversation. When people don't know each other, they should expect roughly equal turns; this extended turn implicates a power differential between the expert and the group. Despite confirming the speaker ("you are quite correct"), he initiates a frame shift through a topic shift from discussing timing to introducing a new topic of how cities can be a better watershed. In doing so, he presents a new goal for the conversation: "that's where your conversation has to go is how can a city be a better watershed through its return flows." By attempting to guide the direction of the conversation and the appropriate topics for talk, he offers interactional expertise, the type of process expertise that you might expect from the facilitator. He ends his this shift saying "I don't know if I'm making sense to you or not." This is not phrased as a question, but instead functions as a distancing statement that seems to reinforce his literal position above the group. This statement implies that if he isn't making sense, the locus of misunderstanding is in the group not his presentation.

Man 3 responds with a challenge about the feasibility of implementing this idea, which is in line with the norms of deliberation designed to consider a wide range of possible solutions and weigh the tradeoffs between them. The expert continues to explain the idea but does not respond to the challenge. Man 3 then reformulates his comment as a question: "are there good examples of that having been done." As an adjacency pair, the question formulation more explicitly calls for a direct response from the expert, even though this question functions as a way to continue to explore the challenges of implementation. By asking for "good examples" the participant does not require the expert to speak against his argument but instead provide evidence of it working. The expert provides confirmation of the question by responding, "yes actually," but then continues "you don't want me going on to that," which is delivered as though it is a joke since he punctuates it with a slight chuckle.

The irony of this exchange is that this is the very sort of question that experts should answer in the "on tap but not on top" model; an expert can present examples that help a group work through the tradeoffs of an idea. Obviously, the "on tap" expert should not be the one advocating for the idea, but the request for an example is reasonable. When the expert declines to answer the question, man 1 says "schedule it," which seems to imply that time should be scheduled for the expert to present this information. The tone is a bit caustic, suggesting that separate time would need to be scheduled for the expert to meaningfully contribute to the group. The expert ends up leaving the group reinforcing his new frame for conversation, which is a return to offering interactional expertise.

4.2 Intervention 2

> *Man 1*: ... you can also say the same thing about ag are they are operating as efficiently as they should well if they operate too efficiently the water doesn't run off to the wetlands and it doesn't go downstream so it's complicated
> *Woman 2*: yes so is water law [laughs]
> *Man 1*: right
> *Expert*: if I may just offer the economists will say yeah MNI [municipalities] is more important than agriculture and I'll tell you why they add more value per acre-foot they add more jobs more money and more income throwing around and tax base and so on but I think you may be right you want to advance your thought about why you would argue for agriculture or wll why should we take on the economists on that in your judgment
> *Woman 2*: um I don't know ummm but when you read this it does say that agriculture if I read it correctly here thoroughly is important and we've depended on that in this county in this whole front range forever but now the big concern is well will municipalities have enough water ard I think an equal concern should be will ag agriculture have enough water
> *Expert*: now does if one gets it does the other have to be without
> *Woman 2*: I don't know but you know they say

Expert: maybe worth exploring

Woman 2: they say you can always buy water from the farmers well you can't farmers are buying water from us

Expert: yeah let's assume that I draw water into my house but I do not put much of a consumptive use on it I cook my carrots a little bit goes off to Iowa as rain through my vent ok well most of it stays in my kitchen I pump it down my sink and it goes in the sewers so it goes out now could we imagine the city as a watershed for agriculture because of this return flow sure we could do that absolutely maybe that's the future

Woman 2: that's a storage though see the water the city would have to put it somewhere where the farmers could use it

Expert: and that is true and so one looks at how that use can be made but in other words instead of seeing agriculture and cities as competitors for water we might want to say that we have to seek partnerships there

Woman 2: I agree that would be the right word.

Expert: uh so that as your city becomes more prosperous it pulls down its consumptive uses in a way and releases them in a way useful for agriculture and other living things on the river who have nothing to do which have nothing to do with agriculture

Woman 2: very well said

The expert intervenes with a politeness token, "if I may," and again affirms the previous speaker "you may be right" only to introduce a new frame (importance of agriculture) for the discussion. The expert poses questions to a previous speaker, woman 2 (not the people speaking when he interrupted). His Socratic line of questioning introduces the notion that cities and agriculture might be in tension with each other. Eventually the expert takes over and argues for seeking "partnerships." At the end of the exchange, woman 2 affirms his position "very well said."

In this exchange, the expert not only introduces a new frame for the conversation, but he takes on the role of facilitator guiding the conversation by posing questions designed for one individual. Eventually, he switches from interactional expertise to normative expertise as he presents his ideal relationship between cities and agriculture. The woman reinforces this move by saying he is "well said," which seems to confirm agreement about the ideological content of his argument as well. Unlike her earlier concern that both agriculture and cities need water, the expert has reoriented this conversation to focus on the nature of their relationship (partnership) instead of who is legally entitled to water (water law) or what their water needs.

4.3 Intervention 3

Woman 1: but water rights in Colorado are one of those like most restricted like kind of arbitrary rules in like a lot of ways because like you can't even put you can't collect rain that comes on falls onto your house because of water rights in Colorado I think that it's like the water rights in general need to be changed and redone because of like time changing times

Expert: if I may speak just to the question of fact there

Woman 1: yes sir

Expert: they did change that here within the last year

Woman 1: they have

Expert: yeah

Woman 1: OK

Expert: meaning you can now collect water off the roof

Man 2: aren't they doing a study though

Expert: and my opinion is simply this we want to watch it that rule made a lot of sense not to be able to connect collect water uh or greywater but that's not for me to say here you know I'll just simply rule number one for me protect your return flows because I always live off the waste of the person above me and I've become the source of waste for the person below me and if we start collecting and putting my return flows to a reuse to grow my petunias or whatever it might be but another consumptive use then I'm depriving you of your water so we have to understand that we use each other's water when I use my water I don't destroy it the water is still there a certain portion goes off to Iowa as rain we call that a consumptive use but most of the water and most uses stays in our system and goes to the next user or to other living things in a=

Woman 2: =to a junior=

Expert: =it goes to junior partner=

Women 2: =partner=

Expert: and so when we interfere with that return flow by collecting it right here and putting it on my petunias and it doesn't go downstream to him we are we are undercutting the sustainability of our civilization

Woman 1: but couldn't you question the use of the greywater and how you use the greywater

Expert: oh you could

Woman 1: because I would say that there's a sustainable way to use like my water that I've collected and use it in a sufficient like self-sustained way

Expert: and if you are using it in a manner in which you're not depriving her of her source you're you're you're giving the waste up so she it's her supply

Woman 1: and it's closing

Expert: then I have no problem but if you are becoming so efficient that you're depriving the return flows to other users that's where we get in trouble

Man 2: if you use your water efficiently and then the next student moves in and uses the water that you didn't use now we've dried up the river more right

Expert: that's right so we have to

Woman 1: but couldn't you say the greywater isn't taking away from the like river water and f- river f- or water farther down the flow because greywater's just basically using like I don't know I would say that it's like a closed area it's like=

Facilitator: so

Woman 1: =using the closing the loop on yourself

Man 4: it's not though your use is=

Facilitator: I think um sorry can I stop you there sorry so this is um kind of one of the contentions of this=

Expert: =there's=

Facilitator: =this issue

Expert: factually there's no contention

Facilitator: ok well that

Expert: factually it is return flow

Facilitator: ok well um that's that's your opinion and she=

Exert: what is cont=

Facilitator: =has her opinion=

Expert: =what is contentious=

Facilitator: =but=

Expert: =is how we manage our greywater but there's no contentiousness about one person's waste being another person's supply

Facilitator: ok um

Expert: that's not a matter of opinion

Facilitator: all right well thank you [laughs] um I just want to in the last couple minutes that we have I want to hear some appreciation because we already talk- talked a lot about the concerns of this approach but could we hear a few appreciations before we fill out our yellow sheets

The outside expert justifies this intervention: "if I may just speak on the question of fact." In the interaction, there was no question of fact. The participants did not raise a question nor was there expressed uncertainty. Instead, the expert hears something he knows is factually wrong so he uses that as a reason to interject. When he corrects woman 1, she defers to his expertise about whether or not the law has been changed (the factual issue). But the expert goes on to provide a justification for the law, including principles for water management. As he starts this turn, he indicates that he knows that this is not his role, "but that's not for me to say here you know." But then he proceeds to offer his opinion anyway. He ends his justification for ending greywater use, noting otherwise "we are undercutting the sustainability of our civilization." This extreme case formulation functions to label the problem with the new greywater law and, thus, challenge woman 1's use of greywater as an example of a problem.

Although woman 1 accepts the factual correction, she challenges this justification for the law, noting that you can "question" whether greywater can be used sustainably. She and the expert go back and forth about the principles being used to evaluate greywater. Woman 1 is focused on sustainability while the expert focuses on altruism and return flows. Towards the end of this interaction, the facilitator interjects to identify this as a contentious issue. Here the facilitator is using a strategy from her training that can be used to bracket fact disputes that are not likely to be resolved in deliberation, particularly without outside research. The expert pushes back on the framing, noting that "factually there is no contention." He continues; what is contentious is "how we manage our greywater but there's no contentiousness about one person's waste being another person's supply." Separating facts from values, the expert correctly identifies the nature of the dispute. But by stating the factual issue as "one person's waste being another person's supply," he attempts to ground his argument as "not a matter of opinion," which suggests that it cannot be contested. This formulation opposes the exchange he had with woman 1 who questioned whether there were sustainable ways for someone to use greywater without damaging someone else's supply, which complicates the expert's "factual" characterization. This exchange demonstrates the sometimes slippery relationship between facts and values during a discussion and the rhetorical advantages of framing something as fact rather than opinion.

5. CONCLUSION—DESIGN INSIGHTS

Our analysis of how an invited expert intervenes in deliberation suggests several lessons for future deliberative design and theorizing the role of experts and expertise in public deliberation. First, having experts intervene during deliberation to correct factual errors may be counterproductive. The third intervention demonstrates an extended tangent created when an expert intervened to clarify a small factual error. Of note, this error was only offered as an example—the factual error had no real consequence to the broader issue under discussion (the complexity of Colorado water laws). From a design perspective, it seems likely that technical experts will hear lots of things during deliberation that they might take issue with. Intervening to correct these issues is likely to fundamentally change the issues under discussion. Rather than intervening during the discussion, deliberative design might benefit from increased engagement with technical experts in developing background materials and identifying factual issues that need clarification in future deliberation.

Second, the relationship between facts and values during interaction is tricky; technical experts may provide normative advice even though they may assume they are simply providing facts. Pellizzoni (2011) forecasts this possibility in his discussion of the transgressive nature of expertise to synthesize knowledge and cut across boundaries, moving from technical rationality to moral judgment and back again. If wicked problems require citizens to work through value conflicts because these choices cannot be made by science alone, deliberative design must separate experts laying out issues from presuming that the public should also follow expert ideologies, principles, and values for working through these dilemmas. Expecting experts to distinguish facts and values during deliberation is not practical, and it may even become strategic. One design response to these problems is to provide experts training on the nature of wicked problems and the ideal roles of experts and publics within them. Since experts likely operate from an epistemological viewpoint that privileges data, training on wicked problems can help provide an alternative epistemology for understanding their role in deliberation.

Finally, this analysis illustrates a technical expert providing interactional guidance by asking questions and introducing new frames for discussion. Considering many technical experts may also be professors accustomed to leading discussions or guiding students, this discourse is not a surprise. But when a technical expert also functions as an interactional expert it can create at least two problems. First, it can end up undermining other interactional experts, namely the facilitator. This occurs in the third excerpt where the facilitator struggles to reframe the interaction only to be challenged—and flustered—by the expert. Beyond this interpersonal problem, conflating technical and interactional expertise can result in the expert offering a proper *telos* of the conversation, which may be interpreted as stemming from their technical expertise. In this situation, the expert's authority encroaches on the democratic polity (Mahdik & Keith, 2011). Here again, training technical experts on the epistemology of deliberation— and the role of the facilitator—may limit technical experts trying to guide interaction away from the norms of deliberation.

REFERENCES

Aakhus, M. (2007). Communication as design. *Communication Monographs, 74*, 112–117. doi: 10.1080/03637750701196383

Aakus, M., & Jackson, S. (2005). Technology, interaction, and design. In K. Fitch and R. Sanders (Eds.), *Handbook of language and social interaction* (pp. 411–436). Mahwah, NJ: Lawrence Erlbaum Associates.

Abelson, J., Forest, P., Eyles, J., Smith, P., Martin, E., & Gauvin, F. (2003). Deliberations about deliberative methods: Issues in the design and evaluation of public participation processes. *Social Science & Medicine, 57*, 239–251.

Bäckstrand, K. (2003). Civic science for sustainability: Reframing the role of experts, policy-makers, and citizens in environmental governance. *Global Environmental Politics, 3*, 24–41. doi: 10.1080/0964401042000274322

Boyte, H. (2009). Civic agency and the cult of the expert. A study for the Kettering Foundation.

Collins, H. and Weinel, M. (2011). Transmuted expertise: How technical non-experts can assess experts and expertise. *Argumentation, 25*, 401–413. doi: 10.1007/s10503-011-9217-8

Coupland, J., & Williams, A. (2002). Conflicting discourses, shifting ideologies: Pharmaceutical, 'alterantive' and feminist emancipatory texts on the menopause. *Discourse & Society, 13*, 419–445. doi: 10.1177/0957926502013004451

Fischer, F. (1993). Citizen participation and the democratization of policy expertise: From theoretical inquiry to practical cases. *Policy Sciences, 26*, 165–187. Retrieved from http://www.jstor.org/stable/4532286

Fischer, F. (2000). *Citizens, experts, and the environment: The politics of local knowledge.* Durham, NC: Duke University Press.

Fischer, F. (2003). *Reframing public policy: Discursive politics and deliberative practices.* New York, NY: Oxford University Press.

Ferkany, M., & Whyte, K. P. (2012). The importance of participatory virtues in the future of environmental education. *Journal of Agricultural Environmental Ethics, 25*(3), 419-434. doi: 10.1007/s10806-011-9312-8

Funtowicz, S. O., & Ravetz, J. R. (1993). Science for the post-normal age. *FUTURES, 25*, 739–755. doi: 0016-3287/93/07739-17

Goodwin, J. (2011). Accounting for the appeal to the authority of experts. *Argumentation, 25,* 285–296. doi: 10.1007/s10503-011-9219-6

Hartelius, E. J. (2010). *The rhetoric of expertise.* Boulder, CO: Lexington Books.

Jacoby, S., & Gonzales, P. (1991). The constitution of expert-novice in scientific discourse. *Issues in Applied Linguistics, 2,* 149–181. doi: http://escholarship.org/uc/item/3fd7z5k4

Lindblom, C. E., & Cohen, D. K. (1979). *Usable knowledge: Social science and social problem solving.* New Haven, CT: Yale University Press.

Mahdik, Z. P., & Keith, W. M. (2011). Expertise as argument: Authority, democracy, and problem-solving. *Argumentation, 25,* 371–384. DOI 10.1007/s10503-011-9221-z

Pellizzoni, L. (2011). The politics of facts: Local environmental conflicts and expertise. *Environmental Politics, 20,* 765–785. Retrieved from http://dx.doi.org/10.1080/09644016.2011.617164

Petts, J., & Brooks, C. (2006). Expert conceptualizations of the role of lay knowledge in environmental decisionmaking: Challenges for deliberative democracy. *Environment and Planning A, 38,* 1045–1059. doi: 10.1068/a37373

Rittel, H. W. J., & Webber, M. M. (1973). Dilemmas in the general theory of planning. *Policy Sciences, 4,* 155–169. Retrieved from http://www.jstor.org/stable/4531523

Yankelovich, D. (1991). *Coming to public judgment: Making democracy work in a complex world.* Syracuse, NY: Syracuse University Press.

Scrambling on Defense: An Anatomy of Anthropological Responses to the Mead/Freeman Controversy

ROBERT STRIKWERDA

Political Science and Women's Studies
St. Louis University
3750 Lindell Blvd. 125 McGannon Hall
St. Louis, MO 63108
USA
rstrikwe@slu.edu

ABSTRACT: In 1984 Derek Freeman launched a crusade against Margaret Mead's *Coming of Age in Samoa*, claiming to have definitively falsified her central claims. Anthropologists responded in a proliferating variety of ways, while failing to project scientific coherence. Underlying anthropology's inability was its changing, and increasingly fissuring, conception of itself as a discipline. Was it a humanistic field or a science, and if a science what kind of science? Most saliently revealed was the discipline's failure to recognize its own tacit acceptance of Mead's impact on the American reading public without clearly—and earlier—advancing critical views of Mead's earliest work.

KEYWORDS: Margaret Mead, Derek Freeman, anthropology, scientific controversy.

1. INTRODUCTION

> This is more than just another academic teapot tempest; anthropology is a science often accused of being a haven for social theorists manipulating facts to prove their preconceived points . . . Mead . . . made major contributions to U. S. social attitudes. Her reputation is secure. The real loser may be anthropology's reputation as a science. If its methods haven't made quantum leaps forward since Mead's day, the whole discipline might find a better home in creative literature. (*Denver Post* as cited in Rappaport, 1986, p. 316)

Margaret Mead has a curiously divided standing today among Americans. For some she is one of the inspirational leaders of the 20th century. Many walls and websites display as a proud motto her proclamation stating "Never doubt that a small group of thoughtful, committed, citizens can change the world. Indeed, it is the only thing that ever has" (Stover, 2005).[1] For some in the new field of evolutionary psychology, however, Mead is close to a *bête noire*; her first book *Coming of Age in Samoa* (hereafter, *COAS*) is excoriated for misleading American social science for many years while being based on a dreadfully misleading hoaxing. Steven Pinker's comments in *The Blank Slate: The Modern Denial of Human Nature*, are representative. "Mead's descriptions of peace-loving New Guineans and sexually nonchalant Samoans were based on perfunctory research and turned out to be almost perversely wrong. As the anthropologist Derek Freeman later documented, Samoans may beat or kill their daughters

---

[1] Stover claims that it is not clear where Mead first made this statement.

Strikwerda, R. (2012). Scrambling on defense: An anatomy of anthropological responses to the Mead/Freeman Controversy. In J. Goodwin (Ed.), *Between scientists & citizens: Proceedings of a conference at Iowa State University, June 1-2, 2012* (pp. 365-377). Ames, IA: Great Plains Society for the Study of Argumentation. Copyright © 2012 the author(s).

if they are not virgins on their wedding night, . . ." (2002, p. 56).[2] And as reflected in the opening quote, this discredits the field of cultural anthropology as well as Mead.

Yet among American cultural anthropologists, whatever their opinions of Mead's work, their general view seems to be that they have quite rebutted Freeman's attacks. I wish to explore, and partly explain the divergence between these differing views. In particular, why hasn't anthropology been able to put Freeman to rest? At the beginning of his 2010 review of Paul Shankman's *The Trashing of Margaret Mead*, Robert LeVine asks: "Can there ever be an end to the 27-year-old controversy over Margaret Mead's Samoan fieldwork of 1926?" Similar plaints can be found throughout the literature, such as this comment from Morton (1996):

> I was amazed to find that yet another contribution to the so-called 'Mead-Freeman controversy' had been published, . . . It is even more unfortunate that authors cannot resist making judgements on this issue and trying to resolve the issues involved, insisting that there is and was a definitive, 'real' Samoa to be discovered. (p. 166).

American cultural anthropology seems to have been continually playing catch up with Mead's critics.[3] Like a monster that one tries to bury, the repercussions of Freeman's attacks still echo in certain venues.

I will focus on one major factor in the controversy: anthropology's inability to frame the criticisms of Freeman in a fashion that made their various defenses of Mead – and repudiations of Freeman's position—persuasive to its inter-field and extra-field audiences. All their responses to Freeman did not accumulate into a cogent overall response. Indeed, as I will argue, anthropologists could not make their attacks persuasive without simultaneously undercutting their field's epistemic standing.

I structure my analysis using a helpful distinction taken from John Lyne (1983) between three audiences: the intra-field audience of other anthropologists, the inter-field audience, scientists and academics in other disciplines, and the extra-field audience of academia—the general public, or reading public. Given, as I will argue, anthropology's history of not taking its extra-field audience seriously, it seems to have lost credibility here, seemingly having lost ground to its inter-field competition, sociobiology and, more recently, evolutionary psychology. Another issue that I emphasize is the responsibility for a discipline to publicly criticize itself, to exemplify what is supposedly one of the hallmarks of a science, being a self-correcting enterprise (Laudan, 1981).

Today we may have lost sight of Margaret Mead's standing in America. She was the best known American anthropologist, one of the most prominent social scientists of the time, a major cultural figure. And Freeman attacked her fieldwork, the defining mark of anthropology as a distinctive discipline: "Ethnography has been, and is, the sine qua non of cultural anthropology. It accounts for our initial status and networks within our profession, legitimizes us as "real" anthropologists . . . and provides us with the means to survive the publishing dictates of the academy" (Farrer, 1996, p. 170). At the same time, the controversy ranged over such a range of issues, from very specific empirical issues of fact, to issues of method—techniques for gathering evidence, to methodology, theories of how research should proceed,

---

[2] For more examples, see Harris (2009), p. 172 and Simpson, Stich, Carruthers, & Lawrence (2006). Also see Buss (1999), p. 26.

[3] For simplicity I will here use anthropology to stand for what more precisely is "American cultural anthropology."

to more abstract philosophic or epistemological issues of the grounds of anthropological knowledge and justification (Harding, 1987, p. 2). Given this range, let me emphasize that this is a very limited look at just certain aspects of the Mead–Freeman controversy.

The notion of controversy itself deserves some attention. First, this is not a "manufactured scientific controversy" in Leah Ceccarelli's sense of one centered around an issue about "there is actually an overwhelming scientific consensus" (2011, p. 196) such as whether HIV causes AIDS or about the reality of global warming. I take Martin Orans to be basically correct in his general conclusion in *Not Even Wrong* that given what we have of Mead's records it is basically impossible to say about the truth of many of Freeman's claims.[4] Further Orans and Shankman are probably correct on the specific issue of whether Mead was hoaxed about female adolescent sexuality—no reason to think she was. If she was wrong here, it was on the basis of no single such incident. Ceccarelli points out that much work in the rhetoric of science criticizes "the world-defining hegemony of scientific discourse" (2011, p. 199). She is looking at cases where there is no more uncertainty than in any other area of science. I am expanding that focus by investigating a messy instance outside of cases of manufactured uncertainty and ones where uncertainty needs to be generated.

It should also be noted that American anthropology has had a goodly number of controversies in recent years. The prominent anthropologist George Marcus (2010) discusses these as one way of tracking "Developments in US Anthropology Since the 1980s." He lists the Mead/Freeman controversy, "Gananath Obeyesekere's critique of Marshall Sahlins' account of the murder of Captain Cook in the Hawaiian Islands, and Sahlins' refutation (early to mid-1990s)" David Stoll's deconstruction of Nobel Prize winner Rigoberta Menchu's testimonial about atrocity in Guatemala, and journalist Patrick Tierney's 2002 attack on the work of Napoleon Chagnon with the Yanomami.[5] As he says, "these controversies were about the modes and the effects of anthropological representations of traditional subjects and the real social stakes involved in representation itself" (p, 206). To this one could add the Tasaday "hoax" in which a small band of people in the Philippines was wrongly claimed to be a paleolithic people[6] (Headland, 1992).

2. COUNTERATTACKS

Certain initial responses to Freeman amounted to counterattacks on him such as those that focused on "debate etiquette." While certainly valid, these responses did not address the issue of the reliability of Mead's work. For example, Freeman never satisfactorily explained why he waited as long as he did to publish his findings, finally publishing them five years after Mead's death. Mead was a certainly a prolific writer, and likely would have mounted a spirited defense. He was also charged with resorting to ad hominem attacks and with seriously selective use of some of his sources (Holmes, 1983). Freeman's personal style clearly was a

---

[4] Orans (1996). Also see Orans's review of Freeman (1999). Anthropologists who agreed with much of his critique at the time include George N. Appell (1984), pp. 133–135 (as cited in Caton, 1990, pp. 275–276). Appell coedited a festschrift for Freeman in Appell and Madan (1988). See also Bock (1983), a review of MSS. Bock "found Freeman's main substantive criticisms convincing" (p. 336), but criticizes the form of the book, and does not "think that his extreme reversals of Mead's overstatements will be very helpful" (p. 340).

[5] In Robin (2004), three of the seven major cases he discusses are from anthropology.

[6] One can also note in sociology, the criticism by W. A. Marianne Boelen of Whyte's classic Cornerville study. See Whyte (1993).

factor; he was characterized as abrasive, on a vendetta, and unprincipled (Caton, 1990, p. 268). An undercurrent in some attacks was a concern that Freeman's work would support sociobiology, or "feed social conservatism and racial backlash in the U. S" (Fields, 1990, p. 231).[7]

Further, as several anthropologists working on Samoa have noted, Freeman never did publish what would have been a valuable contribution, an ethnographic monograph of Samoa (Levy, 1983, pp. 831–832). Freeman had done excellent fieldwork elsewhere; having his considered findings would have been valuable. Rather *Margaret Mead and Samoa* was an extremely a wide-ranging book, not just discussing Mead, but also providing a reassessment of the history of Boasian anthropology, and sketching a new synthesis of the interrelation of biology and culture in anthropology.[8] These responses did not rehabilitate Mead, they only served to detract from Freeman's credibility. And thus they do not particularly vindicate anthropology as a discipline.

3. DEFLECTIONS

Another range of responses I see as trying to deflect or defuse Freeman's criticisms. A first variety focused on the legitimacy of the comparison between Freeman and Mead. Can one compare the 1940s Western Samoan village, where Freeman taught, within a bus ride of the capital of Apia, to the much more isolated island of Ta'u in the Manu'a islands of American Samoa in the 1920s? Could the differences in their accounts be understood in terms of Mead working as 23-year-old woman with adolescent women and Freeman working with older Samoan men, and becoming a chief? I will take up below the claim that he was inappropriately applying 1983 standards to 1920s work.[9]

A second deflecting defense—one I find especially curious—was to claim that though Mead might have been wrong about Samoa in a number of ways, her general conclusions were nonetheless well-taken. Levy (1983) argued that even if Samoans were aggressive, the Tahitians were as the Samoa were supposed to have been.[10] Thus Mead was right, but for inadequate reasons. Obviously such a defense also does not serve to shore up anthropology's status.[11]

Another deflection asserted that ethnography, by its very nature, was inherently impressionistic, and thus the differences between accounts were very understandable, and in sum did not reflect back on any ethnographer's reliability, or indeed the discipline. A good

---

[7] See for example the summary of "The Barnard College Symposium of April 1983," by Fields (1990), p. 231. This was certainly a major concern in the early responses; interestingly it seems to have been quite quickly superseded by other concerns, as attention shifted from theoretical issues such as nature/culture, or nature/nurture, to more methodological, disciplinary ones.

[8] Note that Freeman was a critic of sociobiology per se, offering what I would call a moderate account trying to reconcile culturalist and biological claims in anthropology.

[9] Paul Radin's 1933 critique would seem to indicate that at least this charge against Freeman was invalid.

[10] See also Marcus (1983).

[11] Shankman cites (2009, p. 222) Alice Schlegel and Herbert Barry (1991) who in their review of over 170 anthropological studies touching on adolescence found that the "widespread belief that adolescence in tribal or peasant communities flows smoothly, without completion for resources (which can include a desirable spouse and powerful in-laws) and without areas in which choice must be exercised . . . is belied by the data from this study" (pp. 41–42).

example of this is from Nancy Scheper-Hughes, whose own ethnography in Ireland seemed to conflict with the previous work of Arensberg and Kimball. She argues:

> [W]hen we are talking about Samoan culture or Irish culture we are talking about an interpretation that is the result of a complex series of interactions between the anthropologist and his or her informants. . . . Ethnography is a very special kind of intellectual autobiography, a deeply personal record through which a whole view of the human condition, an entire personality, is elaborated. . . . And the knowledge that it yields must always be interpreted by us, by the particular kind of complex social, cultural and psychological self that we bring into the field. . . . Hence there can be no "falsification" of a 1925 ethnography by a 1940 or a 1965 "restudy" because the particular ethnographic moment in the stream of time that Mead captured is long since gone. (1984, p. 90)[12]

Attribution of ethnographic differences to personal perspectives or interpretive points of views was not limited to this controversy. It was part of a concurrent movement toward seeing ethnography as a much more complicated endeavor than previously held. A greater sense of the personal nature of ethnography, and of the rhetorical construction of ethnography developed in the years after 1983. As Brady points out, these developments

> which we lump under the heading of 'post-modernism', [influenced] . . . a common perception (but very little said in print) that even if Mead was wrong, Freeman didn't have . . . the answer to what was right . . . The 'meta-issues,' in other words, seem to have carried the day against Freeman, against closure on multiple interpretations of Samoan ethnography. (1988/1990, p. 44)

However, while anthropology's internal, or intra-field, audience was not especially interested, its inter- and extra-field audiences may have been drawing different conclusions.[13]

A serious problem that could be drawn from this "defense" is that it renders otiose one major strand of anthropological writing. Atkinson (1982) writes of "what cultural anthropologists do best, namely, it heads full tilt at culture bound assumptions in our own thinking" (p. 257). And this, of course, is something that Mead had made a major part of her own work, make implications for American life. Just as Freeman presented his work as a falsification of Mead's account of adolescence, Mead thought of her work in Samoa as a falsification or refutation of G. Stanley Hall's account of adolescence. This "personal perspectival" defense of Mead, and of ethnography, defuses Freeman's attack, but at the cost of framing ethnography not as a scientific grounding for critique, but a simple personal opinion. Thus this defense did not really salvage Mead, however true it might be of ethnography as a practice.[14]

---

[12] See also Morton (1996).

[13] Though really a matter for another day, I do not believe that postmodernism in any stricter sense than Brady's is really involved. The issues pre-date its rise; it serves more to provide a strawman to criticize.

[14] One complicating factor in assessing Mead's work was that an American anthropologist, Lowell Holmes, had in 1954 done a "methodological restudy" of Mead's work in Ta'u. He had summarized his work as finding that despite some more or less important deficiencies, "the reliability of Mead's account is remarkably high." This had led to letters between Freeman and Holmes, starting in 1966, which Freeman reads as indicating that Holmes's dissertation advisor, a Boasian, made him tone down his original criticisms of Mead's Samoan research; Holmes denies this. See Bargatzky (1988/1990).

4. DISTANCING

Another category of responses to Freeman was to distance contemporary American anthropology from Mead, and certainly from Mead's 1920s work on Samoa. This tacitly admits the legitimacy of Freeman's criticisms, but frames them as old hat, and thus inconsequential for current science. In the review mentioned above, LeVine writes:

> Although Mead became a celebrity through her books and talks popularizing cultural anthropology, her work was often sharply criticized by her anthropological colleagues. Within the discipline, she had, particularly after the 1950s, little theoretical influence. By 1978 (when she died at age 77), however, she was appreciated as a founding figure of American anthropology whose flaws represented an early era of the field. (LeVine, 2010, p. 1108)

However, LeVine's claim that Mead's work was frequently and extensively criticized is not substantiated. *Coming of Age in Samoa* was criticized by some but it was generally well received after its publication (Freeman, 1999, ch. 15; Shankman, 1999, pp. 113–115). Nor is it the case that Freeman was anachronistically applying later-day standards in his critique of Mead's Samoan research. Paul Radin, a well-known student of Boas from a generation prior to Mead, in his 1933 *Method and Theory in Ethnology*, was quite critical of Mead, describing her as a "journalist, though of the best sort" (1965, p. 255). He questioned whether the period of time Mead was in Samoa was at all sufficient to substantiate her claims. But since he does not think a period of even five years would be sufficient, he may have been seen as overly critical. And as Arthur Vidich says in his introduction to the reissue of Radin's work in 1966, the book does not seem to have made much impact (pp. viii, cxiii). [15]

After the first round of reviews there are few published critiques. Among them are a critique by O. F. Raum in his book on Africa (1940) and an article by Peter Worsley in an English journal *Science and Society* (1957). [16] But both of these are by non-American authors: Worsley is English and Raum South African. Now there may be a "fear factor" behind this paucity of critique. In commenting on the controversy Worsley writes that after publication of his article Mead wrote him attacking the piece:

> Taken aback by the virulence of this language, I soon discovered that it evidently was not unusual, for I received several communications from anthropologists in the United States who told me that they had been treated to similar withering counterattacks when they had dared, especially in public situations, to say anything critical of her work. (1992, p. xi).

Another distancing tactic is to place Mead as an exponent of a very particular approach within American anthropology. For example, Conrad Kottak implies that part of the problem is Mead's participation in the "impressionistic" and methodologically less scientific "culture and personality" (1987, p. 282) school of anthropology.

Another point made is that *Coming of Age in Samoa* was written in a time when there was great awareness of the disappearance of many tribes. This was particularly so for Boasians who were very aware of the dying off of the last generations of displaced and decimated North American tribes who had lived in times prior to the massive influence of white American

---

[15] For a substantive methodological critique of Mead's work, see Marvin Harris's *The Rise of Anthropological Theory* (1968). Harris was himself a controversial figure, and a proponent of "cultural materialism," a quite divergent approach to that of Mead.

[16] Raum is mentioned by Bargatzky (1988/1990), p. 255.

society. There was a norm of "one ethnographer/one tribe" (Stocking, 1992, p. 282). In consequence, there was little emphasis on critique of ethnography. On the question of whether there was a culture of criticism in American anthropology, George Stocking, the premier historian of anthropology, notes that in the generation after *Coming of Age*, there was criticism of certain ethnographies, for example, the famous restudy by Oscar Lewis of a Mexican community earlier studied by Robert Redfield and a critique of interpretations of Pueblo culture by John Bennett (1946). Stocking writes:

> The problem of the reliability of ethnographic data—which might perhaps have been suggested by the laboratory metaphor and by the frequent self-identification of the new academic professionals as "scientists"—was largely forestalled by the archetypal distinction between the ethnographic amateur and the academic profession. (1992, p 283.)[17]

When one looks at the decades before Freeman published in 1983, criticism of Mead on Samoa are not at all prominent. Instead she is cited as a paragon of American anthropology. McDowell wrote that "Most significant is [Mead's] concern for the precision and accuracy of the data she gathered. . . . In presenting her material accurately and precisely, Mead is a careful and exceptionally honest ethnographer" (1980, p. 127). An examination of surveys published before Freeman's book in 1983 does not show any signs of this supposed widespread knowledge of Mead's weaknesses. For example, Agar (1980) lists a number of disputes over fieldwork, but does not mention Mead's work as one of these. Edgerton and Langness (1974) discuss a number of cases where ethnography had been questioned—Ruth Benedict's Pueblo work, the Redfield-Lewis divergence—in a chapter where they also mention Mead, but make no indication of any reservations about her work. Indeed the strength of the defenses of Mead after Freeman suggests that he was far from simply rehearsing or amplifying commonly held suspicions, albeit in an objectionably antagonistic fashion.

5. ANTHROPOLOGY AND ITS AUDIENCES

But if Mead's work was either so shoddy or passé, why did it for so long continue to be used as a textbook in college and other courses? Reflection on this, I believe, raises deeper issues concerning the relation of American cultural anthropology to its audiences.

A telling example of the disconnect between the discipline and its audiences is in one of the best early overviews of the dispute by Roy Rappaport in his "Desecrating the Holy Woman: Derek Freeman's Attack on Margaret Mead" (1986). He sets up a category of "Myth" as a third area of discourse separate from the realms of the "necessary truth of logic" and "the empirical truth of science"—that of fact and law. The realm of myth has the "truth of sanctity" (1986, p. 321). Rappaport seems to take over the positivist philosophic tenet that there is some sort of logical gulf between the three. He asserts,

> Anthropology is no more capable of establishing the mythic status of narratives than is chemistry. All anthropology can do is to offer to a public accounts from which that public can select some (as it can from other sources) to establish as myth, leaving the rest to anthropologists' arcane in-house conversations. (1986, p. 322)

---

[17] Bennett states his critique is of interpretations, not data, though how he distinguishes these is questionable.

In dealing with mythic truth, "Where there is disparity between fact—observable states of affairs—and the truth of myth, it is fact that is wrong" (1986, p. 320). (Though he suggests the eugenics 'myth' was somehow "in violation of discovered fact or law" (1986, p. 347).) This last realm is where Rappaport places the true significance of *COAS*, and he insightfully discusses how it both supported and coincided with the American mythos. As science, *COAS* is "not so much incorrect as thin and in need of enrichment, it did make a modest contribution to Samoan ethnography." As myth he suggests it served Americans well insofar as it is "humane and liberating" (1986, p. 347).

But this separation is questionable in a number of ways. First it is not obvious that fact and myth are so entirely distinct. Facts can be taken not simply as "observable states of affairs," but as "substantiated or proven claims." What we take to be factual is influenced by our myths, but also the reverse, especially in the American mythos with its emphasis on facts and science. Even conservative religion is defended via *scientific* creationism.[18] Thus Rappaport's attempt to move the debate away from issues about the facts ("even poor ethnography usually gets its facts straight" (1986, p. 344)) is questionable.

Second, he implies this myth works at the level of American culture, independent of the discipline of anthropology. This is unpersuasive on his own account. As Rappaport mentions, *COAS* "enjoyed substantial classroom adoption for decades" (1986, p. 324).[19] Of course, this was predominantly because anthropologists were ordering it for their classes, a ritual academics are quite used to. As he also states, "participation in a ritual is a formal act of acceptance of whatever is represented in that ritual" (1986, p. 321). Insofar as anthropologists used *COAS* and others of Mead's books in their introduction to anthropology courses, because as Goodenough claims, "they turned students on" (1983, p. 906), it would seem they are implicated in maintenance of Mead's place in the American myth, which Freeman attacked, generating controversy. Both Rappaport and Goodenough want to claim that the "crisis is in the public's view of a public idol" (Goodenough, 1983, pp. 906–907), not in anthropology itself.

This ex post facto exculpation implies a lack of responsibility for one's extra-field audience, and amounts to another version of the distancing strategy, as neatly pointed out in a letter commenting on Rappaport's piece by S. D. Cornell. Cornell writes that as he read,

> I kept waiting for a resume of critical publications by those American Anthropologists who were said to have early recognized the true character of *COAS*. Did they try to warn a deluded public that it was largely myth, not science? Surely the least they owed to those swallowing the story and joyfully accepting it as gospel (good news!) was such a warning. Failure in that seems irresponsible of professionals in a field as closely related to human behavior as cultural anthropology. (1987/1990, p. 253)

Rappaport's rather disingenuous reply was

> First, I don't think that anthropologists have been aware that a few of their texts have played a mythic role in public discourse. The mythic significance of *Coming of Age in Samoa* and Mead's publicly sanctified status were matters that escaped us until Professor Freeman forced us to reflect upon them. (1987/1990, p. 254)

---

[18] The work of the philosopher Willard Van Orman Quine on 'the web of belief', and the impossibility of sharply demarcating necessary truths and empirical truths is relevant here. See Quine (1953) and Quine and Ullian (1968).

[19] See also Marcus and Fischer (1986/1999), p. 160 and Kuper (1989), p. 453.

This raises questions about how a profession polices itself, how it questions itself, and pursues these questions—and portrays itself to its inter- and extra-field audiences. For example, Marcus claims that disputes such as between Mead and Freeman, involving "radically different interpretations from fieldwork in the same culture" are "commonplace" (Marcus, 1983, p. 22) in anthropology; Heider asserts that "ethnographers rarely disagree with each other's interpretations of a culture" (1988, p. 73). A discipline that disagrees about how often it disagrees is not a pretty sight.

6. UPSHOT

Given this range of responses to Freeman, no clear anthropological view emerged as to why Freeman was clearly wrong, and Mead was right, or even any more nuanced account of how one might split the difference. I speculate that the failure to come to some disciplinary consensus left open the door, so to speak, for Freeman in his second 1999 book *The Fateful Hoaxing of Margaret Mead* to advance the very suggestive claim that Mead's closest friends had engaged in the Samoan equivalent of pulling her leg by spinning tales of their sexual exploits with village teenage boys. Even though Orans (1999, pp.1649–50) and Shankman (2009, ch. 13) give quite plausible reasons why this is unlikely, in some inter-field circles this "myth" has quite taken hold. (Of course, as Stover (2005) indicates, this has had little impact extra-field.)

Interestingly the American Anthropological Association, in the Tasaday case and the Tierney–Chagnon dispute, has convened taskforces to investigate and pronounce on a controversy.[20] But closure has not come easily in the second case. After accepting a committee report on the Chagnon case, the association rescinded its acceptance in 2005. Its board stated that this was because of "the members' belief that the El Dorado investigation and the resulting report violated the association's ban on adjudicating claims of unethical behavior and that the El Dorado investigation did not follow basic principles of fairness and due process for the accused" (American Anthropological Association, 2004). Nothing with any approaching such an official imprimatur emerged concerning Mead–Freeman (Shankman's book may be able to do so.)

This disconnect from its extra-field audience is poignantly displayed in the paperback edition of James Clifford and George Marcus *Writing Culture: The Poetics and Politics of Ethnography* (1986). On the inside of the front cover of the book, one of the major works in recent American anthropology, the blurb reads:

> Why have ethnographic accounts recently lost so much of their authority? Why were they ever believable? Who has the right to challenge an "objective" cultural description? Was Margaret Mead simply wrong about Samoa as has recently been claimed? Or was her image of an exotic land a partial truth reflecting the concerns of her time and a complex encounter with Samoans? Are not all ethnographies rhetorical performances determined by the need to tell an effective story?

The book itself contains stimulating and thoughtful essays. But it does not address these questions, much less answer them. Perhaps they did not intend to. Freeman's critique gets a brief, less than two-page mention, and Mead not much more. Very likely the blurb was written

---

[20] For a comprehensive archive of many documents on this see the site of Douglas Hume. Also see Headland (1992).

by some editor at the University of California Press, partly to entice purchasers, but also as a way of indicating the issues raised by the controversy. It is as if a member of the extra-field audience is speaking up from the auditorium, but not getting an answer.

Nor are the deeper theoretical and methodological questions answered much more fully. In his introduction to the book, Clifford poses questions I have tried to highlight here. "These contingencies—of language, rhetoric, power, and history—How are the truths of cultural accounts evaluated?" (1986, p. 25). But he resorts to the relatively weak rhetorical ploy of rhetorical questions, in place of answers or assertions.

Anthropology in the 1980s was in a period of flux, older paradigms being questioned, without successors clearly emerging to succeed them.[21] Feminist anthropologists were focusing critical attention on questions of gender.[22] Much attention was being paid to the nature of fieldwork or ethnography and to how ethnographic experience was channelled, created, and consumed by anthropologists, a concern which drew heavily upon literary criticism (Sass, 1987; Marcus & Fischer, 1986; Clifford & Marcus, 1986). And as Kuper claims, the Mead/Freeman "debate has had a special resonance within anthropology, for it has fed the contemporary unease about ethnographic fieldwork and writing" (1989, p. 454).

For those in anthropology's intra-field audience the various responses catalogued above had genuine persuasiveness; they were more problematic for others in the inter-field audience. Many likely expected responses more along the lines of a natural science. But instead much depended on understanding the less familiar lines of an interpretive human science, such as undergirds the "perspectivalist" responses, such as that of Scheper-Hughes. Although this tactic seems a plausible resolution of some of the 'factual' issues in dispute, even some anthropologists have balked at accepting it as resolving the controversy. In his defense of "Anthropology as an Empirical Science," O'Meara points out that it seems quite implausible to take such a reconciling tactic on questions such as "the probability that 12 menstruating and thus presumably fertile girls could have engaged in sex for periods ranging from two months to four years without a single pregnancy occurring, as Mead claimed" (1989, p. 359).[23] This I suspect is involved in the reactions of the evolutionary psychologists.

7. CONCLUSION

Describing a scientific controversy as a debate or a conversation is a tempting metaphor, but how does one evaluate it in retrospect? When a field does not address an issue, why does it not? When can one say it does not address the questions someone in an audience wants answered? Obviously, the intra-field audience of other anthropologists was most salient, and they can write replies. Inter-field relations do not seem as prominent. Certainly, the perceived standing of anthropology with regard to other social sciences and to biology was an issue. And it seems that anthropology's standing as a source of reliable findings has suffered for many practitioners in these fields.

---

[21] See, for example, Hoebel, Currier, and Kaiser (1982).

[22] There is also a significant "materialist" approach, broadly speaking, influenced by Marx. Participants in it do not seem to have involved themselves in the controversy, as far as I can tell.

[23] Given O'Meara's meta-theoretical focus in the article, it is interesting that he begins by referring to his recent return from the field, apparently to validate his standing to speak on theory by referring to his fieldwork experience (1989 p. 354).

How does a discipline relate to its extra-field audience? American cultural anthropology today has no spokesperson or popularizer anywhere near the stature of Mead in her heyday. And it is unlikely that it could have one with the general public. And to what extent is this its responsibility? The scholarly isolationism supported by Sir Edmund Leach, a prominent British anthropologist, does not seem viable now, if it ever was. Commenting on the controversy and Mead's "impressionistic anthropology," he asserts, "Many of Mead's professional associates both in her own country and elsewhere have taken the view that they have other scholarly duties besides playing to the gallery" (1985/1990, p. 246). Such a stance seems as likely to have engendered problems as to have avoided them.

Indeed, George Marcus's comment, similar to those of Goodenough and Rappaport cited above, apparently intended to downplay Freeman's critique, that "outside of introductory courses, [Mead's] work has not generally been read in recent years" (1990, pp. 232–233) is revealing. It is in precisely such courses that anthropology has its greatest opportunity to educate its audience about itself. As the philosopher Philip Kitcher has suggested in his analysis of the conflicts between evolutionists and scientific creationists, the use by biologists of slogans, simplistic dichotomies ('proven fact' vs. 'only a theory'), and naïve philosophies of science provide readily exploitable starting points for creationists (1983, ch. 2). The extra-field audience for anthropologists, like that for evolutionary biologists, is in part a reflection of how scientists have educated it, including their critics.

It is among feminist anthropologists that the relation of scientists and their audiences is at all addressed. One explicitly feminist response to the issues raised by the Mead/Freeman controversy differs from *Writing Culture* precisely in its orientation to its audience. Frederick Errington and Deborah Gewertz frame their book, *Cultural Alternatives and a Feminist Anthropology*, by discussion of the controversy, explicitly mentioning the effect that *COAS* had on Gewertz's mother and many others (1987, p. 6). They conclude their re-examination of Mead's work on New Guinea with the comment, that despite its flaws,

> Mead was, thus, quite correct in her perception that Chambri women conducted their lives with more assurance than do American women. . . . Through comparing the Chambri with ourselves we can, thus, become assured that male dominance is not inevitable and become more clear-sighted. (1987, pp. 140–141)

Though a piece of scholarship, their book does not place itself outside of politics. It is cultural critique of a conservative strand in American thought, but not simply that. They consciously reflect on the importance of *COAS* in promoting a liberal American movement, and direct their attention to how their work might *support* and *sustain* that tradition, or "myth" according to Rappaport (1986). Rather than distancing anthropology from its effects on American culture via its findings, Errington and Gewertz acknowledge this audience, an audience Mead, more than any other anthropologist, helped create.

ACKNOWLEDGEMENTS: My thanks to Clarke Rountree for help with precursors of this paper, and Penny Weiss for her comments on this paper.

REFERENCES

Agar, M. (1980). *The professional stranger: An informal introduction to ethnography*. New York, NY: Academic Press.

American Anthropological Association (2004). Referendum to rescind the AAA's acceptance of the report of the El Dorado task force. Retrieved from http://www.aaanet.org/committees/nom/05comments/05_ref_eldorado.htm

Appell, G. N. (1988). *Choice and morality in anthropological perspective: Essays in honor of Derek Freeman*. Albany, NY: SUNY.

Appell, G. N. (1990). Freeman's refutation of Mead's *Coming of Age in Samoa*: The implications for anthropological inquiry. In H. Caton (Ed.), *The Samoa reader: Anthropologists take stock* (pp. 79–80). Lanham, MD: University Press of America. (Reprinted from *Eastern Anthropologist*, 1984, *37*, 133–135).

Atkinson, J. M. (1982, Winter). Anthropology: Review essay. *Signs, 8*(2), 236–258.

Bargatzky, T. (1990). Examination of Holmes' quest. In H. Caton (Ed.), *The Samoa reader: Anthropologists take stock* (pp. 255–263). Lanham, MD: University Press of America. (Original work published 1988).

Bennett, J. W. (1946). The interpretation of Pueblo culture: A question of values. *Southwestern Journal of Anthropology, 2*(4), 361–374.

Bock, P. K. (1983). The Samoan puberty blues. *Journal of Anthropological Research, 39*(3), 336–340.

Brady, I. (1990). Letter. In H. Caton (Ed.), *The Samoa reader: Anthropologists take stock* (p. 44). Lanham, MD: University Press of America. (Original work published 1988).

Buss, D. (1999). *Evolutionary psychology*. Needham Heights, MA: Allyn & Bacon.

Caton, H. (Ed.). (1990). *The Samoa reader: Anthropologists take stock*. Lanham, MD: University Press of America.

Ceccarelli, L. (2011). Manufactured scientific controversy: Science, rhetoric, and public debate. *Rhetoric & Public Affairs, 14*(2), 195–228.

Clifford, J., & Marcus, G. E. (Eds.). (1986). *Writing culture*. Berkeley, CA: University of California Press.

Cornell, S. D. (1990). Letter to editor. In H. Caton (Ed.), *The Samoa reader: Anthropologists take stock* (p. 253). Lanham, MD: University Press of America. (Original work published 1987).

Edgerton, R. B., & Langness, L. L. (1974). *Methods and styles in the study of culture*. San Francisco, CA: Chandler & Sharp.

Errington, F., & Gewertz, D. (1987). *Cultural alternatives and a feminist anthropology*. Cambridge, UK: Cambridge University Press.

Farrer, C. R. (1996). Orthodoxy and heterodoxy: Locating ethnography. *American Anthropologist, 98*(1), 170–172.

Fields, C. (1990). The Barnard College Symposium of 1983. In H. Caton (Ed.), *The Samoa reader: Anthropologists take stock* (pp. 230–233). Lanham, MD: University Press of America.

Hume, D. (2012, January 14). Darkness in El Dorado. *AnthroNiche*. Retrieved from http://anthroniche.com/darkness-in-el-dorado.html

Freeman, D. (1983). *Margaret Mead and Samoa: The making and unmaking of an anthropological myth*. Cambridge, MA: Harvard University Press.

Freeman, D. (1999). *The fateful hoaxing of Margaret Mead: A historical analysis of her Samoan research*. Boulder, CO: Westview Press.

Goodenough, W. (1983, 27 May). Letter. *Science*, 220 (4600) 906, 908.

Harding, S. (1987). *Introduction to feminism and methodology*. Bloomington, IN: Indiana University Press.

Harris, J. R. (2009). *The nurture hypothesis: Why children turn out the way they do*. New York, NY: Simon and Schuster.

Harris, M. (1968). *The rise of anthropological theory*. New York, NY: Crowell.

Headland, T. N. (Ed.). (1992). *The Tasaday controversy: Assessing the evidence*. Washington, DC: American Anthropological Association.

Heider, K. G. (1988). The Rashomon effect: When ethnographers disagree. *American Anthropologist, 90*, 73–81.

Hoebel, E. A., Currier, R. L., & Kaiser, S. (Eds.). (1982). *Crisis in anthropology: View from Spring Hill, 1980*. New York, NY: Garland.

Holmes, L. (1983). A tale of two studies. *American Anthropologist, 85*, 929–935.

Kitcher, P. (1983). *Abusing science*. Cambridge, MA: MIT Press.

Kottak, C. (1987). *Cultural anthropology* (4th ed.). New York, NY: Random House.

Kuper, A. (1989, April 8). Coming of age in anthropology? *Nature, 338*, 453–455.

Laudan, L. (1981). Peirce and the trivialization of the self-corrective thesis. In L. Laudan, *Science and hypothesis: Historical essays on scientific methodology*. The University of Western Ontario Series in Philosophy of Science (Vol. 19). Dordrecht, Holland: D. Reidel.

Leach, E. (1990). Middle American. In H. Caton (Ed.), *The Samoa reader: Anthropologists take stock* (pp. 245–246). Lanham, MD: University Press of America. (Original work published 1985).

LeVine, R. A. (2010, May 28). Cutting a controversy down to size. *Science, 328*, 1108.

Levy, R. I. (1983, May 20). The attack on Mead. *Science, 220*, 829–832.

Lyne, J. (1983). Ways of going public: The projection of expertise in the sociobiology controversy. In D. Zarefsky, Sillars, M.O., & Rhodes, J. (Eds.). *Argument in transition* (pp. 400-415). Annandale, VA: Speech Communication Association.

Marcus, G. (1983, March 27). One man's Mead. *New York Times Book Review*, pp. 3, 22, 24.

Marcus, G. (2010). Developments in US anthropology since the 1980s. In A. Boskkovic (Ed.), *Other people's anthropologies: Ethnographic practice on the margins* (pp. 205–209). Oxford: Berghahn Books.

Marcus, G. E., & Fischer, M. J. (1986). *Anthropology as cultural critique*. Chicago, IL: University of Chicago Press.

McDowell, N. (1990). The oceanic ethnography of Margaret Mead. In H. Caton (Ed.), *The Samoa reader: Anthropologists take stock* (pp. 125–127). Lanham, MD: University Press of America. (Original work published 1980).

Morton, H. (1996). Review of *Adolescent Storm and Stress* by James E. Coté. *Oceania, 67*, 166–167.

O'Meara, J. T. (1989). Anthropology as an empirical science. *American Anthropologist, 91*(2), 354–369.

Orans, M. (1996). *Not even wrong: Mead, Freeman, and the Samoans*. Novato, CA: Chandler & Sharp.

Orans, M. (1999, March 12). Mead misrepresented. *Science, 283*(5408), 1649–50.

Pinker, S. (2002). *The blank slate: The modern denial of human nature*. New York, NY: Viking.

Quine, W. V. O. (1953). *From a logical point of view*. Cambridge, MA: Harvard University Press.

Quine, W. V. O., & Ullian, J. (1968). *The web of belief*. New York, NY: Random House.

Radin, P. (1965). *The method and theory of ethnology: An essay in criticism*. New York, NY: Basic Books. (Original work published 1933).

Rappaport, R. (1986). Desecrating the holy woman. *American Scholar, 55*, 313–369.

Rappaport, R. (1990). Letter to editor. In H. Caton (Ed.), *The Samoa reader: Anthropologists take stock* (p. 254). Lanham, MD: University Press of America. (Original work published 1987).

Raum, O. F. (1940). *Chaga childhood*. Oxford, UK: Oxford University Press.

Robin, R. (2004). *Scandals and scoundrels: Seven cases that shook the academy*. Berkeley, CA: University of California Press.

Sass, L. A. (1987, May). Anthropology's native problems: Revisionism in the field. *Harper's Magazine*, 49–57.

Scheper-Hughes, N. (1984). The Margaret Mead controversy: Culture, biology, and anthropological inquiry. *Human Organization, 41*(1), 85–93.

Schlegel, A., & Barry, H. (1991). *Adolescence: An anthropological inquiry*. New York, NY: Free Press.

Shankman, P. (2009). *The trashing of Margaret Mead: Anatomy of an anthropological controversy*. Madison, WI: University of Wisconsin Press.

Simpson, T., Stich, S., Carruthers, P., & Lawrence, S. (2006). Introduction: Culture and the innate mind. In P. Carruthers, S. Lawrence, & S. Stich (Eds.), *The innate mind: Volume 2, culture and cognition* (pp. 3-19). New York, NY: Oxford University Press.

Stocking, G. (1992). The ethnographic sensibility of the 1920s and the dualism of the anthropological tradition. In G. Stocking, *The ethnographer's magic and other essays in the history of anthropology* (pp. 276-341). Madison, WI: University of Wisconsin Press.

Stover, M. (2005). Tales from the Internet: Margaret Mead's legacy in American culture. *Pacific Studies, 28*(3/4), 142-161.

Vidich, A.J. (1965). Introduction. In Radin, P. *The method and theory of ethnology: An essay in criticism* (pp. vii-cxv). New York, NY: Basic Books. (Original work published 1933).

Whyte, W. F. (1993). Revisiting "Street corner society." *Sociological Forum, 8*(2), 285–298.

Worsley, P. (1957). Margaret Mead: Science and science fiction. *Science and Society, 21*(2), 122–134.

Worsley, P. (1992). Foreword. In L. Foerstel & A. Gilliam (Eds.), *Confronting the Margaret Mead legacy: Scholarship, empire, and the South Pacific* (pp. ix–xviii). Philadelphia, PA: Temple University Press.

Analyzing GM Food Risk Arguments through an Online, Multi-media Case Study

TOSH TACHINO

University of Winnipeg
Winnipeg, Manitoba R3B 2E9
Canada
t.tachino@uwinnipeg.ca

DAVID R. RUSSELL

Iowa State University
Ames, Iowa 50011
U.S.A.
drrussel@iastate.edu

ABSTRACT: We constructed an online, multi-media simulation of an environmental debate, and analyzed the uses of scientific information from it that 41 college students included in arguments about the issue. Analysis of the ways students appropriate information to reason about science suggests they do so much as scientists do in public policy debates.

KEYWORDS: argument, case study, multi-modality, online, rhetoric of science, risk, teaching.

1. INTRODUCTION

Rhetorical studies of science have charted what Fahenstock (1986) calls the "rhetorical life of scientific facts" from laboratory through popular media. These studies have generally contradicted the "canonical view" (Myers, 2003) that states that popularization is simply a readable translation of the scientific research; instead there are ample evidence to show that science facts undergo various transformations in their rhetorical lives, and there exists a great deal of differences between scientific research and the popularized account (Charney, 2003; Fahnestock, 1986; Myers, 1991).

However, insufficient attention has been paid to examine a crucial further reinscription, and we know relatively little about how these popularized accounts of science are rhetorically taken up by various members of the public and policy makers. Studies that examined the uptake of popularized accounts so far have been generally limited to surveys that measure superficial knowledge of arbitrarily selected scientific information (e.g., Miller, 1998; Sturgis & Allum, 2004) and experimental studies that isolate scientific information from the naturally-occurring discourse and reinscribe it in an artificial experimental discourse.

Such studies provide useful information on potential behavior, but do not tell us much about why people value and devalue GM products or why they might change their views, in other words, the reasoning used—how decisions are made based on scientific information. In addition, the hypothetical questions and experimental auctions must greatly simplify the discourse as they frame the survey questions or prompts. Whereas for most people, scientific information comes integrated into (some would say buried in) a complex circulation of

Tachino, T., & Russell, D.R. (2012). Analyzing GM food risk arguments through an online, multi-media case study. In J. Goodwin (Ed.), *Between scientists & citizens: Proceedings of a conference at Iowa State University, June 1-2. 2012* (pp. 379-392). Ames, IA: Great Plains Society for the Study of Argumentation. Copyright © 2012 the author(s).

discourse from a range of stakeholders in various media and genres, which consumers or the public must take up in order to use of science for reasoning and/or decision-making. Lusk, Jamal, Kurlander, Roucan, & Taulman (2004) in "A Meta Analysis of Genetically Modified Food Valuation Studies" suggest that "areas for fruitful research lie in explaining *why* consumers have particular a valuation estimate."

This raises the problem we address in this research. How do people use scientific information to make arguments (reason) about the risks of GM foods? Policy makers and consumers alike assess the risks of and make decisions about GM foods based on both scientific/economic data and reasoning or *arguments* about that data that involve value judgments. Facts rarely speak for themselves in policy deliberations, even when the facts are not disputed. People may agree that certain foods contain certain substances at certain levels as established by scientific studies, but they may then *use* that information as part of chains of argument that justify increasing *or* lowering allowable levels, based on differing value judgments and chains of argument. Or they may accept certain scientific information and discount other scientific information in their arguments (and decisions) based on their reasoning and value judgments (including level of trust of the source of the scientific information).

In the past decade, a few researchers have attempted to get at the relationship between scientific information and values in risk decision making through rhetorical (argument) analyses of discourse used by various stakeholders in public deliberations on environmental policy (Cook, 2004), notably case studies of the discourse of particular deliberations. One problem such studies pose is that they are, of course, unrepeatable and limited to particular group of people, those participating in the particular public deliberations studied. Researchers cannot manipulate the situation to introduce different information or stakeholders/participants. To address this methodological problem we turned to an online multi-media case study, in hopes that it would allow us to model the complexities of discourse on a GMO policy issue and elicit discourse for more detailed analysis and larger numbers of participants than case studies of actual deliberations. Our research follows Macoubrie (2003) who constructed an online forum for public discussion of biotechnology policy to elicit arguments for analysis. We also analyze the *arguments* or reasoning people use when presented with a range of information in different media and are asked to make an argument to justify a policy decision.

2. METHODS

We adapted an online, multi-media case study to represent an environmental debate on Golden Rice (which Hessler first developed and used in an extension Biotechnology Ethics course). Golden Rice (GR) is a genetically modified food that contains higher levels of vitamin A. Its chief developer, Ingo Potrykus, argues that it can help prevent VAD, which causes 500,000 cases of blindness and contributes to over 1,000,000 deaths per year (VAD means a 25% greater risk of dying from measles, malaria, or diarrhea) and is the leading cause of blindness in children. But GR has been highly controversial. Proponents in the biotech industry have promoted it as a wonder cure for developing countries, the shining example of biotechnology's promise. Opponents among environmentalists have argued that it does not contain enough VA to address the problem, that it may cause environmental and economic harm, and that existing, effective VA supplement programs are capable of solving the problem and would be defunded to support what is, in their view, a "Trojan Horse" or "poster child" for multinational agriculture corporations.

The Golden Rice case models the complexities of the GR debate by providing textual and video information from pro-GR sources, anti-GR sources, and sources that do not explicitly take a stand. Roughly equal numbers of each were included. The sources that do not take a stand are grouped under "background." Those that take a stance are under "opinions," represented geographically, with no indication to the students whether each opinion was pro or con. In addition, there is a list of links to other information, listed thematically rather than by stance, (and again roughly balanced pro and con no position).

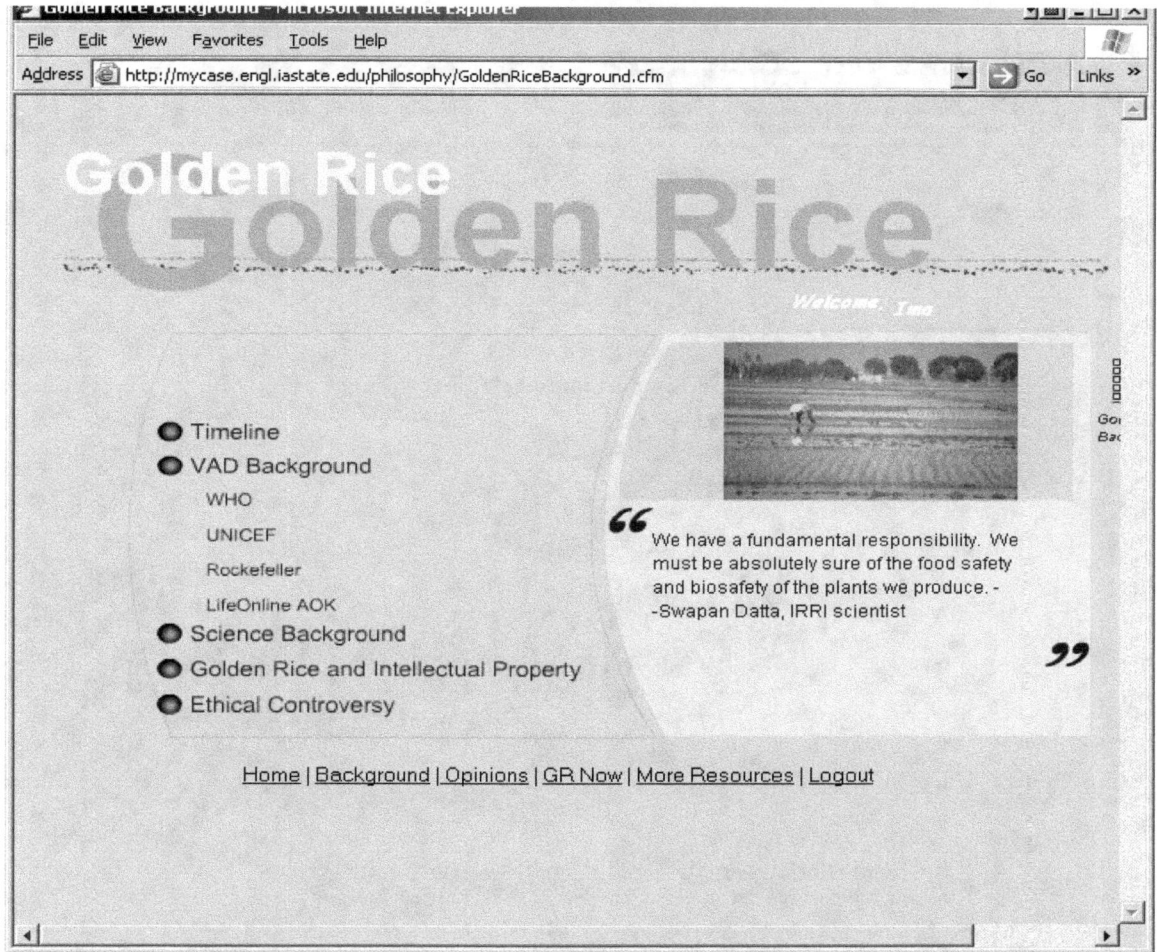

Fig. 1: The Golden Rice online case, main page

In a three week unit, students are given the assignment of writing a recommendation to the Rockefeller Foundation, the leading funder of Potrykus's GR research for over a decade, on the level of funding, if any, that Rockefeller should devote to GR research in the future. During the unit, students first (1) research the case and (2) discuss it in class and in threaded online discussions. As the instructor in the course, I did not intervene in the classroom or online discussions and did not, to the best of my ability, push them in any direction, pro, con, or in between. Finally, they (3) each make a decision about the level of funding and explain that decision in a written argument (about 1000 words), in a letter addressed to the Rockefeller Foundation.

The case was used in two sections of a first-year course in general academic writing, required of all students in a large Midwestern U.S. university of science and technology (N=41). The students represented ten majors (disciplinary curricula), with 10 in the natural sciences, 5 in agriculture, 13 in engineering, 8 in business, and 5 in humanities. Two were international students. This was a homogeneous group of mainly Midwestern U.S. students, in terms of sex, ethnicity, and age, close to the Midwestern University student population as a whole. No students had prior knowledge of Golden Rice, though five had knowledge of GMO crops, firsthand from their farming backgrounds.

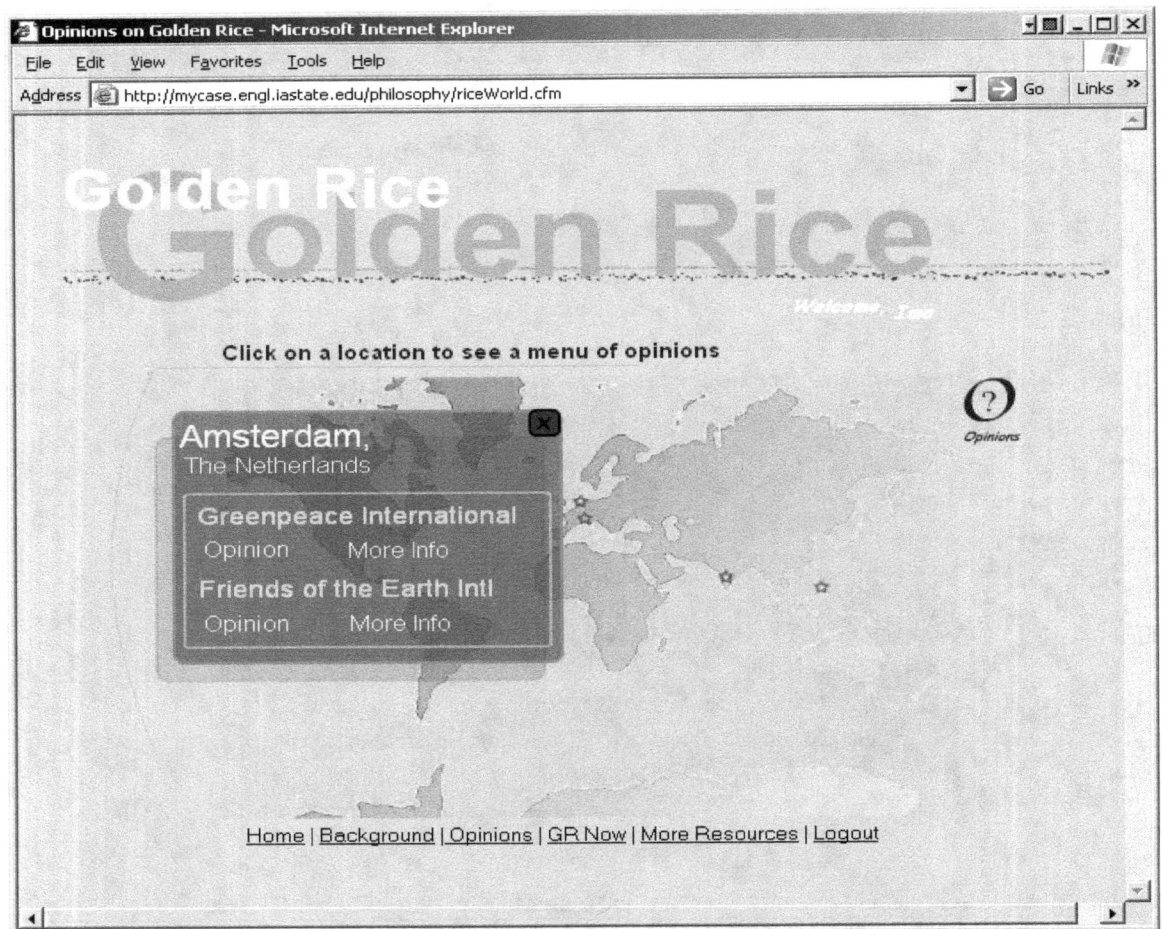

Fig. 2: The Golden Rice online case, clickable world map showing one location

We gathered survey data and student texts for rhetorical analysis. We will give the survey methods and results first, then move to the textual analysis methods and results, and finally try to triangulate the two.

3. SURVEY METHODS AND RESULTS

We administered a 14-question pre-post survey (six point Likert scale) to measure the students' perception of their knowledge of genetic engineering and their attitudes toward biotechnology, which we adapted from one Hessler had used at a biotechnology conference.

Students' perception of their level of knowledge increased $p = < .011$ This is clearly what we would expect as they had just spent three weeks studying it.

Table 1. Survey Results

Question 1. Knowledge of Biotechnology: Summary Statistics

	N	Mean*	S.D.	*T*-Value	*P*-Value
Pre	39	2.72	1.19	-2.688	< .011
Post	39	3.18	1.00		

Questions 2-14. Attitudes toward Biotechnology: Summary Statistics

	N	Mean	S.D.	*T*-Value	*P*-Value
Pre	39	2.48	0.47	5.759	< .001
Post	39	2.15	0.42		

*N.B.: A lower experimental mean = anti-biotechnology

This might be explained by the treatment. Though we tried very hard to control for bias in the case and the pedagogy, we may not have succeeded. But this result is consistent with NSF national survey results, which show "a slight, gradual decline in the American public's support for genetic engineering between 1985 and 2001. The shift can be seen most clearly among college-educated respondents and those classified as attentive to science and technology issues." Huffman, Shogren, Rousu, & Tegene (2003) also found in their experimental auction research that "Participants who claimed to be informed about GM technology in pre-experiment survey were significantly more likely to be out of the market for GM products." We now turn to the qualitative data and our central question. Later we'll return to a more specific analysis of the pre/post survey results to triangulate the two.

4. RHETORICAL ANALYSIS METHODS AND RESULTS

Before we look at the methods and results of the qualitative, rhetorical analysis, we need a bit of theory. This study is based on argument theory and the rhetoric of science (a field about 25 years old).

The version of argument theory we are using suggests that people choose arguments and evidence from the rhetorical resources available, given the constraints of the context, on the basis of their usefulness in accomplishing their goals, not on the basis of some universal principles of rationality or correctness (Bazerman & Paradis, 1991). The fundamental unit of analysis for resources for argument is the topos (plural topoi), from the Greek for "places" (as in topography). Topoi are the common means of persuasion in some community or domain (Aristotle). And the topoi vary with the rhetorical context (or discipline or profession or legislative body, etc.) and the subject, as do the rules or norms for argument (Toulmin, 1979).

Studies over the past 20 years in the rhetoric of science have shown that scientists argue using two very different repertoires of rhetorical resources that scientists use in different social contexts, called the empiricist and the contingent repertoire (Gilbert & Mulkey, 1984?).

> The empiricist repertoire, found in research papers and other formal documents, is characterized by lexical and syntactical arrangements that convey the realist view of science, the ideology that scientists' practices are required by the nature of the physical world. In this [repetoire], scientists are depicted as conduits for the realization of natural phenomena, their own agency deleted. The contingent repertoire, on the other hand . . . is marked by the presence of accounts that attribute influences on behavior to factors external to the physical phenomena under investigation. (Janillo, 2004)

The contingent repertoire is utilized by scientists in their informal talk and—importantly for this study—in most public discussions of science when scientists are speaking as advocates.

EMPIRICIST REPETOIRE	CONTINGENT REPETOIRE
Experimental articles	Editorials, opinion pieces, hall talk, lab talk
Factors contingent on human agency disallowed in persuading colleagues	Factors contingent on human agency allowed to in persuading.
Closed community: experts only (Ph.d. in speciality + original data set required to speak)	More open community: non-experts allowed; no original data set
Data as agents: humans reduced to citations; data assumed to speak; passive voice	Humans as agents
Arguments constrained by methods and norms of field	Arguments open
No appeals to emotion	Appeals to emotion common
No impugning of motives	Impugning motives common
No name calling	Name calling common
Claims highly qualified and hedged	Claims less qualified and hedged (or not at all)

For example, here is Potrykus, the developer of GR in an interview: "If some people decide that they want blind children and white rice, the decision is theirs. I am offering the possibility of yellow rice and no blind children. But the decision about what people want to eat is theirs." This appeal to emotion (rather passive/aggressive at that) would of course never appear in an experimental article. It is not a simple choice between GR and blindness. There are many synthetic and natural supplements available and being used today around the world, as Potrykus is well aware. Rather it's a policy choice about what part if any GR could play in combating VAD, a very complex cost/benefit analysis that micro-biologists do not do. Economists and others do this. But in the absence of a full cost/benefit analysis we are left with arguments from the contingent repertoire.

Similarly, Greenpeace emotionally dramatized the central problem of GR with this photo with the caption: "The amount of Golden Rice that must be eaten every day to obtain the daily RDA of VA." Nine kilos a day. Potrykus experimentally established that current strains of GR provide 8-10% of the RDA and it would take nine kilos a day of GR for a person to get the US RDA of VA. But of course Potrykus argues that with further funding that level could be increased greatly, and furthermore, that the US RDA is "luxurious" and much smaller amounts of VA will prevent enough blindness and other diseases to justify the cost. In the complex circulation of discourse about GM policy, the contingent repertoire is what is useful and appropriate, for scientific experts as well as non-experts.

Fig. 3: Greenpeace photo

4.1 Rhetorical Analysis Methods

First we coded each of the sources and the students' letters pro/con/middle (IRR =.84). Then we located their use of scientific information in the papers. Because the students used so much scientific information in their arguments, we narrowed our analysis to the use of one form of quantitative information, percentages, which figured importantly in their arguments. Current strains of GR contain only about 8-10% of the RDA of VA. Thus students' arguments and decisions often turned on whether these relatively small amounts were enough to justify further research. We then traced each use of a percentage in the student letters to the source it came from. We coded the arguments in which students used a percentage from a source according to whether the student:

- Uses the *same* argument as source
- Cites source to *refute* it
- Uses the information as *evidence* for an argument *not made directly in source*
- Uses the information as *background* information
- Was *untraceable* (due to poor documentation)

With the help of a concordance program, we then looked at each use of "Percent" (in all its forms) to identify the topoi in that the students had used. We were guided by previous linguistic and rhetorical research on common topoi in the GM foods (Cook 2004).

4.2 Rhetorical Analysis Results and Discussion

Students clearly used scientific information to make their arguments. They used Percent 124 times, an average of three times per letter. All but two of the 41 letters used Percent and those two used other quantitative information. The supporters of GR research were no more likely to invoke Percentages than the opponents (47% pro, 49% con). We are now ready to move to our central question. How did these students use scientific information to make arguments (reason) about the risks of GM foods?

4.3.1 What Positions Did Students Take?

Though the assignment specifically invited students to explore middle positions, only two took a middle position. The rest almost exactly evenly divided between pro and con (20 pro, 19 con).

Recalling our theoretical premise that people choose arguments and evidence from the rhetorical resources available, given the constraints of the context, on the basis of their usefulness, one might explain this polarization as an effect of the students perceiving it as a school exercise in debate, like many others they may have encountered, where one is expected to take a clear stand. A second and compatible explanation is that the models available in the case (and in the wider discussions of GM foods) are very much polarized. Within the case materials, there were very few models for discourse that carefully weighed scientific evidence to reach a decision. There were journalistic or background information articles that presented "both sides" (but took no position explicitly). But the materials that took a position, whether from environmentalists, corporations, or—and perhaps most importantly—scientists from academic and humanitarian organizations, did not take a middle position. The closest we can come are the replies that Gordon Conway and Potrykus wrote to environmental activist organizations. There, given the constraints of talking to those who disagree, they produced arguments that conceded points, qualified their statements, and so on. But even in these letters, their positions were clearly and overwhelmingly pro.

4.3.2. How Did Students Use Scientific Information?

Because we focused on how the students used Percent to make arguments compared to how their sources used the same information, we did not focus on the 10 instances of Percent where the students used them as background information or the 15 instances where we could not track down their sources. This reduced the total instances of Percent from 124 to 99 and yielded the following results.

Table 2. Uses of percent information in student letters

	# OF USES (n=99)	# OF LETTERS (n=41)
Uses the *same* argument as the source makes using the Percent information	58	29
Cites source using Percent information to *refute* the source's argument	13	11
Uses the information as *evidence* for an argument not made directly in source	28	15
From source that *doesn't take a position* on GR	21	11
From source that *does not mention* Golden Rice	7	4

4.3.2.1 Same Argument as Source:

Students used scientific information in the form of percentages primarily by appropriating the same argument as the source that contained the percentage (58%). This is not surprising in one sense. If one means to persuade, then it is useful to have ready-made arguments, topoi. And in every raging debate there are topoi available, constructed by others who have thought about it and made good arguments, appropriated and shared among participants. We suggest as a hypothesis that in policy debates, students use scientific information as part of *argument units,* not as individual facts that speak "for themselves." This is also the way scientists primarily use scientific information, whether in experimental articles (the empiricist repertoire) or in private talk (hall talk) or in public debates (the contingent repertoire). Experimental articles are specifically built on the arguments other scientists have made. Each new experimental article tries to push slightly forward (or backward) the arguments other scientists have made ("concept simple as unit of experimental article in the sciences). Students mostly used scientific information not as background information (discrete facts, as science is often presented in textbooks) but rather as evidence for arguments, their goal.

The particular argument units (topoi) the students borrowed the most were also the most salient, the ones that the experts are using when they are most engaged with the problem: the problem of whether and to what extent and how soon Golden Rice could prevent VAD *in comparison to existing supplements.* The fact that the students used Percent arguments in the ways sources did suggests they were able to get at the heart of the issue (what argument theory calls the *stasis* point). These students were able to find the crux (stasis point) of the public policy debate. This suggests that other non-experts may also be able to as well (see Macoubrie, 2003).

In the absence of full economic risk-benefit analyses of the impacts of various supplementation programs, both experts and non-experts are in the same rhetorical boat. But it is nevertheless important for people, whether experts or non-experts, to appropriate and rehearse the argument units, the topoi, because this is how a person makes arguments one's own. When words come out of one's mouth, one become committed to them, though rarely finally. In addition, rehearsing others' arguments can affect one's attitudes (as the discussion below of the survey data will suggest).

The next results, Refutation and Evidence, are more interesting because they suggest ways that students transform the scientific information to make arguments, instead of merely appropriating an argument unit wholesale from a source.

4.3.2.2 Refutation

One way is to take information from a source that disagrees with one's own position in order to refute it. One has to engage with the arguments of the opposition. In 13% of the instances (11 students, six pro, four anti, 1 middle) students took an argument with a percentage from one of the sources and then refuted it. In the empiricist repertoire of experimental articles, scientists very rarely cite others to refute them, less than one percent of the time, according to Hyland (1999). One risks making enemies in a small community. But in the contingent repertoire— public discourse on science policy—the risks of refutation are not so great, and it is very useful in marking off the arguments to understand them for one's self and for persuading others.

All four of the anti-GR students who refuted employed the topos of risk/benefit. All four took the figure that GR currently contains 8-10% of RDA and argued that this is insufficient to justify the investment compared to synthetic and natural plant VA supplements. Five of the six pro-GR students took the 8-10% figure from the then-current research and used it to argue that it is sufficient as a supplement. In both cases, they are not debating the scientific figures. These are taken as agreed (both by the students and by their sources). The arguments concern the interpretation of the numbers, value judgments made without much data. In this, the students are doing what scientists do in public arguments, where the issue is not the results of empirical research, but the meaning of it for policy decisions. However, students engaged in relatively little of the name-calling and impugning of motives that is common among those debating GMO policy, including scientists. Only five students used such tactics: Here is one:

> Greenpeace . . . claims that Golden Rice provides at best 8%01.1 of the RDA of Vitamin A. He then says that an adult would need to eat nine kilos of cooked rice in order to receive their daily dose.[7] This is ridiculous because Golden Rice was never intended to fulfill the daily requirements of vitamin A; it was designed to be a supplement, nothing more. It is also a much better supplement than Greenpeace would have the public believe. A realistic estimation of the amount of Vitamin A Golden Rice imparts is around 10-20% of the RDA when the average 300g is ingested.[3] The lengths that opponents of Golden Rice will go to in order to prevent the spread of GMO's is outrageous. . . .

Compare Potrykus (2004), writing an editorial in a scientific journal *Plant Physiology*, presumably to a friendly audience:

> Thus, the opposition has argued that there is no need for "golden rice" because distribution of synthetic vitamin A works perfectly, or that nobody wants it because it tastes awful, or that people who eat "golden rice" will lose their hair and sexual potential! If you are interested in further misinformation of this kind, please consult various anti-GMO Web sites on the Internet. . . . In my judgment, hindering a person's access to life- or sight-saving food is criminal. To do this to millions of children is so criminal that it should not be tolerated by any society. . . . In my view, the Greenpeace management has but one real interest: to organize media-effective actions for fund raising. The "golden rice" case hopefully may help to unmask the true and shameful face of Greenpeace . . . (p. 23)

Perhaps students used so little of the invective common among experts because they were in a classroom context and worried the teacher would penalize them for it. Or perhaps they were not as invested in the outcome. But perhaps it was because they were arguing in a context in which they had to encounter those with different views directly, as Potrykus does when he is writing not for the relatively sympathetic audience of *Journal of Plant Physiology* readers, but a reply to Greenpeace (Potrykus, 2004).

4.3.2.3 Evidence

Perhaps the most interesting way students used scientific information beyond simply rehearsing the same argument as a source using the same information, was to take information from sources that did not take a stand. This occurred in 21% of the instances, and 11 of the 39 students did this (6 pro, 5 con) (pro 14 instances, con 13, middle 1). Students were synthesizing information. This is particularly interesting because it seems to be a way of using information from sources that don't take a stand in order to come to a decision.

The anti-GR students primarily used information from sources that did not take a stand in order to demonstrate that alternatives to GR are available. They cited studies of food fortified with VA, native (non-GMO) plants high in VA, and VA pill supplements.

> Research done by WHO found that high-dose supplements "produced remarkable results, reducing mortality by 23% overall and by up to 50% for acute measles sufferers." Supplements are effective and cost efficient. If the Rockefeller foundation is serious about fighting VAD they should start funding organizations who provide and distribute supplements.

The pro-GR students primarily used information from sources that did not take a stand to point out difficulties with supplements, in absorption, distribution, and so on.

> Vegetables, even though they contain high quantities of beta carotene, are actually poor sources of vitamin A because only two to four is actually absorbed. Fortification has been shown to be beneficial and costs a little more than the regular product. The problem with this method is that many poor populations do not buy processed foods, thus not receiving the vitamin A intended for them.

Four students used information from sources that did not mention GR, to draw analogies, a common topos in public policy arguments where there is little direct evidence (no field studies of environmental or economic effects). Two pro-GR students used studies of other GM crops to make a case by analogy: *bt* corn, soy, and cotton showed no adverse environmental effects (and positive economic effects); GR will do the same. An anti-GR student used statistics on the Green Revolution's impact on biodiversity to argue that GMO GR would do the same.

In citing information from sources that do not take a position, students were not making original arguments. They were doing what their sources that took a side did: finding evidence for pre-existing arguments that appeared more credible. But they were wrestling with the complexity of the arguments, pushing toward a deeper analysis of the issues, the sort of analysis that would be included in a complex risk-benefit analysis done by scientists (though not yet on GR).

5. TRIANGULATING SURVEY AND RHETORICAL ANALYSIS RESULTS

We now return to the survey results to suggest ways that the rhetorical analysis of arguments might speak to them. As noted earlier, the students' post-test responses showed a statistically significant move in the direction of anti-biotechnology. But when we examined the specific questions that showed statistically significant pre/post difference in light of the rhetorical analysis of their letters, we began to see the results not simply as a change in positions, but, perhaps, as a change of their understanding of the ways science is used to come to policy decisions. The five questions where the students' responses moved in the direction of anti-biotechnology were:

- #2. "Biotechnology is unnatural and should, therefore, be treated with great caution." $p = .005$
- #3. "Better scientific information on biotechnology will lead to greater acceptance of food produced with biotechnology." $p < .001$
- #10. "Technology can solve most of our most pressing human problems." $p = .033$
- #12. "Biotechnology could benefit millions of people." $p < .001$

- #14. "I am suspicious of scientific information on biotechnology that comes from environmental groups." p = .044

Even supporters of biotechnology GMO crops may well have felt, after reading, discussing, and making arguments on GR, that great caution is necessary; that better scientific information is not necessarily what determines acceptance; and that technology is not in itself a solution but must engage with a range of complex and contingent human factors—a rhetorical process. The students' discussions were experiential examples of the complexity and difficulty of arriving at consensus, acceptance, and solutions.

On one question the responses moved in the direction of what we thought, at least when we designed the survey, was pro-biotechnology (although the result is barely significant at the .05 level).

- #7. "We do not need full scientific certainty that biotechnology products are safe before biotechnology products are released." *p = .046*

The question was intended to tease out attitudes against opponents of biotechnology that point to fears as a common topos. Yet here again, the change might suggest that their engagement with the complexities of policy debates made students less confident in the ability of scientific information to settle disputes in the face of the social and rhetoric complexity of coming to a decision, whether they were for or against biotechnology.

6. CONCLUSION

What have we found then about how non-experts use scientific information to make arguments (reason) about GM food issues? What happens when information from experimental articles (empiricist repertoire) enters into the complex circulation of discourse in public debate on policy issues (contingent repertoire)?

Like experts in public forums, students almost always took either a pro or con position, with rather little attempt to find common ground. They mainly reiterated the existing topoi of the debate, rather than introducing new arguments. They made arguments using scientific information largely by appropriating the arguments of the source in which the information appeared. The argument and the scientific information—evidence—makes a unit of argument. When they do not appropriate wholesale, they use scientific information to refute arguments and to support arguments. When students used information from a source that did not take a stand, they did so mainly to support an argument. But when they appropriated information from a source that did not take a stand, they also appear to delve more deeply into the arguments, getting at the complexity of the arguments.

The experts' use of scientific information within public policy debates seems to have served as a model for the students. The assumption on the part of many that the public will change its attitude toward GMOs through increased exposure to scientific information per se may bear further examination. Research on the uptake of scientific information within the complex networks of communication in which they typically receive that information may suggest ways to bring information (facts) from the empiricist repertoire of experimental articles into the wider circulation of information in the contingent repertoire of public discourse.

For example, there were few instances (models) in the sources of extended engagement between those with different views, where arguments could be developed in detail, under the rhetorical pressure of direct rhetorical engagement. It might be helpful to have forums for such engagement, and websites such as this one might be adapted to provide such a forum, even including live chats or threaded discussions that include experts who take various positions. Sites like this might allow experts and non-experts might engage issues more deeply and provide involvement not only of citizens with other citizens, and citizens with experts, but also experts with other experts. (Macoubrie, 2004).

ACKNOWLEDGEMENTS: We thank Iowa State University's Science and Society Project for funding this work. We also thank David Fisher, Kristen Hessler, and Janice Freeman for their invaluable help.

REFERENCES

Albe, V. (2007). When scientific knowledge, daily life experience, epistemological and social considerations intersect: Students' argumentation in group discussions on a socio-scientific issue. *Research in Science Education, 38*, 67–90. doi:10.1007/s11165-007-9040-2

Bazerman, C., & Paradis, J. G. (1991). How Natural Philosophers Can Cooperate: The Literary Technology of Coordinated Investigation in Joseph Priestley's History and Present State of Electricity. *Textual dynamics of the professions: Historical and contemporary studies of writing in professional communities* (pp. 13–43). University of Wisconsin Press. Retrieved from http://wac.colostate.edu/books/textual_dynamics/

Charney, D. (2003). Lone geniuses in popular science: The devaluation of scientific consensus. *Written Communication 20*(3), 215–241.

Consigny, S. (1974). Rhetoric and its situations. *Philosophy & Rhetoric, 7*(3), 175–186.

Cook, G. W. D. (2004). *Genetically modified language: the discourse of arguments for GM crops and food.* London: Psychology Press.

Gilbert, G. N., & Mulkay, M. J. (1984). *Opening Pandora's box: A sociological analysis of scientists' discourse.* Cambridge: Cambridge University Press.

Fahnestock, J. (1986). Accommodating science: The rhetorical life of scientific facts. *Written Communication, 3*(3), 275.

Fisher, D. (2007). CMS-based simulations in the writing classroom: Evoking genre through game play. *Computers and Composition, 24*(2), 179–197. doi:10.1016/j.compcom.2006.06.004

Goodwin, J. (1999). Good Argumentation without Resolution In F. H. van Eemeren, R. Grootendorst, J. A. Blair & C. A. Willard (Eds.), *Proceedings of the Fourth International Conference of the International Society for the Study of Argumentation* (pp. 255-257). Amsterdam: SicSat.

Goodwin, J. (2003). Students' perspectives on debate exercises in content area classes. *Communication Education, 52*(2), 157–163.

Hessler, K., & others. (2003). Case study: Golden rice. The Biotechnology Outreach Education Center at Iowa State University. Retrieved from www.public.iastate.edu/~ethics/GoldenRiceCaseStudy.pdf

Huffman, W. E., Shogren, J. F., Rousu, M., & Tegene, A. (2003). Consumer willingness to pay for genetically modified food labels in a market with diverse information: Evidence from experimental auctions. *Journal of Agricultural and Resource Economics, 28*(3), 481–502.

Juanillo, N. K. (2001). The risks and benefits of agricultural biotechnology. *American Behavioral Scientist, 44*(8), 1246-1266.

Lusk, J. L., Jamal, M., Kurlander, L., Roucan, M., & Taulman, L. (2005). A meta-analysis of genetically modified food valuation studies. *Journal of Agricultural and Resource Economics, 30*(1), 28–44.

Macoubrie, J. (2003). Conditions for citizen deliberation. *Department of Communication, North Carolina State University.*

Miller, C. R. (1987). Aristotle's "special topics" in rhetorical practice and pedagogy. *Rhetoric Society Quarterly, 17*(1), 61–70.

Myers, G. (1991). Lexical cohesion and specialized knowledge in science and popular science texts. *Discourse Processes, 14*(1), 1–26.

Myers, G. (1993). Making enemies: How Gould and Lewontin criticize. In J. Selzer (Ed.), *Understanding scientific prose* (pp. 256–275). Madison, WI: University of Wisconsin Press.

Potrykus, I. (2003). From "golden" to "nutritionally optimized" rice—and from a scientific concept to the farmer. Presentation delivered at the conference *In the Wake of the Double Helix: From the Green Revolution to the Gene Revolution* (pp. 27–31). Bologna, Italy.

Robins R. (2001). Overburdening risk: Policy frameworks and the public uptake of gene technology. *Public Understanding of Science, 10,* 19–36.

Russell, D. R., & Fisher, D. (2009). Online, multimedia case studies for professional education: Revisioning concepts of genre recognition. In J. Giltrow & D. Stein (Eds.), *Genres in the Internet: Issues in the theory of genre* (pp. 163–191). Pragmatics & Beyond New Series. Amsterdam ; Philadelphia: John Benjamins Publishing Company.

Toulmin, S. E. (2003). *The uses of argument.* Cambridge: Cambridge University Press.

Dismantling Expertise: Disproof, Retraction, and the Persistence of Belief

CHRISTOPHER W. TINDALE

Department of Philosophy & Centre for Research in Reasoning, Argumentation and Rhetoric
University of Windsor
401 Sunset Avenue, Windsor, Ontario
Canada
ctindale@uwindsor.ca

ABSTRACT: This paper involves an extended study of the Andrew Wakefield vaccine case as an illustration of three ethotic phenomena: (i) the creation of ethos in the face of attack; (ii) attacks on that expert by other members of the relevant scientific community; and (iii) the extension of ethos to experts from a lay audience. This last is particularly fascinating insofar as it demonstrates the durability of trust that a specific audience continues to have in Wakefield's expertise. The nature and ground of that trust will be explored.

KEYWORDS: autism, ethos, ethotic argument, rhetorical argumentation, testimony, Wakefield.

1. EXPERT ARGUMENTS AND THE POSITION TO KNOW

Many scientific controversies arise and erupt around cases of disputed expertise (Collins & Evans, 2007). Such cases are characterized by one or more of three related, what I will call, 'ethotic' phenomena: self-construction of ethos in the face of attack as a form of defense; the extension of ethos within a scientific community as a means of cooperative reinforcement of positions on which there is some consensus; and the extension of ethos to experts by a lay audience as a natural concomitant to their cognitive response. That is, with respect to the third point, lay audiences consider more than just the argument they receive; they consider the source. The Andrew Wakefield case involving autism and vaccines that is investigated at the core of this paper serves to illustrate all of these phenomena, although most attention is given to the last.

When science is revised through the retraction of claims, by the scientists themselves or the publishing medium that communicated the findings, questions about the control and authentication of ideas are sure to arise. When scientific claims are revised, the field follows certain established protocols in order to return as soon as possible to business as usual. But when public opinion resists such moves, the attempt to "wipe the slate clean" (Harmon, 2010, p.1) falters in a surprising fashion. After all, what knowledge can the public possess that counterweighs the experts' authoritative claims to truth? And when rogue or maverick scientists within the field refuse also to yield, questions must arise about how expertise is constructed, dismantled, and acknowledged.

The typical way in which expert arguments[1] have been treated by argumentation theorists, and especially informal logicians, is in terms of a set of standards or checks to

---

[1] I talk here and throughout the paper of expert arguments rather than arguments from authority. 'Expert' has a narrowness that in cognitive and related argumentative matters has a clarity more useful than the broader 'authority.' Experts do not possess the wide authority of an Aristotle that transcends fields, but invariably have

Tindale, C.W. (2012). Dismantling expertise: Disproof, retraction, and the persistence of belief. In J. Goodwin (Ed.), *Between scientists & citizens: Proceedings of a conference at Iowa State University, June 1-2, 2012* (pp. 393-402). Ames, IA: Great Plains Society for the Study of Argumentation. Copyright © 2012 the author(s).

determine the expert's credentials. There is no thought to completely detach the quality of the reasoning from the source. Douglas Walton's account is symptomatic in this respect: "An expert in a particular domain of knowledge is in a special position to know about propositions in that domain, and therefore the expert's opinion on some propositions of this kind generally has a weight of presumption in its favor" (Walton, 1996, p. 64). That weight can be tested through a series of relevant questions directed at ascertaining the relation of the expert to the domain, the viability of the domain involved, and any circumstantial features of the case that may undermine the expert's assertions.

Such accounts offer serious advances on earlier treatments of the *ad verecundiam* like that found in Hamblin (1970). But, while the focus here is not entirely on the statements themselves—an expert's vested interests must also be considered, for example—the thrust of these accounts is to explore the expert and the argument. Little attention is given to the audience who is addressed and what *further* reasons that audience might have for accepting or rejecting the claims put forward. When the focus shifts to the audience in this way different considerations come to the fore. We start to look at experts not as repositories of knowledge but as communicators, and to consider the full range of what is communicated. Indeed, we start to think about how audiences experience both expert appeals and the sources of those appeals.

Moreover, sets of critical questions for assessing expert argumentation typically fail to deal adequately with a lack of consensus among experts. Consider the Walton account again: We are given six questions with which to explore any argument, one of which asks: "Is the assertion of the expert consistent with what other experts assert?" (Walton, Reed, & Macagno, 2008, p. 15).

It is often this last question that proves most difficult. In his fuller treatment of the argument scheme, Walton suggests that what is important is that an expert's opinion is representative of what is generally accepted in a field (1997, p. 221). Where this is not the case, a further set of critical sub-questions is brought forward, exploring the reasons for the non-representativeness. Has the expert failed to be thorough, or is he or she a maverick or crank? These are strikes against the opinion being based on knowledge, but only presumptively so. If an appropriate explanation can be given, it may be sufficient for the disagreement to be noted and a more qualified conclusion drawn.

We should not discount the figure of the maverick, who often shifts ideas from the margins into the mainstream. But from the perspective of receiving arguments, audiences have fewer or different resources for dealing with the underlying disagreements than they do on other occasions, as the case to be examined may suggest.

2. THE WAKEFIELD CASE

The case focused on here is that of Dr. Andrew Wakefield and the controversy surrounding the relationship between the MMR vaccine (measles, mumps, and rubella) and the onset of autism. Wakefield was a lead author in a February 1998 paper published in the *Lancet* that explored a link between the measles vaccine, irritable bowel syndrome in children and the onset of autism. That paper has since been retracted. While the paper itself did not assert a clear link (in fact, it

---

a more specific or local knowledge of a limited field of inquiry or experience. For a fuller discussion of this distinction and of the argument from authority see Tindale, 1999.

was conceded in the paper that no link had been established on the basis of the twelve cases studied), Wakefield did make such an assertion in a related news conference,[2] and that assertion was seen as instrumental in the subsequent drop of the numbers of children being vaccinated. Many parents acted on what they judged to be Wakefield's expert testimony, even though the majority of relevant experts rejected his claims (John, 2011). Subsequently, serious concerns about the original research came to light, resulting in the retraction not just by the editors of the *Lancet* but also by some of the paper's authors, and in Wakefield and another author being removed from the United Kingdom's medical register in 2010. Among the charges deemed proven by a lengthy inquiry were unethical behavior on the part of the researchers in dealing with children, fraud in the reporting of their results, and in Wakefield's case that he patented a single-measles vaccine from which he stood to profit in the absence of MMR. Wakefield portrays himself as the victim of a conspiracy (Wakefield, 2011, p. 50,73) and continues to operate in the United States (See Dominus 2011).

3. ETHOS *WITHIN* THE SCIENTIFIC COMMUNITY

As Prelli (1989, p. 48) points out, scientific ethos is both technically and morally binding. In the latter case, a community coheres around what they take to be right and good. The reaction to the Wakefield case illustrates this. It would seem to be the *consequences* of Wakefield's claims that prompted the review of his data and procedures and the concern to then counter those claims. According to a CNN Heath Report (January 5, 2011), the alarms Wakefield was believed to have raised in the general populace were deemed to have resulted in a serious drop in the number of vaccinated children, with a commensurate increase in cases of measles and several deaths. Thus, it is the 'good' of the vaccination program that needed to be protected. To this end, the relevant community rallied around, expelling Wakefield (and several colleagues) and closing ranks against him. On the one hand, their justification is scientific insofar as they correct what is technically bad. But on the other hand—and this is the stronger public message—it is moral. They organize themselves to promote a public good that has become at risk. On these terms, the consensus of the experts opposes Wakefield. But as importantly, in making such an opposition and publically explaining their justification they act so as to give themselves weight in the debate, to reinforce their own ethos.

In his recent study of the Wakefield case in *The Philosophical Quarterly*, Stephen John essentially takes the consensual position as given. Hence, the case *then* exhibits a serious instance of the failure to defer to expert testimony (2011, p. 497), since the prevalence of opinion opposed Wakefield's conclusions. The value of Johns' account in my consideration lies mainly in his attention not to the experts and their claims but the audience's reception of those claims. He is concerned with those agents' reasons for acting as they have. Parents "should have deferred to the experts who claimed that MMR was safe. However, a large proportion of parents did not do so" (2011, p. 501), hence, something 'went wrong.' This is because, generally, non-experts ought, in some objective sense, to defer to expert testimony, and here he is only considering the consensual view as representing expertise. He does support this judgment, however.

---

[2] As Stephen John points out in his discussion of the case, Wakefield claimed that the triple MMR vaccine might cause autism in a small number of cases (2011, p. 496). Dominus (2011) discusses Wakefield's continued belief in this causal relationship, and the same is evident from Wakefield's own account (2011).

John's concern with the apparent epistemic failure leads to the development of a simple diagnostic model for social epistemology:

> When non-experts who want to increase their stock of true beliefs in some domain fail to defer to the testimony of true experts, something must have gone wrong. The failure must be the result of either a failure of evidence acquisition or a failure to grasp the simple testimonial principle. (2011, p. 500)

On these terms, parents had an interest in acquiring true beliefs and a choice between deferring to Wakefield and his small group of experts, or the larger group of experts who denied a causal link between MMR and autism. Numbers alone cannot decide between these conflicting groups of experts or establish correctness unless the opinions of additional experts are independent of the views of the main group, as Alvin Goldman (2001) explains in his treatment of experts. This, John suggests, is exactly the case here, since different data sets and statistical tools were used. So numbers do matter. The concerns raised about the credibility of the Wakefield group thus compound the parents' failure.

The most serious problem John identifies, and the crux of his treatment, is his suggestion that the parents engaged in a type of epistemological free-riding (2011, p. 513). If agents adopt a plan of action which is predicated on the belief that other members of our society will accept a claim that has been established with scientific certainty, and that involves not shouldering their share of the burdens following from that claim (and required for maintaining public good), then those agents are engaging in epistemological free-riding. Thus, "failure to defer to expert testimony can be wrong because it is unfair": agents gain a benefit while not accepting the burden necessary for maintaining that benefit (2011, p. 513).

However, two things are noteworthy about John's treatment: (i) He admits that his account with its general tentativeness is speculative. He writes, "I cannot prove these sociological conjectures" (2011, p. 508). Of course, without access to the minds of the agents involved, any conclusions drawn about this general phenomenon need to be qualified in this way. But (ii) as his discussion develops it is clear that he is concerned with a failure of only one group of parents. There are really two sets of parents at issue: (a) those who have some association with autism (children) or autism groups, and (b) those who refuse vaccines. John is interested in the second group, and hence the concern with free-riding carries more weight. But there is also the first group, and while these groups are not mutually exclusive, this second is arguably the more effected and interesting group. I am interested in the distinct issues presented by this group of parents, which pushes our speculation in a different direction.

The aspect of the case that is most relevant to my discussion, then, is that in spite of the counter-claims of experts, and of the evidence of fraud and self-interest, and the failure of other researchers to corroborate the original findings, Wakefield continues to receive substantial support from autism groups and individual parents.[3] Why is this? Simple failure to accede to expert testimony is not enough.

In a different approach to the Wakefield case, Sorell (2007) suggests that there was a crisis of trust and that parents failed to show epistemic deference where they should have by thinking that this was a matter of ethical 'respect.' That is, they believed they had an ethical

---

[3] The following web sites indicate the nature of the support Wakefield continues to receive. Details of the case against him and the U.K. trial can be found on http://www.timesonline.co.uk/tol/ news/uk/health/ article7009882.ece; an indication of his support is apparent from: http://www.wesupportandywakefield.com/. See also Dominus (2011).

right to live by their own values and confused this with an epistemic right to contest expert testimony. This account points to a prior state in the audience that impacts how they receive information. They are not empty vessels but have values that influence how they assess what they read and hear. Any crisis of trust may be traced not just to a failure of experts to agree, but a failure of expert testimony to fit with prior beliefs.

Trust is a powerful force. How audiences judge the character of experts, where this is available, can be as important when claims are considered as what is actually said. But trust must also be balanced by a degree of vigilance, and audiences seem well able to do this (Sperber et. al, 2010). In what follows, I will consider some of what is behind this. It is the failure of traditional models of argument, including informal logic, to adequately account for the influence of things like character that leads us to consider the rhetorical approach to argumentation and, in that light, its origins in the Aristotelian account.

4. ETHOS: BACK TO ARISTOTLE

Aristotle provides one of the earliest accounts of social argumentation, and we can learn a lot about the nature of arguers and audiences from what is set down in the *Rhetoric*. In particular, one of his rhetorical "proofs" is identified as ethos, and a further idea relevant to this discussion is that of the engaged audience. 'Rhetoric' itself is, of course, a term that can mean many things to many people. In his recent book *How to Write a Sentence* (2011), Stanley Fish has helpfully defined rhetoric as "the art of argument" (p. 29), and that is a meaning I'm happy to adopt here. Aristotle's *Rhetoric* defines it as an ability to see in every particular case the available means of persuasion (I.2.1), thus encouraging attention on persuasion and the ways it is achieved. Important in Aristotle's definition is the attention it gives to discovery. The *seeing* that rhetoric is, is theoretical (from the common root of *theorein*), a seeing in the mind.

Aristotelian argumentative agents are situated squarely in social contexts suggesting a much broader conception of rationality than what would be attributed to simple logical beings. This is implicit in the accounts given of the three "proofs," logos, ethos, and pathos (associated with the discourse, or argument, the character of the arguer, and the emotions of the audience). Each of these is considered a proof insofar as it facilitates (but does not guarantee) persuasion. Thus, the logos involved in the *Rhetoric* is one which addresses situations of uncertainty.

Arguing in the social domain characterized by such situations is aided by ethos and pathos. Ethos is particularly important because it involves a speaker communicating her or his character through discourse. No prior reputation is assumed by Aristotle:

> the speech is spoken in such a way as to make the speaker worthy of credence; for we believe fair-minded people to a greater extent and more quickly on all subjects in general and especially where there is not exact knowledge but room for doubt. (*Rhetoric,* I.2.4).

"Character," it is suggested, "is almost, so to speak, the most authoritative form of persuasion" (*Rhetoric*, I.2.4). And while Aristotle does not elaborate on this authoritativeness, it reflects the social nature of argumentation, the dependence, generally, that we place on each other for the authority given to our statements. Reflecting the kinds of ethotic constructions I mentioned in the introduction, the kind of weight acquired by a speaker cannot be simply claimed, it must be *given* by others, and this act involves the type of complex reciprocal recognition that will be worked out later in the tradition. Character is integrally connected with the commitments and obligations we acquire, and thus we might consider the importance Aristotle placed on ethos to be confirmed by subsequent, more recent, discussion of commitments and obligations.

The additional treatment of pathos as a proof further confirms the social focus of the account. Aristotle's early cognitive account of the emotions links emotion with judgment. We do not make the same judgments when we are grieving as we do when rejoicing. Emotions like anger, fear, or shame all rely on thoughts about other people. These emotions all find us outside of ourselves in the world, navigating difficult interpersonal matters that can be understood and converted to sources of persuasion. Only a selection of emotions is discussed, but enough is said to meet the stated claim of explaining how emotions are created and counteracted. Aristotle adds, "from which are derived proofs related to them" (*Rhetoric* II.11.1).

The discussion of pathos in the *Rhetoric* also emphasizes that the rationality of this account is not concerned solely with the "logical being." The whole organism is addressed, and the interactions between the parts will influence the outcome of argumentation. The accounts Aristotle gives of individual emotions indicate their social nature—they arise in relation to a person's perceptions of what is expected of them or due to them in specific circumstances.

Social spaces contain arguers constructing their proofs, interacting with those around them, and addressing the whole person. Those addressed—audiences—are not passive recipients of claims and reasons, but active participants in the process of giving and receiving reasons. In this sense, we see at work a conception of audiences as "engaged." The concern behind the rhetorical enthymeme, for example, is the audience being addressed. Enthymemes involve short arguments with the focus on the audience's ability to grasp the ideas involved. Additionally, this goal will be achieved with greater success if the audience already knows part of what is being put forward, if they can contribute to the argument by way of completing it. Thus we see the idea in the tradition of an argument in which one premise is suppressed because it does not need to be stated—the audience already knows what is involved.

There is a more important sense by which enthymemes bring the audience centrally into the picture, since a sense of 'enthymeme' suggests "something in the mind." The *seeing* involved in the above definition of rhetoric is an internal seeing, an *insight*. The arguer must, in some important ways, see into the mind of her audience and compose the speech accordingly. We need not restrict the sense of this to a particular thing (thought) within the mind, rather than understanding the focus to be on the mind generally and its way of seeing. The contrast between seeing with the body's eye and seeing with the mind's eye is an important one that is exploited in several Aristotelian works. Theorizing is an internal seeing, albeit as this has application to some external state of affairs in the world. And even the more limited interpretation of the enthymeme involving just a particular idea still assumes an insight into others' mental awareness in order to appreciate what is or is not already there.

The engaged audience is a further topic of interest in the last book of the *Rhetoric*, the one that seems to have set aside the more discursive matters and turned simply to matters of style and arrangement, introducing tropes and figures of both speech and thought (energeia). The choice of any stylistic device will involve careful consideration of who it is that is being addressed.

Figures of speech, for example, activate common understandings between arguer and audience. The figure discussed at length here is the antithesis, which works (on one level) in terms of balanced cola. Here, the two cola stand in a relation to each other such that having heard the first phrase, the audience is able to supply the second phrase (whether or not it is subsequently uttered by the arguer). Thus, this figure is invitational in the sense that it invites the audience to complete it.

Under energeia is stressed the conception of actualization that is balanced throughout Aristotle's works with potentiality. An associated concept bringing-before-the-eyes [*pro ommaton poiein*], or visualization, captures the immediacy of what is experienced by an audience: "for things should be seen as being done rather than as going to be done" (*Rhetoric* III.10.6). Something comes alive for the hearer through being actualized in such a way. But for something to be actualized in an audience it must already be available potentially. In this sense, we see again the assumption of a common fund of ideas shared between members of a community, which can be assumed and activated in argumentation. The visualization involved encourages attentiveness and provokes, as we later discover, receptivity [*eumatheia*]. If they are not attentive, hearers will not be receptive, "because the subject is unimportant, means nothing to them *personally*" (*Rhetoric* III.14.7, emphasis mine). The subject must be brought alive for each member of an audience, and that involves making it important to each one personally in order for reception to occur.

Thus, there are important elements of reciprocity in Aristotle's account of argumentation. The audience is conceived as an active participant in the processes of persuasion rather than a passive recipient of persuasive arguments. Audiences contribute details of arguments; their beliefs and knowledge form the materials that arguers must use in conjunction with the statements chosen to convey them. And audiences contribute the parts of arguments that are unspoken, because these are already aspects of their belief structures. More personally, to take a concern in the questions that drive rhetoric speaks to a deeper appreciation of the abilities of citizens who comprise the audiences addressed and of wanting to share in common ventures on a societal level. That is, one addresses a society that one is interested in belonging to and which is comprised of members that one appreciates as being worth interacting with and among whom can be found like-minded people with whom the finer aspects of societal life can be shared. These ideas have contemporary illustration in the extended Wakefield case.

5. UNDERSTANDING ETHOS IN THE WAKEFIELD CASE

What Wakefield has attempted *post facto*, is the creation of a scientific ethos. He narrates events leading up to his expulsion from the medical community and the retraction of the *Lancet* paper so as to present himself primarily as an advocate for children (Wakefield, 2011). To a certain degree, his dispute with the scientific establishment is distant from this. The critics construct their case around the emerging crisis of unvaccinated children as a result of Wakefield's "alarmist" claim. In his account, Wakefield never directly addresses this, claiming no more than that the drop of "uptake of MMR" after his publication may have been offset by a "reciprocal uptake in single vaccines" that were not reported (2011, p. 226). Instead, he presents the issue as one in which the medical establishment is striving to protect the MMR vaccine in spite of its suspect effects (and implies a conflict of interest on the part of some authorities who have apparent associations with drug companies).

But there is one part of his dispute with other experts that is relevant to our concerns, and would seemed to be part of his ethos as this is attributed to him by segments of his audience. He constructs his defense around the stories that he was initially told by the parents of the *Lancet 12* (the twelve subjects of the notorious paper). And he carries this over to a claim about the nature of disease:

> [T]here is a different way of looking at disease . . . This alternative approach does not *just* start with the parental narrative—it is truly invested in it, using it as the navigation system without which the disease is condemned to forever wander in the wilderness. (Wakefield, 2011, p. 160)

This no doubt can account in part for his continued acceptance by the relevant segment of the public. His very perspective is rooted in the beliefs and concerns of his engaged audience, and communicated to them in his statements. And it would appear that this construction was not only accomplished in retrospect, but had an earlier origin in his initial media releases, thus accounting for the original response to him and his ideas.

An audience's experience of expert communication includes several components: their listening to what the expert says (or is reported as saying) and forming ethotic judgments; their emotional response to the expert's ethos with an associated degree of trust; and their understanding of what is said in relation to others things they know or believe which provide corroboration. The last point is important because it can explain why audiences would hold to the claims of "experts" like Andrew Wakefield in the face of conflicting claims. We considered earlier the value of having an engaged audience that already has information that can contribute to the argumentative exchange and complete the reasoning. It is not simply a matter of the audience here weighing competing claims in a logical fashion; it is a matter of taking those claims back into their lives and matching them against their wider experiences. Autism groups and individual parents of autistic children did not just hear Wakefield's claims and come to believe them; they heard those claims and related them to what they already believed or used them to make sense of phenomena that needed to be explained. In their limited way, they are making what Gelfert (2010) calls testimonial inferences to the best explanation, where "best" is decided on their terms and not those of external experts. The attempts of other experts to discount Wakefield's testimony must then do more than uncover fraudulent practices, they must also provide a competing and *better* explanation for the apparent causal relationship which was already suggested in the experiences[4] of those for whom these things are relevant (that is, the parents). It is not simply a matter of continuing to trust Wakefield; it is the extension of that trust to the ideas with which he is associated, ideas that provide the most plausible explanation available to them. In these senses, the ethotic and pathotic elements in the argumentation have epistemic outcomes.

Moreover, this experience is one shared within the affected community. The Wakefield case, at least as philosophers like John approach it, seems to raise the spectre of people exercising epistemic autonomy to the detriment of others. David Coady (2006), arguing along these lines, explains that "we don't want people being epistemically autonomous when they could make their views dependent on others who either have much more information or a much greater ability to make rational inferences from their information" (p. 77). But in the MMR case the relevant dependence is not so much on the experts, at least not *directly*. Nor is it a matter of isolated audience members reasoning on their own. Rather, their dependence is on each other within the affected community. They reason together, corroborating each other's ideas, where such reasoning acts as a corrective and a confirmation. If and when the ideas

---

[4]   At stake here is how information relates to the cognitive environments of an addressee or audience. See Tindale, forthcoming.

change, they will change within the group as the understanding shifts within that corroborative community, perhaps as members leave and join, but more generally through the mutual modification of the group. An important feature of rhetorical argumentation is the interdependence that reasoners enjoy, and that interdependence is suggested here. It is not so much isolated individuals who respond supportively to Wakefield as it is an active community that corroborates the individual experiences of its members. Of course, the fact that a view is widely shared is not a strong reason in favour of that view unless people come to hold it independently (Sperber et. al, 2010, p. 38). But it is the communal holding of this view that reinforces and corroborates it. Epistemic autonomy, perhaps surprisingly, is not even an obvious value here. In a world of non-Cartesian selves, each individual finds personal value only against the backdrop of being with others. When it comes to evaluating expert claims, we do not function well as isolated knowers who must think for ourselves. In such situations, the lesson is to think together, and to situate experts' statements *within* this context, as part of an integrated knowledge base complemented by trust in character assessment. This is a lesson Wakefield's critics would do well to observe.

6. CONCLUSION

We have moved beyond Walton's account in terms of our understanding of expert arguments, adding to that understanding a consideration of how they may be experienced in light of the suggested role played by rhetorical proofs like ethos and pathos in the Wakefield case and the parents' response. We have also moved beyond the largely logos-centric discussion of the Wakefield case produced by John. On both fronts, the research continues. But I hope to have suggested the value of that research and the lessons that might eventually be drawn from it.

Argumentation theorists aspire to understand arguments of all varieties in their natural environments. Those environments are largely social and involve more than interaction between people on what we might think of as a purely logical level. And the full understanding of those arguments must include some consideration of what is involved in the *experience* of being addressed by them.

REFERENCES

Aristotle. (2007). *On rhetoric: A theory of civic discourse.* (G. Kennedy, Trans., 2d Ed.) Oxford: Oxford University Press.
Coady, D. (2006). When experts disagree. *Episteme, 3,* 68–79.
Collins, H. & Evans, R. (2007). *Rethinking expertise.* Chicago, IL: University of Chicago Press.
Dominus, S. (2011, April 24). The denunciation of Dr. Wakefield. *The New York Times Magazine,* p. MM36.
Fish, S. (2011). *How to write a sentence and how to read one.* New York, NY: HarperCollins Publishers.
Gelfert, A. (2010). Reconsidering the role of inference to the best explanation in the epistemology of testimony. *Studies in History and Philosophy of Science, 41,* 386£–396.
Goldman, A. I. (2001). Experts: Which ones should you trust? *Philosophy and Phenomenological Research, 63,* 85–110.
Harmon, K. (2010, March 4). Impact factor: Can a scientific retraction change public opinion? *Scientific American,* 1–4. Retrieved from http://www.scientificamerican.com/article.cfm?id=retraction-impact-lancet.
John, S. (2011). Expert testimony and epistemological free-riding: The MMR controversy. *The Philosophical Quarterly, 61,* 496–517.
Prelli, L. J. (1989) .The rhetorical construction of scientific ethos. In H. W. Simons (Ed.), *Rhetoric in the human sciences* (pp. 48–68). London: Sage Publications.

Sperber, D., Clément, F., Heintz, C., Mascaro, O., Mercier, H., Origgi, G. & Wilson, D. (2010). Epistemic vigilance. *Mind and Language, 25*(4), 359–393.

Sorell, T. (2007). Parental choice and expert knowledge in the debate about MMR and autism. In A. Dawson & M. Verweij (Eds.), *Ethics, prevention and public health* (pp. 95-110). Oxford: Oxford University Press.

Tindale, C.W. (1999). The authority of testimony. *Proto Sociology: An International Journal of Interdisciplinary Research, 13*, 96–116.

Tindale, C.W. (2011). Character and knowledge: Learning from the speech of experts. *Argumentation, 25*, 341–353.

Tindale, C.W. (forthcoming). The words of other people: The fundamental role of testimony in rhetorical argumentation. In H. van Belle, et al (Eds.), *Verbal and visual rhetoric in a media world*. Leiden: University of Leiden Press.

Wakefield, A. J. (2011). *Callous disregard: Autisms and vaccines—The truth behind a tragedy*. New York, NY: Skyhorse Publishing.

Walton, D. (1997). *Appeal to expert opinion: Arguments from authority*. University Park, PA: The Pennsylvania State University Press.

Walton, D. (1996). *Argumentation schemes for presumptive reasoning*. Mahwah, NJ: Lawrence Erlbaum and Associates.

Walton, D., Reed, C., & Macagno, F. (2008). *Argumentation schemes*. Cambridge: Cambridge University Press.

Signatures and Spinoffs: Sequences of Ignorance in the Theory/Practice Split of the Ecological Society of America, 1917–1950

KENNY WALKER

Department of English
University of Arizona
1423 E. University Blvd, Room 445
Modern Languages Building
PO Box 210067
Tucson, Arizona 85721
USA
kcwalker@email.arizona.edu

ABSTRACT: In this paper, I demonstrate the rhetorical life of ignorance as a special *topos* of ecological science. I accomplish this through a topical survey of the genre systems of the Ecological Society of America (ESA), and show how ignorance sequences into larger argumentative signatures that define its disciplinary epistemology. I argue that Ecology is a science with a social praxis and ignoring public application not only underserves the discipline, but is inconsistent with the history of Ecology.

KEYWORDS: rhetoric of science, ignorance, topoi, genre systems, Ecology.

1. INTRODUCTION

> "Disciplines lose the capacity to learn from the work that they chose to ignore." ~ Thomas Miller

> "An intensive topical survey within a discipline could establish characteristic signatures of eide [or special topoi] that may typify the argumentation of that discipline. These signatures could then serve, much like strands of rhetorical DNA, to reconstruct rhetorical inheritance and influence among technical disciplines. . . . By attending to changes in topical signatures over time, we could develop a flexible method for describing and predicting the rhetorical life of STEM disciplines."
> ~Lynda Walsh

The above quotes are about rhetoric within disciplines: one remarks on the products of a discipline's attention, the common argumentative structures of a discipline; and the other remarks on the products of inattention, a loss of growth potential through a constructed ignorance. This paper begins with an inquiry into why we might study ignorance in a discipline, both to watch a strand of its argumentative structure unfold, and to note what it chooses to ignore. Tracing ignorance arguments through a discipline can be useful because they help describe the parameters of a shifting body of knowledge, and allow scholars to question the consequences of those constraints. In other words, the study of ignorance arguments helps rhetorical scholars study disciplinary epistemology. The quote from Miller (2011, p.148) suggests that in an attempt to establish expertise over a body of knowledge, a discipline necessarily chooses to ignore other bodies of knowledge and consequently, it loses the capacity to grow from this work. On the other hand, the quote from Walsh (2010, p. 147) suggests the common topics of a discipline are its rhetorical signatures—they fingerprint the argumentation strategies of a discipline and in this way act like rhetorical DNA. Together these

Walker, K. (2012). Signatures and spinoffs: Sequences of ignorance in the theory/practice split of the Ecological Society of America, 1917–1950. In J. Goodwin (Ed.), *Between scientists & citizens: Proceedings of a conference at Iowa State University, June 1-2, 2012* (pp. 403-411). Ames, IA: Great Plains Society for the Study of Argumentation. Copyright © 2012 the author(s).

quotes suggest a strategy and a method for reconstructing a portion of the rhetorical life of a discipline through a study of its active construction of ignorance. By tracing the argumentation strategies of ignorance, and by watching how these strategies are inherited, or not, over time, rhetoricians of science might better describe and perhaps predict how a discipline takes on a characteristic rhetorical life.

In order to construct a rhetoric of ignorance, I will frame ignorance as a special *topos* of ecological science. Topical analyses have proven useful for the analysis of scientific discourse over the last twenty years because they are sufficiently reductive to be a manageable analytic tool over a large body of work (Miller, 1987; Miller & Selzer, 1985; Perelman & Olbrechts-Tyteca, 1969; Prelli, 1989; Walker & Walsh, 2012; Walsh, 2010; Winterowd, 1973). Topical analyses allow for a depth of insight into one particular argumentation strategy, rather than a broad, encompassing perspective. Walsh (2010) likens an author's selection of a *topos* to "selecting a stance from which to view a landscape" or "a slice through a data set" (pp. 125–126). In either conception, conducting an analysis through a *topos* is a way to foreground chosen thematic elements while leaving others to the margins or outside the field of view altogether. The issue of reliability is also easier since the frame of analysis is only one term, in this case ignorance, and its sequences into larger argumentative structures. Using a topical methodology within a diachronic reception study, Walker and Walsh (2012) identified how the special *topos* of uncertainty in Rachel Carson's *Silent Spring* created a site for public engagement in popular environmental science. But this conception of scientific uncertainty restricted ignorance to a disciplinary *topos* which simply articulated "the unknown," and as Proctor and Schiebinger (2008) suggest, ignorance is much more complicated than this. A topical study of how ignorance changes over time in a scientific discipline might help reveal a larger capacity for ignorance arguments.

In this short paper, I seek to complicate the term *ignorance* beyond "the unknown" and use a diachronic topical analysis to demonstrate the sequences of ignorance in ecological science. I accomplish this by tracing ignorance through the genre systems of the Ecological Society of America (ESA) 1917–1950, particularly the genres that led to the creation of the "Ecologist's Union." By tracing this *topos* through various ESA genres, I show how ignorance sequences from technical to public to disciplinary spheres of argument to deliberately shape the ethos of a professional ecologist as willfully ignorant of public application. First, this analysis reveals how the public application of ecological science is inherent in its technical arguments, and second, it reveals how a discipline argues from a technically-based ignorance to an ethos-based ignorance. I argue that Ecology's active ignorance of public application not only underserves the discipline, but is inconsistent with the history of ecological science. To start, I briefly construct a rhetoric of ignorance, and review how topical analyses can trace ignorance arguments as shifting social constructs.

2. A RHETORIC FOR AGNOTOLOGY: TOPICAL ANALYSES OF DISCIPLINARY IGNORANCE

Agnotology, or the study of ignorance, has been directly evoked at least since Scottish Philosopher James Frederick Ferrier (1856) wrote his treatise on "Agnoioloy or Theory of Ignorance" in his *Institutes of Metaphysic: Theory of Knowing and Being*. Social theorist Michael Smithson (1989) was one of the first to synthesize various and competing definitions of ignorance through his work in cognitive science in the late eighties. His work described ignorance and uncertainty as "emerging paradigms" largely ignored by western epistemologies (p. 3). Since then science communication scholars such as S. Holly Stocking and Lisa Holdstein (1993; 2009) have written about the social construction of ignorance between scientists and journalists for over a decade. A large portion of their work contends that ignorance is socially constructed through a process of claims making. Those who wield arguments based around what is unknown often do so "in ways that reflect and contribute to their own interests" (1993, p. 189). In this way ignorance claims can allow journalists to fulfill their roles as watchdogs and thus meet the needs of their audiences (2009, p. 37). For rhetorical scholars, Stocking (1998) argues "if you concentrate on the spaces around the knowledge, if you focus on what you don't know, on ignorance, you may do a better job of knowing" (p. 177). More recently, historian of science Robert Proctor (2008) has expanded on this work in his edited collection, *Agnotology: The Making and Unmaking of Ignorance*. He argued for a broad approach to ignorance, conceptualizing it as a structural production with various manifestations (p. 2). Though Proctor is cavalier in his recognition of Ferrier as an antecedent, it is clear that Ferrier's conceptions of ignorance as a correlative of epistemology and as an active construct are a central part of both Smithson and Proctor's definitions.

The definition of ignorance I start with is Proctor's (2008), who divides ignorance into three phases: ignorance as a resource; ignorance as a lost realm; and ignorance as an active construct (pp. 4–18). I understand these phases to equate with sphere of argument (technical, public, and disciplinary) and also with the rhetorical appeals (logos, pathos, and ethos). In order to code for signatures of ignorance I worked inductively from each text surveying for articulations of ignorance that were consistent with the definitions laid out in Proctor's first chapter in *Agnotology: The Making and Unmaking of Ignorance*. Finally, I analyzed this data set according to spheres of argument and their rhetorical appeals.

Conducting a topical analysis of a single *topos* like ignorance across the genre systems of a specific scientific community allows scholars to trace how this *topos* sequences into larger rhetorical signatures. This combination can help rhetorical scholars trace the development of a specific rhetorical strategy across genres and, as Paul, Charney and Kendall (2001) note, watch acceptance or rejection unfold over time. This not only allows scholars to link specific rhetorical acts to institutional and disciplinary contexts, it allows one to study the social intentions of a group over time. In my example below, locating and tracing the sequences of ignorance should provide insight into the ESA's contested epistemology. It should help locate and theorize where and when the discipline chose to stop paying attention.

2.1 Ignorance as Logos (Technical): Research Articles 1920/1950

The journal *Ecology* was originally founded in 1920 to publish research papers "of ecological interest from the entire field of biological science" (as cited in Burgess, n.d., p. 17). Burgess notes in his history, "At some point, as yet undetermined, *Ecology* instituted a policy strongly favoring original research, and with few exceptions, opposing theoretical or review submissions" (p. 17). This development belies a more fundamental shift in the workings of a scientific research organization toward less social, political, and public engagement through research and, what Russell (2002, p. 5) calls, "the triumph of specialization." Research and only research became the ESA's specialty. Given this trajectory, it comes as no surprise that within these technical genres ignorance primarily functions as a logical appeal to call for further research. In my survey of the eighteen research articles from the first issue of *Ecology* in 1920, and the last issue in 1950, there is an overwhelming trend toward ignorance as a resource for further scientific research. In a few hundred pages of scientific research there are 41 ignorance quotes, all of which use the *topos* as a resource to either describe what has not yet been researched, and/or to call for more research. Certainly the use of the ignorance *topos* for the identification of a gap is not new. John Swales (1990) identified that such moves were common to at least 14 academic fields. Clearly ignorance is a signature of many disciplines that use it to create a research space and call for the perpetuation of their academic research. Framing what is unknown, but suitable for research is a move every academic researcher learns to use effectively, and ecologists of the early 1920s were not an exception.

2.2 Ignorance as Pathos (Public): Preservation Reports 1921 and 1933

The Ecological Society of America was founded as a group of scientists whose purpose was not only to conduct research on the relationship between organisms and their environment, but to also actively work to conserve those environments where their subjects of study live. In the early years, these two activities were inseparable. One of the ESA's first committees to serve this function was the *Committee for the Preservation of Natural Conditions*. The driving force behind the preservation committee was Professor Victor E. Shelford of the University of Illinois. Considering it his life's work to use scientific arguments to preserve landscapes for future generations, in 1921 his preservation committee produced *Preservation of Natural Conditions* (Croker, 1991, p. 123), a report to the National Research Council, outlining the rationales for preservation.

This 29-page report contains 19 articulations of ignorance which demonstrate how the logical appeals that call for future research in technical genres directly sequence into pathetic appeals that call for conservation action in public policy genres. For example, in the policy report you can find the signature of ignorance as you might find it in a research article but it comes with an active sense of urgency:

> The science of ecology, for example, depends upon *undisturbed patches of nature* as its 'material.' More important still, all that we have learned of geographic distribution and geographic variation has come from the study of native species taken in their original habitats. *The work is far from being practically completed* . . . We must study the actual products of evolution as they have arisen in nature. (1921, pp. 10–11, emphasis mine)

The signatures of ignorance which demonstrate a need for future research (the work is far from complete) here also sequence with signatures of ignorance which have more of a sense of urgency because of the need for "undisturbed patches of nature as its 'material.'" Rather than emphasize what could possibly be known with more research, these signatures of ignorance emphasize what could possibly be lost if these spaces are not saved. Thus, the argument from ignorance sequences to stake a claim for what we "ought" to do: "Without [concerted action] we shall *certainly lose the greater part of the material* upon which our sciences of ecology, geographical distribution, taxonomy, etc. are based" (p. 11, emphasis mine). Losing the materials of scientific research provides the emotional pull of landscapes "now almost gone" (p. 6), or under the stress of "needless destruction" (p. 7). The threat of lost material for research provides a spark to identify the destruction of precious resources and mitigate it: "Early action is the only effective kind of action" (p. 9). Quite seamlessly, the *topos* of ignorance argues for the perpetuation of research and the creation of sanctuaries of ignorance where the unknown facts of ecology can be preserved, and later, studied. This sequencing shows how the arguments for the preservation of land are laden with the social and cultural values of an ideology that links conservation with the potential benefits derived from scientific progress.

Throughout the early 1930s, the preservation committee of the ESA was heavily involved in conservation activities with the National Park Association and the National Park Service (NPS). Many of the activities of the committee directly involved lobbying Congress by writing letters, attending meetings, and organizing conferences. The NPS was by far the most powerful conservation agency during the early twentieth century. They viewed the ESA as a natural partner because they provided the scientific arguments for establishing sanctuaries for research inside National Parks that primarily functioned as national recreation spaces. The result of all these activities for Victor Shelford, still the head of the ESA preservation committee, was that he spent most of 1932 crafting "a final ESA policy statement on nature sanctuaries" (Croker, 1991, p. 129). This statement would go on to have a large influence on various NPS policy reports, which eventually led to the adoption of a conservation policy for the NPS in 1961.

Shelford published his five-page policy document, "The Preservervation of Natural Biotic Communities," a year in the making, in April of 1933 in the *Bulletin of the Ecological Society of America*. The report is remarkable for how it uptakes the two forms of ignorance I discussed in the previous report and how it hints at the research/practice split coming in the postwar period. Here in a policy statement, ignorance sequences into arguments for what scientific societies can choose to actively ignore and what they cannot. The policy document allows ignorance to sequence from a technical logos, to a pathos-based call for action, and finally to a claim about disciplinary ethos. Taken together, the 1933 policy statement demonstrates just how the topos of ignorance sequences into larger discourses about the social construction of disciplinary knowledge.

Shelford (1933, p. 241) begins his policy document by admitting that an original state of nature "never can be known," but because "in the United States and Canada areas of nearly natural vegetation are larger . . . fewer of the animals have been lost. It is possible, therefore, to recognize several classes of Nature Sanctuaries in North America (1933, p. 242). In addition to framing ignorance as a lost realm and a resource needing protection, Shelford uses ignorance to make arguments about trends in the specialization and education of ecologists:

> The whole trend of research and education is toward specialization on particular objects or particular organisms. These are stressed while the assemblage to which they belong is ignored or forgotten, together with the fact that they are to be regarded as integral parts of the system of nature. (Shelford, 1944, p. 240)

In arguing that "assemblage[s]" and "the system of nature" were ignored, Shelford makes an argument about what is lost in the products of a disciplines' inattention. In short, the trend toward the particular and away from the assemblage was a disturbing one for an applied researcher like Shelford. Actively ignoring or forgetting the system as an integral part of each study was a dangerous disciplinary construct. In this report, he sequences the signatures of ignorance to argue about the importance of paying attention to "assemblages" and the "system of nature" in need of preservation (See table 1).

2.3 Ignorance as Ethos (Disciplinary): Letters and Commentary 1944

By 1944 there was a growing sense that the politics of conservation were becoming increasingly problematic for a scientific organization such as the ESA. A new leadership had emerged in the ESA hierarchy which "strongly questioned the nature preservation role of the ESA" (Croker, 1991, p. 139). Anticipating this move, Shelford wrote an open letter to all ESA members in the *Bulletin*. The open letter reported on a questionnaire Shelford (1944) had sent out, proving "the vast majority of members of the Ecological Society of America wish to see it carry out a strong program for the preservation of natural areas, and the maintenance of natural conditions in existing reservations" (p. 12). Shelford included testimonies from the responses he received. These testimonies directly reflected the attitudes from the ESA preservation reports and policy documents, and he noted these themes were often repeated. Shelford picked them up and placed them into these open letters:

> A considerable number of correspondents stressed the need for more attention and especially sympathy for applied ecology. "It has disturbed me considerably that the society has displayed so little interest in problems of this sort or in anything which smacks of the application of ecological methods and principles . . . Incoming officers have had little knowledge of the work which lies outside the field of the usually constituted scientific society. The Ecological Society is not of the ordinary type." (p. 12)

The accusation that the ESA had actively ignored application of its research illustrates the contested and shifting nature of disciplinary epistemology. For the ESA leadership, application was not the point. Research for research's sake is what would keep them in the upper echelons of the academic hierarchy. A few pages later, Shelford again takes up the arguments laid out in the preservation reports and states: "But this time there should be haste, as the war is and the post-war period will be characterized by unusual destruction of nature, the basic material for the research of many of the members" (p. 13). These claims allow Shelford to argue with the full armature of ignorance: not only do ecologists need resources for research, but they are being destroyed, and the society has "little interest" or "little knowledge" about these occurrences. By sending out a questionnaire showing the majority of members supported the preservation efforts by the ESA, Shelford essentially drew the battle lines along the theory/practice split. At the annual business meeting of the ESA that September, the executive committee, wielding its own form of active ignorance, recommended that the preservation committee be abolished (Croker, 1991, p. 140). They thought the Society was better off ignoring the activities of conservation because, in Shelford's words, "they believed they shouldn't get mixed in politics" (as cited in Croker, 1991, p. 140).

Meanwhile Shelford wrote a commentary in the journal *Science* about the inevitable consequences that occur when scientists ignore the practical applications of their research. It was a statement not only to the ESA, but to postwar science-policy integration broadly. It amounted to a defense of the linear model (where scientists directly recommend policy to policy-makers) and a damning critique of the postwar "is/ought" divide (see Pielke, Jr.). Not surprisingly the argument is framed around ignorance as an active construct. Shelford (1944) begins by identifying how the public contact committees of various scientific organizations had been abandoned, then states:

> Agencies representing special fields of knowledge, some of it technical, cannot make presentations through another less scientific agency. To minimize misconstructions and misrepresentations,

public application of scientific principles and the needs of future research should be urged by the specialists themselves. Human society, which supports research, will hold scientific men and the societies which they constitute responsible for failure to urge the application of their knowledge. (p. 451)

In this passage, Shelford articulates what he views as the two essential functions of a scientific discipline—its ability to call for more research, and its ability to urge "public application of scientific principles." In other words, a scientific discipline's ability to pursue knowledge also means it must pursue, not ignore, public application. The "misconstructions and misrepresentations" Shelford worries about, are notions taken up in the conclusion.

In July of 1945, the ESA voted on an amendment to the bylaws defining the duties of the preservation committee. The amendment stated: "It shall encourage the preservation of natural conditions by providing information and advice to those interested in securing sound legislation for this purpose, but shall not have the authority to take direct action designed to influence legislation on its own behalf" (as cited in Croker, 1991, p. 143). Under pressure by the executive committee and rumors of their singular ability to secure research funds, the amendment was approved 213 to 115, and by the end of 1945, the ESA would not have any organization which took up direct action to influence legislation (p. 143). In the next five years, Shelford organized an Ecologist's Union, and by 1950 he secured enough money to start a new non-profit agency directly involved with the politics of Ecology. It went by the name *The Nature Conservancy*.

3. CONCLUSION

In its early years, Ecology engaged the politics of conservation as one of its central functions for the continuation of scientific research, but by mid-century it disengaged from the social and political implications of its activities and constructed itself as a society for the research and writing of ecological papers. Over half a century later, the consequences of this are ever-present. Shelford's (1944) warning of the "misconstructions and misrepresentations" of scientific principles still plays out in the multiple debates about the lack of public understanding of science (p. 451). While I don't wish to rehash those debates here, this study supports the notion that Ecology as an academic discipline is complicit in creating a space for misrepresentations. On one hand, this ignorance has meant Ecology has been slow to learn from public application and political engagement; and on the other hand, framing itself as strictly a research-based discipline has not necessarily served Ecology well. Many critics both inside and outside the discipline view Ecology as a marginal science, overly reliant on fuzzy terms, without proper reduction, or experimentally rigorous methods (Phillips, 2003). Acknowledging that Ecology is a science with a social praxis, with an inherent application which requires public engagement, might help remedy many of these problems. A study of ignorance argumentation at the end of the 20th century when Ecology began to pay much more attention to application and public engagement is an interesting area for future research on this subject. Such a study might not only describe more arguments from ignorance but also demonstrate how it may have been possible to predict that Ecology's arguments from ignorance would resurface to play a role in reshaping disciplinary ethos.

I hope the strategies outlined here will be of use to rhetorical scholars who seek to contribute to agnotology as a field of study. Tracing the special topos of ignorance over time and across genres can be a useful method to study a rhetorical life of a discipline. Studies of ignorance in this manner might give us more insight into the rhetorical nature of many academic disciplines.

Table 1—Sequences of Ignorance in ESA Genres 1920-1950

	Ignorance as Logos (Technical) Research Resource	Ignorance as Pathos (Public) Perceived Loss	Ignorance as Ethos (Disciplinary) Active Construct
Research Articles (1920)	22	0	0
The Preservation of Natural Conditions (1921)	7	11	1
The Preservation of Natural Biotic Communities (1933)	2	2	3
Open Letter and Commentary (1944)	1	1	5
Research Articles (1950)	19	2	0

ACKNOWLEDGEMENTS: I wish to acknowledge Drs. Ken McAllister, Thomas Miller, and Lynda Walsh for their guidance during this project. I also wish to thank Drs. Greg Wilson and Kevin deLaplante for their insightful comments.

REFERENCES

Bazerman, C. (1994). Systems of genres and the enactment of social intentions. In A. Freedman & P. Medway (Eds.), *Genre and the new rhetoric* (pp. 79–101). Bristol: Taylor and Francis.

Berkenkotter, C. (2001). Genre systems at work: DSM-IV and rhetorical recontextualization in psychotherapy paperwork. *Written Communication, 18,* 326–49.

Bulletin of the Ecological Society of America. (1919). *5,* 1–4.

Burgess, R. L. (n.d.). The ecological society of America: Historical data and some preliminary analyses. Oak Ridge, TN: Oak Ridge National Laboratory.

Croker, R.A. (1991). *Pioneer ecologist: The life and work of Victor Ernest Shelford, 1877–1968.* Washington, D.C.: Smithsonian Institution Press.

Ecology: Continuing the Plant World. (1920). *1*(1), 6–55.

Ecology: All Forms of Life in Relation to Environment. (1950). *31*(1), 1–146.

Ecological Society of America: Committee on the Preservation of Natural Conditions. (1921). The preservation of natural conditions. Springfield, IL: Schnepi and Barnes. Retrieved from http://books.google.com/books?id=3xFHAAAAYAAJ&pg=PP3&dq=the+preservation+of+natural+conditions&hl=en&sa=X&ei=WsnYT5idNoym8gS_79DpAw&ved=0CEAQ6AEwAA#v=onepage&q=the%20preservation%20of%20natural%20conditions&f=false

Fahnestock, J. (2005). Rhetoric of science: Enriching the discipline. *Technical Communication Quarterly, 14,* 277–286.

Ferrier, J. F. (1856). Agnoiology or theory of ignorance.*Institutes of metaphysic: The theory of knowing and being* (2nd ed., pp. 405–447). Edinburgh: William Blackwood and Sons. Retrieved from http://books.google.com/books?id=WLMVAAAAYAAJ&printsec=frontcover&dq=Institutes+of+metaphysic&hl=e

n&sa=X&ei=n8nYT8LnLJGu8QTgl7HLAw&ved=0CEwQ6AEwAA#v=onepage&q=Institutes%20of%20met aphysic&f=false

Gross, A. (2006). *Starring the text: The place of rhetoric in science studies.* Carbondale, IL: Southern Illinois University Press.

Gross, A., Harmon, J., & Reidy, M. (2002). *Communicating science: The scientific article from the seventeenth century to the present.* Oxford: Oxford University Press.

Miller, C.R. (1984). Genre as social action. *Quarterly Journal of Speech, 70,* 151-67.

Miller, C.R. (1987). Aristotle's 'Special topics' in rhetorical practice and pedagogy. *Rhetoric Society Quarterly, 17,* 61–70.

Miller, C. R., & Selzer, J. (1985). Special topics of argument in engineering reports. In L. Odell & D. Goswami (Eds.), *Writing in non-academic settings* (pp. 309–341). New York, NY: Guilford.

Miller, T.P. (2011). *The evolution of college English.* Pittsburgh, PA: Pittsburgh University Press.

Paul, D., Charney, D., & Kendall, A. (2001). Moving beyond the moment: Reception studies in the rhetoric of science. *Journal of Business and Technical Communication, 15,* 372–399.

Perelman, C., & Olbrechts-Tyteca, L. (1969). *The new rhetoric: A treatise on argumentation.* (J. Wilkinson & P. Weaver, Trans.). Notre Dame, IN: University of Notre Dame Press.

Phillips, D. (2003). *The truth of ecology: Nature, culture, and literature in America.* Oxford: Oxford University Press.

Pielke Jr., R. A. (2007). *The honest broker: Making sense of science in policy and politics.* Cambridge, Cambridge University Press.

Prelli, L. J. (1989). *A rhetoric of science: Inventing scientific discourse.* Columbia, SC: University of South Carolina Press.

Proctor, R., & Schiebinger, L. (2008). *Agnotology: The making and unmaking of ignorance.* Palo Alto, CA: Stanford University Press.

Russell, D. (2002). *Writing in the academic disciplines: A curricular history* (2nd ed.). Carbondale, IL: Southern Illinois University Press.

Shelford, V. E. (1944). The conflict between science and biological industry. *Science, 100*(2603), 450–451.

Shelford, V. E. (1944). Two open letters. *Bulletin of the Ecological Society of America, 25,* 12–15.

Shelford, V. E. (1933). The preservation of natural biotic communities. *Ecology, 14,* 240–245.

Smithson, M. (1989). *Ignorance and uncertainty: Emerging paradigms.* New York,NY: Springer-Verlag.

Stocking, S. H. (1998). On drawing attention to ignorance. *Science Communication, 20,* 165-178.

Stocking, S. H., & Holstein, L.W. (1993). Constructing and reconstructing scientific ignorance claims in science and journalism. *Science Communication, 15,* 186–210.

Stocking, S.H., & Holstein, L.W. (2009). Manufacturing doubt: Journalists' roles and the construction of ignorance in a scientific controversy. *Public Understanding of Science, 18,* 23–42.

Swales, J. (1990). Research articles in English. *Genre analysis: English in academic and research settings* (pp. 110–176). Cambridge: Cambridge University Press.

Walker, K., & Walsh, L. (2012) 'No one yet knows what the ultimate consequences may be': How Rachel Carson transformed scientific uncertainty into a site for public participation in *Silent Spring. Journal of Business and Technical Communication, 26,* 3–34.

Walsh, L. (2010). The common *topoi* of STEM discourse: An apologia and methodological proposal, with pilot survey. *Written Communication, 27,* 120–156.

Winterowd, R. (1973). 'Topics' and levels in the composing process. *College English, 34,* 707–708.

Balancing Substance and Style on a Budget: How North Carolina Sea Grant Communicates Science (Part 1)

HEATHER WARD

Elm City, NC
heatherward63@gmail.com

ABSTRACT: North Carolina Sea Grant College Program coastal extension and communication specialists share a strong belief in applied science and pursue outreach much like agricultural extension agents. These papers document the challenges inherent in communicating science by this community of practice through 12 months of participant observation and in-depth interviews.

KEYWORDS: boundary organization, communities of practice, discourse analysis, science communication.

1. INTRODUCTION

Sea Grant's mission statement reads, "NOAA's National Sea Grant College Program enhances the practical use and conservation of coastal, marine and Great Lakes resources to create a sustainable economy and environment." Congress established the program in 1966, administered through the National Oceanic and Atmospheric Administration (NOAA) within the U.S. Department of Commerce. In some ways, 'Sea' Grant resembles the 'Land' Grant program, created in 1862 to accelerate U.S. agricultural development.

Today, thirty-two Sea Grant programs exist in every coastal and Great Lake state and Puerto Rico. The national program headquarters is in Silver Spring, Maryland. Each program is a unique partnership between federal and state government, participating universities and community colleges, and private organizations. Research funding and extension priorities reflect NOAA's national priorities as well as state interests. Areas of concern common to all the state programs include rapid coastal population growth, habitat degradation, limited coastal literacy, impacts from climate change, stressed fisheries, the loss of maritime cultures, and how to incorporate science into decision-making processes at all levels of government.

In recent years, the National Sea Grant's College Program's annual federal appropriation ranged from 55 to 65 million dollars. The non-profit Sea Grant Association, whose membership is comprised from academic institutions that participate in the program, lobbied Congress for $72 million dollars in the fiscal year 2009 Commerce, Justice, Science Appropriations Act (SGA 2012).

North Carolina Sea Grant (NCSG) became the nation's 12th National Sea Grant College Program in 1976 and is an inter-institutional center within the University of North Carolina (UNC) system. Scientists from 16 UNC campuses and any college, university, or community college in the state may apply for research grants. The administrative headquarters is at North Carolina State University in Raleigh. Extension specialists also work out of the Coastal Studies Institute in Manteo (administered by East Carolina University), NC State's Center for Marine Sciences and Technology (CMAST) in Morehead City, and UNC-Wilmington's Center for Marine Sciences (CMS). Coastal Carolina University in Conway,

Ward, H. (2012). Balancing substance and style on a budget: How North Carolina Sea Grant communicates science (Part 1). In J. Goodwin (Ed.), *Between scientists & citizens: Proceedings of a conference at Iowa State University, June 1-2, 2012* (pp. 413-426). Ames, IA: Great Plains Society for the Study of Argumentation. Copyright © 2012 the author(s).

South Carolina hosts a regional climate extension specialist, shared with South Carolina's Sea Grant program.

NCSG's total annual budget (state and federal) was around $3 million at the time of this research. Several extension specialists come to NCSG with partial "special purpose" funding from outside organizations, such as the N.C. Division of Environment and Natural Resources and Cooperative Extension Service.

2. COMMUNITIES OF PRACTICE AND CLAIMSMAKERS

This analysis considers North Carolina Sea Grant as a 'community of practice' and extension agents as important, yet under-analyzed social mediators and scientific 'claims-makers.' This paper is the first in a series documenting the challenges inherent in communicating science and the coping strategies of this community of practice through 12 months of participant observation and in-depth interviews. This first paper aims in particular to deliver a thick description of NCSG activities and discuss identity. How does this community of practice define itself institutionally and how do individual communication and extension specialists see their roles in communicating science? Subsequent studies will discuss specific communication and outreach techniques, knowledge production, clientele perceptions and overall communication effectiveness.

The sustained pursuit of a shared enterprise creates 'communities of practice' (Wenger, 1998) everywhere. Schoolteachers, military members, and North Carolina Sea Grant officials find meaning together and practice their trade in ways that distinguish them from other groups. Members do work in a historic and social context that offers both identity and structure. Community members engage in 'joint enterprise' and share a 'repertoire' of language, tools and regulations (Wenger, 1998, pp. 72–85), but more importantly embedded and often hidden understandings—shared sensitivities, worldviews and rarely questioned assumptions about what counts as knowledge. Communities of practice are not always benign or 'benevolent' (Wenger, 1998, p. 132). They also can become exclusionary with reactionary tendencies that create tension between newcomers and old-timers (Jewson, 2007, p. 72).

Casual conversations with Sea Grant coastal extension specialists and communicators suggest that they face a wide variety of communication challenges. They need to quickly grasp complicated scientific concepts, then translate and package that information for multiple audiences that include the general public, fishermen and crabbers, small business owners, government regulators, educators, and local elected officials. They must maintain relationships with researchers and the media and often mitigate disputes between environmental advocates, property owners, and commercial interests. Applied scientific research like that practiced by NCSG goes through a process of mutual construction and negotiation as researchers, policymakers, bureaucrats and coastal citizens interact and challenge each other's evidence and perspectives. North Carolina Sea Grant's scientific discourse shapes perceptions of North Carolina's coast through grant awards, articles in the organization's flagship magazine, *Coastwatch*, and the presentation of research results to state agencies and other groups.

Contemporary geographers are interested in critically evaluating the processes and politics that produce environmental knowledge (Meindl, 2002, p. 684) and "the notion of claims-making is a useful way of conceptualizing the influence that applied scientists wield in constructing and disseminating environmental knowledge" (Meindl, 2002, p. 685). NCSG researchers and extension specialists make 'cognitive' and 'interpretive' claims as a matter of course, both describing reality and establishing the relevance of research findings to other experts, decision-makers and the public (Aronson, 1984, p. 14; Meindl, 2002, p. 685).

Applied science in particular is a social enterprise and research is needed both to better understand how scientific information is constructed and communicated and to improve information dissemination from scientist to citizen and vice versa. Yet, little mention is made in academic literature of coastal extension services and nowhere a thick description of coastal extension, education, and outreach activities. Case studies about coastal and ocean policymaking are few and far between. Those that discuss communication processes are even rarer. The ability of coastal managers and related academics to generalize and build grand theory will improve as more case studies are investigated, documented, and compiled.

3. DATA AND ANALYTIC METHODS

This paper makes use of information collected during the author's one-year service as the science communications fellow at NCSG between June 2007 and June 2008. The fellowship is ideal for scientists interested in learning to communicate their work to audiences outside academia or for aspiring science writers. The fellow summarizes and promotes research projects funded by North Carolina General Assembly's Blue Crab and Fishery Resource Grant programs and writes articles for NCSG's flagship publication, *Coastwatch* magazine.

3.1 Research Questions

Primary research questions included:

(1) How do coastal extension agents perceive their role and the purpose of the organization?
(2) What philosophy underpins educational, outreach, and extension work?
(3) How do NCSG officials define coastal extension, education and outreach services?
(4) What clients and audiences do they serve?
(5) What concepts do extension specialists consider the most important to convey to the public, fishers, teachers, and other clients?
(6) What outcomes do they desire for coastal communities and public policy, if appropriate?
(7) Does NCSG see itself as a boundary organization? What problems and opportunities does this present?
(8) What are the communication challenges? Are any unique to coastal extension?
(9) What techniques are most useful in advancing their efforts? What additional tools and training are desired?

3.2 Data Set and Collection Methods

In-depth, personal interviews with NCSG extension specialists and communicators followed six months of participant observation. The survey instrument was tested in advance of the interviews with the cooperation of two extension specialists and two communicators outside of NCSG, but within the national Sea Grant network. Interview questions included:

(1)   What do you see as North Carolina Sea Grant's (NCSG) purpose (or mission)?

(2)   What are the values of this organization? Has the organization determined these values, or are they subjective?

(3)   What is the role of an extension specialist/communicator? Institutional or personal definition?

(4)   Who are your major clientele groups or audiences?

(5)   How did you determine these?

(6)   What percentage of time do you spend working with them?

(7)   What are the particular challenges in communicating with each group?

(8)   What are your desired outcomes?

(9)   What clientele behaviors most need to change?

(10) What techniques and tools do you find most useful in overcoming these communication challenges and advancing your efforts?

(11) Have you received any particularly useful training (from govt agency, university training, other group)?

(12) What additional training or tools would help you do your job more effectively?

(13) Do you ever find yourself wanting to express a stronger opinion?

(14) What is the greatest obstacle to your success?

(15) Does the public or your clientele possess any misperceptions about your role? If so, describe them.

(16) What are some challenges that may affect how you interact with audiences in the future?

(17) How have email and Internet communication techniques changed your job, if at all?

(18) Tell me a success story. What are you most proud of in your time with Sea Grant? Have you had the opportunity to follow up on these efforts?

Pat Corcoran, an extension specialist with Oregon Sea Grant, helped the author create a Communication Success Continuum. The Continuum is a matrix on an Excel spreadsheet. The X-axis displays science communication goals beginning with 'Awareness' → 'Buzz' → 'Advocacy' → 'Engagement' → 'Understanding' → 'Using Products and Tools' → 'Adoption' → 'Policy Change' → through 'Behavior Change.' Respondents highlighted communication tools listed within the matrix that they used routinely, such as 'posters,' blogs 'workshops,' 'public comment sessions,' or 'commission recommendations.'

Textual data collected included NCSG's strategic plan and other planning documents, director and staff meeting notes, research proposals and list of awards granted, a content analysis of *Coastwatch* articles from 2000 to 2007, a sampling of NCSG products, and monthly reports submitted by extension specialists and communicators.

3.3 Analytic Techniques: Discourse Analysis and Thick Description

Findings here are informed by the field of discourse analysis. Discourse analysis identifies the ideas that count as knowledge or are privileged as 'truth' within communities and investigates 'discursive structures'—the unwritten assumptions that produce an individual or group's authoritative account of the world. The validity of discourse analysis depends on the richness of the information collected and the analytical skills of the researchers rather than numerical validity (Patton, 1990).

Discourse analysis focuses on the situational and contextual nature of communication as it happens in the environments that shape it and are shaped by it. Theories and techniques from several types of discourse analysis are borrowed and applied to this case study, including social discourse analysis (Van Dijk, 1997b) and Fairclough's (2001) critical discourse analysis. Social discourse analysis is an appropriate method here because the techniques work at various scales and target interactions, the more general 'discourse-society interface.' According to Van Dijk (1997b), discourse should be studied as action, showing the social, political and cultural functions of discourse within institutions, groups or society writ large. In social discourse analysis,

> we find that social reality may be constituted and analysed anywhere between a more micro and a more macro level of description, for instance as (details of) acts and interaction of social actors, and as what whole institutions or groups 'do,' and how both thus contribute to the production and reproduction (or challenge) of social structure. (Van Dijk, 1997b, p, 6)

According to Fairclough (2001, p. 91), DA occurs in three stages: "*description* of text, *interpretation* of the relationship between text and interaction, and *explanation* of the relationship between interaction and social context." Interpretation through discourse analysis is a complicated and intuitive process. Underlying relationships, institutional characteristics and key ideas only emerge when the researcher becomes absorbed in the 'texts' fully, and then reviews the material several times over with fresh eyes and ears. Below are fundamental discourse analysis considerations used to approach this case study's data set, organized according to Fairclough's three stages, and derived from Chilton and Schaffner (2002), Fairclough (2001), Foucault (1980), Scollon (2008), Van Dijk (1997a) and others:

Stage 1: Description of Text
- Contents: activity, purpose, units of analysis
- Subjects: Who are the actors, stakeholders, voices, or identities?
- Role of Language: audience, communicative function, media
- Genealogy: How did this discourse come about, institutionally or culturally?

Stage 2: Interpretation of Interactions
- Interests of Participants: agendas and perceptions
- Networks
- Argumentation or Persuasive Methods
- Institutional Constraints
- Routines and Rituals
- Dominant Discourses
- Mechanisms that Silence
- Validity and Knowledge Claims
- Regimes of Truth
- Tactics and Strategies

Stage 3: Explanation and Effects
- Social Context
- Action-Forcing Devices
- Relationships: between texts and actions taken by social actors
- Episteme or Paradigm
- Beliefs, Values or Underlying Ideologies
- Policy Outcomes

Ideally, common principles or a paradigm will emerge from the collected data and comparisons, inspired by Glaser and Strauss's grounded theory (1967). The case study first offers a 'thick description' of a particular social activity along the lines of Gilbert Ryle's (1968) explanation of a wink in his essay *The Thinking of Thoughts* or Clifford Geertz's (1973) description of a Balinese cockfight in *The Interpretation of Cultures*. The intent of much research is to generalize from the particular, but case study methods are often criticized because sampling sizes are small and results tied to specific temporal and spatial contexts. The overarching goal in this initial phase of analysis is to generate necessary particulars.

4. RESULTS

4.1 Participant Observation Notes

Sea Grant extension and communication specialists engage a broad range of publics, make plain trade-offs faced by coastal decision-makers, and work to generate practical solutions in particular places. Sea Grant clientele includes academic researchers, fishermen and women, industry representatives, coastal managers, local elected officials, city and environmental planners, and non-profits and concerned citizens. Extension specialists do original research and monitor collaborative research projects. They conduct training workshops, facilitate technology transfer, and alongside Sea Grant communicators, perform education and outreach. Communicators inform Sea Grant clientele about scientific results and marine and coastal issues using any means possible—magazines, targeted brochures and newsletters, video productions, education and outreach workshops, and, whenever affordable, short radio and television spots.

NCSG's *Associate Director* manages the state's Fishery Resource Grant (FRG) program. The defining principle behind the program is that people in fisheries-related activities possess compelling ideas for improving and protecting fisheries, but may lack the financial resources or scientific background to conduct experiments, collect data, and analyze results. Sea Grant administrators require academic researchers, members of the fishing community, and occasionally non-governmental groups to collaborate on proposal submission, research design and study execution.

Additional 'core' research projects fall within NOAA and Sea Grant National Office chosen themes and priority areas. Five categories of grants were in place at the time of this research—aquaculture, coastal communities and economies, coastal hazards, urban coasts, and ecosystems and habitats. The four 2009-2013 national focus areas are: safe and sustainable seafood supply, sustainable coastal development, healthy coastal ecosystems, and hazard resilient coastal communities (NSGCP, 2012). NCSG's advisory boards evaluate grant proposals using a competitive, peer-reviewed process that additionally tries to balance awards between coastal regions and state universities. Mini-grants are smaller and more flexible.

NCSG's 15-member extension team manages a multitude of diverse research and outreach efforts from Raleigh and along the coast. The Extension Director organizes specialists by topics, including commercial and recreational fisheries and habitats, seafood technology, coastal hazards, water quality, law and policy, community development and marine education. Programs in other states adopt different structures, depending on their needs and state structures. Florida Sea Grant, for example, works more closely with the state's Cooperative

Extension Service and many of their county marine extension agents operate out of agricultural centers.

Extension highlights from 2007:

- Two *Water Quality Specialists* work to improve coastal waters through land-use planning and urban stream restoration. One is collaborating with other agencies and the University of North Carolina School of Government to develop a training program for local officials on water science and coastal growth strategies. The other is developing an evaluation matrix to help engineers and funding agencies make informed stream restoration design choices and was invited recently to India to help develop their stream restoration program.

- The *Coastal Construction and Erosion Specialist* helps communities understand and resolve issues related to erosion, inlet maintenance, dune enhancement. and weather-resistant construction techniques. He is on a National Academy of Sciences special committee to recommend improvements for Federal Emergency Management Agency flood maps and on a national task force seeking bridge designs that can withstand storm events.

- The *Marine Education Specialist* is developing educational materials and workshops for teachers in all of North Carolina's regions, based on a coastal literacy survey. She runs a 3-credit online graduate oceanography course for teachers, coordinates curriculum materials for *Coastwatch on NC Now*, and works with NOAA's Coastal Services Center on the Southeast Phytoplankton Monitoring Network that involves students in water sampling.

- The *Fisheries Specialists* occasionally mediate disputes between watermen, regulators, and scientists. One "collaborative learning" demonstration project in the Albemarle Sound region brought together all of these groups to design more innovative management strategies. Specialists helped design and implement the North Carolina Coastal Ocean Observation System (NCCOOS) and co-authored a booklet designed to demystify state fisheries management processes for the public.

- The *Law, Policy and Community Development Specialist* is collaborating with the N.C. Division of Coastal Management to study North Carolina's ocean policy and propose revisions. The Coastal Law, Planning and Policy Center is a partnership between NCSG, the UNC School of Law and UNC's Department of City and Regional Planning. Development pressure raises issues that cross federal, state, and local boundaries. The center's biannual newsletter and research papers provide balanced analyses of complicated legal and planning issues.

- Blue crab shedding operations bring millions into the coastal economy. In addition to running the state's Blue Crab Research Program, NCSG's *Blue Crab and Mariculture Specialist* authored *Closed Blue Crab Shedding Systems: Understanding Water Quality*. The pamphlet explains the importance of maintaining and monitoring water quality in tanks that must absorb large numbers of peeler crabs over a short time period.

- The *Seafood Technology and Marketing Specialist* helped to coordinate the Carteret Catch campaign. Commercial fishing families, seafood distributors, restaurant owners, scientists, tourism experts and educators teamed to develop a logo and theme to promote local seafood. He also helps deliver Hazard Analysis and Critical Control

Point (HACCP) training and assistance to small seafood dealers, packers and processors in North Carolina.

- The *Coastal Resource and Enterprise Specialist* helped to change the dynamic between regulatory agencies and industry from conflict to cooperation through the NC Clean Marina Partnership. Pressure washing vessels in marinas introduces a lot of pollution to coastal waters. Seventy gallons of water are needed to clean a 25-foot vessel and some marinas clean 150 to 175 vessels in a season.
- Sea Grant staff provide scientific reports and expertise to state advisory boards and committees in the Department of Environment and Natural Resources, the Marine Fisheries Commission, and the N.C. Coastal Resources Commission, among others.

The Communicator's Role

North Carolina Sea Grant's communication specialists help people understand marine and coastal policy issues, including state policymakers, academics and secondary school educators, students, coastal business leaders, marine resource users, local elected officials and the public. NOAA and Sea Grant research opportunities and results are publicized and new technologies and fishing techniques promoted. The subject matter is complex and the audience broad, perhaps too broad.

In 2007, NCSG's communications staff consists of a director, two full-time communicators, and one administrative support associate/distribution manager. The Fisheries Resource Grant and Blue Crab Research Programs support one science communications fellow, replaced yearly. The communication's budget supports one undergraduate intern per semester. Freelance designers and photographers work on an "as needed" basis.

The office maintains a publications database, produces and distributes informational brochures, pamphlets and books, creates public exhibits and responds to media inquiries about Sea Grant-sponsored research and activities. A Web site provides a template for soliciting research proposals and publishing results. News releases announce scientific findings, workshops and products.

The *Communications Director* supervises the development, editing, and production of all publications, facilitates all education and outreach efforts, supervises media relations, mentors communication staff and interns, and puts together annual reports for NOAA, the N.C. General Assembly and Congress. As a member of the management team, the director works with NCSG's Advisory Board to identify statewide outreach and education needs and to develop partnerships with state agencies and non-governmental groups. With the help of two writer-editors and extension specialists, the director applies for grants that support communications-related research and innovative projects.

North Carolina Sea Grant *Communicators* are not only experts in translating scientific information for the general public, but are necessarily multidisciplinary. Each communicator is assigned extension specialists and university investigators to "cover." They keep track of research topics and progress and, much like journalists, try to choose the most interesting and compelling stories. Communicators are keenly aware of the role the media plays in shaping public policy and bend over backwards not to favor particular universities or state regions and to avoid the appearance of advocacy on the side of commercial fishermen, government regulators, or environmental groups.

The communicators also produce *Coastwatch* magazine, a 32-page, self-cover magazine published six times a year. North Carolina Sea Grant's flagship publication emphasizes the organization's research and outreach efforts, highlighting not only the natural environment, but also the people, places, and culture that define North Carolina's coast.

NCSG's communications director wants to increase the subscriber base to cover more operational costs and expand the transfer of knowledge, especially to educators and policymakers. The magazine has about 1,200 paid subscribers ($15/year) and consistently earns top honors in writing and communication competitions. Some 1400 complimentary copies are distributed to state and local officials, libraries, state welcome centers, and museums. The staff is formulating a marketing and business plan for the magazine and considering branding and social marketing approaches.

North Carolina Sea Grant and UNC-TV are teaming up to present Coastwatch on North Carolina Now. The first segment—focusing on the Rocky Branch stream restoration project—aired in December as part of statewide public television network's nightly newsmagazine. A companion print story appeared in the holiday issue of the magazine. UNC-TV's television spot that corresponds to the Winter 2008 *Coastwatch* story "Finding Fish in Lots of Water" aired in March 2008. North Carolina State University's Office of Extension, Engagement and Economic Development provided a grant for three pilot segments.
Recent communication efforts included:

- An online informational kiosk—www.ncseagrant.org/waterfronts—kept the Waterfront Access Committee, legislators and the public apprised of ongoing discussions, including presentations by members of the N.C. Coastal Resources Law, Planning and Policy Center.

- Consumers seeking seasonal North Carolina seafood can turn to new wallet-size cards. Local Catch: North Carolina Seafood availability cards, developed by North Carolina Sea Grant and the North Carolina Aquariums premiered at the October North Carolina Seafood Festival in Morehead City. The cards highlight commercial fisheries by season, reveal how and where North Carolina seafood is harvested and offer "Quality Counts" tips for selecting seasonal fresh seafood.

- The *Break the Grip of the Rip* public information campaign is a continuing partnership between Sea Grant communicators and the National Weather Service, beach communities, and the National Park Service on the Outer Banks. NCSG's communications director represented the national Sea Grant network, reviewed message content and publication designs, and coordinated the logistics for the national news kickoff in 2004.

Sea Grant communication and extension efforts extend overseas as well. NCSG sponsored a N.C. State engineering/coastal policy graduate student from Indonesia 15 years ago. He is now secretary general for the country's Ministry of Marine Affairs and responsible in part for passage of coastal zone legislation, a new tsunami warning system, and a Sea Partnership Program, modeled after Sea Grant. The communications director performed as a coastal ambassador, visiting Indonesia to support their fledgling program in 2005.

4.2 Interview Results—Identity

What follows are answers to four interview questions (1, 2, 3, and 8) intended to reveal institutional identity and individual perceptions of roles within the organization and as communicators of science.

What do you see as North Carolina Sea Grant's (NCSG) purpose or mission?

NCSG extension and communication specialists define their purpose broadly, but consistently. They "serve coastal communities" by "funding research" and "translating the science for constituencies." Several emphasized the applied nature of their work. One remarked, "Research and outreach, but not from an ivory tower." Another said, "We solve problems through science." Respondents listed multiple and diverse audiences—"fishermen, seafood dealers, property owners, K-12 educators, graduate students, anyone interested or using coastal resources,"—and multiple areas of expertise. "I give advice on a variety of minefields, engineering, marine construction, coastal processes and erosion control alternatives, pointing out boundaries and encouraging best practices." Education proved another common theme. "We expose students to the Sea Grant model and train the next generation of researchers to get outside the lab."

What are the values of the organization? Has the organization determined these values or are they subjective?

Science is the preeminent discourse within coastal management circles, considered the most reliable source of knowledge, and NCSG proved no exception to this rule. 10 of 14 respondents said that it was important to remain "unbiased," "honest," or "neutral" and many married this ability to scientific method. "We promote honest, open discussion, and sound science and practices that are replicable with no bias."

- "Your individual expertise may have drawn you to a conclusion. We remain neutral and present the verifiable facts through science, not the way we'd like it to be. It is not our job to make our clientele environmentalists."
- "I work around regulatory processes and get at the edges of trouble. My work is often open to criticism, so I stick with the science. Comforting."

Most attributed this value to the institution and its leadership. "We check our facts and I was told to be empathetic not sympathetic. We're not taking up the causes of coastal communities, but we want to establish a personal connection." In only one case, did there appear to be any tension between objectivity, resource use, and environmental advocacy. The respondent remarked, "I want to minimize human impact as much as possible," "turn out good stewards," and "affect positive change on the coast."

Several used this question to remark on how well their values coincided with those of the organization. "NCSG is nice for a self-propelled, motivated individual." Or, "Teamwork; I don't think I'd find another worksite where I'd feel so comfortable."

What is the role of an extension or communications specialist?

Every extension and communication specialist answered this question similarly and most said that defining their role was a personal choice not an institutional mandate. All said that their role was to "distribute information," to "translate research for different constituent groups" in terms "anyone can understand," or "get knowledge to state agencies and coastal communities." Respondents used words such as "liaison," "connecting piece," "dialogue," "conduit," or "to be the bridge between science and users, industry, the university, and the public" to describe their role. A few argued that their role involved "identifying emerging issues" and to "Do a plan of work that is proactive, not reactive." Several respondents remarked on their role "translating science" and "bringing researchers back down to earth," adding we are "more in touch with public notions of science."

What are your desired outcomes?

Desired outcomes varied somewhat because each specialist is assigned different areas of responsibilities, however several common themes emerged, notably visibility and sustainability. Four specialists mentioned a desire to increase NCSG's visibility. "I want more people to know what Sea Grant is." "We are judged on publications out the door, new product development and promotional programs to gain visibility." "My goal is broader exposure and greater knowledge of our work so that we get more interesting research proposals and projects." Some of this emphasis could be blamed on concurrent staff meetings to discuss branding and website upgrades. However, many specialists expressed a seemingly sincere desire to be an "information center" or an "objective organization to help." One hoped to be recognized as the 'go to' person in their area of expertise. "I want to be a resource for whatever coastal, marine, or aquatic information they need," said another.

Balancing the needs of fishing communities against coastal ecosystems also figured prominently in respondents answers. "I want to have an impact on the continuance of North Carolina's fishing industry. NC as it was, is, and should be with more money and earnings for those folks." Another aimed to preserve a "sustainable economy without impacting the environment."

4.3 Communication Success Continuum

Respondents were asked to complete the Continuum at the conclusion of the interview while reflections on their activities remained fresh. After highlighting communication tools that they used within the matrix, five respondents chose 'Understanding' as their overarching communication goal, located exactly in the center of the Continuum. Three chose 'Awareness' at the very beginning of the Continuum and presumably the easiest communication goal to achieve. Two selected 'Engagement' and one chose 'Using Products and Tools.' Two respondents chose 'Behavior Change' at the far end of the Continuum, but did not indicate that they used any of the tools within the column below. Arguably, 'Behavior Change' is the most difficult communication goal to achieve. Two respondents also added, "We do all of these things." Most respondents highlighted tools underneath goals at the beginning or middle of the Continuum, such as 'brochures,' 'workshops,' teacher curricula,' 'conferences and training,' 'gray press,' 'event calendar,' 'publicize scientific reports' or 'press releases.'

5. DISCUSSION

Sea Grant could be described as a boundary or bridging organization, attempting to narrow the divide between science and public policy as well as academic research and local practical knowledge. Sea Grant encourages the sustainable development of marine resources informed by scientific research. The organization aims to facilitate the transfer of useful knowledge between academic researchers, decision-makers, and the coastal public, improving policymaking by doing and publicizing applied research relevant to local and state communities. Associated education and extension efforts attempt to temper and mitigate ever-increasing development pressures in the coastal zone, reminding the public of the economic and environmental importance of the region's fragile ecosystems.

That North Carolina Sea Grant is a community of practice is clear from the analysis of the first few interview questions. The group shares three characteristics of a community of practice—mutual engagement, joint enterprise, and a shared repertoire (Wenger, 1998, p. 73). All are mutually engaged and share a spirit of belonging. Extension specialists and communicators understood the mission or purpose of the organization in much the same way as the national organization's definition that emphasizes practical use of the coast and a balance between sustainable economies and the environment. They are engaged in and held accountable by a joint enterprise that is defined by the participants even as they pursue their goals (Wenger, 1998, p. 77). Finally, the respondents clearly share a repertoire. All took great pains to frame their role as 'bridging,' rather than 'advocacy,' proponents of 'applied research' drawing on 'sound science.' The repertoire of a community of practice includes processes, tools, words, and concepts produced or adopted by the group and incorporated into their practice (Wenger, 1998, p. 83).

Similarly, discourse analysis proved a helpful analytical tool. Sea Grant officials take pride in their role as "honest brokers" in coastal communities because their organization is non-regulatory and science-based. The language of several specialists further demonstrates that they wield public perceptions of this role for tactical advantage. One specialist encouraged locals to get involved in research and gather their own data, "rather than allow government regulators to control all of the science." Reiterating NCSG's goal to fund research driven by societal need ahead of researcher curiosity encourages demand for scientific information and coastal extension services. Terms, such as 'unbiased,' 'honest,' or 'bridging' become points of focus around which meaning is negotiated both inside and outside the organization.

Additional analysis of interview questions should shed light on a few emergent questions and suggest ways to improve NCSG's science communication. Lack of visibility is a recurring theme. Extension and communication specialists listed a plethora of audiences/clients, casting doubt on how effective they can be accommodating the needs of such diverse groups.

Participant observation, however, revealed specific priority clientele. The groups the extension specialists work with are in line with the talents and interests of the specialist combined with the organization's goal to support applied coastal research. Academics and Principal Investigators (PIs) are important clientele, followed by targeted groups—seafood dealers, marina owners, city managers, coastal homebuyers, landscape architects, law students, and educators and so on.

The lack of visibility and need to serve specific clientele suggest several possible changes in how NCSG communicates science.

1. Modify *Coastwatch*'s format to free up time and talent for other avenues of communication and make the publication a more effective tool for the organization. Extension specialists love the look of the magazine, but expressed doubt about how useful it is to their specific clientele. Because the magazine targets the general public and broad geographic areas, they do not find a compelling reason to distribute any one particular issue widely in their communities.

Consider publishing four times a year—three regional editions (Upper Coastal, Southeast, and an Island issue) and a 4th 'Emerging Issues' edition. This format involves the extension specialists more fully in determining *Coastwatch* content, makes planning articles ahead of time easier, and frees up time for communication tasks beyond writing. Free issues could be distributed in featured regions and saturate state agencies and the legislature with an annual forward-looking Emerging Issues edition. Corporate or group sponsorships will be easier to obtain with a regional focus as well, if that ever becomes a funding necessity.

Each regional edition would highlight NCSG-sponsored research conducted in the area or about the area and feature the local coastal extension specialist—their projects, role as an unbiased community resource, and contact information. Once a year, every specialist could use their regional *Coastwatch* issue as a calling card in the community, 'working' that region to enhance visibility immediately after the issue is delivered.

The 'Emerging Issues' edition would cover broader state-wide concerns, target the legislative and policy audiences and feature Raleigh-based specialists and Triangle and Sandhill projects. Municipal and county officials and the public will look forward to their region's issue each year. Crafting the 'Emerging Issues' edition forces NCSG to think well forward about the coast once a year, serving in a way as an organizational planning tool. It will allow the revamped NCSG website to handle more time-sensitive topics. Paying subscribers from the general public will enjoy the geography tour. Providing more in-depth local details will better connect them to their place and people and, when their region is not the focus, inspire travel.

2. Create an 'outreach communications' position—someone in the field explaining what North Carolina Sea Grant does and why science-based knowledge matters to relevant constituencies. This position would be dedicated fully toward boosting the organization's visibility, canvassing one region at a time and policymakers once yearly—an ongoing roadshow with NCSG's flagship publication in hand.

The outreach communicator's work plan would bridge communications and extension and marry publications to extension work plans. He or she will operate in the coastal area most recently featured in a regional *Coastwatch* edition and work with extension specialists to determine content for the regional *Coastwatch* editions and write four articles yearly. The outreach communicator would visit target audiences—identified by extension specialists and the NCSG management team—to explain and promote NCSG applied research opportunities and to present research results at public meetings, in universities, and to state agencies. The position could conduct science communication scholarship and monitor similar NCSG-funded research projects.

3. Over time, locate a communication specialist at every NCSG coastal office (Manteo, Morehead and Wilmington) to work closely with the extension specialists. Relationships, inspiration, writing and communicating about coastal issues will flow more smoothly 'in place.'

All of the above would likely build momentum and access to groups that most reliably support NCSG research and outreach and go a long way toward alleviating the visibility problem identified by the extension specialists. Face-to-face outreach communication efforts alongside upgrades to the website and a proven publication create marketing synergies and a powerful campaign. Visiting policymakers, academic departments, and non-profits regularly will generate more impressive research proposals to choose from, higher quality results, and ongoing demand for NCSG scientific expertise and extension services.

ACKNOWLEDGEMENTS: The author thanks everyone at North Carolina Sea Grant for their dedication to the coast, generosity with their time, and candor and Oregon Sea Grant extension specialist, Pat Corcoran, for his help creating the Communication Success Continuum.

REFERENCES

Aronson, N. (1984). Science as claims-making activity: Implications for social problem research. In J.W. Schneider & J.I. Kituse (Eds.), *Studies in the sociology of social problems* (pp. 1–30). Norwood, NJ: Ablex Publishing Corporation.

Bocking, S. (2004). *Nature's experts: Science, politics, and the environment.* New Jersey: Rutgers University Press.

Carrada, G. (2006). Communicating science: A scientist's survival kit. European Commission Directorate-General for Research. Retrieved from www.cordis.europa.eu.int/nanotechnology

Chilton, P., & Schaffner, C. (2002). *Politics as text and talk: Analytic approaches to political discourse.* Philadelphia, PA: John Benjamins Publishing Company.

Corbett, J. (2006). Communicating *nature: How we create and understand environmental messages.* Washington, D.C.: Island Press.

Fairclough, N. (2001). *Language and power* (2nd ed.). London: Pearson Education Limited.

Foucault, M. (1980). *Power/Knowledge: Selected interviews and other Writings 1972–1977.* New York, NY: Pantheon.

Geertz, C. (1973). Thick description: Toward an interpretive theory of culture. In *The interpretation of culture: Selected essays* (pp. 3-30). New York, NY: Basic Books.

Glaser, B., & Strauss, A. (1967). *The discovery of grounded theory: Strategies for qualitative research.* Chicago, IL: Aldine Publishing.

Gregrich, R. (2003). A note to researchers: Communicating science to policymakers and practioners. *Journal of Substance Abuse Treatment, 25,* 233–237.

Guston, D.H., Clark, W., Keating, T., Cash, D., Moser, S., Miller, C., & Powers, C. (2000). Report of the workshop on boundary organizations in environmental policy and science. Belfer Center for Science and International Affairs.

Healey, M., & Hennessey, T. (1994). The utilization of scientific information in the management of estuarine ecosystems. *Ocean and Coastal Management, 23,* 167–191.

Jewson, N. (2007). Cultivating network analysis: Rethinking the concept of 'community' within 'communities of practice'. In J. Hughes, N. Jewson, & L. Unwin (Eds.), *Communities of practice: Critical perspectives* (chap. 6). New York, NY: Routledge.

Meindl, C., Alderman, D., & Waylen, P. (2002). On the importance of environmental claims-making: The role of James O. Wright in promoting the drainage of Florida's Everglades in the early twentieth century. *Annals of the Association of American Geographers, 92*(4), 682–701.

National Sea Grant College Program (NSGCP). Retrieved from http://www.seagrant.noaa.gov/

Patton, M.Q. (1990). *Qualitative evaluation and research methods* (2nd ed.). Beverly Hills, CA: Sage.

Ryle, G. (1968). The thinking of thoughts, What is 'Le Penseur' doing? *University Lectures, 18.* University of Saskatchewan. Retrieved from http://lucy.ukc.ac.uk/CSACSIA/Vol14/Papers/ryle_1.html

Scollon, R. (2008). *Analyzing public discourse: Discourse analysis in the making of public policy.* New York, NY: Routledge.

Sea Grant Association (SGA).Retrieved from http://www.sga.seagrant.org/

Van Dijk, T. (Ed). (1997a). *Discourse as structure and process—Discourse studies: A multidisciplinary introduction* (Vol. 1). Thousand Oaks, CA: Sage Publications.

Van Dijk, T. (Ed). (1997b). *Discourse as social action—Discourse studies: A multidisciplinary introduction* (Vol. 2). Thousand Oaks, CA: Sage Publications.

Weber, J., & Word, C. (2001). The communication process as evaluative context: What do nonscientists hear when scientists speak? *BioScience, 51,* 487–495.

Wenger, E. (1998). *Communities of practice: learning, meaning, and identity.* New York, NY: Cambridge University Press.

Expertise and Inauthentic Scientific Controversies: What You Need to Know to Judge the Authenticity of Policy-Relevant Scientific Controversies

MARTIN WEINEL

Centre for the Study of Knowledge Expertise Science (KES)
School of Social Sciences
Cardiff University
Glamorgan Building
King Edward VII Avenue
Cardiff
CF10 3WT
Wales
weinelm@cardiff.ac.uk

ABSTRACT: The paper explores how non-experts can assess whether a policy-relevant technical or scientific issue is subject of a genuine controversy amongst experts or not. A criteria-based approach is suggested. While a number of criteria are introduced, the focus is on the expertise that one needs to employ the criteria appropriately. It is suggested that this expertise, which is called 'sociological discrimination' and which refers to an understanding of the nature of science, is an essential prerequisite for making adequate authenticity judgements.

KEYWORDS: AZT, Mbeki, expertise, inauthentic scientific controversy, nature of science, sociological discrimination

1. INTRODUCTION

In October 1999, South African President Thabo Mbeki surprised many when he claimed that an ongoing scientific controversy about the safety of an antiretroviral drug called AZT made it impossible for his government to implement a country-wide program based on AZT to prevent mother-to-child transmission of HIV (MTCT). Scientific research had found that AZT, when given for a short period to pregnant women living with HIV/AIDS as well as their newborns, was capable of reducing the risk of MTCT significantly. While this fact remained undisputed, Mbeki suggested in a speech to the second chamber of Parliament that AZT might just be too toxic to be used in the public health sector for MTCT prevention:

> [W]e are confronted with the scourge of HIV-AIDS against which we must leave no stone unturned to save ourselves from the catastrophe which this disease poses. Concerned to respond appropriately to this threat, many in our country have called on the Government to make the drug AZT available in our public health system. [. . .]
> There . . . exists a large volume of scientific literature alleging that, among other things, the toxicity of this drug is such that it is in fact a danger to health. These are matters of great concern to the Government as it would be irresponsible for us not to heed the dire warnings which medical researchers have been making. I have therefore asked the Minister of Health, as a matter of urgency, to go into all these matters so that, to the extent that is possible, we ourselves, including our country's medical authorities, are certain of where the truth lies.

Weinel, M. (2012). Expertise and inauthentic scientific controversies: What you need to know to judge the authenticity of policy-relevant scientific controversies. In J. Goodwin (Ed.), *Between scientists & citizens: Proceedings of a conference at Iowa State University, June 1-2, 2012* (pp. 427-440). Ames, IA: Great Plains Society for the Study of Argumentation. Copyright © 2012 the author(s).

> To understand this matter better, I would urge the Honourable Members of the National Council to access the huge volume of literature on this matter available on the Internet, so that all of us can approach this issue from the same base of information. (Mbeki, 1999)

This was a remarkable statement as there was no previous indication that the government was concerned about the safety of AZT. Mbeki invoked "medical researchers," i.e., experts, who warned about the safety of AZT to challenge the seemingly popular belief that using AZT to reduce MTCT was safe. Mbeki also referred to a huge volume of scientific literature that detailed these warnings. These claims proved to be important for the policy-making process as the government stuck to them for several months and blocked any attempts by non-governmental stakeholders such as AIDS activists to force a positive decision on the introduction of AZT-based MTCT prevention programs in South Africa.

The delay in decision making that resulted from Mbeki's invocation of a scientific controversy about the safety of AZT had serious consequences in the South African context. In 1999, South Africa hosted the reportedly biggest and fastest growing HIV/AIDS epidemic in the world. Official figures suggested that about 4.5 million South Africans were living with HIV/AIDS by the time Mbeki gave his speech and that almost 500,000 new infections per annum were fueling the epidemic (Grimwood, Crewe, & Betteridge, 2000; Marais, 2000; UNAIDS, 2008). Given that in South Africa women between 15 and 40 years constitute the social group most heavily affected by HIV/AIDS, MTCT and its prevention was a huge issue. Scientific research suggests that without any intervention there is roughly a 30% chance that women living with HIV/AIDS will transmit the HI virus to their children shortly before, during or after giving birth. Scientific research had established that AZT-based MTCT prevention programs had the potential to reduce the risk of MTCT by about 50% (Newell, 1998; Bulterys, 2001). Given that official estimates suggested that in 1998, when no such programs were in place, about 60,000 children were infected with HIV through MTCT, the potential consequences of the policy decision about the use or non-use of AZT for thousands of children in South Africa become visible (Marseille et al., 1999; Wilkinson, Floyd, & Gilks, 2000; Chigwedere, Seage, Gruskin, Lee, & Essex, 2008; Nattrass, 2008).

2. INAUTHENTIC SCIENTIFIC CONTROVERSIES

An inauthentic scientific controversy can be understood as a degree of publicly visible disagreement about some scientific fact or issue which does not represent uncertainty in the community of relevant experts. According to this understanding, the briefly described episode of a larger policy-making process about using AZT to reduce the risk of MTCT in South Africa is a clear example of an inauthentic scientific controversy. There never has been any disagreement about the safety about AZT when used for MTCT prevention within the relevant expert communities.[1]

Over recent years the issue of inauthentic scientific controversies has been given some prominence within the sociological literature (e.g., Latour, 2004; Oreskes, 2004; Ceccarelli, 2008, 2011; Weinel, 2008, 2009; Michaels, 2008; Collins, Weinel, & Evans, 2010; Oreskes & Conway, 2010). It has been recognized that some actors in the policy arena clearly invoke

---

[1] An extensive analysis can be found in Weinel (2008, 2010). Similar views about the authenticity of this alleged scientific controversy can be found in Nattrass (2007, 2012), Gevisser (2007), Cullinan and Thom (2008), Geffen (2010), and Ceccarelli (2008, 2010).

inauthentic scientific controversies for political purposes. Quoted in an editorial of the New *York Times*, Frank Luntz, a Republican pollster, for example, has been very frank about the usefulness of manufacturing scientific controversies for policy purposes in the debate about global warming: "'Should the public come to believe that the scientific issues are settled,' he [Luntz] writes, 'their views about global warming will change accordingly. Therefore, you need to continue to make the lack of scientific certainty a primary issue'" ("Environmental," 2003).

Michaels (2008), who has analyzed in detail the strategies of tobacco manufacturers in their bid to delay the introduction of anti-smoking legislation, quotes an executive of a cigarette manufacturer, who is clearly aware of the political advantages of being able to point to a scientific controversy: "*Doubt is our product* since it is the best means of competing with the 'body of fact' that exists in the mind of the general public. It is also the means of establishing a controversy" (Michaels, 2008, p. x, emphasis in original).

Michaels, who concentrates mainly on the controversy-creating activities of big industrial players, points out that such actors have learned "that debating the *science* is much easier and more effective than debating the *policy*" (2008, p. xi, emphasis in original). Given the inherently probabilistic nature of scientific knowledge, there is always room to create doubts and to declare uncertainty.

While the recognition and description of inauthentic scientific controversies as an important issue in technological decision making is to be welcomed, a hitherto neglected aspect is how to distinguish systematically between genuine and inauthentic scientific controversies. Recognizing inauthentic scientific controversies is important since it can be argued that they should not have an impact on technological decision making. One reason for keeping inauthentic scientific controversies out of decision-making processes is that they can have real consequences. Manufactured scientific controversies are usually used to delay decision making, thereby delaying potentially life-saving legislation as the South African example, introduced above, suggests. Another reason is that it denies publics proper political debates by hiding political agendas behind invented scientific concerns. In the South African example, a proper debate about the pros and cons of using a drug-based intervention to reduce the risk of MTCT was denied by Mbeki's suggestion that there would have to be an agreement about the facts first within the scientific community before policy makers could act.[2]

To make these authenticity judgements as transparent and as independent of the personality of judges as possible, an approach based on criteria that can be applied by anyone to judge the authenticity of scientific controversies seems to be most promising route. Elsewhere, four criteria have been suggested that might help outsiders to make judgements about the authenticity of scientific controversies (Weinel, 2008, 2009, 2010). These criteria are: (1) explicit argument, (2) expertise of claim makers, (3) constitutive work and (4) conceptual continuity with science.

The first criterion refers to the existence of an ongoing and meaningful disagreement about a scientific fact or issue, which is a constitutive feature of any scientific controversy. While the presence of a disagreement is necessary, not every discernible disagreement can be

---

[2] Fred Kauffeld (personal communication, 2 June 2012) has suggested two more reasons for why it is important to keep inauthentic scientific controversies out of policy making. First, the "misrepresentation" of science in the context of inauthentic scientific controversies might discredit and undermine public support for science. Second, inauthentic scientific controversies might also result in tighter regulations due to attempts to stamp them out. This might then limit the freedom of scientific research.

regarded as being ongoing and/or meaningful as will be explained below. The task for the analyst is thus to check whether a contemporary meaningful disagreement exists. The second criteria suggests, however, that only disagreements between "scientific experts" can constitute a scientific controversy. The analyst has therefore to check whether those who disagree are genuine experts in the field under consideration. The third criterion states that a controversial claim, even if it has been made by a genuine expert, is not enough if it is not backed up by some sort of scientific work as otherwise any speculative utterance by an expert would constitute a scientific controversy. The analyst has to check whether claims are supported by scientific work such as theorizing or experimentation.[3] The fourth criterion, conceptual continuity with science, suggests that to constitute a scientific controversy, allegedly controversial claims have to fall into the realm of science.[4]

3. EXPERTISE AND THE OPERATIONALIZATION AND APPLICATION OF DEMARCATION CRITERIA

The focus in the present paper is not on discussing or testing the suggested demarcation criteria—this has been done elsewhere (Weinel, 2010)—but on a different aspect, namely on what sort of expertise is required to use the criteria "appropriately." Returning to the South African case study, it is interesting to note that Mbeki has implicitly used the first two of the four criteria to justify his verdict that the safety of AZT was the subject of a genuine scientific controversy. With regard to the first criterion, Mbeki appeared to interpret the existence of a "huge volume of scientific literature" which challenged the view that the benefits of using AZT for the prevention of MTCT outweighed its risks, as a serious disagreement about the "scientific facts." With regard to the second criterion, Mbeki claimed that those claims were made by "medical researchers," i.e., experts, and had therefore to be taken seriously.[5]

This contravenes the judgement reached by the present author (Weinel, 2008, 2010), who found that there was virtually no disagreement whatsoever in the relevant scientific communities that dealt with the issue of safety of AZT in the context of MTCT prevention. The discrepancy between the judgement reached by Mbeki and the author might shed some doubt upon the suitability of criteria to determine the authenticity of scientific controversies. After all, it is not promising when two parties use the same criteria and derive mutually exclusive judgements about the authenticity of the controversy about the toxicity of AZT.

Far from being a cause for worry, the discrepancy between Mbeki's and the author's assessments is instructive and productive as it directs attention to the way the criteria are

---

[3] The term "constitutive work" draws on the idea of a "constitutive forum," a concept developed by Collins and Pinch (1979). They define a *constitutive forum* as an abstract "space" which comprises "scientific theorising, and experiment . . . [with or without] corresponding publication and criticism in the learned journals and, perhaps, in the formal conference setting" (Collins & Pinch, 1979, pp. 239–240).

[4] For the sake of avoiding misunderstandings it has to be repeatedly emphasised that establishing the authenticity of scientific controversies is a problem relevant to the policy arena. The criteria are not useful to "play" science police, i.e., to try and dictate who can do research on what and when.

[5] The perspective taken in this paper requires taking Mbeki's comments on face value. As Clark Wolf pointed out during the discussion at the GPSSA "Between Scientists & Citizens" conference in Ames on June 1 and 2, 2012, another interpretation of the episode is to assume that Mbeki only used scientific arguments to disguise a political agenda. This requires, however, inside knowledge of Mbeki's motives, which my research was unable to generate.

operationalized and applied as well as to the type of expertise that underpins their operationalization and application. The argument put forward is that Mbeki did not adequately operationalize and apply the criteria because he did not have an adequate understanding of the nature of sciences and how they work in practice. In turn, it is argued that an adequate operationalization and application involves a type of expertise which might be called "sociological discrimination" (Weinel, 2010; Collins & Weinel, 2011). To explore the meaning of "sociological discrimination" in more detail, a comparison is undertaken between Mbeki's and the author's operationalization and application of one of the criteria, namely the "explicit argument" criterion.

4. EXPLICIT ARGUMENT

As pointed out above, the criterion "explicit argument" turns on the fact that any controversy implies the presence of a "contemporary disagreement" about some scientific fact or issue. While indentifying arguments can be fairly simple, knowing when one encounters a contemporary meaningful disagreement is not a straightforward task. There are at least two difficulties to consider here. First, historical studies of scientific controversies show that it is rarely the case that those who oppose a view that becomes the scientific mainstream position change their minds. Rather, it is far more common that they hold on to their views but become marginalised to an extent that their views are ignored. For example, Peter Duesberg's belief that HIV does not play a causal role in the aetiology of AIDS was rejected in the late 1980s and does not play a role in the scientific discourse. Nonetheless, Duesberg has not changed his mind and still argues the same position, but he finds it now very hard to get his views published or get heard at conferences, etc. (e.g., Epstein, 1996; Nattrass, 2012). The difficult part for an outsider is to decide whether an argument is still explicit in the sense of contemporary or whether a particular controversial claim has already been debated and rejected within the scientific community.

Second, an issue that has become particularly acute in the age of the Internet is that one can find a myriad of potentially controversial claims with regard to almost any scientific fact or issue. Whether it is the carcinogenicity of coffee consumption, the influence of planetary constellations on the mutation of viruses, the ability to build machines that supposedly ignore the second law of thermodynamics, the health benefits of obscure substances—there always appears to be someone who has published a claim that challenges the scientific mainstream. Taking each and every controversial claim seriously in the context of technological decision making is impossible. First, there are not enough resources to investigate every claim and to establish whether there is something to it or not. Second, even if this would be possible, it wouldn't resolve the problem since extreme skepticism cannot be refuted, which means that a determined character can always justifiably uphold a position that disagrees with any finding. Third, and most importantly, it would lead to a breakdown of decision-making processes as any such process could be stopped by simply "making up" controversial claims.

Having argued that identifying an explicit argument is difficult, how can a scientific outsider establish whether one exists or not? The most fruitful avenue for outsiders to judge whether an explicit disagreement is affecting a particular policy-relevant scientific domain seems to be to scrutinize the scientific literature of relevant technical domains and look for disagreements.[6] STS research indicates that this literature can reflect the state of a scientific discourse, but only when the literature is looked at in a wider context (Collins & Pinch, 1998; Collins, 2004; Shwed & Bearman, 2010). The contextualization itself can take two forms. One way in which a social analyst can contextualize particular pieces of literature is by immersing herself into the oral discourse of an expert community. By doing so, it is possible to scrutinize the standing and acceptance of particular pieces of published scientific work.[7] This is, however, a highly impractical way for outsiders as immersion into an expert community requires access, resources and time. Other sociological research indicates that there is another way in which literature can be used to assess whether a fact or an issue is regarded as controversial or consensual within scientific communities. This research does not involve immersion into an expert community but relies on the analysis of certain quantities of peer-reviewed scientific literature (Shwed & Bearman, 2010). The detection of consensus or controversy works on the basis of pattern recognition. In a situation of genuine controversy, a steady output of publications respectively supporting two or more sides in a dispute can be expected. Once a controversial matter has been settled, experts tend to devote their energies to work within an established paradigm. The support for the "defeated" positions in form of publications tends to fall away.

The lesson for an operationalization of the criterion "explicit argument" based on sociological discrimination is that it is possible to concentrate on the literature to establish whether a meaningful contemporary disagreement among scientists exists. One can reasonably claim that an explicit disagreement is ongoing if one can show that there is a reasonable amount of contemporary literature that supports diverging positions.

Applying this operationalization, a good starting point is to look through the peer-reviewed scientific literature and to try and find evidence for an explicit argument about the safety of AZT. Accordingly, a search of the PubMed database, limited to the period between January 1, 1994 and December 1, 1999 and searching for the terms "AZT" or "Zidovudine"

---

[6] While deep immersion into a technical field in the mould of an ethnographer might yield more reliable judgements, this route is not open to most people. Outsiders have to make do with accessible scientific literature. This literature ought to be published either in peer-reviewed scientific journals or by widely recognized technical institutions such as the World Health Organization, the International Labour Organization, UNAIDS, World Bank, etc. While this is a crude approach, fraught with all sorts of problems, it helps to keep outsiders away from literature that is published without any attempt to engage with a certain domain. On the Internet, a realm where anyone can publish almost anything they like, one can find "literature" that supports almost any imaginable claim such as that HIV was invented by the CIA or that the "bird flu virus" mutated due to particular constellations of planets. If one starts to take such literature into account, one can simply do away with policy making as any technological topic would have to be classified as being "controversial."

[7] Collins (2004) has shown that this can work. For example, he recognized that claims in a certain paper by Joe Weber did not lead to a scientific controversy despite contradicting widely held beliefs in gravitational wave physics. Curious about the lack of debate within the domain-specific journals, Collins discovered by talking to practitioners in the field that Weber's published claims had regularly been rejected during conferences and workshops. Given that Weber was regarded as a maverick and given that his claims had been refuted by leading figures within the field, no one had deemed it to be necessary to respond to his claims in writing. As Collins (2004) points out, a written rebuttal of Weber's claims was only published in the early 1990s, almost ten years after Weber's initial claims had been published. The reason for this belated response was that Weber, on the basis of his findings, lobbied politicians to refuse the funding of laser interferometers. Faced with this threat to the funding of new equipment, a refutation of Weber's claims was published to satisfy outsiders, such as politicians, that Weber was wrong.

and "toxicity" in the title and abstract, finds 318 publications.[8] Not all of them are relevant as they do not all deal with the use of AZT in the context of MTCT prevention. Looking through the abstracts of those that do, the vast majority either recommend the use of AZT for MTCT prevention explicitly or do not offer an explicit opinion. Only one article, a literature review by Papadopulos-Eleopulos et al. (1999), seems to outright reject the possibility of using AZT as an antiretroviral drug.

The institutional literature is even more in agreement about the safety of AZT. The drug was approved for use in MTCT prevention by all major licensing bodies, among them the very influential U.S. Federal Drug Administration (FDA) and the South African Medicines Control Council (MCC). The MCC, in response to Mbeki's and Tshabalala-Msimangs claims in late 1999, immediately reviewed the safety of the drug and an interim report in mid-November 1999 concluded that the benefits of AZT outweighed the risks when used for MTCT prevention (Nattrass, 2007). AZT was also put on the World Health Organization's (WHO) "Essential Drug List" from 1998 onwards, which meant that its safety and effectiveness had been reviewed by WHO experts (Gray & Smit, 2000). A WHO expert team also published a literature review in January 2000 which concluded that AZT was safe to use for MTCT prevention.

In sum, following the operationalization of the "explicit argument" criterion, one has to conclude that there was no explicit argument about the safety of AZT in the context of MTCT prevention. The bigger picture in late 1999 was that dozens of peer-reviewed scientific articles as well as the institutional literature recommended the use of AZT, while exactly one peer-reviewed scientific article argued against the use of AZT. Using his sociological discrimination, the author can point to several features that shed doubt about the credibility of this singular article. First, the article claims to be a literature review, but nonetheless comes to conclusions that are not supported by the literature. Rather than reviewing the literature, the article uses selected claims from a wide range of literature to arrive at a novel claim, namely that AZT cannot function as an antiretroviral agent. Second, the article only makes indirect and theoretical claims about the toxicity of AZT. It does not actually demonstrate that AZT, when used for MTCT prevention, is too toxic as no experiments are done or reviewed. Instead, the article claims that AZT cannot possibly work and that the drug is toxic; thus, exposing pregnant women to it involves only risks but no benefits. Third, the article does not state a principle that underlies the selection of the literature under review. Published in late 1999, only a small subset of the reviewed literature is recent—the majority of the reviewed literature is pre-1994, i.e., before the time AZT has been used for MTCT prevention. Fourth, the article has been published in a journal on the fringes of the scientific domain—one might expect that an article reporting paradigm-shifting findings would appear in a journal with a higher impact factor than *Current Medical Research and Opinion* (Cherry, 2009; Nattrass, 2007).

It is suggested that if Mbeki or anyone else following the above specified operationalization and using "sociological discrimination" to inform their judgements would have to conclude that there was no explicit argument about the safety of AZT in late 1999. So, why exactly did Thabo Mbeki arrive at a different conclusion? It is argued that Mbeki, instead

---

[8] The publication types in this search have been limited to: Clinical Trial, Meta-Analysis, Randomized Controlled Trial, Review, Case Reports, Classical Article, Clinical Trial, Phase I, Clinical Trial, Phase II, Clinical Trial, Phase III, Clinical Trial, Phase IV, Comparative Study, Controlled Clinical Trial, Corrected and Republished Article, Government Publications, Journal Article, Multicenter Study, Technical Report, Twin Study, Validation Studies.

of basing his judgement on "sociological discrimination," used his common sense, which seemingly did not entail much regard for an understanding of the nature of sciences.

Mbeki was first and foremost a politician and had very little exposure to sciences in practice throughout his life (e.g., Gevisser, 2007). Mbeki himself was aware of this and publicly acknowledged that he was not well-positioned to read and understand scientific literature. Accordingly, he also publicly acknowledged that he never made direct claims about AZT. Rather, he suggested that all he did was to draw attention to the alleged ongoing dispute among scientists about the safety of AZT.

For Mbeki, recognising such an ongoing scientific dispute appeared to be a very simple task which did not really require any specialist knowledge. Indeed, at the end of his speech in October 1999 he suggested that any Parliamentarian (and presumably anyone for that matter) could come to the same conclusion as the president:

> To understand this matter better, I would urge the Honourable Members of the National Council to access the huge volume of literature on this matter available on the Internet, so that all of us can approach this issue from the same base of information. (Mbeki, 1999)

The question is what "huge volume of literature" was he actually referring to given that the above analysis was unable to uncover any significant disagreement in the institutional and peer-reviewed scientific literature.

Unfortunately, Mbeki never properly elaborated what literature he had actually read nor did he specify where found it. Mbeki himself only ever mentioned one article directly and this was the aforementioned literature review by Papadopulos-Eleopulos et al. (1999), published in *Current Medical Research and Opinion* (Robertson, Hartley, & Paton, 2000; Myburgh, 2007, 2009; Cherry, 2009). In contrast to the above assessment by the author, Mbeki attributed great significance to the article. While he did not assess the content of the article, Mbeki claimed that it was an important piece because it was "very lengthy with millions of references" and had been published in a "very senior scientific journal" (Robertson et al., 2000). It seems odd that Mbeki uses the lengths and the (vastly exaggerated) number of references to assess the importance of a scientific article. It is also unclear which standard Mbeki used to refer to *Current Medical Research and Opinion* as a "very senior" scientific journal. The most commonly used and widely accepted marker of importance or "seniority" of a journal is usually the impact factor and the impact factor of this journal, as shown above, did not suggest great importance (Cherry, 2009; Nattrass, 2007).

An indirect way of getting to the supposedly critical literature that is likely to have informed the government's view is to scrutinize a "book" called *Debating AZT*, which was written by Anthony Brink (2001), a South African lawyer and self-proclaimed expert on AZT. Brink repeatedly claimed that it was his "research" into AZT that alerted Mbeki to the dangers of the drug (Brink, 2001; Myburgh, 2007, 2009; Cherry, 2009; Nattrass, 2007, 2012; Geffen, 2010). *Debating AZT* claims to be an extensive review of the critical scientific literature on AZT and at some stage Brink deals with the toxicity of AZT in the context of the drug's use for MTCT prevention. Brink repeatedly stresses the importance of four papers—Olivero et al. (1997a, 1997b), Blanche et al. (1999), and The Italian Register for HIV Infection in Children (1999)—that apparently support his claim that AZT is unacceptably toxic in the context of MTCT prevention. In the context of the whole body of literature, these four papers, if truly represented by Brink, would still represent a minority view on AZT. A closer look at the original publications, however, reveals that Brink's representation of the content of the articles is problematic. Specifically, Brink quotes selectively from this set of papers and fails to mention that the respective authors actually endorse the use of AZT for PMTCT.

For example, with regard to a paper by Olivero et al. (1997a), Brink gives the impression that these researchers recommend not to use AZT for PMTC by quoting selectively from the paper:

> Since "AZT is unequivocally a transplacental genotoxin and carcinogen [and] given transplacentally to mice, benzopyrene produced lung and liver tumour multiplicities similar to those observed [with AZT]," the researchers recorded their concern that "the current practice of treating HIV-positive women and their infants with high doses of AZT could increase cancer risk in the drug-exposed children when they reach young adulthood or middle age." (Brink, 2001, p. 43)

Readers, who might not have access to the full text of the publication, are not made aware that the researchers state in the next sentence: "The remarkable effectiveness of AZT in preventing fetal HIV infection indicates that the immediate need for treatment of a potentially fatal disease should outweigh the potential cancer risk" (Olivero et al., 1997a, p. 1607).

Brink repeats the trick of selectively quoting from scientific papers with regard to the other papers and creates the impression that medical researchers argue against the use of AZT in peer-reviewed scientific journals.[9]

If those scientific papers do not argue against the use of AZT for MTCT prevention, what other literature is there that might constitute a "huge volume"? By his own admission, Mbeki got most of his literature from the Internet (Mbeki, 1999). Indeed, a day after Mbeki's speech in October, a media liaison officer working for the presidency confirmed that the president had indeed accessed a lot of literature on AZT on the Internet: "'The president [has] got a thick set of documents. He went into many sites, including the World Health Organisation's one. The president goes into the Net all the time,' she said" (Sulcas & Randall, 1999)

According to one South African scientist, the "thick set of documents" comprised at least 1,500 pages (Cohen, 2000). While one can access peer-reviewed scientific literature via the Internet, this is apparently not what Mbeki has done. Instead he appears to have accessed a variety of "critical" websites, among them websites such as virusmyth.com which contain a number of popular, non-peer-reviewed or otherwise unpublished texts disputing, among other things, the safety of antiretroviral drugs (Cherry, 2009). The only peer-reviewed scientific article on virusmyth.com with limited relevance for the use of AZT for MTCT prevention is the aforementioned Papadopulos-Eleopulos et al. (1999) review.

In sum, Mbeki seemed to operationalize the explicit argument criterion in an over-simplified way. While rightly looking for disagreement, Mbeki seemingly did not pay too much attention to where exactly the claims about the excessive toxicity of AZT were made. The vast majority of his "huge volume" of literature comprised texts downloaded from various Internet portals associated with people denying that HIV caused AIDS. As shown further above, only one peer-reviewed article supported the claim Mbeki was making, but this article made only indirect claims about the toxicity of AZT and represented an opinion on the very fringes of the scientific discourse. Interestingly, Mbeki, by stating that this article appeared in a "very senior scientific journal," appeared to be aware that the quality of a journal reflects on

---

[9] A longer discussion can be found in Weinel (2010).

the articles published in it, but he then failed entirely to recognise that *Current Medical Research and Opinion* was a rather undistinguished and unimportant journal.

Mbeki's claims about the literature that supported the view that AZT was too toxic suggest that he did not understand how sciences work. He appeared to be unaware that identifying an ongoing scientific controversy is actually harder than just finding some disagreement. He also failed to realize that "scientific literature" is predominantly published in scientific journals. He also appeared unable to adequately assess the quality of journals and individual texts, the latter being demonstrated by his assessment of the Papadopulos-Eleopulos et al. article.

5. CONCLUSION

The case study of decision making about the use of AZT to reduce the risk of MTCT in South Africa aims to illustrate the importance of establishing the authenticity of scientific controversies. Recognizing inauthentic scientific controversies and ensuring that they do not affect technological decision making has significant practical value. The South African case study indicates that there was at least some chance that PMTCT might have been introduced earlier if President Mbeki had been unable to influence the technological decision-making process by claiming that the safety of AZT was the subject of a scientific controversy. As well as the potential saving of lives and reduction of suffering (Chigwedere et al., 2008; Nattrass, 2007, 2008, 2012) there is also a political imperative that makes the recognition of inauthentic scientific controversies important. Actors who try to use inauthentic scientific controversies effectively camouflage political arguments as scientific arguments. De-politicizing matters that are political, as happens when certain issues are declared to be matters for expert inquiry and therefore beyond the lay public's grasp, denies citizens a proper political debate.

While four criteria to determine the authenticity of a policy-relevant scientific controversy have been cursorily introduced, the main point of this paper is to draw attention to the claim that operationalizing and applying these criteria requires a particular type of expertise. This type of expertise has been referred to as sociological discrimination (SD). It has been claimed that in contrast to Mbeki the author was able to draw on an additional understanding of the nature of sciences which he acquired partly due to his immersion into the field of science studies and partly due to his fieldwork on the topic which included talking to experts.

SD, in its broadest sense, refers to an understanding of the nature of sciences or to "social intelligence about science" as Kutrovátz (2009) has put it. The knowledge that might be associated is not "well-defined" or "canonical," at least not at the present state of thinking about it. While the idea of SD appears to be plausible to the author at least, the extent of it or what exactly falls into it remains largely unclear and fuzzy. To get a better understanding of the idea of SD, it is worth taking a closer look at the specifics of the SD that played a role in analyzing the South African case (Table 1).

Table 1: Sociological Discrimination

Criteria	Operationalization	Relevant SD
Explicit argument	Analyze seemingly "controversial claims" in larger context of published and /or oral discourse of a domain	• Awareness that identification is not 'easy' (i.e., involves more than simple identification of two or more opposing points of views) • Awareness that if judgement is made on the basis of literature alone, it needs to involve "scientific literature" which is sufficiently contextualized (i.e., individual pieces are not looked at in isolation) • Ability to perform competent literature search (i.e., knowledge of where and how to look for relevant literature) • Awareness of limitations and "power" of different types of scientific literature • Ability to recognize the limitations of individual pieces of literature • Awareness of quality differences of scientific journals (i.e., "impact factor" as rough guide • Awareness of differences in relevance and quality of individual pieces of literature • Ability to recognize differences in relevance and quality (requires knowledge about standards, i.e., a "good" literature review sets out criteria that guide literature choice and includes up-to-date literature)

What appears to be the case is that it is highly unlikely that one person can gain "complete" SD as this requires an understanding of all sciences and their respective *modi operandi*. As such, SD is necessarily collective knowledge distributed among a multitude of actors. It might, however, be the case that one person has the relevant SD to make adequate judgements about specific cases. For example, the author does not claim to fully know how a particular science works, let alone how all sciences work. Rather, he happens to know useful and relevant bits that set him apart from Mbeki's understanding of sciences when it comes to judging the specifics of the alleged controversy about the safety of AZT.

This raises the question of how SD can be acquired and be made widely accessible within societies. The author has had the chance to acquire SD in the course of his (fairly shallow) immersion into the field of science studies. This, to be clear, should not be taken to mean that a formal degree in science studies is sufficient or even necessary for the acquisition of SD. The formal aspect of the studies only contributed to part of the author's understanding of the nature of sciences. More useful and relevant appears to have been his fieldwork related to the case study of technological decision making around the use of antiretroviral drugs, which involved reading about AZT and talking to scientists, doctors, activists and other observers of the policy process.

437

While direct immersion into a scientific field or into a range of scientific fields is a good way of acquiring an understanding of the nature of sciences, there are aspects of SD that can be made explicit and formal. There is thus the potential for wide distribution within societies and there is nothing in principle that prevents anyone from gaining knowledge that falls into the SD category. This can and possibly ought to start at school level. At the moment, it seems that children are predominantly taught specific content of science, e.g., they learn domain-specific specific knowledge in biology, chemistry, physics and so on.[10] While this, to a certain degree, is important, education on the nature of science should be taught in a way that reflects contemporary understanding. The output of science studies in general and post-positivist science studies in particular is relevant here and it ought to be incorporated into contemporary school curricula. This would, of course, require a considerable effort as teachers would have to be trained, courses would have to be designed and accessible text books would have to be written, but that such an effort can be undertaken is demonstrated in Brazil, where sociology has been a mandatory part of the curriculum of secondary schools since 2007. Interestingly, some pilot projects specifically teach some of the key findings of the sociology of scientific knowledge in schools.[11]

ACKNOWLEDGEMENTS: I would like to thank the Cardiff School of Social Sciences for funding the research on which this paper is based. Harry Collins and Rob Evans have supported the research intellectually. This paper was presented at the "Between Scientists & Citizens" conference at Ames, Iowa, on June 1 and 2, 2012, which was brilliantly organized by Jean Goodwin, Michael Dahlstrom and Kevin deLaplante and numerous other who cannot all be named here, but whose contribution to a fantastic conference has been noted. I am grateful to Jean Goodwin for inviting me and even more for "forcing" me to write this paper. Thanks also to Jean Wagemans, who has taken the time to extensively comment on the paper in the context of the presentation. I would also like to thank those who have been present during the presentation of the paper for their comments and critiques.

REFERENCES

Blanche, S., Tardieu, M., Rustin, P., Slama, A., Barret, B., Firtion, G., . . . Delfraissy, J. F. (1999, September 25). Persistent mitochondrial dysfunction and perinatal exposure to antiretroviral nucleoside analogues. *The Lancet, 354*, 1084–1089.

Brink, A. (2001). Debating AZT: Mbeki and the AIDS drug controversy. Pietermaritzburg, South Africa: Open Books. Retrieved from http://www.tig.org.za/pdf-files/debating_azt.pdf

Bulterys, M. (2001). HIV during pregnancy: Preventing vertical HIV transmission in the year 2000: Progress and prospects—A review. *Placenta, 22*(Supplement A, Trophoblast Research 15), S5–S12.

---

[10] In contrast to earlier and largely ill-fated attempts to educate the wider public about the *content of science* under the banner of "public understanding of science," an educational effort ought to concentrate on relaying insights about the *nature of science*. The short-comings of the Public Understanding of Science and its association with the deficit model are well-established (e.g., Irwin & Wynne, 1996; Miller, 2001; Burns, O'Connor, & Stocklmayer, 2003; Sturgis & Allum, 2004). Given the societal usefulness of sociological discrimination, a case can be made for a new programme that leads to the "public understanding of the nature of science" (PUNS).

[11] Thanks to Rodrigo Ribeiro for information on this issue that formed the subject matter of his presentation at the 4S conference in Washington in 2009.

Burns, T. W., O'Connor, D. J., & Stocklmayer, S. M. (2003). Science communication: A contemporary definition. *Public Understanding of Science, 12*(2), 183–202.

Ceccarelli, L. (2008, April 11). Manufactroversy: The art of creating controversy where none existed. *Science Progress*. Retrieved from http://www.scienceprogress.org/2008/04/manufactroversy/

Ceccarelli, L. (2011). Manufactured scientific controversy: Science, rhetoric and public debate. *Rhetoric & Public Affairs, 14*(2), 195–228.

Cherry, M. (2009). The president's panel. In K. Cullinan & A. Thom (Eds.), *The virus, vitamins & vegetables: The South African HIV/AIDS mystery* (pp. 16–35). Johannesburg, South Africa: Jacana.

Chigwedere, P., Seage G. R., Gruskin, S., Lee, T., & Essex, M. (2008). Estimating the lost benefits of antiretroviral drug use in South Africa. *Journal of Acquired Immune Deficit Syndrome, 49*(4), 410–415.

Cohen, J. (2000). AIDS researchers decry Mbeki's views on HIV. *Science, 288*(5466), 590.

Collins, H. M. (2004). *Gravity's shadow: The search for gravitational waves*. Chicago, IL: University of Chicago Press.

Collins, H. M., & Pinch, T. J. (1979). The construction of the Paranormal: Nothing unscientific is happening. In R. Wallis (Ed.), On the margins of science: The social construction of rejected knowledge (pp. 237–270). *The Sociological Review Monograph, 27*. Keele, UK: Keele University Press.

Collins, H. M., & Pinch, T. J. (1998). *The golem: What everyone should know about science* (2nd ed.). Cambridge, UK: Cambridge University Press.

Collins, H. M., Weinel, M., & Evans R. J. (2010). The politics of the Third Wave: New technologies and society. *Critical Policy Studies, 4*(2), 185–201.

Cullinan, K., & Thom, A. (Eds.). (2009). *The virus, vitamins & vegetables: The South African HIV/AIDS mystery*. Johannesburg, South Africa: Jacana.

Environmental word games. (2003, March 15). *The New York Times*. Retrieved from http://www.nytimes.com/ 2003/ 03/15/opinion/environmental-word-games.html?ref=frank_luntz

Geffen, N. (2010). *Debunking delusion: The inside story of the treatment action campaign*. Johannesburg, South Africa: Jacana.

Gevisser, M. (2007). *Thabo Mbeki: The dream deferred*. Johannesburg, South Africa: Jonathan Ball Publishers.

Gray, A., & Smit, J. (2000). Improving access to HIV-related drugs in South Africa: A case of colliding interests. *Review of African Political Economy, 27*(86), 538–590.

Grimwood, A., Crewe, M., & Betteridge, D. (2000). HIV/AIDS—Current issues. In A. Ntuli, N. Crisp, E. Clarke, & P. Barron (Eds.), *South African health review 2000* (pp. 287–300). Durban, South Africa: Health System Trust.

Irwin, A., & Wynne, B. (Eds.). (1996). *Misunderstanding science? The public reconstruction of science and technology*. Cambridge, UK: Cambridge University Press.

Italian Register for HIV Infection in Children. (1999). Rapid disease progression in HIV-1 perinatally infected children born to mothers receiving zidovudine monotherapy during pregnancy. *AIDS, 13*(8), 927–933.

Kutrovátz, G. (2010). Trust in experts: Contextual patterns of warranted epistemic dependence. *Balkan Journal of Philosophy, 2*(1), 57–68.

Latour, B. (2004). Why has critique run out of steam? From matters of fact to matters of concern. *Critical Inquiry, 30*(2), 225–248.

Marais, H. (2000). *To the edge: AIDS review 2000*. Pretoria, South Africa: University of Pretoria.

Marseille, E., Kahn, J. G., Mmiro, F., Guay, L., Musoke, P., Fowler, M. G., & Jackson, J. B. (1999, September 4). Cost effectiveness of single-dose Nevirapine regimen for mothers and babies to decrease vertical HIV-1 transmission in sub-Saharan Africa. *The Lancet, 354*, 803–809.

Mbeki, T. (1999). Address of President Mbeki, at the National Council of Provinces. Retrieved from http://www.info.gov.za/speeches/1999/991028409p1004.htm

Michaels, D. (2008). Doubt is their product: How industry's assault on science threatens your health. New York, NY: Oxford University Press.

Miller, S. (2001). Public understanding of science at the crossroads. *Public Understanding of Science 10*(1), 115–120.

Myburgh, J. (2007). The Virodene affair. *Politicsweb*. Retrieved from http://www.politicsweb.co.za/ politicsweb/ view/politicsweb/en/page71619?oid=83156&sn=display

Myburgh, J. (2009). In the beginning there was Virodene. In K. Cullinan & A. Thom (Eds.), *The virus, vitamins & vegetables: The South African HIV/AIDS mystery* (pp. 1–15). Johannesburg, South Africa: Jacana.

Nattrass, N. (2007). *Mortal combat: AIDS denialism and the struggle for antiretrovirals in South Africa*. Scottsville, South Africa: University of KwaZulu Press.

Nattrass, N. (2008). AIDS and the scientific governance of medicine in post-apartheid South Africa. *African Affairs, 107*(427), 157–176.

Nattrass, N. (2012). *The AIDS conspiracy: Science fights back.* New York, NY: Columbia University Press.

Newell, M. L. (1998). Mechanism and timing of mother-to-child transmission of HIV-1. *AIDS, 12*(8), 831–837.

Olivero, O. A., Anderson, L. M., Diwan, B. A., Haines, D. C., Harbaugh, S. W., Moskal, T. J., . . . Poirier, M. C. (1997a). Transplacental effects of 3'-Azido-2',3'-dideoxythymidine (AZT): tumorigenicity in mice and genotoxicity in mice and monkeys. *Journal of the National Cancer Institute, 89*(21), 1602–1608.

Olivero, O. A., Anderson, L. M., Diwan, B. A., Haines, D. C., Harbaugh, S. W., Moskal, T. J., . . . Poirier, M. C. (1997b). AZT is a genotoxic transplacental carcinogen in animal models. *Journal of Acquired Immune Deficiency Syndromes & Human Retrovirology, 14*(4), A29.

Oreskes, N., & Conway, E.M. (2010). *Merchants of doubt: How a handful of scientists obscured the truth on issues from tobacco smoke to global warming.* New York, NY: Bloomsbury Press.

Oreskes, N. (2004). Beyond the ivory tower: The scientific consensus on climate change. *Science, 306*(5702), 1686.

Papadopulos-Eleopulos, E., Turner, V.F., Papadimitriou, J.M., Causer, C., Alphonso, H., & Miller, T. (1999). A critical analysis of the pharmacology of AZT and its use in AIDS. *Current Medical Research and Opinion, 15*(Supplement 1), S1–S45.

Robertson, H. Hartley, R., & Paton, C. (2000, February 6). Face to face with the president: Sunday Times Interview with President Mbeki. *Sunday Times,* p.6.

Shwed, U., & Bearman, P. (2010). The temporal structure of scientific consensus formation. *American Sociological Review, 75*(6), 817–840.

Sturgis, P., & Allum, N. (2004). Science in society: Re-evaluating the deficit model of public attitudes. *Public Understanding of Science, 13*(1), 55–74.

Sulcas, A., & Randall, E. (1999, October 30). Mbeki sparks row over Aids drug. *The Sunday Independent.* Retrieved from http://www.iol.co.za/news/south-africa/mbeki-sparks-row-over-aids-drug-1.17874.

UNAIDS. (2008). *South Africa: Epidemiological fact sheet on HIV and AIDS—Core data on epidemiology and response.* Geneva, Switzerland: UNAIDS.

Weinel, M. (2008). Counterfeit scientific controversies in science policy contexts (Cardiff School of Social Sciences Working Paper No. 120). Cardiff, UK: Cardiff School of Social Sciences.

Weinel, M. (2009, March 19). Thabo Mbeki, HIV/AIDS and bogus scientific controversies. *Politicsweb.* Retrieved from http://www.politicsweb.co.za/politicsweb/view/politicsweb/en/ page71619?oid= 121968&sn=Detail

Weinel, M. (2010). *Technological decision-making under scientific uncertainty: Preventing mother-to-child transmission of HIV in South Africa.* (Doctoral dissertation). Cardiff, UK: Cardiff School of Social Sciences.

Wilkinson, D., Floyd, K., & Gilks, C. F. (2000). National and provincial estimated costs and cost effectiveness of a programme to reduce mother-to-child HIV transmission in South Africa. *South African Medical Journal, 90*(5), 794–798.

World Health Organization. (2000). Safety and tolerability of Zidovudine. Geneva, Switzerland: World Health Organization. Retrieved from http://www.who.int/reproductive-health/publications/archive/zidovudine.htm

The Explanatory Value of Cognitive Asymmetries in Policy Controversies

FRANK ZENKER

Lund University
Department of Philosophy and Cognitive Science
Kungshuset, Lundagård
22 222 Lund, Sweden
frank.zenker@fil.lu.se

ABSTRACT: Citing an epistemic or cognitive asymmetry between experts and the public, it is easy to view the relation between scientists and citizens as primarily based on trust, rather than on the content of expert argumentation. In criticism of this claim, four theses are defended: (1) Empirical studies suggest that content matters, while trust(worthiness) boasts persuasiveness. (2) In social policy controversies, genuine expert-solutions are normally not available; if trust is important here, then a clear role for cognitive asymmetry is wanting. (3) Social policy controversies pivot on values, so that biases and ideologies may explain participant behavior. (4) Few experts communicate perfectly; rather than cognitive ones, one might cite social differences.

Keywords: *ad hominem, ad verecundiam*, deficit model, *ethos*, expert, lay audience, *logos*, trust, values

1. INTRODUCTION

It is standard to claim that laypersons, policy makers included, are often deficient in the sense of being unable to fully understand expert argumentation, and must therefore—on the grounds of an epistemic or cognitive asymmetry—base policy decision exclusively or to a large extent on trust. This claim features at least implicitly in explanations why some audiences receive an expert's reason-claim complex as a mere appeal to authority (catchphrase: only the conclusion travels, but the reasons receive no uptake).

Furthermore, the cognitive asymmetry claim serves to explain the observation that expert argumentation may change upon "traveling" across contexts (Rehg, 2010), for instance "from a technical argument to an appeal to authority" (Goodwin & Honeycutt, 2010, p. 22). The above claim may also feature in explanations why the public context regularly witnesses attempts at lowering the expert's trustworthiness (*ad hominem* argumentation) rather than counter-argumentation engaging critically with the cognitive content of the expert's reasons.

As it were, participants to public context discussions are unable to engage at the level required for an expert context. Yet, these participants regularly align their discourse to, and therefore presumably understand, the role of (implicit) trust.[1] As Weingart (1999) contends, when it comes to policy making, some evaluative verdicts are inevitable:

---

[1] For Hardwig (1991) "[i]f A *trusts* B *implicitly*, she will often not have or even feel the need to have good reasons to believe what B says" (p. 699, italics mine). The importance of trust is recognized in science communication and elsewhere. "Key to the relationship between science and the public is trust" (Gregory & Miller, 1998), cited after Borchelt (2008, p. 153) who treats trust as an amalgam comprising competence, integrity, and dependability of scientists.

Zenker, F. (2012). The explanatory value of cognitive asymmetries in policy controversies. In J. Goodwin (Ed.), *Between scientists & citizens: Proceedings of a conference at Iowa State University, June 1-2, 2012* (pp. 441-451). Ames, IA: Great Plains Society for the Study of Argumentation. Copyright © 2012 the author(s).

If scientific knowledge is linked in any way to 'interests' (in policy-making), it is evaluated as supportive, contradictory, or even dangerous. Knowledge inevitably comes under these evaluative verdicts once it enters the public arena and is considered politically relevant. This is, again, an aspect of the politicisation of science inseparable from the scientification of politics. (Weingart, 1999, p. 56)

Our purpose is to critically examine the explanatory value of the cognitive asymmetry claim. In particular, we question its usefulness in explaining (what appears to us as) a regular pattern in expert-layperson communication: cognitive message-contents originating with an expert meet with a primarily non-cognitive (or affective) response by non-expert audiences. Our thesis is that an appeal to epistemic or cognitive audience-deficits explains too much, and potentially hides alternative explanations from analysts' views.

Section 2 provides a sketch of the explanation which grounds the public uptake of expert argumentation as a mere authority argument, such that audiences are more aligned to considerations of ethos rather than logos (see below). To express this in a more precise manner, Bayesian terms prove helpful and—paired with empirical results—lead to a rival explanation that is open to empirical testing (Sect. 3). We then present four challenges with respect to the cognitive asymmetry claim (Sect. 4), and close with a brief summary (Sect. 5).

2. EXPLANATION SKETCH ON A DEFICIT MODEL

The explanation sketch is as follows: Complete understanding of an expert's reason-claim-complex is not possible, unless message recipients possess domain-specific information and/or skills comparable to that of the expert-sender. Non-experts are assumed to lack as much. In this respect, they may be considered deficient—hence the term *deficit model*.[2] As it were, lay folk cannot evaluate the expert's reason-claim-complex *on a cognitive basis*. Here, cognition contrasts and indeed competes with emotions or affects, widely understood, and crucially with (dis)trust.

To illustrate, assume that an expert, EXP, forwards reasons $R_1, \ldots, R_n$ to support a policy P, and to undermine a range of alternative policies, P* (which to support would entail that P is a suboptimal choice vis-à-vis P*). Assume further that a non-expert hearer, H, understands at least P (and possibly also P*), while the same does not hold for each R_n presented by EXP, or for the manner in which EXP construes the support between $R_1, \ldots, R_n$ and P (and, respectively, construes undermining P*). Finally, assume that H desires to

---

[2]   The deficit model has been criticized and is sometimes rejected as inadequate for science communication. Alternatives include the *dialogue* and the *participation* model, both of which recognize citizens' active role as discourse partners and providers of local knowledge, respectively. See Bucchi (2008) for an overview. As Jones and Salter's (2003) discourse analytical study demonstrates, the deficit model nevertheless "retains a foothold in the discourse arena" (p. 30). More than that:

> It can be argued that the transparency and openness frame [which, for instance, includes scientist-citizen consensus meetings] is in fact a revised version of the deficit model. ... [T]he assumption that a deficit of information suggests a deficit in ability has been discarded, but the assumption that, if the public had the appropriate information, they would support the new technology has been retained. This is where transparency reinvents the deficit model. (Jones & Slater, 2003, p. 34)

personally evaluate P vis-à-vis EXP's reason-claim-complex, rather than withhold judgment (*epoché*).[3]

On the deficit model, it now sounds natural that H will seek recourse to other cues as the basis for an evaluation of EXP's argument.[4] Social psychological experiments can demonstrate that, under controlled conditions, evaluations may exclusively build on "atmospheric" factors, i.e., cognition may be replaced by affect (see the empirical study by Witte & Boy, 2007, reported in Zenker, forthcoming).[5] To say that, in such cases, a hearer's evaluation is comparatively more attuned to the speaker's *ethos* than *logos* (and using these terms in a wide sense) renders the phenomenon in traditional terms.

By finding it remarkable that evaluations of argumentative messages may be attuned also to a speaker's *ethos*, one implicitly suggests that—as far as the evaluation of expert argumentation is concerned—considerations of *ethos* and *logos* need not always cohere. Particularly, two cases are of interest: *ethos* overrides/replaces *logos*; *ethos* influences *logos*. In both cases, H's evaluation of the expert's argument (or justification) for a policy P may be negatively affected vis-à-vis one not coming about under conditions of a cognitive deficit.

The above sketch can be used to suggest a reason why *ad hominem* argumentation is with some frequency reconstructable in public discussions of social policies: personal characteristics of the message source, i.e., the expert, become more important than the cognitive contents of the message. Irrespective of whether our currently best measures of the public understanding of science in fact support the deficit model (see Sturgis & Allum, 2004 for a criticism), our ability to account for the occurrence of *ad hominem* argumentation lets the deficit model appear attractive. It suggests *why* empirically validated verbal participant behavior comes about.[6]

3. BAYESIAN EXPRESSIONS

To make the above more precise, one may use standard terms from the Bayesian approach to argumentation (Hahn & Oaksford, 2006). One would like expressions for the comparative contributions of cognitive and affective factors to the evaluative result of a reasoning claim complex. Let E (for evidence), particularly E_{logos} and E_{ethos} abbreviate the cognitive and the

---

[3] This last condition may easily go unnoticed. *Pace* those with high hopes for participatory democracy, it is not a given that a lay person desires to form an opinion on a policy option, unless personal reasons motivate. Moreover, the issue would be precluded negatively, unless one assumes audiences to be unbiased in the sense of not already *fully* endorsing a policy *alternative* to that being argued for. I return to this in Sect. 5.

[4] Goldman (2001) explicitly recognizes the reasonability of evaluating experts *also* on the basis of how they handle dialectical obligations in debate settings:

> The idea of indirect argumentative justification arises from the idea that one speaker in a debate may demonstrate dialectical superiority over the other, and this dialectical superiority might be a *plausible indicator* for [the non-expert or novice] N of greater expertise, even if it doesn't render N directly justified in believing the superior speaker's conclusion. (p. 95, italics mine)

[5] Zenker (forthcoming) contends that the cognition/affect distinction, although "rough and ready," deserves to be developed (rather than rejected), and should come out as a limiting case of any distinction replacing it.

[6] Compare how Pragma-dialecticians explain the occurrence of fallacies, namely in a *functional* way as a conflict in balancing a rhetorical and a dialectical goal (van Eemeren, 2010). In contrast, the deficit model suggests what in Durkeim's terms is a *real explanation*, purporting to name the efficient cause of a social phenomenon.

affective components of a message. Further, let H (for hypothesis) abbreviate the policy in question, and $P(H)$ the subjective degree of belief in this hypothesis (P for probability).

Then the ratio $P(H|E_{logos})$ over $P(H|E_{ethos})$ equals 1 *if and only if* (*iff*) considerations of *ethos* and *logos* contribute equally; it is <1 *iff* ethos considerations outweigh logos considerations; and it is >1 in the reverse case. Furthermore, it may seem natural to assume that the sum of *logos* and *ethos* considerations yields a measure for the overall evaluation of a policy (or, in our terms, a hypothesis). This would yield (1):

(1) $P(H|E_{logos}, E_{ethos}) = P(H|E_{logos}) + P(H|E_{ethos})$

However, the overall evaluation may well be a more complex function (abbreviated *f*) than mere summation, yielding (2)—to which we return below:

(2) $P(H|E_{logos}, E_{ethos}) = f[P(H|E_{logos}), P(H|E_{ethos})]$

Another standard assumption is that messages received from experts are more reliable than messages received from non-experts. This is normally expressed by demanding that, for experts, the probability of (receiving) the evidence(-report) is greater if the hypothesis is true than if it is false, as in (3).

(3) $P(E|H) > P(E|\textit{not } H)$ [likelihoods of receiving evidence from reliable sources]

Adjusting (3) to the distinction between evidence based on the expert's *logos* and evidence based on considerations of the expert's *ethos* yields:

(4) $P(E_{logos}, E_{ethos}|H) > P(E_{logos}, E_{ethos}|\textit{not } H)$

4. SOME EMPIRICAL RESULTS

Empirical studies suggest that, when asked to compare messages, participants tend to rate the following, in this order, as increasingly *less persuasive* evidence types: statistical evidence, expert report, causal evidence, anecdotal evidence (aka non-expert report) (Hornikx, 2008). Pornpitakpan (2004, p. 243) finds "[t]he main effect studies of source credibility on persuasion seem to indicate the superiority of a high-credibility source over a low-credibility one."

Moreover, when comparing participants' degree of belief in a conclusion vis-à-vis evidence received from a comparatively more reliable source (read: expert) and from a comparatively less reliable source (read: non-expert), Hahn, Harris, & Corner (2009) and Hahn, Oaksford, & Harris (2012) report the earlier to boost the posterior belief. Put more succinctly:

(5) $P(H|E_{EXP}) > P(H|E_{nonEXP}) \geq P(H)$ [comparison of prior and posterior belief in a hypothesis, given evidence from an expert or a non-expert source]

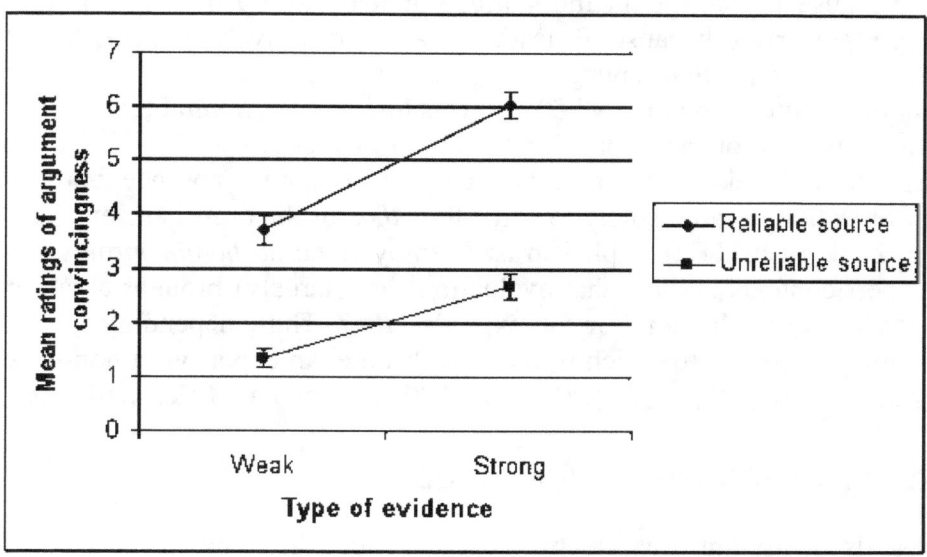

Fig. 1. Crossing weak and strong evidence types
with reliable and unreliable sources (Hahn et al., 2012)

Hahn et al. (2012) further report that, when crossing weak and strong evidence types with reliable and unreliable sources, $P(H|E)$ turns out to be sensitive to both the persuasive strength of the evidence and the reliability of the message source (see Fig. 1). A reliable source communicating a strong (or persuasive) type of evidence correlates with a comparatively greater increase in the posterior degree of belief than a reliable source communicating a weak type of evidence.

These and similar studies being externally valid would suggest that some audiences can and do distinguish types of evidential strength. Such results create difficulties for the view that, provided only the conclusion travels across contexts, *then for cognitive reasons* (see below). Moreover, for the overall evaluation of the message, audiences seem to treat the *reliability* of a message source literally as a contributing *factor*, and not as a summand.

Provided one identifies the intended effect of an *ad hominem* (which aims at discrediting the expert's ethos) as the lowering of the perceived reliability of the message source, it is natural to ask: as low as what? Obviously, it is not clear to what degree of belief audiences change if a presumably reliable source is attacked on the basis of *ethos* considerations. One might think that attacking the expert's *ethos* lowers an audience's degree of belief *at most* to that applicable if a non-expert (i.e., a comparatively less reliable source) presents an identical argument (or: the same evidence). In this case, the degree of belief would be that which is effected when only content merits are tracked, i.e., $P(H|E_{logos})$.

However, it is at least possible that an *ad hominem* affects a degree of belief at values below that assigned to a comparatively less reliable source communicating the same evidence. Thus, just as source reliability *boosts* the posterior degree of belief (in ways modeled by a factor rather than a summand; see Fig. 1), an *ad hominem* could influence audiences' degrees of belief in a similar factorial way—but in the negative direction.

If so, then *ad hominem* argumentation may be frequently reconstructable not (so much) because audiences are unable to engage with the argument content (in a cognitive manner), but because the attack on the source's ethos may be the more powerful suasory strategy[7]—more powerful, because it leads to a comparatively greater reduction in the audience's degree of belief than engaging cognitively *as a non-expert* with an expert-claim might. It is another matter altogether whether this is how things *should* be.

Vis-à-vis this hypothesis, consider the (ethno-methodologically valid) observation that a necessary condition for identifying an expert-context is—not the absence, but—the presumed irrelevance of objections geared solely towards the *ethos* of the arguer (or expert). This stance is defended in Zenker (2011). If implicit trust is enjoyed, an *ad hominem* move would hardly arise. Else, expert contexts assume that evaluations are (largely) brought about and (largely) remain independent of evaluations of the expert's *ethos*. Thus, depending on the context in which the argument arises or to which it may travel (here: an expert vs. a non-expert context), the function f in $P(\text{H}|\text{E}_{logos}, \text{E}_{ethos}) = f[P(\text{H}|\text{E}_{logos}), P(\text{H}|\text{E}_{ethos})]$ may differ markedly.

5. FOUR PROBLEMS FOR THE DEFICIT MODEL

The above yields a straightforward empirical question: In a public context, does an *ad hominem* attack lead to comparatively greater reductions of an audience's degree of belief in the expert's claim than engaging with that claim cognitively as a non-expert? The question is presumably not so much in need of discussion, but demands empirical investigation.

What may need discussion is the claim that an empirical investigation is at all worth pursuing. To support this latter claim, we now present four problems for the deficit model. Above, we had taken this model to suggest why the public context is "ripe" with *ad hominem* attacks against expert argumentation, namely: non-expert participants to the public context cannot engage cognitively with the expert's argumentation. We would like to suggest that this may be true (or relevant) in far fewer cases than readily assumed.

Although participants may be able to engage cognitively, they may chose not to do so whenever cognitive reasons do not significantly favor one policy option over others (Sect. 5.1), or for strategic reasons, i.e., when arguing *ad hominem* is the more persuasive strategy (Sect. 5.2), or for moral reasons, i.e., when policy controversies pivot on moral differences (Sect. 5.3). Finally, we suggest that some cases allegedly indicative of cognitive differences between expert and lay audience might be viewed as cases of communicative incompetence (Sect. 5.4).

5.1 When Content Matters but Trust Counts

The literature harbors case studies of so-called controversies, including scientific ones (e.g., on the "nature" of light), and those at the boundary between science and society (e.g., nuclear energy, stem cell research, global warming). Here, it is characteristic that *unique* expert

---

[7] Martin (2000, p. 203) observes for the case of *science debating*:

> It is extremely difficult, though not impossible, for audience members to seriously challenge or expose a speaker. Experienced speakers have heard nearly every question many times before and rehearsed their responses. Furthermore, they have the authority of being the speaker, with a presumption of more time to speak and having the last word. Occasionally, though, a speaker may lose a joust with an audience member, usually by being caught off guard or ill prepared by someone who is extremely knowledgeable and well prepared.

solutions are not known at the time one may legitimately speak of a controversy.[8] With the "[q]uasi-experimental implementation of new technologies" (Weingartner, 1999, p. 158) being the current normal case, analysts may therefore expect various mutually exclusive policy options to be available also because the technological alternatives may not yet be worked out such that *exactly one* alternative may deemed to be objectively best.

> [A] multitude of scientific–technical issues has captured public attention: the safety of recombinant DNA, the ethics of reproductive technologies, the application of biotechnology in agriculture, the ethical, political and economic implications of sequencing the human genome, the depletion of the ozone layer, the implications of CO_2 emissions for anthropogenic climate change, the transferability of mad cow disease (BSE) to humans, and other less visible ones. In all these controversies, it has become commonplace that the adversarial parties, be they governmental or non-governmental groups, engage scientific experts to present evidence which supports their respective views. (Weingartner, 1999, p. 156)

Thus, without adopting some material assumptions that expert-discussants disagree over, (discourse-)analysts face difficulties in justifiably claiming that a policy option is de facto favorable over another. In such cases, it may *not* be unreasonable for public context audiences to align themselves to the expert's *ethos* in order to decide for or against one or the other policy, rather than come to no evaluation.

Therefore, in the absence of (what are normally called) *decisive reasons*, it can be legitimate to personally endorse the opinion of experts found to be (more) trustworthy, and attack those not deemed worthy of one's trust.[9] Perceptions of trust may thus be legitimate reasons for favoring one policy over another. Hence, provided with a choice among equally (un)supported policy options—in the sense of a tie between the respective argument-contents—otherwise identical verbal behavior becomes explainable independently of cognitive deficits or asymmetries.

Conversely, one way of adjusting the explanation on the deficit model consists in the demand that the policy in question (for or against which an expert has raised an argument claim complex) is not part of a controversy featuring equally unsupported policy options. Other than for cases already "shelved" (according to our best historical accounts), it may prove difficult to ascertain as much.[10]

---

[8] However, unique solutions may become available later. Presently, for instance, inoculation is widely (though not universally) accepted as the best policy. When introduced in 18th century England to fight smallpox, at first, the procedure proved controversial throughout modern Europe.

[9] Relatedly, one may also explain *ad hominem* attacks when this condition is flouted. Thus, assume that EXP supports policy P with reasons $R_1, \dots, R_n$, while citizen C rejects P for reasons unrelated to $R_1, \dots, R_n$—say, because P will literally change C's backyard, which C perceives to threaten her entrenched ways of life. Then E will not appear trustworthy, because C is already "sold" on P being a false policy. See the next subsection.

[10] In specific cases (e.g., risk estimates or the distribution of funds to competing projects), analysts may find it reasonable to side with the results of elicitation techniques which make us of performance weighed solutions to aggregating (or "pooling") expert opinion such as the Cooke Method, which "generally produces uncertainty spreads that are narrower than the 'democratic' pooling approaches [e.g., the equal weight view], but wider than those provided by single experts" (Aspinal, 2010, p. 294).

5.2 Controversies Pivoting on Moral Values

It is characteristic of some long-standing disagreements that participants diverge over a (set of) moral value(s). Jackson's (2006) study of the "science of race" may be read as a case in point. Values may provide the perhaps most decisive reason for or against some policy. Thus, moral values and the biases (Gigerenzer & Brighton, 2009) or ideologies (see Rehg, 2010, pp. 57–80) sustaining them may explain behavior otherwise accounted for on a deficit model.

Provided that the moral values endorsed differ between expert and audience, audiences may consider some value-differences to support a legitimate personal attack— namely on the grounds of the expert not endorsing a particular value subset. To give extreme examples, think of those who do not endorse the free speech principle in political debates or do endorse an unconditional free-rider strategy in economic games. Perhaps less extreme, think of those who reject human embryonic stem cell research on religious grounds, or not (see Zenker 2010).

Such value divergences between experts and their non-expert audiences leave hardly any explanatory role for cognitive asymmetries, unless some values are considered morally true, yet cognitively comparatively more costly to recognize *as true* than others. In fact, in the case of a controversy at the scientist-citizen interface, such value differences may preclude (some, perhaps all) cognitive grounds from having relevance. As an example, consider that freedom of research constitutes a value that, presumably, is strongly endorsed by many scientists, and is presumably not strongly endorsed by many citizens.

Conversely, one way of adjusting the explanation on the deficit model consists in the additional demand that the policy in question is not part of a controversy pivoting on moral values. It may prove difficult to find too much in the remainder set. Would the discussion on the potential harm of electromagnetic radiation (mobile phones, microwaves, etc.) qualify as a controversy or a scientific debate?

5.3 Ad hominem as a Strategic Move

To postulate a cognitive deficit on the part of the non-expert audience is strictly insufficient to yield the desired *explanandum*: the occurrence, at a high frequency, of *ad hominem* attacks. A complete explanation based on this postulate would include at least two additional factors: the desire or preference (ascribable to the non-expert) to somehow evaluate the expert, her argument, or both. Moreover, a complete explanation would also include a desire or preference to engage in argumentative verbal behavior. After all, why not rest silent, content with judgment suspended?

Generally, an *ad hominem* attack implies a cognitive deficit only if participant behavior maps strictly into one side of the presumed cognitive asymmetry, namely into a "lack of understanding." However, strategic attempts at deception ("misleading signals") provide a counterexample. Put differently, when exploiting a cognitive asymmetry to explain public-context data, one supposes that public-context audiences do not play strategies.

It is not clear if this constitutes a sound assumption, especially as the media impress upon the public context various forms of "spin doctoring." As outlined above, a personal attack may be the "better" strategy even if a non-expert fully understands the expert's reasoning-claim complex, and could perhaps also engage with that complex on the basis of its content-

(de)merits. At any rate, participant behavior to-be-explained is not without further assumptions indicative of a cognitive deficit or asymmetry.

Conversely, one way of adjusting the explanation on the deficit model consists in the demand that these conditions are met. Yet, it is not clear when we can safely say they are.

5.4 What Did You Say?

Without special training, few communicate in readily understandable terms, even to experts in other fields.[11] At times, the overt message-content may be so rudimentary that a correct uptake requires having understood it previously. Very short talks at very large conferences make for a "good" example. Politeness standards, a long standing tradition of disregarding ("playing over") communicative infelicities, and—perhaps more widespread among academics than other professions—fear of appearing ignorant by asking for clarification sustain this condition.

Apart from the above, experts regularly draw from a specialized lexicon normally not shared outside of their domain. They also tend to endorse ways of construing support relations between reasons the domain-relative validity of which precludes them from being shared widely. Think of *inference to the best explanation*, or a logical calculus rule such as *negation as failure*. Empirical studies suggest that, without special training, lay arguers may normally not be assumed to correctly understand classical argumentation figures such as *reductio ad absurdum* (van Eemeren, Garssen, & Meuffels , 2009, p. 80) which are frequent in academic settings.

Finally, when speakers intentionally favor "code" (aka esoteric expressions) over more "pedestrian" ways of putting matters, then the relevant asymmetry may not so much implicate a cognitive, but a social level—that at which status is perpetuated.

> [C]ognitive boundaries between technical and lay knowledge are, at least to a significant degree, rhetorically constructed, and are crucial to the maintenance of social and political boundaries. Thus, interrogating the cognitive distinction challenges the social distinction, and the power produced by that distinction, as well. (Kinsella, 2002, p. 199)

One need not restrict this view to expert-citizen interaction, but may apply it also to discourse across scientific fields. Even the unavoidable division of labor typical of large scale research projects—which is said to incur a constant reliance on others' expertise—might be interpreted as evidence in favor of unsuccessful communication between scientists.[12]

---

[11] Hardwig (1991) points out that (implicit) trust rules most ways of coming to know *even between scientists*. After all, experiments are mostly not repeated in identical setup; the review process does normally assume that data are not faked; in large scale research, scientists can at best reliably evaluate the competences of a few of their collaborators.

[12] For the medical domain, Fallowfield and Jenkins (1999) trace such communication deficits to both institutional and educational shortcomings (e.g., too little time; disciplinary training vis-à-vis interdisciplinary work-life), and report these to negatively affect communication between medical researchers, practitioners, and patients, measured, for example, by information recall. Perhaps one reason in support of this general view is the almost complete absence, on most of our campuses, of "old fashioned" ideas such as a *studium generale* which could, in principle, enable scientists—*before* they acquire and further develop specialized knowledge—to learn to communicate "with one voice," rather than be socialized into the disparate jargons of various fields.

6. SUMMARY

We have tried to undermine the claim that citing cognitive deficits between experts and audiences provides a good explanation for the observation that the public context regularly witnesses attempts at lowering an expert's trustworthiness, rather than engaging with the content of an expert's argument. While such audiences may be able to engage cognitively, we have suggested that they may chose not to do so provided that cognitive reasons do not significantly favor one policy option over others, or when arguing *ad hominem* is the more persuasive strategy, or when policy controversies pivot on differences in moral values. Moreover, we have suggested that cases which allegedly indicate cognitive differences may also be viewed as cases of communicative incompetence on the part of experts or instances of perpetuating social differences. In particular, the claim that arguing *ad hominem* may be the more persuasive strategy compared to engaging cognitively as a non-expert is open to potential falsification.

REFERENCES

Aspinall, W. (2010). A route to more tractable expert advice. *Nature, 463,* 294–295.

Borchelt, R.E. (2008). Public relations in science: Managing the trust portfolio. In M. Bucchi & B. Trench (Eds.), *Handbook of Public Communication of Science and Technology* (pp. 147–157). Oxon, UK: Routledge.

Boy, R., & Witte, E. (2007). Do group discussions serve an educational purpose? *Hamburger Forschungsberichte zur Socialpsychologie, 79.* Retrieved from: http://psydok.sulb.uni-saarland.de/volltexte/2008/2342/pdf/HAFOS_79.pdf

Bucchi, M. (2008). Of deficits, deviations, and dialogues: Theories of public communication of science. In M. Bucchi & B. Trench (Eds.), *Handbook of Public Communication of Science and Technology* (pp. 57–76). Oxon, UK: Routledge.

Fallowfield, L, & Jenkins, V. (1999). Effective communication skills are the key to good cancer care. *European Journal of Cancer, 35*(11), 1592–1597.

Gigerenzer, G., & Brighton, H. (2009). Homo heuristicus: Why biased minds make better inferences. *Topics in Cognitive Science, 1,* 107–143.

Goldman, A.I. (2001). Experts: Which ones should you trust? *Philosophy and Phenomenological Research, 63* (1), 85–110.

Goodwin, J., & Honeycutt, L. (2010). When science goes public: From technical arguments to appeals to authority. *Studies in Communication Sciences, 9*(2), 19–30.

Gregory, J., & Miller, S. (1998). *Science in public: Communication, culture, and credibility.* New York, NY: Plenum Press.

Hahn, U., & Oaskford, M. (2006). A Bayesian approach to informal reasoning fallacies. *Synthese, 152,* 207–223.

Hahn, U., Harris, A.J.L, & Corner, A. (2009). Argument content and argument source: An exploration. *Informal Logic, 29*(4), 337–367.

Hahn, U., Oaksford, M., & Harris, A.J.L. (2012). Testimony and argument: A Bayesian perspective. In F. Zenker, (Ed.), *Bayesian argumentation: Contributions to a workshop at Lund University, October 2010.* Dordrecht: Springer (forthcoming).

Hardwig, J. (1991). The role of trust in knowledge. *The Journal of Philosophy, 88*(12), 693–708.

Hornikx, J. (2008). Comparing the actual and expected persuasiveness of evidence types: How good are lay people at selecting persuasive evidence? *Argumentation, 22,* 555–569.

Jackson, Jr., J.P. (2006). Argumentum *ad hominem* in the science of race. *Argumentation & Advocacy, 43*(1), 14–27.

Jones, M., & Salter, B. (2003). The governance of human genetics: Policy discourse and constructions of public trust. *New Genetics and Society, 22*(1), 21–41.

Kinsella, W.J. (2002). Problematizing the distinction between expert and lay knowledge. *New Jersey Journal of Communication, 10*(2), 191–207.

Martin, B. (2000). Behind the scenes of scientific debating. *Social Epistemology: A Journal of Knowledge, Culture and Policy, 14*(2–3), 201–209.

Rehg, W. (2009). *Cogent science in context: The science wars, argumentation theory, and Habermas.* Cambridge, MA: The MIT Press.

Pornpitakpan, C. (2004). The persuasiveness of source credibility: A critical review of five decades' evidence. *Journal of Applied Social Psychology, 34*, 243–281.

Sturgis, P., & Allen, N. (2004). Science in society: Re-evaluating the deficit model of public attitudes. *Public Understanding of Science, 13*(1), 55–74.

van Eemeren, F.H. (2010). *Strategic maneuvering in argumentative discourse: Extending the pragma-dialectical theory of argumentation.* Amsterdam: John Benjamins.

van Eemeren, F.H., Garssen, B., & Meuffels, B. (2009). *Fallacies and judgments of reasonableness: Empirical research concerning the pragma-dialectical discussion rules.* Dordrecht: Springer.

Weingart, P. (1999). Scientific expertise and political accountability: Paradoxes of science in politics. *Science and Public Policy, 26*(3), 151–161.

Zenker, F. (2010). Analyzing social policy argumentation: A case study on the opinion of the German National Ethics Council on an amendment of the Stem Cell Law. *Informal Logic, 30*(1), 62–91.

Zenker, F. (2011). Experts and bias: When is the interest-based objection to expert argumentation sound? *Argumentation, 25*(3), 355–370.

Zenker, F. (forthcoming). In support of the weak rhetoric as epistemic-thesis: On the generality and reliability of persuasion knowledge. In H. van Belle, P. Gillaerts, B. Van Gorp, D. Van de Mieroop, & K. Rutten (Eds.), *Verbal and visual rhetoric in a mediatised world, Vol. 2: Proceedings of Rhetoric in Society (RIS) III, January 2011, Antwerp.* Antwerp: University of Antwerp Press.